ENVIRONMENTAL SCIENCE FOR ENVIRONMENTAL MANAGEMENT

ENVIRONMENTAL SCIENCE FOR ENVIRONMENTAL MANAGEMENT

Second Edition

Edited by

TIMOTHY O'RIORDAN

School of Environmental Sciences
University of East Anglia
Norwich

PRENTICE HALL

Pearson Education Limited
Edinburgh Gate
Harlow
Essex CM20 2JE
United Kingdom
and Associated Companies throughout the world

Visit us on the world wide web at:
http://www.awl-he.com

© Pearson Education Limited 2000

First published 1995
Reprinted 1995 (twice), 1996
This edition 2000

ISBN   0 582 35633 4

British Library Cataloguing in Publication Data
A catalogue record for this book is available from the British Library

Library of Congress Cataloging-in-Publication Data
A catalog entry for this book is available from the Library of Congress

Typeset by 30 in 9.5 Garamond
Printed in Great Britain by Henry Ling Ltd., at the Dorset Press, Dorchester, Dorset.

Dedication

*This book is dedicated to HRH The Prince of Wales,
who is Patron of the School of Environmental Sciences at the University of East Anglia.
His vision and courage in promoting the cause of holistic environmental science
provide the inspiration that created this second edition.*

Contents

List of contributors

All of the lead authors are at the University of East Anglia. They are presented below in order of contribution, together with their main areas of research and email contact addresses.

Professor Timothy O'Riordan, School of Environmental Sciences
(environmental policy, international environment agreements, sustainability and locality)
t.oriordan@uea.ac.uk

Dr Andrew Jordan, Centre for Social and Economic Research on the Global Environment
(environmental politics, European environmental politics)
a.jordan@uea.ac.uk

Dr W. Neil Adger, School of Environmental Sciences
(agricultural economics, development and sustainability, environmental economics)
n.adger@uea.ac.uk

Dr Paul Dolman, School of Environmental Sciences
(ecosystem management, biodiversity)
p.dolman@uea.ac.uk

Dr Michael P. Meredith, School of Environmental Sciences
(physical oceanography)
m.meredith@uea.ac.uk

Dr Julian Andrews, School of Environmental Sciences
(coastal geomorphology, geochemistry)
j.andrews@uea.ac.uk

Dr Andrew Lovett, School of Environmental Sciences
(geographical information systems)
a.lovett@uea.ac.uk

Professor Michael Stocking, School of Development Studies
(development policy, soil degradation, agro-environment management)
m.stocking@uea.ac.uk

Dr Richard Hey, School of Environmental Sciences
(river geomorphology, gravel bed movement)
r.hey@uea.ac.uk

Dr Kevin Hiscock, School of Environmental Sciences
(environmental geohydrology)
k.hiscock@uea.ac.uk

Professor Tim Jickells, School of Environmental Sciences
(ocean chemistry, biogeochemistry)
t.jickells@uea.ac.uk

Dr Alastair Grant, School of Environmental Sciences
(marine biology, marine sediment chemistry)
a.grant@uea.ac.uk

Professor Peter Brimblecombe, School of Environmental Sciences
(environmental chemistry at all scales)
p.brimblecombe@uea.ac.uk

Dr Robin Haynes, School of Environmental Sciences
(epidemiology and health sciences)
r.haynes@uea.ac.uk

Dr Simon Gerrard, Centre for Environmental Risk
(environmental risk management, risk communication)
s.gerrard@uea.ac.uk

Dr Jane C. Powell, Centre for Social and Economic Research on the Global Environment
(waste management, life cycle analysis)
j.c.powell@uea.ac.uk

Preface

Science generally is under fire. Both politicians and articulate interests among the public are beginning to doubt the authenticity and advocacy of science, especially when funded and researched in government and corporate settings. This is particularly the case in areas of risk, where scientific analysis is couched in uncertainty, and the precautionary principle may have to hold sway. Yet environmental science does have a vital role to play, not just in its conventional form as an analyser of processes and a forecaster of modelled outcomes. Environmental science is also a legitimate part of the pricing process and of actual decision taking in that it can help to blend the feelings and judgements of interests in ways that might not otherwise be accommodated. Participatory or stakeholder consultation is an emerging approach to planning and resource management. This is the 'soft' end of environmental science at work. It cannot succeed unless the 'hard' end of testing for fundamental laws and of creating plausible models runs sympathetically and in parallel, so as to provide valuable contextual information.

This new edition of the book is set in the context of a changing environmentalism and a challenged science. It makes the case for interdisciplinarity and adaptability (learning). And it shows the need for consensus-seeking stakeholder round tables and similar informal consultative approaches to public identification and problem solving. It retains as its theme the highly variable relationships between globalism and localism, yet always seeks to promote a global perspective.

Environmental science is still an oddball in the higher education curriculum. Although nearly every institution of higher learning contains an environmental dimension in the course structure, the composition of these offerings varies as much as the cloud patterns over Table Mountain in Cape Town. Indeed, the notorious Research Assessment Exercise, the five-yearly audit of departmental research which determines the ranking of every academic unit, cannot agree on a single panel to oversee the environmental science family. The distinction between Earth, life and social sciences remains pretty insulated from the temptations and trials of interdisciplinarity. The main research councils, both in the UK and elsewhere, flirt with interdisciplinarity in varying degrees of seriousness and earmarked cash allocations. Even the most determined of these efforts rarely create true interdisciplinarity. Scientific training, scientific principles and career pathways, to say nothing of conventional peer review procedures, tend to corral the young and budding interdisciplinarian into one of these three comfort zones of like-minded colleagues.

Interdisciplinarity does not come easily, or cheaply. It is easy to carp and to despair at the apparent lack of progress. Such a perspective would mislead and misrepresent the very real commitment to interdisciplinarity that is now being made by all institutions of higher learning throughout the world. This is precisely why this book is reproduced with such a major rewrite of key chapters compared with its original predecessor of five years ago.

There is an exciting age ahead for interdisciplinary environmental science. All the major issues of global environmental change encompass the coupling of physical systems and human intervention. It is increasingly necessary to connect science not just to policy, but actually to policy processes. Almost all key environmental matters nowadays are to do with sustainability, the bonding of people to the planet in a placenta of care, equity, justice and progress. This is why the major new chapters to this second edition come at the front – looking at the reunification of science to policy and to the formation of knowledge in a democratic and localising society, reassessing the political dynamics of sustainable development in a post-Rio global age, and outlining just how politics works in response to global–local relationships, scientific indeterminacy, and a civil society capable of forging all kinds of coalitions.

The editorial introductions to each chapter set the scene for the book as a whole. To assist the reader, cross-referencing is widespread. This is both a textbook and an odyssey. The texts are designed to be read as introductions, not conclusions. There is no possibility that a book such as this can be fully comprehensive of such a vast and expanding topic. The readings and references are sincerely provided to assist the reader to be a companion in this exploration of interdisciplinarity, and joyously to travel the byways of imagination and opportunity that hopefully this book will offer and excite.

Environmental management is presently regarded as an arena overflowing with excellent graduates, but underprovided by employment and innovation. This is only a part-truth. Government and business are exploring creative partnerships that link good environmental practice to secure co-operative investment and trust-building participatory procedures. Here is a vast arena for the true interdisciplinarian. The management of whole catchments, of coasts and of cities requires these creative partnerships, often involving North to South funding, but combining local knowledge and social identity. It is this arena of co-operative analysis, social judgement and creative governance that will become the hallmark of sustainability in the decades to come. The interdisciplinarian will never work alone. It is the constructive opportunity of working within, then creating, liveable futures for the whole of the planet and its human family, and at all scales of attention, that will become the trademark of the forthcoming generation of environmental scientists. I can only hope that this book goes even a little way towards achieving that vision.

Timothy O'Riordan

ST. JAMES'S PALACE

Environmental science plays an essential part in any discussion of the relationship between mankind and the natural world which sustains us. Without reliable data, informed analysis and the ability to forecast, we could not begin to make the increasingly complex judgements on which our future depends. But I am not convinced that it is either sensible or necessary to adopt an approach which is based solely on dissecting the totality of any situation into ever smaller and more specialised parts. Looking at the bigger picture, and using all the insights available from different disciplines, is far more likely to produce the genuinely sustainable solutions which are our ultimate goal.

Taking a broader environmental perspective should also enable us to incorporate a more reflective, ethical dimension into our thinking and decision making. Fundamental concepts, such as the need to respect and work in harmony with Nature, are all too easily forgotten or derided, but long experience shows that there are some rather basic rules which govern our existence on this planet. Attempts to get something for nothing are simply not sustainable in the long run. There are lessons, too, from traditional knowledge and wisdom which can be integrated with the best of modern science, if only we can muster enough humility to accept the prevalence of things we don't understand, as well as those we do. And if we can muster that humility we will begin to discover that at the very heart of what we call "sustainability" lies an understanding of a sense of the Sacred. For, ultimately, it is only that sense of the Sacred – which I believe is lodged deep in the soul of Mankind – that can remind us of the need for balance and harmony and prevent us from treating this precious planet as a gigantic, experimental laboratory.

I am delighted that Tim O'Riordan, who understands – and is able to explain as well as anyone I have ever met – the wider role of science in the context of society, has produced, together with his colleagues, a text that explores environmental science in an admirably interdisciplinary fashion. I hope that a substantial readership will be inspired to adopt an approach that combines scientific precision with sound judgement, strong commitment and a broad, thoughtful perspective on humanity.

Acknowledgements

This is the second, and extensively revised, version of a book that has sold 10 000 copies over five years. To make a revision happen is a huge task. All my colleagues responded magnificently to the opportunity to display their wares for a new age. Such a commitment does not come lightly, as I am all too aware. Their co-operation was invaluable, and also a source of pride. My special gratitude goes to Pauline Blanch, who created a magnificent final copy out of endless drafts and constant changes, and always with a smile. Matthew Smith and his colleagues at Addison Wesley Longman provided all the support required, and more, and waited so patiently while this ugly duckling came of age.

We are grateful to the following for permission to reproduce copyright material;

Academic Press (London) Ltd. for Figure 14.2 (Hamilton & Clifton, 1979); American Society of Civil Engineers for Figure 12. 10 (Paice & Hey, 1993); Annual Reviews Inc. for Figure 17.4.2 (Rasmussen, 1990); R P Ashley for Figures 13.1 and 13.2; Blackwell Scientific Publications for Figure 14.3 (Ruivo, 1972); Butterworth Heinemann Ltd for Figure 17.4.1 (Carter, 1991); The IPCC Scientific Assessment for Figures 7.2.1 and 7.1 (Houghton, Jenkins and Ephraums, 1990), Tables 7.1, 7.2 and 7.3, and Figures 7.2, 7.3 and 7.4; Elsevier Science for Figure 14.8 (Fonselius, 1982), 16.3 (Cobb *et al.*, 1959); Helgolander Meeresunterchungen for Figure 14.4 (Ernst, 1980); HMSO for Figure 18.3 (HMSO, 1952); Institute of Fisheries Management for Figures 12.13, 12.14, 12.16 and 12.17 (Hey, 1992); NHPA for Figure 5.1; Research Institute for Agrobiology and Soil Fertility for Figure 14.1 (Salomens & De Groot, 1978); The Royal Society for Figure 14.6 (The Royal Society, 1983); Springer Verlag GmbH 7 Co. KG for Figure 12.12 (Brookes, 1987); John Wiley & Sons Inc for Figure 12.5 and Table 12.1 (Winkley, 1982); World Health Organization for Figure 16.4.1 (Beagleholem Bonita and Kjellstrom, 1993).

Whilst every effort has been made to trace the owners of copyright material, in a few cases this has proved impossible and we take this opportunity to offer our apologies to any copyright holders whose rights we may have unwittingly infringed.

List of journals

Environmental management advances through research. Peer review journals report the best of new ideas and fresh perspectives. If you want to keep up with this rapidly changing field, it would be helpful if you look regularly at the following list of journals identified by the contributors. Access to these journals is commonplace through websites.

Environmental policy and science

The standard scientific magazines *Nature*, *Science* and *New Scientist* carry regular features and news reports on environmental policy issues and the significance of new scientific findings. *Environment* is a monthly magazine that covers a wide range of environmental science reports and is well worth reading. Similarly, *Ambio* maintains high-quality science reports of interest to the general reader. *The Ecologist* covers both North and South environmental themes: it is cheerfully radical but always stimulating. *Global Environmental Change* is the lead journal for global change issues with an interdisciplinary perspective. *Environmental Politics* covers a fine range of contemporary political analysis. It is also a repository of green environmental thinking, as is another newish journal, *Environmental Values*. This journal and *Environmental Ethics* provide excellent coverage for ideological and ethical issues. *The Journal of Environmental Management and Planning* covers an international range of planning and policy issues, as do the *Journal of European Environmental Policy and Management* and *European Environment*. For the multidisciplinarian, *Progress in Environmental Science* is worth a look, as is *Environmental Assessment, Policy and Management*.

Environmental economics and resource management

Ecological Economics and *Environment and Development Economics* are interdisciplinary journals representing the key outlets for the emerging discipline of ecological economics. The primary environmental economics journals are the *Environmental and Resource Economics* and the *Journal of Environmental and Economics and Management*, representing the European and North American professional societies, respectively. They publish the main theoretical and empirical developments in the field, concentrating on the core areas of market solutions to environmental issues, resource use and scarcity, and economic valuation of environmental assets. *Land Economics*, the *American Journal of Agricultural Economics* and the *Journal of Agricultural Economics* all focus on environmental economics issues, and key developments in environmental economics methods are often discussed in these journals. Also important are *Environment and Planning A*, *Journal of Environmental Management* and

Environmental Management, publishing both environmental and ecological economics as well as resource management research. *World Development* is a key journal for development issues, often touching on environmental issues such as forestry, conservation and broader environment and development conflicts. Other mainstream economic journals such as *Journal of Public Economics* and *American Economic Review* and economics journals dealing with particular issues are important. For example, the highly accessible *World Economy* has covered major developments in the trade and environment debate. *Energy Policy* and *Resource and Energy Economics* are key journals for the energy field. Resource management issues are covered in *Ecological Applications, Environmental Conservation, Environmental Values, Global Environmental Change, Environment* and many geography journals. *Population and Development Review* and *Demography* are important in terms of demography and environment.

Ecosystem management and environmental ethics

The main applied ecological journals are *Conservation Biology*, the *Journal of Applied Ecology*, the *Journal of Biological Conservation, Biodiversity and Conservation* and *Conservation Biology*. Each of these has case study material of great interest. At a popular level, especially countryside management, read *Ecos, Land Use Planning* and the *Journal of Rural Studies*.

Oceans and coasts

The best journal for coastal research is *Marine Policy*, though this also has fine coverage of fisheries issues. *Marine Pollution Bulletin* is a lively journal with many topical articles. *Estuarine* covers coastal themes. At the basic level, *Ocean Challenge* is a readable magazine, while *Oceanus* is a colour magazine suitable for anyone interested in the oceans. For a journal covering new oceanographic theories and observations read *Deep-Sea Research*.

Soil erosion and land degradation

The three most useful journals here are *Land Degradation and Development*, the *Journal of Soil and Water Conservation* and *International Agricultural Development*. Each offers good case studies from all over the globe. The more specialised journal, *Soil Use and Management*, also carries articles of interest.

Hydrology and hydrogeology

The principal journals that publish research in the fields of groundwater contamination and remediation include *Ground Water*, the *Journal of Hydrology, Journal of Contaminant Hydrology* and *Water Resources Research*. Other journals to consult include *Applied Geochemistry*, the *Hydrology Journal*, which contains interesting international case studies, and *Water Research*, which includes published articles relating to surface water and groundwater treatment processes.

Air pollution, energy and risk

Atmospheric Environment is a most important scientific journal. Subsidiary but useful journals are *Environment and Technology* and the *Journal of Air and Waste Management Association. Risk Analysis* is by far the most comprehensive publication in the area of health and safety, as well as social perception. *The Journal of Risk Research* is a valuable newcomer, as is the *Journal of Risk and Health. Energy Policy* is the most reliable source for energy matters.

Environmental law

The *Journal of Environmental Law* is the most useful review of planning and law matters, but the *Journal of Environment and Planning Law* is invaluable as a source of comment on recent legislation. *Environmental Policy and Practice* contains plenty of topical articles on environmental policy and management from a legal perspective. For international environmental law, two publications are suggested: the *Review of European Community and International Environmental Law* and *European Environmental Law Review* provide valuable European coverage of European and international issues.

Chapter One

Environmental science on the move

Timothy O'Riordan

Science does not need qualifiers like 'good' or 'green', or suffixes like 'ism'. Adding the -ism is designed simply to bring science down to the level of the pseudo sciences such as Marxism or Creationism. People who do so think it a ticket of entry: actually it is a rejection slip.

(Alex Milne writing in the New Scientist, 12 June 1993)

Do not underestimate the difficulties of interdisciplinarity. The more specialities we try to fit together, the greater are our opportunities to make mistakes – and the more numerous are our willing critics. Science has been defined as a self-correcting system. In this struggle, our primary adversary should be 'the nature of things'.

(Garrett Hardin on the 30th anniversary of his 'Tragedy of the commons' article, writing in Science, 1 May 1998, 683)

Topics covered

- Science and culture
- Science and policy
- Disciplinarity, multidisciplinarity and interdisciplinarity
- Precautionary principle
- Civic science
- Knowledge and feeling

A challenge to science

Alex Milne speaks for many scientists who see their culture threatened by a wave of populist criticism. The claim of the non-conformists is that the established ways of conducting science act against sensitive and precautionary environmental management by reflecting and reproducing the elitist and exploitative aspects characteristic of all instruments of power. For example, Brian Wynne and Sue Meyer (1993) echo the environmental activist when they argue that research seeking a high degree of control over the system being studied, and which enables precise observations of the behavioural correlations between a small number of variables, draws the regulator into examining only those phenomena where cause and effect can be either proved or shown to be reasonably unambiguous.

This practice tends to place the regulator on the defensive. There may be a great number of intercorrelating factors that are not measured with equivalent diligence, due to lack of resources or inadequate recording equipment. But in democratic political cultures, the regulator has to justify the level of protection being sought. A challenge by a prosecuted discharger can result in a costly and time-consuming appeal. The courts tend to operate on provable and substantive evidence, so clever and wealthy defendants can afford to wheel in a scientist who can throw doubt on many alleged chains of cause and effect. It is tempting for the regulator to play safe and determine the environmental standard or the permitted level of discharge on the basis of the evidence that can stand up in court. That in turn will rely on the conventional scientific method. Hence the very essence of the scientific technique becomes a political weapon in the legal culture of appeal and ministerial determination of environmental quality.

Science is value-laden, as are the scientists who practise their trade. That is to be expected, though it is not always recognised. We shall see that the process of peer review is designed to iron out any obvious ideological wrinkles. More important is the belief that the practice of science may reinforce a non-sustainable economic and social culture. Because we do not know where the margins of sustainability are, as pointed out in the next chapter, the scientific approach may provide a justification for pushing the alteration of the planet beyond the limits of its tolerance. Even by playing safe, the scientific approach may, quite unintentionally, create a sense of false security over the freedom we have to play with the Earth. The critique is therefore directed at the role and self-awareness of science in a world that is grappling for the first time with seeking to restrain human aspiration and imposing global obligation of self-fulfilling private and public enterprise. Until now those qualities have been the very essence of progress and material security. To challenge them requires boldness and a cast-iron justification.

Science as culture

In a revealing public opinion poll, MORI (Worcester, 1998) asked a representative sample of the British population how much confidence they had in scientists associated with particular organisations. The results are presented in Figure 1.1. Over half said that they had either no confidence or not much confidence in scientists working for government, and almost the same proportion felt the same way about scientists working for industry. More to the point, the trust rating for government-funded and government-based scientists has fallen steadily over the past three years. Yet the same sample had almost twice as much confidence in scientists working for environmental organisations. In a separate study, Marris *et al*. (1998) found that about 60% of a sample of Norwich (UK) residents trusted scientists linked to independently financed

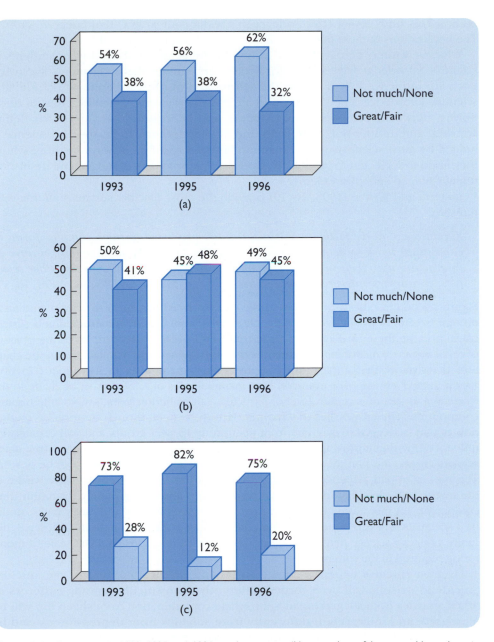

Figure 1.1: Responses, in 1993, 1995 and 1996, to the question 'How much confidence would you have in what each of the following have to say about environmental issues? (a) Scientists working for government; (b) scientists working in industry; (c) scientists working for environmental groups.' Responses: A great deal/A fair amount/Not very much/None at all/Don't know (omitted).

Source: MORI Business & The Environment Study.

bodies. This study reviews this evidence in a wider context in O'Riordan *et al.* (1997). The implications of these findings will also be analysed in Chapter 17.

What we witness here is the recognition by the general public of what students of scientific knowledge have argued for many years, namely that judgements about science are influenced by views on the setting and purpose of their work, and that scientific knowledge is not 'objective' at all. It is socially constructed by a host of rules, norms, networks of bias and expectation, and peer group pressures that define approval and disapproval, conformity and non-conformity. To be specific, science is judged by most scientists in one manner, but in very different ways by those whom they hope to serve and to inform. We shall see that processes of judgement are institutionally framed and value-laden.

Jasanoff and Wynne (1998, 16–19) provide the most succinct summary of this relationship:

> *Knowledge and the technology it sustains are produced through complex forms of communal work by scientists and technical experts who engage with nature in interaction with their multiple audiences of sponsors, specialist peers, other scientists, consumers, interpreters and users.* (Jasanoff and Wynne, 1998, 16–17)

Figure 1.2 outlines this view, known as the *constructivist approach* to science and technology. No fact is independently verifiable outside some set of constructed and agreed rules. Because these rules change, no scientific endeavour is absolute. Even the practice of replication is a matter for agreement and consensus-seeking. We shall see in the chapters that follow that apparently straightforward matters as measurement involve 'norm-framed' choices. For example, how many samples of rainfall must be made in the North Sea to 'prove' a certain level of particulate air pollution from, say, volatile organic compounds? The answer lies in what peer groups expect, though the precise methodology that is acceptable may vary from peer group to peer group, and will certainly be influenced by the improving quality of instrumentation and measurement. How many sensors are required on the ocean surface before the character of a gyre can be proven? The answer, in part, depends on the size of the research grant, and the relative wealth of the research funders. Big international programmes absorb big money for monitoring equipment. Yet monitoring is not attractive as a research activity and rarely picks up the best scientists and core funding. As Meredith and his colleagues note in Chapter 8, the future of ocean science may be influenced by political decisions over the scale of monitoring required in the coming decades.

Data on suicides depends on the social and political context of the suicide. Suicide in prisons will be recorded differently from suicides among sufferers of AIDS. Suicides from alleged organo-phosphate contamination from sheep dip will be measured, and interpreted, differently from suicides from depression caused by unemployment. All phenomena being measured are not independent of a context, a meaning, a statistical rule or a peer review as to acceptable methodological norms. If a researcher reaching the end of PhD fieldwork finds a run of 'quirky' data, what does he or she do? Ignore them? Suppress them? Ask for more funding but lengthen the PhD process? Recalibrate the instruments? The answers will reveal a lot about the inner pressures of scientific conformity or unconformity. For the young researcher, these are important observations. We shall discuss the application of 'third dimension power' in Chapter 3. This is the encapsulation of the outlooks regarding 'good science' by subordinates by those who wish to order such young values in their own interests. The young researcher seeks to please and impress peer groups: to step out of line too early is not deemed wise. Conformity may, however, be regretted, especially if interdisciplinarity is the goal.

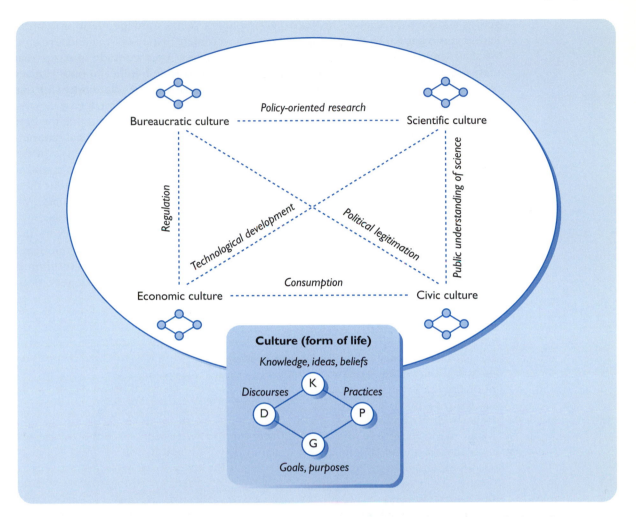

Figure 1.2: Knowledge generation and institutional framing. Any knowledge is mediated through a series of cultures, but most notably scientific, bureaucratic, civic and economic cultures. These frame both 'facts' and 'interpretations' of calamity and hence responsibility. Culture, in this context, involves knowledge, discourse (i.e. agreed interpretations of contentious issues), practices and goals (or idealised reasons for being and acting). The dashed lines indicate the nodes of scientific interaction with the civil society, in policy, regulation, public understanding and attitudes to lifestyles. This constantly changing set of relationships is continually being reconstructed, hence the term 'constructionist' view of science.
Source: Jasanoff and Wynne (1998, 17).

Science and policy

In a controversial article, Representative George Brown Jr (1997, 13), the ranking Democratic member of the US House of Representatives, claimed that environmental science 'has been distorted to serve political purposes'. His point was that ideological critics of tough and costly regulatory rules sought to prove to a responsive Republican majority that the science justifying many of these high regulatory standards was promoted by the mindset of ideological liberals. These people, he commented, together with 'an unholy alliance of scientists eager to reap the benefits of federal research funding and environmentalist groups determined to support a political agenda that systematically exaggerated environmental problems', are regarded by the Right as dangerous subversives. Science, he warns, is simply not seen as independent or neutral. Rather, it is regarded by those who dislike 'big government' as self-serving and arrogant. We shall see in Chapter 7 that it has taken the US Congress over a decade to accept that there is a global warming 'problem', and even now, a sizeable number in both Houses of Congress remain unconvinced for many of the reasons that Congressman Brown cites. This does not make constitutional ratification of the Kyoto agreement easy.

Brown argued that there was no evidence of scientific misconduct, and that the best environmental science was 'being conducted in an objective and apolitical manner, consistent with the traditional norms of scientific integrity' (Brown, 1997, 14). On the other hand, the critics, often funded by lobbies of special interests, failed to confront scientists on their own ground of published peer review, and sought to present their views in opinion pieces aimed at policy makers, the media and the general public. We shall see in Box 7.7 that this is a particularly controversial point in climate science.

Figure 1.3 looks at the changing interpretations of science and policy. Note the steady shift from a 'hands off' approach to one which increasingly involves ethical and judgemental issues, formerly the domain of politicians acting alone. This evolution is by no means universally shared amongst the scientific and policy practitioners. Both prefer to 'educate' the public from their particular perspectives.

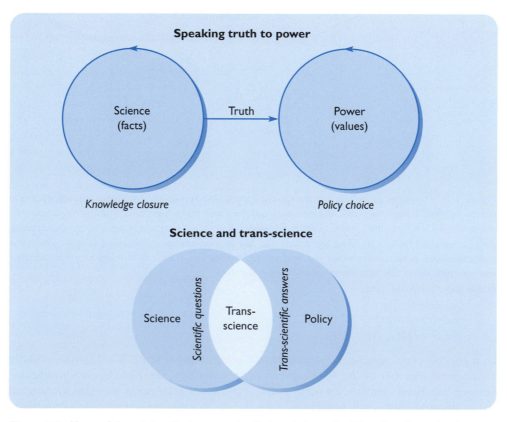

Figure 1.3: Views of the relationship between scientific knowledge and public policy. Conventionally science is seen as providing the necessary facts for policy makers to select and interpret. Subsequently scientists recognised that facts and judgemental bias are interconnected, and the notion of trans-science was born. Subsequently a more interactive version, namely civic science, became the vogue.
Source: based on Jasanoff and Wynne (1998, 8).

At stake here is who decides what is 'sound science', or 'good grounds' for early political action. According to Brown, the politicians on the libertarian Right see in science a justification for predetermined policy positions. Productivity in science effectiveness terms is regarded as the ability to alter public opinion by persuasion rather than by knowledge. Science could become funded by political fad rather than genuine enquiry. Congressman Brown advocated even more rigorous peer review of all scientific evidence, full disclosure of funding sources, and the adoption of science 'courts' where evidence is genuinely contested. Box 1.1 looks at the relationship between science and policy in more detail. What seems clear is that new forms of science–policy relationships will be needed to meet the general public concern over the integrity of science in an age of uncertainty and discontinuity.

Box 1.1 Scientific advice and political decisions

Figure 1.1.1 provides one characterisation of the relationship between scientific advice and political decisions. Before we analyse these relationships, the role of scientific advice depends enormously on the openness of information to wide public access, the broadening of the membership of advisory committees to include a wide array of interests whose views are designed to be heard, and some sort of science and technology review office in the national legislature, the reports of which are commissioned and run by the legisla- tors themselves. The reports of this office should also be available to the public (Jasanoff, 1990).

The diagram chooses only two axes, namely the perceived urgency of the issue, and its relative novelty. Politicians love to buy time, because this puts off the discomfort of making a decision, takes the heat out of a controversial issue and allows for some sort of scientific review to legitimise any final decision taken. Where urgent action is desperately needed, then scientific advice is often subservient to political expediency. For example, the reputed carcinogenic properties of dioxin are probably less serious than originally claimed. But any dioxin scare (around a chemical plant, linked to pulp and paper and printing industries because of the bleaching process) tends to create a high-risk adverse reaction. Only the very highest standards of dioxin removal are now permitted on incinerators. Similarly, if a problem is seen to be urgent and a country is forced by international agreements to treat it in the same way as other countries are, then science becomes subservient to these commitments. A good example in the US is the grudging recognition of the need to remove sulphur dioxide and nitrogen oxides from coal-burning stations in order to save Canadian forests and wildlife. A similar example is the decision by the UK government to stop dumping of sewage sludge in the North Sea even when the scientific evidence suggests that the UK contribution is both economically inefficient and ecologically marginal.

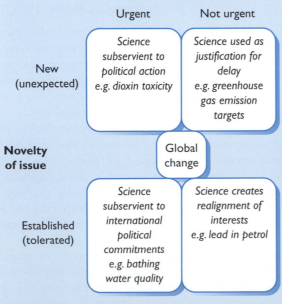

Time required to act

	Urgent	Not urgent
New (unexpected)	Science subservient to political action e.g. dioxin toxicity	Science used as justification for delay e.g. greenhouse gas emission targets
Established (tolerated)	Science subservient to international political commitments e.g. bathing water quality	Science creates realignment of interests e.g. lead in petrol

Novelty of issue

Global change

Figure 1.1.1

Box 1.1 (continued)

Where a problem is not perceived as urgent, scientific advice is welcomed as a justification for delay. The function of the Intergovernmental Panel on Climate Change was influenced in part by the need for formal justification of the international measures inflicted on nation states to get rid of excess CO_2 and methane emissions. To do this is both costly and potentially labour shedding. So a science justification was considered vital.

Finally, under the right conditions of time and established public acceptance, scientific advice can be used to realign interests. This is the active role of interdisciplinarity and science in the 'open classroom' that is very much the aim of applied environmental sciences. The conditions noted at the outset of this box have to hold for this to be a successful arrangement.

Science and uncertainty

The Congressional debate highlights a more general theme, namely that scientific methods are used to validate judgements implicit in assessment and valuation procedures such as cost–benefit analysis, environmental auditing, life-cycle analysis and risk management. All of these tools, and a number more, are used in this book. They are the standard repertoire of the environmental scientists, and are used to make their reputations. Yet, as Lemons (1996, 1) points out, if the practitioners do not make explicit their assumptions and interpretations of uncertainty, 'environmental decision makers and managers often accept scientific analysis of environmental problems as being more factual than warranted'. Professor Roy Anderson, a prominent member of the UK government's Spongiform Encephalopathy Advisory Committee (on BSE), emphasised this point:

> *The scientific community has to think more carefully about presentation to policy-makers. There is sometimes a tendency to resort to jargon or technical language. We are not blameless.*
> (reported by Meikle, 1998, 19)

Schrader-Freschette (1996, 13–17) explores the commonly adopted 'two-value frame' of *hypothesis falsification* and *provisional acceptance*. She points out that the reasons for its adoption are the application of rigorous testing against scepticism, and pure pragmatism. It is better to accept, for the time being, at least some hypotheses, rather than none at all. The consequence of this is that where there is no evidence of a particular causality, or outcome from some initiating event, it is easier to assume, at least provisionally, that there is no link. This means that, in general, scientists like to play safe. It is better to avoid the embarrassment of suggesting a causality that may not be there, but where the cost of avoidance would be huge (Type I, or false positive errors), and then to continue to investigate. This is normally the case even when highly indeterminate evidence may be on offer, but not strong enough to pass the test of falsification (Type II or false negative errors). It is therefore more likely that the burden of proof falls on those who wish to show a causal link than on those who can more readily show why it may not exist.

This is a crucial point for the integrity of environmental science, and for its relationship to the policy machinery. In many aspects of risk, for example the so-called oestrogen-disrupting chemicals (organochlorines for the most part), causality is all but impossible to prove, yet public anxiety is high. This is fertile ground for environmental groups and consumer bodies, as well as 'non-conformist' scientists. The issues of trust and accountability of science are also terribly important for the manner in which influential groups within the body politic play their cards. We shall see in Chapter 17 that in the area of risk management this problem is now under attack. The delay in being sure of the scientific link between bovine spongiform encephalopathy (BSE) and Creutzfeldt–Jakob disease (CJD) is a classic example of the wish to avoid a false positive result. What was particularly revealing about the BSE–CJD connection was the unwillingness of most scientists associated with the ministry responsible for regulating food safety, and for promoting the food industry generally, to accept any provisional hypothesis of a link. Though they were acting in a long scientific tradition, such scientists were inevitably accused of colluding with political and commercial interests. We may never know for sure what factors influenced them, and it is this doubt over institutional integrity that finally blew the compromising link between food safety regulation and food production promotion out of the water. More to the point, the evidence is growing that those scientists expressing misgivings over this initial position of 'no link' were not supported to speak out, or, in some cases, were denied further funding to pursue their assumptions.

Shackley and Wynne (1995) and Irwin (1995) look for an informal 'knowledge-sharing' arrangement between regulator and regulated. Many senior officials of the major regulatory agencies in the UK are former business scientists who understand both the business and regulatory culture. Regulatory discussions in the UK are characterised by personal friendships and close encounters. This regulatory guidance role is now a matter for more serious attention, with efforts to distance the regulatory body from the regulated, at least in the perception of the interested public. Whether this will change the common culture of regulatory science remains to be tested.

One way out is to adopt a more equity-laden, or justice-driven, mode of interpretation. This focuses on who is likely to be affected by a possible causal chain, and especially how far such people, or natural systems, are likely to be *vulnerable*. We shall discuss vulnerability in a number of ways in this book, notably in Chapter 6. What is important here is the so-called *maxi-min rule*. This is the basis of assuming the worst outcome by concentrating on those likely to be made worst off, and then ensuring that, at the very least, they are no worse off at the end. To do this involves principles of precaution, citizens' rights and responsibilities, property rights, and procedures for deliberation and inclusive power sharing. It is to all these that we now turn.

Civic science

Civic science is a phrase that Alex Milne, cited in the introduction, would abhor. It means a form of science that is deliberative, inclusive, participatory, revelatory and designed to minimise losers. Its purpose is to recognise that groups in society have to be involved if fairer and more comprehensive decisions are to be made. It also accepts that certain types of scientific uncertainty cannot be handled by traditional peer review procedures. A more widely based validation arrangement is required. In Chapter 3, we shall see that environmental politics is riddled with bias, manipulation, suppression, power generally, and the mobilisation of coalitions. Nevertheless, in active democracies, where maxi-min approaches are being demanded, some form of inclusive and deliberative science is necessary.

> *Managing large ecosystems should not rely merely on science, but on* civic *science; it should be irreducibly public in the way responsibilities are exercised, intrinsically technical, and open to learning from errors and profiting from successes.*
>
> (Lee, 1993, 161)

The American scholar Kai Lee coined the term civic science to denote a science that is explicitly technical and moral, as well as middle term (i.e. extending beyond politicians' terms of office). This means that it is adaptive, visionary, learning and promoting through a constructive engagement with active social networks.

Ortwin Renn and his colleagues (1995) look at how democratic procedures should be developed on the basis of trust. This includes ensuring representativeness, generating non-distorting communication and reaching open consensus. The key is *inclusiveness* and *consensus building*. This is a complicated process, but the nub of the issue is as follows:

- *Trust* is possible only if interests feel connected, respected, listened to, indispensable and included in all parts of decision making.

- *Representativeness* is possible only if all participants are networked to their constituent interests. This may best be achieved by the training of local 'facilitators'. These are invaluable individuals who link the science process to communal interests by translating it into the vernacular of local values and cultures to ensure that authentic views are aired and appreciated.

- *Inclusiveness* is possible only when all the relevant parties are connected and trusting of one another. Inclusiveness is an outcome, not an input. It cannot be designed in: it is created by a successful process.

- *Fairness* is the most difficult quality to ensure in a democracy. Fairness comes out of *empowerment*, which is the process of treating individuals with respect and sincerity, and enabling them to form cohesively and to participate effectively. To achieve fairness, there needs to be agreement as to what principles underlie justice and appropriate treatment among the various social groupings involved. This is unlikely to be reached without a process of consensus building, of creating confidence in the process and in one another, and of the appropriate use of liability and compensation rules.

In an important innovation, the Royal Commission on Environmental Pollution (1998, 101–112) examined the case for novel approaches to incorporating values into standard setting and environmental protection. The Commission endorsed the principle of face-to-face inclusive processes, recommended that all forms of government adopt such techniques where the issues are 'complex and broad in scope', and urged monitoring of experiments so that the 'full potential' of such approaches could be realised. This is a helpful and timely intervention that should promote the cause of such approaches in many aspects of environmental management in future years.

Inclusive deliberation for a participatory civic science requires a set of conditions that are difficult to fill, even in the most smoothly functioning democracies. Examples of this approach are presented in Box 1.2, and the various kinds of scientific uncertainty for which these procedures may be invoked are presented in the discussion that follows, as well as in the introduction to Chapter 7. Here are the conditions, many of which are proving very difficult to achieve.

- *The elected politicians agree to share power*. Should they do this, elected politicians actually enhance their legitimacy to act. But few believe this, for they cherish the power of patronage and representation. Hence, most politicians remain deeply sceptical about real community democracy.

- *Regulatory and executive agencies are fully transparent*. This means that such agencies allow all legitimate interests to enter fully and openly into their deliberations. We shall see in Chapter 17 that many risk regulatory bodies nowadays include representation from consumer, environmental and social justice organi-

Box 1.2 Innovations in public participation

Attempts to widen the basis for consultation are becoming more confidently pursued in an age of more active civic involvement, and a recognition that expertise has to be linked to interest group judgements and biases if politically acceptable decisions are to be reached. Some of the ways for incorporating inclusive and deliberative participation are listed below, but the examples are by no means exhaustive.

Judgements informed by deliberation

- *Citizens' juries*: a group of representative citizens listen to presentations and argument amongst experts and informed public opinion, and come to a consensus conclusion. They may be assisted by a facilitator, or a 'judge', though this is not always required.

- *Consensus conferences*: a large group of representative citizens hear argument from a range of opinion on a matter of contested science. They vote, usually electronically, on a series of points raised by a moderator, but according to an agreed agenda.

- *Referendums*: a voter-wide poll, triggered by a 'question' and informed by a variety of debating arrangements. In some countries (e.g. Switzerland, Sweden) and some states (e.g. in the USA) these have the force of law. Elsewhere, they are advisory to a legislative, but in effect politically binding, process. The level of discussion is variable.

Discourse techniques

- *Issues forums*: local forums that may meet over many months to discuss a range of issues, often hearing evidence. Many of these become round tables for particular themes, for example Local Agenda 21.

- *Study circles*: groups of people in social settings aiming to be informed and to reach consensus on a range of issues.

- *Citizens' panels*: a mix of the two above, more formal and representative than the circles, but not so coherent as the forums generally. These have value at the local level, and are the focus of a variety of participatory techniques such as Future Search, Visioning Exercises and Integrated Assessment Modelling.

Source: Stewart (1996, 3–14).

---

sations. Furthermore, they seek to operate on a consensus basis, through which all members have a *de facto* veto, but also the huge obligation of reaching agreement. Increasingly, the minutes and the policy papers of such agency advisory bodies are on the Internet, so that all interested parties can see what decisions have been reached.

- *All parties accept the responsibility to reach agreement.* Rights to participate are not absolute. There is an obligation to respond to the interests of other parties when inclusive procedures are invoked. Social networks usually throw up 'mini-elites' who do not always want to share the power they have recently attained. Inclusive procedures suggest that they must do so, but in a manner that is dignified and fair.

In a whole host of environmental sustainability areas, inclusive and deliberative processes are being recognised for what they can offer for effective, legitimate and fair decision making and action. This is an area in environmental science that is still underdeveloped, notably in the skills of negotiation and facilitation. Designing the appropriate procedures from the mix of science, policy, interest involvement and accountable, but cost-effective, outcomes is still in its infancy. We still see that in areas such as water futures, coastal management, risk assessment, biodiversity action plans, climate futures, transport options, sustainability indicators and future states of liveable cities, this approach will become increasingly adopted. For readers of this book, the key is to understand the policy process (Chapter 3), recognise how science and policy interrelate (this chapter), perceive the coupling effect of physical criticality and social vulnerability (Chapter 6), and devise an interdisciplinarity where knowledge and feeling intertwine (also this chapter).

Inclusive and deliberative processes are much more difficult to make successful than many practitioners generally realise. For one thing, the people involved require skills of absorbing complex information, converting this to clear verbal argument, and being ready to accommodate to the interests of others. For another, many of the techniques summarised in Box 1.2 require a political process that is geared to deliver on the outcomes of the deliberations. Failure to do so may increase subsequent alienation and distrust. Third, unless groups are trained in a sensitive and caring manner to participate in exercises such as focus group discussions on visioning of possible future states, many simply will be excluded as a consequence of their perceived inadequacy. So for such procedures to be successful, both democracy and educational skills need to be made appropriately sympathetic to the new requirements. This is proving to be a tall order.

Environmental science and interdisciplinarity

It has long been held that environmental science equates with 'Earth science', namely that it is the study of the atmosphere, the land and the oceans and the great chemical cycles that flow through the physical and biological systems that connect them. For the most part this is still the case in the content of environmental science courses in higher education in the UK. It is becoming less so nowadays as we see that modern environmental science is increasingly becoming *interdisciplinary*, preparing people for *global citizenship* and training them to be flexible yet competent *analysts and decision takers*. These are not objectives that are easily reconciled in one degree course, and certainly not in one subject area.

At stake here is an unusual form of scientific training and investigation. The trigger for this, according to the American philosopher Richard Tarnas (1993, 352–359), was the emergence of genuine uncertainty in quantum physics, the interdependence of observer (in the Heisenberg principle), and the 'irreducible irrationality, already recognised in the human psyche, being recognised in the structure of the physical world itself' (*ibid.*, 358). With its problematic link to the adverse social and distributional consequences of growth and technology, together with its links to political, military and corporate establishments, science had to be captured for humanity's sake, and for its appropriate stewarding and trusteeship roles for the planet (see Chapter 19).

Tarnas (1993, 430–434) proceeds to argue that the necessary relationship between the human mind and the world is not dualistic in the Newtonian and Cartesian sense of an observable planet and an autonomous intellect. Rather the relationship is participatory. The human mind is ultimately the organ of nature's own process of self-revelation. Nature's reality only comes into being through the very act of human cognition. Nature is thus contested as is the human logic.

This point is particularly well made by Macnaughton and Urry (1998). But we might go further. Nature is a metaphor for our changing souls, relationships, feelings and energy fields. If we choose to be open to such experiences, nature is enriched by our openness and acceptance. If we choose to remain closed, so too will the reality of nature's interpretation in our minds and actions.

Let us try to define interdisciplinarity in the context of disciplinarity, pluridisciplinarity and multidisciplinarity. Box 1.3 also takes up this theme.

Box 1.3 Multidisciplinarity and interdisciplinarity

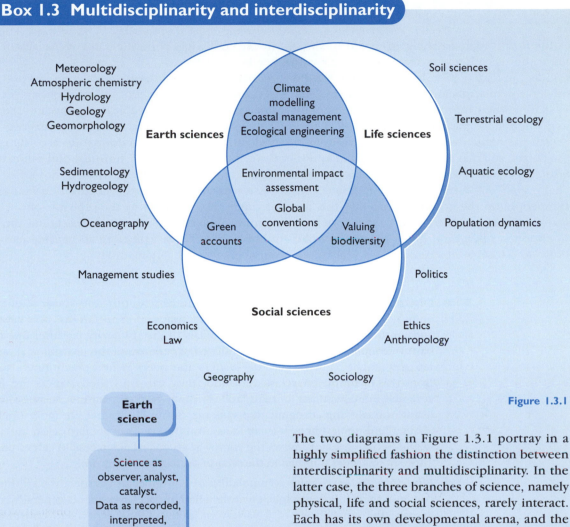

Figure 1.3.1

The two diagrams in Figure 1.3.1 portray in a highly simplified fashion the distinction between interdisciplinary and multidisciplinarity. In the latter case, the three branches of science, namely physical, life and social sciences, rarely interact. Each has its own developmental arena, and the attractive research areas are by no means found only on the margins, where an element of joint disciplinarity takes place. But it is possible for the climate system to be studied in this way, with much success, and for the emerging field of environmental economics to flourish in the interstices between economics and ecology. Where a more formal multidisciplinarity is often found lies in the ever-widening area of environmental impact assessment. Admittedly even today, after the Americans first introduced the approach in the 1969 National Environmental Policy Act, this technique is more an amalgamation of mini reports based on predetermined approaches rather than a truly integrative document.

Box 1.3 (continued)

But for EIA to flourish it needs not just to integrate the disciplines. It also needs to provide regular guidance over how a project should evolve. Interdisciplinarity recognises that power is shared when information exchange is open and when those likely to be affected by change actually negotiate their values and reactions from the outset, and that data are a function of knowledge, experience and power. In other words, the vital ingredients for the EIA have to come from the people on the ground whose interests are directly affected. This is not just a matter of well-intentioned liberalism: it is a recognition that those who have used a resource for generations are the best judges of how impacts of change can best be evaluated.

This interdisciplinarity goes well beyond environmental assessment, even of policy, let alone projects. It covers a fresh approach to the gathering and interpretation of data, it recognises the need to weigh information according to various parameters of political and ethical norms, and it creates an extension of power for those who are not always recognised as being of importance. This is why the principle of precaution, the practice of civic science and the ethos of knowledge as feeling all became part of the interdisciplinary method.

- *Disciplinarity* is the regime of a discipline of scientific and social scientific enquiry as laid down by common practice, untested expectations, peer review and methodological acceptance. The basis of disciplinarity is rather like the basis of common law. It is the accumulated experience of the profession as written and taught by its intellectual leaders and peers. When practitioners stray from the common paraclassic ground, they can find their papers are not accepted in mainstream journals, and they may not be guaranteed to be short-listed in qualifying academic departments. Some disciplines are more prickly about all this than others, it is fair to say. But the confines are fairly clear and the would-be practitioner knows the routeway to ensure a successful career path. These points were outlined in the section on science as culture.

- *Pluridisciplinarity* is the accumulation of a number of perspectives around a common theme. For example, environmental impact assessments generally congregate the specialities of a variety of analysts and experts to convene a corporate picture, if not an integrated whole.

- *Multidisciplinarity* involves a procedure that is different, as well as a purpose and an outcome. The procedure is to examine *at the outset* how a topic should be characterised, say as a problem or as an opportunity, so that the actual *design of the study* is agreed by practitioners beforehand. In essence, multidisciplinarity *locks in* a common objective and a co-ordinated (but not integrated) methodology.

Since multidisciplinarity is closer to interdisciplinarity than to the other two, it is worth giving an example. Consider an attempt at a 'holistic' evaluation of the environmental and economic (but not social) effects of a switch from inorganic to organic methods of production (Cobb and O'Riordan, 1999). The task was to examine how the shift in regime of management:

- altered the nutrient status of the soil, and hence reduced off-farm nutrient losses to the atmosphere and to groundwater;

- invigorated a more diverse plant life in the margins of the fields, and in 'reservoir' unfarmed areas;

- increased the species numbers and varieties of animals and insects so that the farm was aesthetically more attractive and also more adaptable to pest control;

- strengthened the 'money' economics of profitability, and also improved the 'social' economics of a more resilient, self-perpetuating and ecologically diverse enterprise.

Despite the presence of an overseeing steering group, and irrespective of a vast number of researcher meetings, this study could not fully integrate the evidence into the analysis. To 'fully integrate' means a change in the basis of both data collection and interpretation to ensure that certain 'holistic' perspectives are in place. Two of these holistic perspectives were 'soil health' and 'biodiversity'. Both of these came to be examined from a disciplinary perspective as a contribution to a multidisciplinary study. Soil nutrient analysis was modelled on a whole-farm basis, but also at the level of the field. In both cases, the variability of nutrient movement was enormous. It was measured at different scales. But finding a 'mean' for nutrient flows at any scale simply became meaningless as a measure of soil health. Microbiological activity was too unknown to be helpful. Yet the 'feel' of the organic soil, compared with its previous 'non-organic' structure, was really different. It looked, felt and smelt different, and withheld water, nutrients and soil biota far more securely. The soil was more resilient, but no-one could *measure* why, even though the farm manager and the farm staff could *feel* the difference.

Similarly, biodiversity could not be quantified in totality. There are far too many ecological systems and processes to examine, whether on their own or in combination. The consequence was that the team had to select 'emblem' species of insects (butterflies), animals (fieldmice) and pest-controlling agents (carabid beetles) as a symbol of biological 'health'. Of course, this was only a fragmentary picture. In any case, biodiversity is a cultural as much as an ecological phenomenon. This means that the more people with influence *visit and value* farm wildlife from an experimental and educational viewpoint, the more they are likely to appreciate the ecological mix in the end.

Once again, this interaction between human agency and physical system defied the multidisciplinarians. All that could be gained was a second-hand version of aesthetic appreciation via a series of cognate studies done elsewhere in the agri-environmental field, even though there is good evidence from a study in the Yorkshire Dales (O'Riordan *et al.*, 1994) that visitors and residents of a national park do value environmental conditions much more when they are enabled to understand how and why such conditions are important, changing and threatened.

We now conclude our previous series of definitions by turning to interdisciplinarity.

- *Interdisciplinarity* involves a combination of knowledge and feeling, of measurement and judgement, of information and ethics, of explanation and participation. Interdisciplinarity is the outcome of a dialogue of understanding and revelation, of communication and response. Interdisciplinarity accepts that the natural world is an extension of human knowing and feeling, of behaviour and expectation. In short, interdisciplinarity starts from the premise that there is no distinction between a natural system and human interpretation of that system. Nature's 'truth' is only the unfolding of our own appreciation and a manifestation of how we treat our fellow human beings. If we abuse nature, we can be sure that we also abuse our fellow humans (see Box 1.4).

This is not, generally speaking, acceptable stuff to mainstream science. The prestigious Royal Commission on Environmental Pollution (1998, 28) was unequivocal: 'A clear

Box 1.4 On the nature of interdisciplinarity

Children under the age of five, we are told, think holistically. So do many peoples who do not have the advantage of a structured scientific education. Interdisciplinarity is the merging of knowledge into common concepts, and the application of ideas in the round for real-world problem solving. True interdisciplinarity probably has never existed, because, by definition, the phenomenon involves the unification of concepts that are designed to be conceived as separate entities. The simile is trying to create an orange with segments and pith separated out. With skill, and with some adhesive, a plausible orange will appear, but it will be a contrivance, not a living entity. Thus we can distinguish between multidisciplinarity or pluridisciplinarity, and true interdisciplinarity. The first two are essentially synonymous; the last is fusion *ab initio*. This is why there is confusion around the term. The common way out is for multidisciplinarity, the co-ordination of specialisms, based on framework, but with integration occurring only by effort and often by chance. True interdisciplinarity is a fundamentally unique approach to total science.

It is, however, possible to identify some common concepts that embrace both the social and natural sciences. Here are four that look promising.

● *Chaos* is a concept that enables dynamic systems to be understood for their unpredictability and patterns of indeterministic behaviour. Chaos takes place as much in governing institutions as in the worlds of quantum physics or hurricane formation. The point is to learn from the principles involved as they apply to various circumstances.

● *Social learning* is the process by which organisms 'see' their environmental circum-

stances by intelligence gathering, and act with foresight or prepared adjustment. This principle of precautionary but evolutionary adjustment may be a vital one for responding to environmental stress.

● *Dynamic equilibria* abound in nature, but usually in an entropic sense, namely that over time equilibria may shift to new sets of dynamically stable relationships. Just as this may be the case for oxygen or sulphur fluxes, so too it may be for carbon and the human population.

● *Carrying capacity and evolutionary adjustment.* This is another version of 3 above. Carrying capacity is akin to sustainability and involves the application of co-operation, competition, justice and respect for individuals as well as group welfare. Enshrined in these principles are also those of reciprocity, sharing to ensure that adequate conditions of survival are maintained. Carrying capacity must be linked to physical and ecological parameters such as tolerance and adaptability.

This in turn may require the rethinking of approaches to research training and the connections between educational institutions and industry, commerce and government. There should be scope for including the university and college beyond its walls to collaborate its activity into organisations engaged in policy analysis and environmental problem saving. There is also the probability of joint PhDs involving research undertaken in partnerships between universities and a supportive external institution. Such PhD schemes would provide both training and problem-solving skills.

dividing line', it concluded, 'should be drawn between analysis of the scientific evidence and consideration of ethical and social issues, which are outside the scope of a scientific assessment'. That may not be the last word on this matter.

In the organic transition case, soil health would become a totality of nutrient status, interpretative management and 'feel' that enriched an ethical perspective as to what is 'good' about naturalness and nutrition. Biodiversity is a coupling of ecological functioning and human experience of that functioning. The more people sensed the 'health' of a diverse biological system by serving, feeling, smelling, listening and

Box 1.5 Science for sustainability

In their follow-up to the famous *Limits to Growth*, Meadows *et al.* (1992) looked at the conditions for a creative dialogue between the sciences and the community of nations to generate a sustainable future. They identified five key parameters.

● *Visioning or imagining* what you really want, and not what someone has told you you should want, or what you have learned to settle for. Visioning means taking off all the impediments to treasured dreams and uplifting future states. This is the basis for the new institutions of sustainability based on paying one's way, sharing, identifying with the totality of creation and acting out of commitment, not duty or loyalty.

● *Networking* or community groupings. Informal networks convey both information and values, often very effectively. Networks are also connections between equals: they allow everyone to have a legitimate say and to feel genuinely part of the whole, no matter how complex the whole might be. This is the one arena where the modern world of frightening information overload and social alienation is most oppressive.

● *Truth telling* or devising a language that beckons, not frightens or divides. Truth telling generates a sense of commitment and vitality as well as honesty, and it allows communications to function far more effectively. So the language should be open, not closed. For example, do not say change is sacrifice, but change is a challenge that opens up new potentials: sacrifice is a misnomer because giving to future generations is an act of sharing, not loss.

● *Learning* means acting locally but in a global context. Each molecule of CO_2 or CFC saved is a contribution to human well-being, as is the skill to enable others to take such action. Governments can only create the conditions for citizen action: that should be extra-governmental or anarchic.

● *Loving* means extending solidarity of the joy of creation to all humans and to all life on Earth. Loving means friendship, generosity, understanding and a real sense of empathy with life before, during and beyond temporal existence. It is unfashionable to act such ways beyond 'church': but loving is the essence of religion, and religion is the very basis for bonding people together. Sadly, the majority of religious fundamentalists have forgotten the very basis of their faith.

tasting, the more the totality of the sensory and ecological experience entwines. This is not the stuff of science but of an interaction between scientific method and human experience in the round. The truly appreciated organic farm is visited by enquiring citizens, is the source of local jobs and educational curricula, and the basis of local food supply. The 'value' of an organic system is only calculable in a holistic economic, social, environmental, ethical and participatory framework. The science becomes active, not passive, engaging directly with the behaviour and outlook of citizens (see Box 1.5).

This perspective presents a headache for neoclassically trained economists. Say that a fully transformed organic management regime increases the butterfly population by 100 individuals. To the conventional economist, the 100th butterfly should be valued less than the 99th additional butterfly if the well-known law of diminishing marginal returns is followed. But if the fully integrated organic farm is visited and understood for all its purposes by an appreciative public, then, arguably, the 100th butterfly is worth *more* than the 99th. This is because the blossoming of butterflies is associated with the 'goodness' of an integrated management regime through which the public learns to acquire a more holistic perspective on interactive valuation. The increase in butterflies, being a symbol of an increase in sustainability generally, is valued more, because of the direct engagement with the farm as a lived world for the consumer as citizen.

Such a perspective tends to be ridiculed by disciplinarians and others, including even multidisciplinarians. It is also all but impossible to quantify and thus to pass the test of analytical credibility. So interdisciplinarity languishes in the frustrations of would-be visionaries, while various forms of multidisciplinarity and pluridisciplinarity become popular under the false guise of interdisciplinarity. Arguably, this is the norm in environmental management practice.

In summary, interdisciplinarity remains underdeveloped for the following reasons.

- The modern scientific tradition does not readily share its culture with other cultures of knowing and understanding.

- Career advancement in science tends to be accelerated if peer review accepts the basis of research and the consistent high quality of the research method.

- Allthough multiple authorship is now a more recognised basis for career promotion, multiple authorship involving the integration of many disciplines can be psychologically stressful to achieve and time-consuming to produce, even though ultimately rewarding.

- Many established higher education institutions find it difficult to create interdisciplinarity out of long-serving single-discipline departments, especially those with high research ratings and a steady drip-feed of research funding.

- There are too few properly executed training programmes and methodology seminars for interdisciplinarity. These require the stimulus of jointly funded research projects, or realistic career paths for young scientists, and of close co-operation with business, government at all levels and pressure groups, including citizens' associations.

Interdisciplinarity will flourish only when there is a genuine change in the scientific culture towards the relationship between knowledge and feeling, between measurement by collection and measurement by participatory observation, and in the management of research. There is also enormous scope for jointly executed research involving academics and practitioners, and, even, jointly authored MSc or PhD theses. The twinning of academics and community facilitators to joint research degrees would be an important first step. These points are given further airing in Chapter 19. Another angle on 'open science' in introduced in Box 1.6.

Box 1.6 Science and the Internet

Every day over 35 000 physicists browse a website (*http://xxx.lanl.gov*) by Paul Ginsparg and maintained by Los Alamos National Laboratory in New Mexico. The site contains e-prints, papers written before peer review, notes that update research, and 'conversations' with other researchers via email. Peer review is thus decentralised, informed and fun. The final products may be much better than might be the case with two or three academic refer-ees. For interdisciplinary research the website may provide the answer, opening up ideas, methodologies and intriguingly novel interpretations. Another source for Internet advice is the 'bytes of note' column in *Environment Magazine*, written by Tom Parris of the Kennedy School of Government at Harvard University. These monthly columns are an invaluable source of information for scholars of interdisciplinary environmental science.

Broadening the scientific tradition

There has always been a lively debate about the objectives and definitions of science. A flavour of the dispute was provided at the outset. Opinions die hard and views are strenuously defended. Part of the problem lies in the misinterpretation by the various antagonists of each others' position. Science certainly seeks to move forward on the basis of broad principles, theories, laws and hypotheses, namely statements of interpretation that apply to a broad array of circumstances, and which are subject to continuous scrutiny through experiment, observation, falsification, verification and replication.

These are proper procedures. They form the basis for both the social and natural sciences. As we have seen, the problem is not so much whether these approaches are necessary, because they are. The issue is whether they should be extended by other forms of judgement and dialogue to create a partnership with society on a broader front. This would allow science to be more aware of its scope for misdirecting human development even when it is sincerely searching for the truth.

Let us look first at the scientific method as it is commonly understood. (See de Groot, 1993, for a good review.) Science evolves by *theory building*, *theory testing* and *normative evaluation*. The basic theories themselves are examined for their correctness in terms of their internal logicality, and for their consistency, that is their inherent plausibility.

These theories in turn are converted into hypotheses or propositions whose truth or applicability to a given set of circumstances is subjected to analysis. Normally that analysis relies upon *observations* and meticulous recording, *experimentation* also with meticulous recording, or *modelling* through which representations of 'reality' are created to provide a more manageable basis for examination and prediction. Where there is a historical record, the model can be calibrated against measured output to test for its *robustness* and *accuracy*. Where there is no historical record, or where the model is essentially designed to depict the future, then the only test for reliability is *peer group criticism* of the model's assumptions, interactions and sensitivities to relationships between cause and outcome, which are uncertain or simply not known.

Peer review, or the combined judgement of those who are not only knowledgeable and experienced but who sincerely wish to retain the credibility of their collective profession by maintaining the very highest standards of excellence, is the vital basis of predictive science. We shall see in the chapter that follows that all the great global change issues – climate change, ozone depletion, biodiversity loss, tropical forest removal, microtoxicological disturbance of ecosystems – cannot be predicted with absolute certainty. All are therefore subject to networks of peer scientific review with the aim of generating consensus as a basis for political conviction and action. This prediction is followed for both the social and the natural sciences. For interdisciplinary science the task is more difficult and less successful, but the principle of retaining authoritative professionalism remains.

So much for the fundamental principles of *corrections of theory* and *faithfulness of the empirical method*. There is a third aim of science, namely to provide a background of advice as to what is good practice. This is known as its *normative role*, which can only be conducted through evaluative criteria based on socially agreed norms. Such norms are usually controversial, and certainly ambiguous. They apply to principles of justice, fairness, efficiency and what else is deemed to be morally right. Clearly, the definitions of these principles will vary from political culture to political culture, and will even be disputed by scientists themselves. For example, economists regularly battle over whether efficiency or equity should have supremacy. We shall see in Chapters 4, 5 and 7 that this is by no means a clear distinction any longer. Politicians prefer to think in terms of fairness or evenness of treatment, even when this means a more costly (i.e. less efficient) solution.

Consider this point from HRH The Prince of Wales (1998, 252) over the morality of genetic modification:

> *I happen to believe that this kind of genetic modification takes mankind into the realms that belong to God, and to God alone. Apart from certain highly beneficial and specific medical applications, do we have the right to experiment with, and commercialise, the building blocks of life? We live in an age of rights – it seems to me that it is time our Creator had some rights too.*

The Prince of Wales looks for science to inform, and be informed by a social morality that is revealed through open debate in partnership. Some of his thinking is embodied in Box 1.7.

Box 1.7 Two kinds of science

Figure 1.7.1 suggests that there are two approaches to the pursuit of understanding, one active and interventionist, the other integrative and contemplative. Both are necessary for the environmental science of the future. The interventionist mode reflects a view that the mysterious 'beyond' is attended to by the principle of stewardship, or leaving matters better, by improvement and wealth creation, yet viable for the future. The partnership mode looks at humanity as a co-evolutionary trustee in a companionship of stewardship. The first primarily generates knowledge and action, the second tends towards judgement and integration. Both nurture the other in true interdisciplinarity.

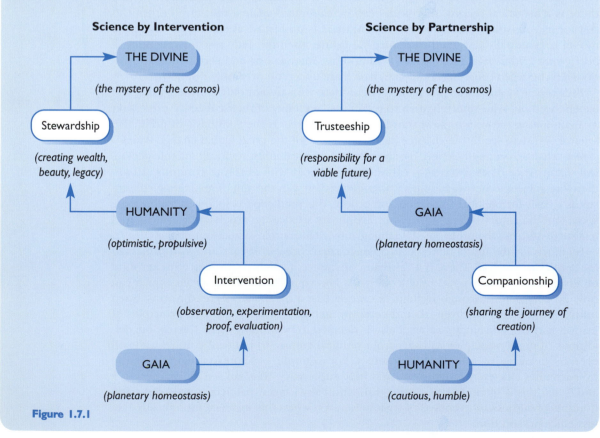

Figure 1.7.1

It is wise not to assume that there is a single normative criterion. Different circumstances will throw up varying yardsticks; some examples follow.

- *Efficiency in the form of least-cost solutions* is fine where there is close to a functioning market which manages to incorporate environmental side-effects into price signals.

- *Fairness or equity* for all concerned tends to operate where rights are universally shared between present and future generations, and where collective action, usually involving many nation states acting together, is necessary to produce a desired outcome.

- *Paying for past debts* (i.e. differential equity) is applied to issues such as the clean-up of contaminated land, the reduction of greenhouse emissions (where some countries have emitted over longer periods), and, increasingly, the reallocation of water rights. However, such a norm is very contentious politically as usually those first in, and/or politically the most powerful, have to be persuaded by a collectivity of weaker (usually victimised) interests to pay up. Yet in modern environmental science this normative principle is an important one.

- *Equivalence of treatment* may not be very cost-effective, but it applies to the principle of burden sharing. This is commonly found in circumstances where a number of countries are contributing to environmental degradation, and even when some are creating more damage than others, everyone is expected to pull their weight simply because it is seen as socially responsible and a statement of collective solidarity. In such circumstances 'scientific' justification of contribution and removal is by no means the basis for negotiation. It is primarily a matter of being part of the whole commitment. Such shared action helps to keep all countries involved in on the act.

We shall see in the chapters that follow how all of these evaluative criteria apply to environmental problem solving. Efficiency issues tend to dominate economic analysis (Chapter 4), while equity considerations strongly influence collective agreements (Chapters 2, 7 and 19). Paying for past debts appears in risk management matters, while equivalence of treatment turns up in ethical approaches to ecosystem management for the good of the planet as a whole (Chapter 5), in health and environment issues (Chapters 15 and 16), and in risk assessment (Chapter 17).

The precautionary principle, critical load and the public trust doctrine

One of many themes of the text that follows is the endless search for better understanding of environmental processes. This is the basis on which to make sound models and reliable predictions of what may happen if various policy measures, including doing nothing, are put in place. Clearly we are nowhere near that state of certainty with regard to all environmental systems. Indeed there is a vigorous debate as to whether we know enough to make any sensible predictions at all, let alone guide and justify political action.

One should distinguish three levels of uncertainty in environmental science:

- *Data shortage.* Often there is neither the historical record nor the comprehensiveness of monitoring to form a reliable picture of what is happening. Take the case of nutrient enrichment of the North Sea (see Chapter 14). There is good evidence of great fluctuation in phytoplankton (microscopic plant life), normally a sign of high nutrient status. But there is no accurate historical trend, so we do not know if these fluctuations are unusual. Also, from time to time, nutrient-rich water from the North Atlantic sweeps into the North Sea, carried by currents and driven by wind. This may be a prime natural factor in altering the nutrient status. But so could be the nutrient-rich discharge from the major rivers fed by municipal sewage treatment waste, industrial discharges and agricultural runoff. It is possible to generate the data over time, but no amount of new data will make up for the lack of historical record in the medium term.

● *Model deficiencies.* Models of global climate change (see Chapter 7) and associated sea level rise (Chapter 9) rely on an understanding of the links between atmosphere, ocean, biota and ice, which are still being generated. Scientists can test their models for the sensitivity to input parameters, but any model will be limited by ignorance of these highly complicated but little understood relationships. Here is where peer review and networks of scientists both collaborating and criticising each other act as the safeguard, but the models are still highly imperfect. Arguably such models can only be refined, not made representationally accurate. In a similar vein, fisheries science is being criticised for being too unwilling to admit fundamental ignorance. The result is to fail to convince policy makers, and to allow overfishing to continue (see also Chapter 8) (*Nature*, 1997, 105)

● *Beyond the knowable.* Both data limitations and model imperfections can be overcome with time and effort. But there is a school of thought which claims that certain natural processes are indefinable and indeterminate because they operate in mysterious ways that can never be fully understood. Just as stars and galaxies undergo seemingly chaotic phase changes and comprehensive chemical transformations apparently without warning or in relation to no known laws, so earthly systems appear to act chaotically (randomly) and catastrophically (flipping into new phase states or exhibiting huge but transitory turbulence). These are not essentially modellable, and by definition, data sets do not provide any great insight. In the foreseeable future, the predictive pathways of such intricate systems seem beyond comprehension. Can anyone *really* tell us just what would happen to climate, rainfall and the Earth's species mix if nine-tenths of the world's tropical forests were to be removed in the next half-century?

Each of these three qualities of uncertainty places potentially very great strains on environmental science. We cannot follow the practice of utilising the natural assimilability of environmental services, the so-called *critical load* approach, because we cannot be certain where these thresholds are. Nor can we use population dynamics, predator–prey relationships or indicator species theories of ecosystem change, because we do not know enough about these phenomena to do justice to our judgement of what is tolerable. We can and must improve our theoretical models and empirical designs, but that will not provide us with robust interpretations of disturbance and consequence.

So we are beginning increasingly to fall back on the *principle of precaution*. This is still a misunderstood notion that is held in deep suspicion by some scientists and developers. This is because the application of precaution changes the balance of power between science and the community, between developers and environmentalists, and between those who exploit the environmental services and the vulnerable who depend on these services for their survival.

The attraction of precaution lies in the aphorisms 'a stitch in time saves nine' and 'better safe than sorry'. Where there is uncertainty of the type that requires extended peer validation, or where there is real indeterminacy, then the notion of 'being roughly right in due time, rather than being precise too late' carries attractions. The principle is invoked under the following circumstances:

- there is an active civic science;
- headline issues of threshold collapse cause much anxiety;
- dislocation at the local level accommodates to a global problem;
- vulnerable groups are championed by citizens' justice organisations;
- business begins to worry about reputation and social responsibility;
- regulation is more transparent;
- direct action groups can communicate on the Internet to plan their forays.

The concept of precaution will emerge throughout the text that follows. Put at its simplest, it can be regarded as having four meanings.

- *Thoughtful action in advance of scientific proof* of cause and effect based on the principles of wise management and cost effectiveness, namely better to pay a little now than possibly an awful lot later. In this sense, precaution is a recipe for action over inaction where there is a reasonable threat of irreversibility or of serious damage to life support systems.

- *Leaving ecological space as room for ignorance.* This means quite deliberately not extracting resources even when they are there for the taking, for example fish or whales, or even mineral ores, simply because we do not know what will be the longer-term consequences of virtually total removal. This also applies to levels of development. It is a moral responsibility of wealthy nations to allow room for the poorer nations to develop unsustainably for a while, just to grant them time to make the transition less painfully.

- *Care in management.* Because it is not possible to forecast all the possible consequences of altering a habitat, or manipulating an ecosystem, or cleaning up a waste dump, so it is necessary to carry the public through the trial and error of experiment and adjustment. This calls for focused and creative participation in the very process of management. It is particularly relevant in the control of hazardous activities such as nuclear power stations, incinerators and the release of genetically modified organisms (see Chapter 17).

- *Shifting the burden of proof from the victim to the developer.* This is by far the most controversial aspect of the precautionary principle, for it signifies the power shift mentioned earlier. In the past, development has operated on the basis of risk taking, environmental assessment and compensation out of the proceeds of growth. Arguably, many innovations and explorations would not have taken place if risks were not accepted as a necessary element of progress. Similarly, where damage has occurred, it is customary to pay the victims out of the gains and profits associated with the damage-creating scheme. Growth is the basis for subsequent redistribution. Environmental impact assessments were designed to list the likely consequences of a proposed course of action in a reasonably formal and public manner and this was deemed sufficient.

The precautionary principle seeks to change all this. It aims to place much more emphasis on the responsibility of those who seek to improve things to show that they will not cause harm, or at least will put up a performance bond to provide a fund for possible compensation to subsequently proven victims. This is the maxi-min rule at work, so it is hardly surprising that such an approach is regarded as deeply normative.

In a wide-ranging review of the principles underlying environmental standard set-ting, the Royal Commission on Environmental Pollution (1998, 60) endorsed the precautionary principle. It pointed out that the application 'should be transparent and subject to review in the light of development of understanding. Relevant data should be collected and reviewed on a continuous basis, and if a standard has been set, it should be revised up or down as necessary.' This is an important statement in the evolving context of regulatory science.

The issue of changing the terms of liability is incorporated into the public trust doc-trine, namely a commitment to put back into the Earth at least the equivalent of what is being removed in any particular development. Clearly this compensatory investment does not necessarily mean replacement. But it does mean providing the equivalent of the environmental services that have been expropriated. So profits from non-renewable resource use should, in part, be directed at providing renewable substitutes. Removal of tropical forests or coral reefs should be limited by punitive requirements on guarantee-ing protection for sizeable remaining amounts of these resources, possibly by payment into a global heritage fund, or via royalty payments for taxonomic research so we know more of what species are remaining, or by tradable development rights whereby rights to develop are denied in sensitive zones but can be purchased in less sensitive spots. These are very much ideals, and quite contrary to established ways of doing things. Hence the controversy and resistance from powerful development interests. The doctrine is never-theless vital if biodiversity is to be guaranteed, and property ownership always contains a responsibility for planetary stewardship (see Chapter 5).

Precaution should be distinguished from prevention in that the latter is applied to eliminating known hazards, such as toxic substances, or at least reducing noxious materials at the point of production and/or use. Thus prevention is simply a regulatory measure aimed at an established threat. Precaution is a wholly different matter. It introduces the duty of care on all actions; it seeks to reduce uncertainty simply by requiring prudence, wise management, public information and inclusive participation, and the best technology, all over the planet. In this sense, it is not determined by the socio-political and economic circumstances of geography. Such a line is highly contro-versial, but it is increasingly being forced upon multinational corporate business.

Applying the precautionary principle also propels action across nations even when some countries are obviously going to be inconvenienced more than others as a result. This is why the international interpretation of precaution suggests that it should be applied *according to capabilities*, that is with due regard for the ability of a country to take advance action, and *in proportion to the likely benefits*. At present, therefore, the concept is held in check by the principles of ability (or reasonableness of treatment) and proportionality (or related cost of advance action in relation to likely benefits: see Figure 1.4). In the latter case, as we shall see in Chapter 19, side payments or induce-ments have to be made to poorer or disadvantaged countries to compensate them for investing in courses of action more expensive than they would otherwise have spent resources on, in order to meet a global benefit (such as reducing or substituting for ozone-depleting products, or greenhouse gases).

Environmental science as seen from the outside

Environmental science is evolving internally by becoming more problem-focused, policy relevant, interdisciplinary and self-critical. These are healthy developments, but in no way do they remove the fundamentals of the scientific tradition. Yet to accommodate the needs of uncertainty, indeterminacy, public outreach and political manipulation

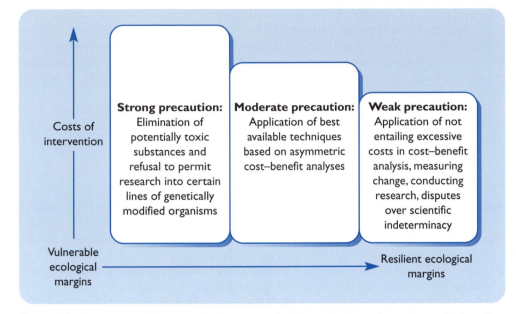

Figure 1.4: Precaution and the principle of proportionality. Proportionality applies to the application of costs and benefits where part of the valuation of benefit is the avoidance of unnecessary risk by playing safe. Proportionality is defined by attitudes to environmental and social resilience, vulnerability and periodic, possibly irreversible, thresholds. These attitudes are measured in turn by attitudes to science, expertise, international obligations and rights of nature.

requires a wariness, an openness and a refreshing frankness that will stand this science in good stead (see Box 1.4).

Interestingly, both business and non-governmental organisations are also adapting to the new look of environmental science. Business has become overwhelmed by the plethora of international treaties, regulations and codes of practice that is demanded of it by international government organisations such as the United Nations, the North Sea Basin States, the European Union and the various trading blocs. In addition, much modern national legislation is influenced by these international agreements as well as by forceful public opinion.

Business is beginning to forge alliances with environmental scientists across a wide array of themes – cost–benefit analysis, environmental impact assessment, risk management, eco-auditing, life-cycle analysis and troubleshooting of new regulations, especially those covering the use of new products and the responsibility for cleaning up waste. This is allowing environmental scientists to find new job opportunities as managers and public relations specialists, as ethical investment analysts and as environmental assessors. At the same time, business is providing very fertile ground for environmental science research. It creates case studies, good data sets and interesting problems that can only be solved by creative interaction. We shall see that nowhere is this more evident than in the aftermath of the Kyoto agreement on climate change (Chapter 7).

Environmental pressure groups are also hiring scientists and liaising with research institutions. All the major international groups such as Greenpeace, Friends of the Earth and the World Wide Fund for Nature have a science group responsible for reviewing scientific information and spotting gaps or flaws. Campaigners in these organisations are usually found at major science congresses, and are actively connected to computerised science information flows. It is not wise for the modern pressure group to be scientifically illiterate. Equally, it is not wise for the group to be too conventional and uncritical over the role of science in environmental problem identification, analysis and solution. Nowadays there is a formative debate between the NGOs, industry, government and academia that is pushing the whole of environmental science forward. The net result is an extension of science into the world of politics, commerce and social change.

This is a wonderful time to be an environmental scientist. You can be sure that, whatever your views on the subject matter now, you will have changed your stance in a decade or so. Such is the dynamism of the subject matter and its methods of enquiry.

References

Brown, G. (1997) Environmental science under siege in the US Congress. *Environment*, **39**(2), 12–20, 29–31.

Cobb, R. and O'Riordan, T. (1999) On evaluating integrated science for sustainable agriculture. *Journal of Agricultural Economics* (in preparation).

de Groot, W. T. (1993) *Environmental Science Theory: Concepts and Methods in a One-world, Problem-oriented Paradigm*. Elsevier, Dordrecht.

Hardin, G. (1988) Extensions to 'The tragedy of the commons'. *Science*, 1 May, 682–683.

HRH The Prince of Wales (1998) Seeds of disaster. *The Daily Telegraph*, reprinted in *The Ecologist*, **28**(5), 252–253.

Irwin, A. (1995) *Citizen Science: a Study of People, Expertise and Sustainable Development*. Routledge, London.

Jasanoff, S. (1990) *The Fifth Branch: Science Advisers as Policy Makers*. Harvard University Press, Cambridge, MA.

Jasanoff, S. and Wynne, B. (1998) Science and decisionmaking. In S. Raynor and E. Malone (eds), *Human Choice and Climate Change: the Societal Framework*. Battelle Press, Columbus, Ohio, 1–87.

Lee, K. (1993) *Compass and Gyroscope: Integrating Science and Policy for the Environment*. Island Press, New York.

Lemons, J. (1996) Introduction. In J. Lemons (ed.) *Scientific Uncertainty and Environmental Problem Solving*. Blackwell, Oxford, 1-11.

Macnaughton, P. and Urry, J. (1998) *Contested Nature*. Sage, London.

Marris, C., Langford, I. and O'Riordan, T. (1998) A quantitative test of the cultural theory of risk perceptions: comparison with the psychometric paradigm. *Risk Analysis*, **18**(3), 190–205.

Meadows, D. H., Meadows, D. L. and Randers, J. (1992) *Beyond the Limits: Global Collapse or a Sustainable Future*. Earthscan, London.

Meikle, J. (1998) Luck to be a lady to me: analysis of risk. *The Guardian*, 7 October, 19.

Milne, A. (1993) The perils of green pessimism. *New Scientist*, 12 June, 31–37.

Nature (1997) Fisheries science: all at sea when it comes to politics? *Nature*, **386**, 105–110.

O'Riordan, T. and Jordan, A. (1995) The precautionary principle in contemporary environmental politics. *Environmental Values*, **4**(2), 191–212.

O'Riordan, T., Marris, C. and Langford, I. (1997) Images of science underlying public perceptions of risk. In Royal Society, *Science, Policy and Risk*. Royal Society, London, 13–30.

O'Riordan, T., Wood, C. and Shadrake, A. (1994) Landscapes for tomorrow. *Journal of Environmental Planning and Management*, **9**(1), 41–57.

Renn, O., Webler, T. and Wiedermann, P. (eds)(1995) *Fairness and Competence in Citizen Participation: Evaluating Models for Environmental Discourse*. Kluwer, Dordrecht.

Royal Commission on Environmental Pollution (1998) *Setting Environmental Standards*.

Twenty-first Report. The Stationery Office, London.

Schrader-Freschette, K. (1996) Methodological risks for four classes of scientific uncertainty. In J. Lemons (ed.), *Scientific Uncertainty and Environmental Problem Solving*. Blackwell, Oxford, 12–39.

Shackley, S. and Wynne, B. (1995) Global climate change: the mutual construction of an emergent science policy domain. *Science and Public Policy*, **2**(4), 218–230.

Stewart, M. (1996) *Deliberative Procedures in Local Agenda 21*. Local Government Management Board (Local Government Association), London.

Tarnas, R. (1993) *The Passion of the Western Mind: Understanding the Ideas that have Shaped our World View.* Pimlico Press, London.

Worcester, R. (1998) Greening the millennium: public attitudes to the environment. *Public Opinion Quarterly*, **69**, 73–85.

Wynne, B. and Meyer, S. (1993) How science fails the environment. *New Scientist*, 5 June, 33–35.

Further reading

For those who like to start at the beginning there are two major books on the changing nature of science, namely

Kuhn, R. (1970) *The Structure of Scientific Revolutions*. University of Chicago Press, Chicago.

Merton, R. (1973) *The Sociology of Science*. University of Chicago Press, Chicago.

Other works that are well worth consulting are

Ezrahi, Y. (1990) *The Descent of Icarus*: *Science and the Transformation of Contemporary Democracy*. Harvard University Press, Cambridge, MA.

Irwin, A. (1995) *Citizen Science: a Study of People, Expertise and Sustainable Development*. Routledge, London.

Irwin, A. and Wynne, B. (eds) (1996) *Misunderstanding Science: the Public Reconstruction of Science and Technology*. Cambridge University Press, Cambridge.

Jasanoff, S. (1990) *The Fifth Branch: Science Advisers as Policy Makers*. Harvard University Press, Cambridge, MA.

Latour, B. (1987) *Science in Action*. Harvard University Press, Cambridge, MA.

For a review of the precautionary principle, see

Costanza, R. and Cornwell, L. (1992) The 4P approach to dealing with scientific uncertainty. *Environment,* **34**(9), 12–20, 42.

O'Riordan, T. and Cameron, J. (eds) (1994) *Interpreting the Precautionary Principle*. Cameron and May, London.

For a broader look at the possible role of environmental science, see

O'Riordan, T. (1998) Civic science and the sustainability transition. In D. Warburton (ed.), *Community and Sustainable Development.* Earthscan, London, 96–115.

Chapter Two

The sustainability debate

Timothy O'Riordan

If current predictions of population growth prove accurate and patterns of human activity on the planet remain unchanged, science and technology may not be able to prevent either irreversible degradation of the environment or continued poverty for much of the world.

(Statement by the National Academy of Sciences and the Royal Society, 1992)

Topics covered

- Global environmental change
- Excluding the commons
- The Rio process
- Sustainable development
- Perverse subsidies
- Socio-ecological footprints

Editorial introduction

The global environmental debate is a mixture of anxiety over the capacity of the planet to absorb the changes wrought by human activity, and concern for the billions of future, less fortunate people, whose plight will be by no means of their choosing and whose children may well face a possible future of tragic choices. This, at its worst, may lie between dying soon or staying alive only to undermine further the viability of the existence of their progeny. The tragedy may be all the more poignant, for the capacity of the very poor to cope with and to survive through the adoption of many forms of livelihood keeps many in some sort of survival (see Chapter 6). The guarantee of that survival should be better served by the global family.

There are hundreds of books and reports which argue persuasively and with supportive evidence for a greater sharing of livelihood opportunity for the planet. The most prominent is the Brundtland Report (1987) prepared by the World Commission on Environment and Development, established by the UN and chaired by the former Norwegian prime minister, Gro Harlem Brundtland. The most recent are those by Ayres (1998), McLaren and his colleagues (1997), Carley and Spapens (1998) and

Sachs *et al.* (1998). All these reports provide extensive evidence of the changing physical and economic circumstances of the planet, together with assessments of the ecological and social 'burdens' or 'footprints' of particular activities and consumption patterns. We will look at this area of analysis later in the chapter. The point here is that improved analyses of socio-ecological 'rucksacks' are being made nowadays, so it is a little easier to look at the broader picture, as well as gain a better basis for incorporating taxes or levies into economic activity. 'Paying our way' is now a significant policy arena, as outlined by O'Riordan (1997) and developed by Neil Adger in Chapter 4.

Despite improved measurement, far too little is actually being done to promote the passage of the transition to sustainable development. We shall see in the chapter that follows that this transition, or pathway, is enormously problematic in political terms even to embark upon, let alone traverse. There is no clear agreement as to what sustainable development is, every pathway begins and ends at different points, and almost inevitably the current pattern of economic gain and political power is institutionally ensnared in non-sustainable development. Arguably, it is the non-sustainability that retains this institutional order, so one can hardly expect it to write its own epitaph in the interests of a contradictory and ambiguous goal.

Any talk of transferring resources, capabilities or technology from the rich North to the poorer South raises doubts as to how such transfers are to be guaranteed in such a way as not to increase corruption or dependency, or both. Of much interest is the changing perception of trade, democracy and security across the North and South. The pattern of trade is assisting non-sustainable competitive advantage between countries, while for many political economists free trade is the precursor for better democracy. The Group of 77 (actually over 125 poor countries) feels that the wealthy nations recognise their responsibility for keeping the wells of life intact. This is the so-called *natural fabric* of life-retaining processes that form the biogeochemical fluxes of the 'Gaian' bonding that is this amazing planet. In Chapter 4, we look at how far these critically important attributes can be valued and placed in the global economic account. The science and educational implications of the Gaia theory are covered in Chapter 19. This natural fabric cannot be taken for granted. Indeed, its integrity depends on the co-operative responsibility of *social fabric*, namely the cultural norms and accepted behaviours of a society that is motivated at the personal and the collective level to care for and to tend both the human family and its precious ecological support base.

It is no wonder that global ecological security is high on the international agenda. At any one time, there are over 40 local conflicts that displace people and destroy livelihoods (see also the final section of Chapter 6). The maintenance of civic liberties in a society that also recognises the intrinsic rights of the natural fabric is becoming more and more central in the interpretation of sustainable development. Humanity does have the livelihoods of its members on its hands. This is why global environmental change will dominate international economic and political affairs, as well as strategic military analysis, for many years to come. We take a look at the coupling links between security and insecurity throughout the book, but especially in Chapters 4, 5 and 6.

This chapter also looks at the Rio Process, both the causes of the UN Conference on Environmental and Development held in Rio de Janeiro in June 1992, and its very mixed and contradictory aftermath. Sustainable development is an unavoidable concept these days. It may be impossible to define, in part because it is intensely 'geographical' in its manifestation. After all, it is a combination of culture, history, land, people and institutions, so its character and conduct will be as varied as the geography of the globe. However, there are common themes and shared experiences, and these will be reported on in this chapter. The most important point to grasp about sustainable development is the paradoxical observation that it will only succeed by capturing and redirecting social and economic change, yet it also has to act as an organising focus independent of that capturing process. It requires both an identity and an accumulative role in a myriad of circumstances.

The state of the planet

According to the conclusions of a distinguished meeting of a cross-section of the world's scientists,

> *Highest priority should be given to reducing the two greatest disturbances to planet Earth: the growth of human population and the increase of resource use. Unless these disturbances are minimised, science will become powerless to assist in responding to the challenges of global change, and there can be no guarantees of sustainable development.*   (Dooge *et al.*, 1992, 7)

This statement, together with the one that introduces this chapter, both from the pinnacle of the scientific profession, conclude that humanity has the power to disturb but not destroy the life support systems of a globe that may be unique in the universe, and which has maintained life for at least 3.5 billion years. We shall see that what is really threatened is humanity itself, or at least the one-third of the population which is vulnerable even to relatively modest reductions in basic resources such as fertile land, fuel, water and waste disposal capability.

Evidence from the fossil record suggests that the Earth itself is staggeringly resilient. The biosphere – the living envelope of air, water and land that maintains life – has endured volcanic 'blackouts' and huge meteorite collisions which have drastically altered climates and caused catastrophic species kills (Wilson, 1989). But, given time, the Earth has not only restored its equilibrium, it has actually increased the diversity and complexity of species, not least evolving *Homo sapiens*. (For a fine review, see Lenton, 1998.)

However, these periods of restoration have always been very long in human time scales, and in the interim there were undoubtedly wholesale shifts in the chances of survival at a regional scale. The Earth itself has no feeling for its species: species reproduce and die according to the mysterious processes we call creation and evolution. Until humans appeared there was no sentient being to mourn loss or welcome birth. It is we, as a species, who care about life on the Earth as a whole. As the historian Arnold Toynbee (1976) observed, humanity is both good and evil; it destroys, but it can also create and restore. Alone amongst all living beings, humans have rational minds and emotional souls: humans have a conscience so are capable of reflective judgement. Throughout history, humans have been troubled by knowledge of their ability to undermine their livelihood and an awareness that only they can stop themselves from doing so.

Report after report, book upon book, conference following conference all conclude that humans are in trouble and that the Earth is patchily stressed. How serious all this is will be reconsidered in Chapter 6.

There are four annual reports that should be consulted if you wish to remain up to date with the scale and extent of global environmental change. These are *World Resources*, published by Oxford University Press and produced by the US World Resources Institute; the *Human Development Report*, also published by Oxford University Press and produced by the UN Development Programme; *State of the World*, published by Earthscan and written by the US Worldwatch Institute; and the UN

Development Programme *Environmental Data Report*, published by Blackwell. Box 2.1 summarises the evidence as it has accumulated over the past 20 years. Such is the ubiquitousness of the media and the widening of the school curricula that much of this is well known to the reader.

Box 2.1 The state of the planet as summarised by various scientific audits

Before the advent of telemetry and remote sensing monitors, much information on global environmental change was anecdotal and subject to cultural biases at a local level. Now the stream of data is almost ubiquitous and could be timeless if sufficient money was made available for monitoring. This is a topic of some controversy in its own right. Monitoring by accurate but remote means is very expensive and potentially open-ended in scale, such is the frenzied rush of new technology. Whether such relatively expensive data are worth all the new evidence is a moot point. In Chapters 11 and 13 we make the case differently for two key themes of global environmental change, namely soil health and ocean well-being. What is of interest in the recent audits is the reversal of the *Limits to Growth* findings (Meadows *et al.*, 1972). The world is not running out of the non-renewables such as oil, gas, coal, bauxite and other minerals. Indeed, commodity prices are so low as to weaken the economies of countries dependent on the export of commodities. It is the renewable resource base that is under siege nowadays.

- World population is 5.8 billion and increasing by 88 million annually, or around 250 000 daily.

- World economic activity has grown by 3% per year since 1950. If this trend continues, total world output will be five times larger than it is today by 2050. That would require a second planet to accommodate it, if ecological burdens remain the same.

- 500 million people exist on marginal lands, already incapable of feeding them.

- 20 countries already suffer from water stress, having less than 1000 m³ *per capita* per year, and total water availability has dropped from 17 000 to 7000 m³ *per capita* per year.

- Since 1970, the world's forests have fallen from 11.4 km² to 7.3 km² per 1000 inhabitants.

- 25% of all fish stocks are currently overfished, and a further 44% are fished to their biological limits. Over 1 million people cannot get adequate fish protein.

- 425 000 ha of mangrove swamps has been lost to shrimp farming in East Asia over the past five years, exposing the coastline to brackish water and high tidal inundation.

- Wild species are becoming extinct 50–100 times faster than they would naturally.

- 2.2 million people die every year from indoor air pollution, mostly in rural areas.

- 30% of the population in developing countries lack access to safe drinking water, and 2 million die every year from associated diseases. Over 90% of all wastewater in the developing world is untreated.

- Air pollution from cars and industrial exhausts kills over half a million annually, mostly in Third World cities.

- Nearly 30% of domestic solid waste is uncollected in Third World cities.

- 11 million poor agricultural workers suffer from pesticide poisoning annually.

- Acid rain has damaged 60% of Europe's commercial forests and is reducing crop yields by 25% in East Asia.

- Household debt is growing at 3% per year. In the USA, households save only 3.5% of their income, half as much as 20 years ago.

- global consumerism has grown by over 350% since 1990, and is rapidly increasing in developing countries. Already total spending exceeds $1.5 billion per day

Source: United Nations Development Programme (1998, 66–85).

The following conclusions reveal the severity of the crisis.

- *About 5000 children per day* die because of avoidable lack of food, water, sanitation and basic health care. This is primarily an outcome of poverty, powerlessness, and the inability of national and international organisations to provide for basic needs.

- *About 900 million people live in circumstances where their established means of producing food and gathering fuelwood and clean water are no longer sufficient* to keep them or their families alive above the bare subsistence level. This is a function partly of poverty and partly of the lack of appropriate food markets and a suitable distribution system.

- *About 15 million people have been displaced* from their homelands because of the inability to keep alive where they once lived, or because of oppression or military insurrection. About another 10 million people are displaced within their own borders into marginal lands or already highly stressed regions.

- *Loss of protective soil cover and forest cover is now so widespread* that erosion of land is beginning to prevent the creation of new food producing areas. The introduction of high yield, and genetically modified, varieties and improved marketing and storage have maintained overall *per capita* food production, but in parts of Africa and in South-east Asia, *per capita* food production has dropped for over a decade as populations catch up with productivity increases. Table 2.1 summarises the most recent estimates of degradation of global land. Annual loss of productivity is estimated at 0.1%, and cumulative loss since 1945 is around 4.5% (Crosson, 1997, 9). Crosson suggests that, overall, these losses are not too significant in light of these productivity increases, but that there are 'hotspots' of real degradation in the Indus and Tigris basins, in north-east Thailand and Brazil, in the steeply sloped areas of China and South-east Asia, and on semi-arid hill slopes in most tropical areas. Michael Stocking provides more information in Chapter 11.

Degradation category	Amount of land affected (million ha)	Lost production (%)
Total land	8735	
Not degraded	6770	0
Degraded	1965	
lightly	650	5
moderately	904	18
strongly	411	50

Source: Crosson (1997, 9).

Table 2.1: Degradation status of global land in crops, permanent pasture, forest and woodlands.

- *Population growth, compulsory migration and landlessness*, forcing people to move, as well as road schemes that improve accessibility to previously uninhabited areas, are all contributing to the loss of tropical forests, at the rate of 2% annually, as are national economic policies that even today subsidise the removal of hardwood despite the calamitous environmental consequences.

● *Possibly as many as two-fifths of the world's peoples live under conditions where small changes to climate, water availability and access to fuelwood will have disproportionate effects on their chances of survival.* Climate change may be a century away in terms of truly noticeable departures from temperature norms. But relatively small fluctuations in the form of precursors to that change will almost certainly adversely affect many millions of people, despite their in-built resilience to environmental change (this matter is also covered in Chapter 6).

● *For the great majority of the world's poor the killer is not environmental degradation on a global scale.* It is the common scourge of disease, inadequate sanitation and nourishment, and localised pollution from cars, household and industrial waste, and domestic fuel burning. For this beleaguered group, global environmental politics are effectively the same as polishing the brass on the *Titanic*.

Environmental stress and social deprivation are with us already. Television footage shows us that virtually every day. The problem, as everyone knows, is to try to move development into an environmentally sustainable path in such a way as to make it as painless as possible for the already deprived, whose poverty and local indebtedness drives them to destroy the very land upon which their survival depends.

Much of the debate so far has emphasised the priority that should be given to food security, land stabilisation, energy availability and civil rights for minorities and women generally, if the basic preconditions for sustainable development are to be met. The richer nations of the Organisation for Economic Co-operation and Development, however, tend to emphasise instead the 'big' global issues that appear to threaten them directly. These are *climate change* with its imponderable implications for agriculture, water supply and coastal protection, *ozone depletion* with its equally uncertain implications for skin cancer, eye cataracts and the functioning of phytoplankton on the surface of the sea, and *loss of species and habitats* with its attendant waste of potentially vital genetic resources for pharmaceuticals, food technology and pest control. Figure 2.1 summarises the framing of global environmental change, and in Chapter 19 we look more closely at the law and politics of multinational environmental agreements.

For the poor South these are potentially devastating dangers, but they are not so clearly a matter of priority as is the day-to-day requirement of survival that affects so many of its peoples. Box 2.2 looks at the particular circumstances of South Africa to illustrate this point, though the theme is more generally applicable to other countries in economic and social transition. We shall see that the poor countries are becoming very dismayed that the high talk of sustainability is not being converted into serious commitment to assist their own particular pathways to sustainability.

It is tempting to conclude that wealthy nations worry about global changes whose consequences may be two or more generations away, while poor nations do not have the cash to provide clean water and proper waste removal in desperately polluted towns filled with toxic dust and fumes. In a global survey of heavy metal pollution, Nriagu (1990) concluded that over 150 million people experienced elevated lead in their blood, 250 000–500 000 suffer renal failure due to cadmium poisoning, and another 500 000 are believed to have skin cancer as a result of arsenic poisoning. In developing countries generally, Nriagu (1990, 32) concludes:

These people are much more predisposed to being poisoned by toxic metals in their environment: poor nutrition and health, high population density, poor hygienic conditions, and a preponderance of children and pregnant women – who are considered to be most at risk – all enhance the susceptibility to environmental metal poisoning.

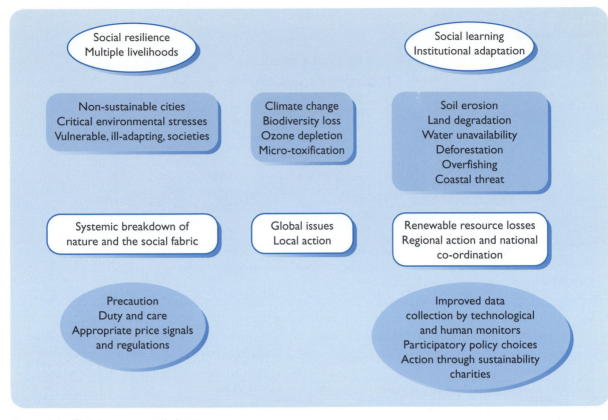

Figure 2.1: Global environmental change.

Box 2.2 Sustainability for survival in South Africa

The emergence of democracy in South Africa brought with it a drive for more justice, equality and economic opportunity for the 22 million formerly oppressed and impoverished by apartheid. This led to the Reconstruction and Development Programme of the African National Congress, and the emergence of the Growth, Equity and Redistribution Programme, both aimed at getting into the job market. To assist in this process, the rise of an already active civil society was fostered by local citizens' empowerment programmes, and the creation of a national participatory process on environmental quality and sustainable development. This is the Consultative National Environmental Policy Process (CONNEPP) that ran into difficulty the more it promoted participation, but the less it managed to incorporate or convince disadvantaged groups. An added element was the framing of the 1996 Constitution, which ostensibly gives citizens rights to an environment not harmful to health or well-being. In practice, however, this is not proving an effective vehicle for civil action, despite the palpable evidence of failure to deliver. Nevertheless, the institutional framework of sorts is in place, and this is a hopeful beginning. Figure 2.2.1 summarises this framework, the so-called 'triple helix'. The following abbreviations are used in the figure:

● RDP: Reconstruction and Development Programme aimed at ensuring that the underprivileged gain access to government and the economy.

● GEAR: Growth, Equity and Redistribution Programme aimed at affirmative action in favour of the less privileged gaining access to jobs and management responsibility.

Box 2.2 (continued)

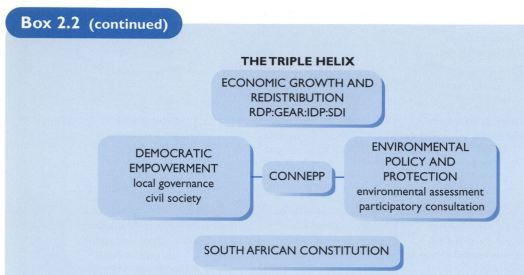

Figure 2.2.1

- IDP: Integrated Development Programmes to create partnerships in development and environmental well-being via participatory governance.
- SDI: Spatial Development Initiatives – zones of 'fast track' economic activity.
- CONNEPP: Consultative National Environmental Policy Process, a nationwide debate that led to the highly innovative National Environmental Management Act 1998.

The South African Constitution, for all citizens, lays down the following:

- a right to an environment not harmful to health and well-being;
- a right to ecologically sustainable development and use of natural resources while promoting justifiable economic and social development;
- a right to information from public and private bodies;
- a right of access to the courts for redress of grievances;
- intergovernmental dispute resolution procedures at national, provincial and local levels;
- reallocation of natural resource property rights for the common good;
- a duty of envrionmental care on all public agencies.

Source: O'Riordan (1998, 102).

So it should not be surprising that a 22-nation poll prepared specially for the Rio Summit (see below) showed a widespread concern over environmental contamination, with a majority of all people sampled placing environmental protection and restoration among the top three issues for governmental attention (Dunlap *et al.*, 1993).

Table 2.2 summarises the details of global environmental concern. When environmental issues are not specifically mentioned, only a few spontaneously mention this item as a priority issue. Nevertheless, over a fifth did so in the Netherlands, Mexico, Finland, India, Switzerland and Chile. Most significant is the relatively comparable responses of anxiety about environmental degradation for all countries, irrespective of their wealth. This suggests a ground base of permanent concern for environmental improvement the world over. Figure 2.2 reveals that, in the UK, concern for environmental quality appears to be treated as a luxury, to be expressed only when the economy is booming, crime is falling, and health and educational services are well

Table 2.2: Comparison of responses to environmental concern from a global survey of 24 nations. In general, environmental matters do not score highly when unprompted, but underlying concern is often quite strong, even in poorer countries. These responses suggest that, in broad terms, policy action in favour of environmental protection will attract public support in almost every country. The table shows the percentages of respondents who say that environmental problems are the 'most important' or a 'very serious' issue in their country, and of personal concern to them 'a great deal' or either 'a great deal' or 'a fair amount'.

	Most important	Very serious	Great deal	Great deal/ fair amount
Industrialised countries				
Ireland	39	32	22	73
Netherland	39	27	16	71
Finland	28	21	16	63
Portugal	25	51	46	90
Switzerland	20	63	12	42
Denmark	13	26	12	53
Japan	12	42	23	66
USA	11	51	38	85
Canada	10	53	37	89
Germany	9	67	14	63
Norway	7	40	18	77
Great Britain	3	36	28	81
Developing countries				
Mexico	29	66	50	83
India	21	51	34	77
Chile	20	56	30	70
Turkey	18	61	12	40
South Korea	9	67	22	80
Russia	9	62	41	78
Uruguay	3	44	38	82
Brazil	2	50	53	80
Philippines	2	37	55	94
Nigeria	1	45	71	87
Poland	1	66	4	25
Hungary	1	52	32	79

Source: Dunlap, R.E., Gallup, G.H., Jr and Gallup, A.M. (1993), *Health of the Planet*, George H. Gallup International Institute, Princeton, NJ.

provided for. This may be very misleading. As we shall see below, sustainable development provides a policy bridge between environmental quality, economic enterprise and social caring. We shall discuss in Chapter 6 that maintaining the natural and social fabric in a creative and progressive manner may well become the focal feature of our economy in the mid-22nd century, as we learn more that enriching livelihoods for all is the routeway to ecological security. Sustainability is not a tradable luxury.

But even more insidious in terms of real livelihoods is the very subtle but persistent erosion of freedom, denying people any vestigial ability to cope with the drip-drip loss of health, nutrition and wealth. This is the issue of creeping powerlessness and exclusion.

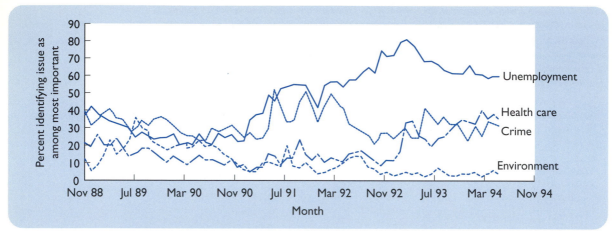

Figure 2.2: Environmental attitudes in the UK, 1988–94.

Source: MORI (Market Opinion Research International). Reproduced with permission.

Excluding the commons

The Ecologist magazine was one of the first to champion the cause of sustainability, long before it was fashionable in UN circles. Its pioneering issue *Blueprint for Survival* (1972) presaged the first UN Conference on the Human Environment in Stockholm in 1972. Much of that text was rubbished by commentators at the time, but the majority of its analysis and prescriptions have subsequently appeared in various official reports, not least the Brundtland Report (1987), of which more later. Let us look at the argument for the restoration of the commons put forward in the 20th anniversary edition of *The Ecologist* in June 1992.

'The commons' is not so much an amalgam of collectively owned resources as a form of communal existence where reciprocity is the order of the day. In the commons, people recognise their dependence on each other and on the Earth that supports them; people aid those who are in distress, because they know they in turn will be assisted when times are hard; wealth is shared as a recognition of social responsibility and veneration; and services are exchanged rather than commodities sold for money.

The commons is therefore a particular form of social and cultural organisation:

local or group power, distinctions between members and non-members, rough parity among members, a concern with common safety rather than accumulation, and an absence of constraints which lead to economic scarcity.

(The Ecologist, 1992, 125)

In this context, 'power' may refer to the right to exclude outsiders or to punish anyone who abuses the commons. It may also mean an additional structure of internal rules, rights, duties and beliefs that allows the commons to become the 'owner' of human behaviour. Thus the commons becomes a metaphor for a sustainable economy in an equitable and attentive society that, almost without the need for recordable communication, knows why and how to co-exist.

According to this analysis, the true global dilemma lies in the steady enclosure of the commons – by multinationals, by nation states, by intergovernmental organisations and by thieves. In the words of *The Ecologist*:

Development can never benefit more than a minority; it demands the destruction of the environment and of peoples. It attempts to dominate, fragment and dispossess – in a word, enclose. The challenge is to reject development and reclaim the commons.

(front cover)

Enclosure means to open up the commons to materialism, individualism, dependency and oppression. It means denying people the rights to survive and to earn a sustainable living in the name of progress and trickle-down profit dissemination. It means demeaning traditional norms by punishing those who cling on to past ways, eventually to eliminate what they stand for. Enclosure is the process of non-sustainability. According to the US commentator Ivan Illich, enclosure is the 'new ecological order': its most insidious quality is to create structures of control that reinforce its authority. For example, the green revolution not only resulted in dependency on fertilisers and pesticides: it created commercial alliances between seed manufacturers, agrochemical firms and pharmaceutical corporations (Shiva, 1991). A similar pattern of commercial hegemony is now being observed in the biotechnology conglomerates which are beginning to dominate global food markets (see Environmental Data Services, 1998, 18–30). For example, in 1997 alone, the US agri-chemical giant Monsanto spent $6.5 billion on acquisitions of seed, pesticide and fertiliser companies.

This analysis is radical but problematic. It is based on an assumption that frugality, self-restraint and the joy of minimalism is somehow an ideal human condition, the natural end-state of a creative culture. It suggests that maintenance of such conditions by barring entry into commons of developmental activities such as more open markets, health care, literacy and credit can become culturally eroding and ultimately destructive. This is dangerous rhetoric: it implies that retaining a self-reliant barrier to external influence is somehow the best path for sustainable development. There is a touch of well-meaning elitism about this perspective, but surely there must be and should be almost as many tracks to sustainability as there are communities. Yet it is troubling in that there is more of a grain of truth to the proposition that the endeavours of those who preach sustainable development are proposing reforms that could well further enclose the commons. We examine this conclusion in more detail in Chapter 6.

The reader wishing to pursue this analysis should consult Sachs (1993). He sees the act of seeking a more sustainable salvation being embraced by a political order that seeks to retain efficiency, progress and the vested interests of acquisitive capitalism.

As governments, business and international agencies raise the banner of global ecology, environmentalism changes its face. In part, ecology understood as a philosophy of a social movement is about to transform itself from a knowledge of opposition to a knowledge of domination ... The purpose of global environmental management is nothing less than the control of the second order; a higher level of observation and intervention has to be installed, in order to control the consequences of control over nature.

(Sachs, 1993, xv)

The UN Conference on Environment and Development: Rio 1992

This famous conference, referred to as UNCED or the *Earth Summit*, produces strong and opposing reactions amongst commentators. Needless to say, much of this has to do with the position on the sustainable development spectrum, the experience of the pre-Rio bargaining period, and the dashing of exaggerated expectations but genuine demands.

Box 2.3 Quotes from UK environmental NGOs at Rio

We have to climb a mountain, and all governments have succeeded in doing here is meander in the foot-hills having barely established a base camp.

(Jeremy Legget, scientific director, Greenpeace, *Sunday Times*, 14 June 1992)

We need a paradigm shift. I saw no sign of that happening in Rio. Of course we have to welcome any progress, but it has been microscopic.

(Jeremy Legget, scientific director, Greenpeace, *The Independent*, 15 June 1992)

The Earth Summit has exposed the enormous gulf that lies between what the public want and what their leaders are willing to do. The North has done little to signal, let alone address, the issue of its over-consumption. Much of the burden of the environment and development crisis has been left on the shoulders of ten of the world's poorest countries in the South.

(Andrew Lees, campaigns director, FoE, *The Independent*, 15 June 1992)

It's all generalities. We need to know specifically what is going to be done and by when.

(Charles de Haes, director general, WWF-International, *Financial Times*, 15 June 1992)

The Earth Summit was a failure. The words were there but the action was lacking.

(Chris Rose, campaigns director, Greenpeace, *The Guardian*, 15 June 1992)

I came here with low expectations and all of them have been met.

(Jonathon Porritt, *The Guardian*, 2 June 1992)

As the text suggests, these were hip-firing reactions partly to meet the needs of media seeking instant and memorable summaries. The trouble with the media craving for simplicity and 'one liners' is that no service is provided for thoughtful and considered debate.

One influence on public opinion was the devastating critique offered by the non-governmental organisations almost before the summit was finished (see Box 2.3). More significant, perhaps, was the difficulty which any but the most involved had with conceptualising the concept of sustainability and subsequently converting that into their own behaviour and political aspirations. After all, the sustainability problematique requires millions of people to undertake hundreds of consumer and leisure choices differently, without necessarily any price signals to guide them, and with no guarantee that their own contribution will be matched by others. This is the classic 'free rider' problem, namely the corrosive thought that one's 'sacrifice' is a gain for the conscienceless. It takes dedication of a high order, or a coalescence of regulatory and economic institutions, to overcome this paradox of converting micro-behaviour to macro-sustainability. We look at this again in Chapter 19.

The conference itself was designed to take stock of the state of the world 20 years after the first major Earth summit, the UN Conference on the Human Environment held in Stockholm in June 1972. That conference was noted for the suspicion, held by many Third World countries, of the motives of the developed world. The view was that somehow the rich nations wanted to limit development for the sake of the Earth (or their aesthetic sensibilities), or at least control future aid and trade on environmentally conditional terms. This led to charges of environmental colonialism and subjugation of the poor to the whim of the wealthy, much of whose riches were earned on the sweated backs of the impoverished.

Box 2.4 The various meanings of sustainable development

Note the distinction in the wording between interventionist and nurturing modes as captured in the language adopted, in general, by the economic ministries on the one hand and the environmental ministries on the other. Note, too, the subtleties between realist and more integrationist language as part of regime creation. Underlying this language is a rich texture of interpretations of power, Gaia, commons and revelation.

Sustainable development is economic and social improvement that is:

● continuous and permanent;

● durable and reliable;

● proactive and just;

● enterprising and sharing.

Three fundamental principles are:

● to maintain and protect the essential life support processes of the planet;

● to utilise renewable resources to the point of precautionary replenishment;

● to price the cost of living according to its natural burdens and social disruption.

Vehicles for implementation are:

● international agreements around the safeguarding of life support;

● compensatory transfers to recognise the legitimate needs of the sacrificers and the vulnerable;

● national integrative policies coupled to formal duties of environmental care;

● comparative economic performance and regulatory equal treatment;

● local commitment via formal and informal community democracy.

Source: O'Riordan and Voisey (1998, 6).

The Stockholm Conference produced an uneasy relationship between North and South, tempered by the establishment of the UN Environment Programme aimed at monitoring environmental change and championing the cause of directed assistance to alleviate environmental damage and resource exploitation. Subsequent conferences on desertification, the social agenda, women's rights, population and urban settlements achieved little of note except plenty of rhetoric and grand programmes of aid, few of which added new money. The UN Conference-go-round was regarded as an expensive circus of little relevance to the needs of those most desperate.

The World Commission on Environment and Development was formed under UN auspices in 1983, under the chairmanship of Mrs Brundtland. This commission was charged with the task of identifying and promoting the cause of sustainable development. This is a phrase that remains mysterious in application (see Box 2.4). Its very ambiguity enables it to transcend the tensions inherent in its meaning. It has staying power, but no-one can properly put it into operation, let alone define what a sustainable society would look like in terms of political democracy, social structure, norms, economic activity, settlement geography, transport, agriculture, energy use and international relations. Arguably the Earth Summit was convened around a mystery.

If nothing else, the Rio Summit was a triumph of organisation. It was presaged by four two-week preparatory conferences, each involving up to 170 countries and each attended by NGOs representing nine 'stakeholder groups' – *indigenous peoples, environmental and developmental organisations, science, local government, business, farmers, trades unions, women, and youth*. These were hard-fought occasions focusing particularly on financing mechanisms, North–South technology transfers, international conventions and overconsumption of resources (i.e. waste or inefficient use). In the event, UNCED attracted 110 world leaders, representatives from 153 countries, 2500 non-governmental groups, 8000 accredited journalists and over 30 000 associated hangers-on. Many leaders had never addressed the twin theme of environment within development before, few countries had embarked on any organised environmental plan, and even fewer had created any sort of mechanism to generate the dialogue among the stakeholder groups that was regarded as a prerequisite for any kind of strategy inching its way towards sustainable development.

The good points about Rio were as follows.

- The excitement of the event itself, the feeling of international solidarity amongst activists, the tremendous surge of self-belief and support for groups who, before Rio, had struggled in an unreceptive culture to try to organise civil rights in depleted commons. That green networking should never be underestimated. Nowadays there are 25 000 NGOs active in the post-Rio scene, and many are coalescing around environment and development issues. There is at least a recognition that the poverty groups in the developed world, and the aid charities in the developing world, should form some sort of accommodation with environment groups as well as civil rights organisations such as Amnesty International. This coalescence of action at international, national and local levels is mobilised by a real momentum of 'grass roots enthusiasm'. Because the groups are formally part of all negotiating parties, Rio has electrified the NGO world. Of course, these are still early days, and even the most effective of NGO co-ordination can do little if governments do not listen, international financing bodies remain unresponsive and unhelpful, and party political ideologies refuse to embrace the social and egalitarian aspects of sustainable development.

    Nevertheless, as democracies spread to over half of the world's nations, so the NGO movement is allowed to flourish. True, there are far too many examples of serious abuse and oppression of environmental rights and social justice groups even in so-called functioning democracies. The Amnesty International website will provide the reader with plenty of evidence here. But, on the whole, the drive towards attending to localism and in formal patterns of governance the world over is a helpful indication of the linkage between sustainability and participatory democracy. This relationship will have to be encouraged by aid donors and by international agreement in the years to come, and there are signs from the World Bank and the development agencies that this is indeed the case (see World Bank, 1997).

- Three global conventions were signed, on climate change, biodiversity and desertification. Chapter 5 looks at biodiversity, Chapter 7 climate change, Chapter 13 desertification and Chapter 19 the wider issues associated with managing the global commons. These are based on 'soft law' or flexible wording which steadily commits all parties (including non-signatories) to international obligation. Each convention is overseen by a Conference of the Parties (the signatory nations) backed by scientific and technological assessment committees, international financing arrangements, and a small but significant secretariat. Each expects all signatory nations to produce annual reports of performance aimed at reaching agreed targets of commitment. Lack of these reports can be subjected to legislative surveillance and NGO politicking. In some countries this will be the case, in others not. But the effort of meeting similar norms of performance will help to spur the laggards. Chapter 19 looks at this aspect in more detail.

● Agenda 21 was the centrepiece of the meeting, a 40-chapter report on what was wrong and what should be done to correct it. Each chapter deals with goals, priorities for action, a programme of follow-up and a cost estimate. These form the basis of national sustainable development strategies that are now sent annually to the UN Commission on Sustainable Development. Chapters 23–32 of Agenda 21 deal with strengthening the role of nine key stakeholder groups. Chapters 35–37 cover science, public awareness and education, and capacity building for both science and technology transfer in developing countries. We shall see below that cash for capacity building is bound up with global financial transfers. But the concept of a global audit as outlined by Brundtland would upgrade the existing database on global change, tie it to models of criticality on a regional basis and connect it to strategies for survival, hopefully with targeted international cash. These are long-distance objectives, but at least the concept is laid down. These chapters are fundamental for the future of active environmental science and the promotion of global citizenship – a community that is both knowledgeable and has the tools to alter their societies and economies towards greater sustainability. The political self-awareness mentioned in the introduction is vital here. The success of Agenda 21, in essence, hinges on the willingness of governments to create both the educational and political conditions for global citizenship to flourish. The test of Agenda 21 is to see how far these take place in the sustainable development strategies that each nation has to prepare as their contribution to meeting Agenda 21. Box 2.5 summarises the main agreements signed at Rio.

Box 2.5 The agreements signed at Rio

The Framework Convention on Climate Change

● Climate change is a serious problem.

● Action cannot wait until resolution of scientific uncertainties.

● Developed countries should take the lead.

● Compensation paid to developing countries for additional costs of implementing the Convention.

● Developed countries aim to reduce by variable amounts above and below 1990 emission levels by 2010.

● Reporting process established, along with a host of implementing mechanisms.

Convention on Biological Diversity

● Conservation and sustainable use of biodiversity and components.

● Affirms states have sovereign rights over biological resources in their territories.

● Benefits of development of biological resources to be shared in fair and equitable way on mutually agreed terms.

● Countries to develop national plans.

● Funding arrangements under GEF.

Agenda 21

● 40 chapters outlining action plan for sustainable development.

● Integrates environment and development concerns.

● Strongly oriented towards bottom-up participatory and community-based approaches.

● Acceptance of market principles, within appropriate regulatory framework.

● Performance targets mostly limited to those previously agreed elsewhere.

Rio Declaration

● 27 principles for guiding action on environment and development.

● Stresses right to and need for development and poverty alleviation.

● principles concerning trade and environment are ambiguous and in some tension.

● Rights and roles of special groups addressed.

Box 2.5 (continued)

Forest Principles
- Represents blocked attempt to negotiate a convention on forests.
- Emphasises sovereign rights to exploit forests.
- General principles of forest protection and sustainable management.

The Convention to Combat Desertification
- Focus on Africa, Asia, Latin America and the northern Mediterranean.
- Involvement of local people and indigenous cultures.
- Co-ordination of agencies and governments.
- But lack of adequate global interest on what is still seen as a regional problem.

Any reader anxious to follow the Rio Process would be advised to look at the books by Bill Adams (1990), Ian Moffatt (1996), Clayton and Radcliffe (1996), Felix Dodds (1997), Osborn and Bigg (1998) and Lafferty and Eckerberg (1998). Two invaluable critiques are provided by O'Keefe *et al.* (1993) and Wolfgang Sachs (1993). What follows cannot do justice to the breadth of detail provided by these books, and by the more theoretical treatment provided in the introduction to a book edited by O'Riordan and Voisey (1998). So what follows is a basic summary, helpfully also provided by Jordan and Voisey (1998, 93–97).

The Rio Conference was always designed to be a process rather than a meeting. It put into place a number of institutional arrangements, of which the most noteworthy are the Commission on Sustainable Development (see Box 2.6), a UN Department of Policy Co-ordination and Sustainable Development, and an agreement, to review progress in UN Plenary Session every five years. The first review took place in New York as a Special Session of the UN General Assembly (UNGASS). This was not a success: it was not well prepared, the necessary preliminary conference and political commitments were not put in place, and the main players in Europe, North America and East Asia were too preoccupied with their own domestic politics to be attracted to global issues of apparently some generations away. Here is a summary of UNGASS (O'Riordan and Voisey, 1998, 27–29).

- The social agenda was at least recognised as an integral aspect of sustainable development. This took the Brundtland agreements on social justice, civil rights and stewardship further than originally developed. But it has yet to appear forcefully in NGO alliances and governmental–NGO partnerships. There are separate budget pools and competing 'image-turfs' to fight for.

- The Convention on Desertification came into force before UNGASS, but did not get the further support its CoP was seeking. A conference on highly migratory fish stocks (see also Chapters 8 and 16) created a forum for assessing the decline in fish resources. But this is very much a talkshop with too little in the way of urgency, given the dramatic levels of overfishing worldwide. The forestry issue remains firmly in touch, with no significant convention in place.

- Freshwater depletion has also moved into the international spotlight, partly because of persistent warnings that up to a third of the world's population will not have access to sufficient water by 2050 (Dodds, 1997, 15). This theme is bound to become more important in future Rio updates. As yet no major new institutional arrangements are in place, though a convention is likely to be signed eventually.

Box 2.6 The Commission on Sustainable Development

The Commission is an interesting example of an evolving institution that has not failed in the labyrinth of competing UN organisations. It is a subset of the main UN ECOSOC Committee, primarily regarded as pro-development. But skilful chairs, plus an able secretariat, assisted by frenzied and persistent lobbying by NGOs from all over the world, have kept the Commission as an active force for sustainable development. It holds regional conferences on key topics, promotes best practice, receives the national sustainable development strategies and comments extensively on their quality and direction, and acts as an important conduit for sustainable development into the whole of the UN institutional system. The CSD is enormously dependent on good leadership, powerful patronage and supportive cabinets across the planet. That support has to be won: it can never be taken for granted.

The fact, however, that government representatives at CSD sessions are predominantly from environment ministries, with little influence over their more powerful colleagues in trade and industry departments, has undoubtedly lessened its impact. So too has the lack of interest shown by the green lobby. Aside from a hard core of environmental activists, some of its meetings are very poorly attended. The chief problem is a lack of clarity on what the whole reporting process is meant to achieve and the best means to get there. In the rush to 'do something' at Rio, many feel the criteria for assessment were neglected. What, for example, does the CSD do when it has received reports from every country? Beyond these procedural issues, critics question who the CSD is actually meant to be representing. Some claim that it remains intergovernmental and that it has made limited efforts to incorporate local environmental groups in its proceedings or even canvass their opinions.

UNGASS proved that sustainable development is not attractive to mainstream and insecure statespeople when herded into an international conference hall without adequate preparation and diplomatic protection. Quite frankly, the main international institutions are not up to the job. The UN system is not renewed or reorganised and remains generally discredited, despite the noble efforts of its new Director General, Kofi Annan. The principal agencies, notably UNEP, the United Nations Environment Programme (Sandbrook, 1998), lie in between the Conferences of the Parties and their secretariats that now cover climate change, biodiversity and desertification. Unless UNEP can wrest away the co-ordinating function of global environmental change from a host of petty bodies that are left, and ensure adequate collaboration at the level of the Convention and Protocol, then the preparation for Earth Summit III in 2002 will be equally flawed. To avoid that, there needs to be a report card for all international institutions dealing with environment and sustainable development, together with a focused secretariat and political commitment of a high order (Osborn and Bigg, 1998, 25–30). This is by no means guaranteed. Possibly, an even more strident presence of NGOs during the run-up to Rio III will prove counterproductive, no matter how exasperated they must feel. The view of the Zaïrese delegate at UNGASS reflects the sense of failure to link sustainability to development:

If this kind of Earth Summit circus continues, then the people of Africa will perish. We need the rule of law. We need democracy, peace with justice, and we need fair terms of trade so we can develop a proper market economy, then we can protect our environment. (Quoted in Jordan and Voisey, 1998, 94)

Financing sustainable development

Various estimates have been made as to how much the transition to sustainability might cost. For the most part this is creative guesswork, because no-one can know the full figure. To begin with we have no idea in detail just what needs to be done – in provision of clean water, in reforestation, in education, in primary health care, in waste treatment, in land remediation, and in human rights. This is why the capacity-building element of Agenda 21 is so important. Local knowledge needs to be amalgamated into local, regional and national accounts to build up a picture of restoration and protection. Sadly, it is precisely in this zone of local knowledge, and the incorporation of locally inclusive participation, so central to the practice of interdisciplinarity (see Chapter 1), that the Rio process has noticeably failed. Despite recent efforts to 'green' aid by targeting it towards NGOs and community uplift in the poorer areas, this total effort is swamped by global economic pressures.

Why is this the case? To begin with, the UN system has no muscle for ensuring comprehensive reporting. Second, there is inadequate donor money targeted to the organisations that could undertake this level of reporting. Third, the actual evidence is difficult to collect and to measure in compatible databases. Fourth, the NGO movement is still stifled in this vitally sensitive arena of human justice reporting. And finally, science in general has still to find an interdisciplinary 'language' to give this kind of activity the prestige it deserves. These are the kinds of institutional failures that plague global environmental change generally, and environmental education in particular (see Chapter 19).

Since 1960, the ratio of income of the top 20% of the global population to the bottom 20% has increased from 30:1 to 80:1. Figure 2.3 summarises the nature of the growing inequality. Equally evident is that globalisation forces up the expectations of 'basic needs'. For example, citizens in Thailand and Korea have three times the number of cars per 1000 people at the same level of income as was enjoyed by Europeans in the 1950s. Some 270 million 15–18-year-olds in 40 countries enjoy the same 'pop culture of clothes and leisure and entertainment' – almost as a right. Yet 300 million people in the richest nations still lack all the basic services for survival and self-esteem (UN Development Programme, 1998, 3–10).

Levels of development assistance to the poor nations have fallen from 0.34% of GDP (gross domestic product) in 1992 to 0.25% in 1997 (Bramble, 1997, 192–193). The UK (–0.6 %), France (–12%), and, above all, the USA (–28%) have retreated in a conspicuous manner as domestic politics intervened to cut international charity. The total debt of the developing countries has risen to over $2 trillion, up by 8% in 1995 alone. Efforts to reduce that debt in the G7 summit in Birmingham in June 1997 noticeably failed (see Box 2.7).

All this adds up to a serious erosion of the capacity of Third World nations to manage their way to sustainable development. One has to be careful here. Sustainable development is about enterprising self-renewal and not huge subsidy or aid transfer. As we shall see in Chapters 6 and 11, community resilience and adaptability are hallmarks of survival, security and livelihood maintenance. But such admirable coping strategies do have to be supported and nurtured. It is this failure to target sensitively and constructively that is the most disturbing feature of the financial failure. UNGASS simply failed to deliver on any new financing mechanisms and targeted capacity building. Chapter 6 provides guidelines as to how capacity building can be determined.

Another controversial aspect of international financing for sustainability concerns how far international money is necessary to pay for the extra expenditure required to meet a global commitment rather than a national priority. This is known as the *incremental*

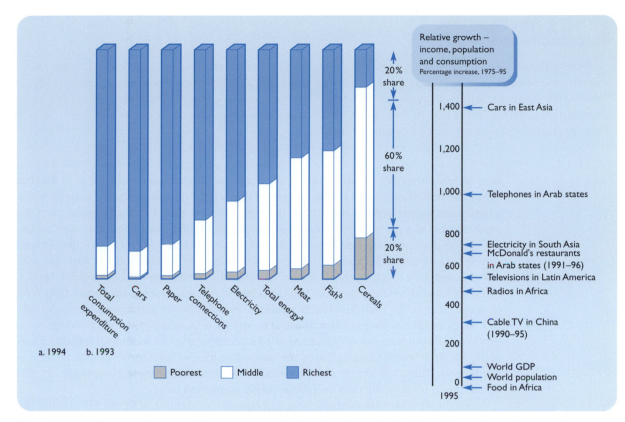

Figure 2.3: Rapid consumption growth for some, stagnation for others, inequality for all – with mounting environmental costs.
Source: United Nations Development Programme (1998).

Box 2.7 The unresolved debt crisis

The poorest countries receive disproportionately little aid: 80% of all the global capital flows are targeted at only 12 countries. Fifty countries spend more than a third of their wealth creation on debt repayments. The Heavily Indebted Poor Countries Initiative, aimed at streamlining debt relief and building in thresholds for sustainability, has all but collapsed since its inception in 1996. Despite compliance with the conditions set by the International Monetary Fund, Uganda and Bolivia had to wait over a year before they became eligible for debt relief. Tanzania will not be eligible until 2002, yet currently it spends three times as much on debt repayments as on education, and nine times as much as on primary health care.

Proposals for improvement include the following:

● Cut the waiting period for eligibility from six to three years, based on economic performance criteria acceptable to donors and aid agencies.

● Reduce the debt service to exports from 20–25% to 10–15%.

● Link debt relief to positive performance targets on human development and poverty reduction, in line with OECD guidelines for 2005.

● Accelerate debt relief for the 25 most needy countries, tied to specific investment in basic education and health, poverty eradication, water and sanitation.

Box 2.7 (continued)

The political resolve of the G8 group of industrialised countries is still desperately needed to remove a mere $2 billion of debt relief for 20 African countries. Yet $100 billion from the International Monetary Fund was pumped into East Asia in six months in mid-1998. Virtually no relief has gone to Africa during that same period of international economic turmoil.

Source: United Nations Development Programme (1998, 100).

cost. In UN terms this figure, which applies only to developing countries, is defined as either the cost of meeting a global obligation that would not otherwise be committed by a nation in its own defined interest, or the extra amount added to a scheme to meet the requirements of a global convention. An example of the first would be the cost of replacing ozone-depleting chemicals with ozone-friendly substitutes. A case of the second would be a switch from, say, coal- to gas-fired electricity production because gas produces about 40% less CO_2 than coal on a heat-for-heat basis, as well as virtually no SO_2 and about half the nitrogen oxides. So the additional costs of using gas rather than coal, or renewables rather than fossil fuels, would be classified as an incremental cost and subject, in principle, to any sustainability account. This is relatively new thinking, so to produce a country-by-country statement of incremental costs would take many, many years. We shall see in Chapter 7 that the innovative idea of a clean development mechanism for the Third World to benefit from carbon trading strategies will be plagued by this definitional problem. This has become a deeply politicised issue.

Nevertheless, various analysts have attempted to calculate a figure for costing the transition to sustainability. Table 2.3 provides one such estimate from the World Watch Institute in Washington. This suggested a figure of around $150 billion annually through to 2000 starting in 1991, so it would be closer to $300 billion in year 2000 dollars if a serious start were to be made in 2000. This figure contrasts with $1000 billion annually spent on arms (Brundtland, 1987) and $125 billion spent on aid to the developing countries in 1996. Agenda 21 came up with a figure of $600 billion annually for the period 1993–2000, but this was estimated sector by sector largely on the basis of special pleading by lobbies. It is grossly exaggerated and probably did more harm than good, because it frightened some potential donors, beset as they are by structural recession. Taking a more dispassionate view, Grubb and his colleagues (1993) recalculated the implementation cost of Agenda 21 at $130 billion per year. The UNCED secretariat's estimate of the incremental costs for the climate and biodiversity conventions range from $30 million to $70 billion. Table 2.4 puts these figures in perspective.

All these estimates should be treated with a pinch of salt. Until the national sustainable development strategies come on stream, and environmental science in the round is deployed in a rigorous manner, calculations of the costs of the sustainable transition have to be regarded as political statements. This is evident from the subsequent battles over any additional financing following Rio.

The one financing institution that has come out of this moderately well is the *Global Environment Facility*. This was created in 1991 as a three-year pilot scheme to pay for selected incremental costs in the area of greenhouse gas emission reductions, biodiversity protection, international water management and energy conservation generally. The GEF is jointly run by the World Bank, the UN Development Programme and the UN Environment Programme and advised by a scientific and technical advisory panel (see Jordan, 1994; 1995).

Table 2.3: One set of estimates for converting the Earth to health. This is a best guess but is based more on standardised and somewhat simplistic calculations than on 'ground truth' evidence. Much now needs to be done in applied environmental science to gather better calculations, related to actual conditions region by region. (Figures in US$ billions).

Year	Stabilising population					Energy conservation			Retiring Third World debt	Total
	Family planning services	Education and health improvements	Financial incentives	Reducing deforestation and conserving biodiversity	Forest and tree planting	Increasing efficiency	Developing renewables	Protecting topsoil on cropland		
1991	3	6	4	1	2	5	2	4	20	47
1992	4	8	6	2	3	10	5	9	30	77
1993	5	10	8	3	4	15	8	14	40	107
1994	5	11	10	4	5	20	10	18	50	133
1995	6	11	12	5	6	25	12	24	50	151
1996	6	11	14	6	7	30	15	24	40	153
1997	7	11	14	7	8	35	18	24	30	154
1998	7	11	14	8	8	40	21	24	20	153
1999	8	11	14	8	8	45	24	24	10	152
2000	8	11	14	8	9	50	27	24	10	161
Total	59	101	110	52	60	275	142	189	300	1288

Source: Brown and Wolf (1993, 183).

Table 2.4: The
world's priorities?
(annual
expenditure).

Basic education for all	$6 billion[1]
Cosmetics in the USA	$8 billion
Water and sanitation for all	$9 billion[1]
Ice cream in Europe	$11 billion
Reproductive health for all women	$12 billion[1]
Perfumes in Europe and the USA	$12 billion
Basic health and nutrition	$13 billion[1]
Pet foods in Europe and the USA	$17 billion
Business entertainment in Japan	$35 billion
Cigarettes in Europe	$50 billion
Alcoholic drinks in Europe	$105 billion
Narcotic drugs in the world	$400 billion
Military spending in the world	$780 billion

Note:

[1] Estimated additional annual cost to achieve universal access to basic social services in all developing countries.

Source: United Nations Development Programme (1998, 37).

The use of the World Bank as the main funding body was highly controversial because most developing countries and the environmental NGOs regard the World Bank as the worst form of non-sustainable capitalism. They claim that it props up non-democratic and corrupt regimes in the Third World with prestige projects for which the recipient countries are expected to pay back the loan with interest. Much of these interest payments are met by exploitative resource mismanagement schemes that leave many destitute and dismayed. It has to be said in fairness to the World Bank that such methods of repayment are primarily for the recipient country to determine: there is no reason in principle why the system should be exploitative. The fact that it is, and that the bank appears to condone this practice when it is quite prepared to impose conditionality in other aspects of its lending programme, is what gives it such a bad press in many countries.

Since 1994, however, the World Bank has begun to take sustainable development seriously. It has initiated community needs surveys and created liveable cities projects across the planet. Its focus is on basic health and sanitation needs, educational uplift and best practice, which is shared on the Internet. The World Bank will never be a true angel in the sustainability transition, as it is primarily a lender to countries for economic and military strategic reasons. But it does at least recognise that building social resilience is good for local economic enterprise, and this is a step forward. The further progress in the World Bank should be monitored by assiduous observers. This progress will be hugely influenced by efforts to reorganise the international economic order following the collapse of the East Asian and Russian economies in mid-1998. It is not yet evident that any of this restructuring is being contemplated with sustainability in mind.

The GEF, for better or for worse, has become the institution around which North–South transfers are occurring. In its pilot phase, it was granted $1.6 billion. From mid-1994 to 1997 it was given an additional $2–3 billion, by far the most significant element of additional funding to emerge from UNCED. But both the climate and the biodiversity conventions demand that the GEF should be restructured to be more accountable to funders and beneficiaries. The South would like an arrangement of one member, one vote; the North prefers voting power to reflect donor investment. This is an important issue, because there is no other effective mechanism for transfers of this magnitude unless either a whole new financing scheme is put in place, which is most unlikely, or existing multilateral banks and aid institutions are regionalised and reoriented to deal with incremental cost funding. The most likely bet is a reorganised GEF, more free-standing with diplomatically created mechanisms for greater accountability. This in turn would be linked to other post-Rio institutional changes requiring a more formal commitment to more sustainable development policies and practices. These points will be given further airing in Chapter 19.

Before we leave the theme of international financing, it is often argued that much of the existing flow of $125 billion from the North to the South could be better spent. This would require policies to target the poorest countries, to the poor in those countries, and to sustainable development schemes to upgrade their livelihoods (Holmberg, 1992, 307). At present only about a third of all aid is free to be targeted in this way, while meeting basic needs for the poorest groups utilises only about 10% of the funds. Sadly, much is either siphoned off to the rich or to the military, or it is used to subsidise energy and raw material prices to keep the urban middle classes relatively content. The majority of developing countries actively subsidise resources that should bear at least a 'normal' market price, and arguably a full environmental cost (Box 2.8). For a variety of domestic political reasons this is not the case, so precious resources are squandered and much collateral environmental damage takes place as a result. Also, as much as 40% of technical aid to developing countries goes to ex-colonial civil servants or non-local consultants. Far too little is directed to indigenous capacity building in the basic environmental sciences of monitoring, mapping, analysing and evaluating resource management schemes.

A study by Norman Myers and Jennifer Kent (1998) estimates that all economic subsidies are worth $1900 billion annually, of which $1450 billion is environmentally

Box 2.8 The pattern of perverse subsidies

Subsidies serve many purposes. One is to maintain economic activity when competition may eliminate it from a region or a nation that wants to retain its own supply and employment structure. Another is to promote an economic investment in some socially desired technology or practice when currently the market does not provide sufficient incentive. A third is to protect an industry or a labour force from change for which it is not adapted. A fourth is to ensure that a particular innovation does not die because it is ahead of its time. So subsidies distort markets, but markets distort economies and societies. Rarely does one set of distortions cancel the other.

Perverse subsidies are those which create definable costs to the natural and social fabric of the planet, or a given locality. Soil erosion caused by price-supported intensive agriculture would be a good example, as would be the underwriting of the social costs of a car for business use. Norman Myers and Jennifer Kent

estimate the total cost of such subsidies to be worth $1.5 trillion annually, of a total global package of subsidies of $1.9 trillion per year. Table 2.8.1 summarises their findings.

This figure of $1.5 trillion is about 5% of the total world product. The figure for Britain could be around $35 billion annually, and for China $90 billion per year. The German lignite industry is subsidised to the tune of $6.7 billion, while $20 billion is pumped into the global fish catch to prop up the market price. But one must remember that subsidies are supported by political lobbies and powerful vested interests, while the losers from the effects of these subsidies are not so mobilised or identifiable. This is at least one good reason why perverse subsidies remain. They are difficult to quantify, historically rooted, protected by client lobbies, and maintain parties in power. Perverse or not, they will be difficult to root out.

Source: Myers and Kent (1998, 135).

Box 2.8 (continued)

	Sector subsidies	Conventional costs (billion dollars per year)	Environmental subsidies	Total Perverse subsidies (brackets indicate range) (billion dollars per year)
Agriculture	325	250	575	460 (390–520)
Fossil fuels	145	n.a.	145	110
Road transport	58	359	917 (798–1041)	
Water	60	175	235	220
Fisheries	22	n.a.	22	22

perverse. By the term 'subsidy' the authors look at both direct financial support for an activity and an indirect failure to tax social and ecological burdens associated with that activity. This figure of $1.5 trillion amounts to around 5% of the total world capital product of around $30 trillion annually. The authors argue that the North accounts for two-thirds of all perverse subsidies, the USA alone one-fifth, and road transportation as a whole swallows up nearly half (Myers and Kent, 1998, xvi). Box 2.8 summarises the principal findings of this intriguing, though incomplete, assessment.

The Myers–Kent study reports on efforts to create new citizens' coalitions to tackle this issue. One such coalition is the 'green scissors' movement of 22 NGOs in the USA which has managed to cut 47 government-subsidised but anti-sustainability projects to the tune of $39 billion over five years. They estimate that $300 billion could be saved by this strategy in the USA alone, more than the current $126 billion federal deficit.

Calculations such as these need to be treated with caution. A subsidy may mean a job and an enterprise of value in the social and spatial realm that is still sustainability. To remove such subsidies is by no means costless either in economic terms or in electoral outcomes. No politician is prepared to tackle head-on the subsidies that maintain the powerful patronage that maintains electoral support. This is why ecological taxation is proving so awkward to implement. Daniel McCoy (1997, 205–212) discusses the actual likelihood of a reduction of a distorting subsidy as a result of ecotaxation. He concludes that one must be sceptical. In theoretical terms, such a mutual gain is unlikely if existing taxes complement or contradict each other, as many do, and if markets do not function perfectly, which they do not. Much depends on the interactive relationships of existing taxation, and the manner in which any additional ecotax revenue is recycled in the economy. In practice, tax regimes are too convoluted, too politically embedded and too manipulable to legal avoidance and special pleading for the deceptively attractive notion of double dividend actually to work. Only when the sustainability transition is more fully in place are these impediments likely to be overcome. The success or otherwise of that transition is our next port of call.

The politics of sustainable development

At the heart of sustainability is self-regeneration and self-reliance – of the soul, as well as of economy, ecology and society. When the soul embraces the recognition of caring and sharing so that enterprising livelihoods are enabled to flourish in millions of cultures and spaces, then the transition might move forward. The concept of sustainable development lies rooted in human supremacy over peoples and places, and emerged with the paternalistic air of decaying colonialism. The wealthy recognised that their future well-being required the better management of those 'Gaian' systems of gyroscopic planetary regulation that lie primarily between the tropics (see Lenton, 1998, 441). The anxiety over biodiversity is certainly influenced by this view, as are to some extent the worries over prolonged climate change destabilising huge numbers of people in water-depleted regions with declining food-producing capability. When these anxieties get deflated by other political considerations, and as the struggle to cope with growing debt and destitution eludes political accommodation every time it is faced, so we get the UNGASS type of outcome: 'sustainable development is all very well, but let us look at it again next year'.

Figure 2.4 summarises two views of the three principal domains for sustainability. The key point to grasp is the 'triple bottom line' of planetary maintenance (ecological sustainability), social equity (ethically loaded sustainability) and economic enterprise (livelihood sustainability) – in that order of priority. But most commentators look at it the other way round. Driving from the bottom in the conventional interpretation is the *technology and economy domain* of greater efficiency in materials and energy usage per unit of product. This is the brave new world of eco-efficiency and clean technology, where growth and consumption are supposed to be free of environmental burdens. The prophets in this wilderness are Ernst von Weizsäcker and his two American colleagues Amory and Hunter Lovins (1998). This is the Club of Rome antidote to its

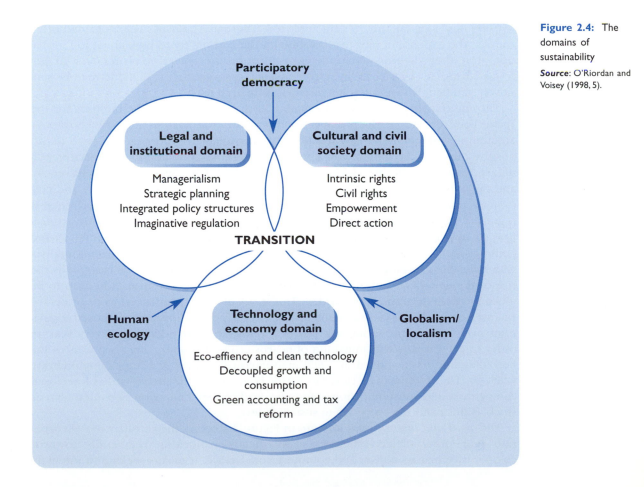

Figure 2.4: The domains of sustainability
Source: O'Riordan and Voisey (1998, 5).

more famous parent *Limits to Growth* (Meadows *et al.*, 1972), which sought to show that it could not be possible to grow and to conserve resources, consequent upon the principles of thermodynamics.

> *Factor Four, in a nutshell, means that resource productivity can – and should – grow fourfold. The amount of wealth extracted from one unit of natural resources can quadruple. Thus we can live twice as well – yet use half as much.*
>
> (von Weizsäcker *et al.*, 1998, xviii)

The curious aspect of all this is that few people appear to believe Factor Four is actually achievable, even though it may be possible. Certainly the technology of resource use efficiency is being developed at a rapid pace, thanks to the microchip, tougher regulations and a steady drip-drip of ecotaxes. And certainly business sees opportunities in resource waste reduction (see Chapter 18), if for no other reason than that international and domestic regulations make it commercially attractive to find ways of recycling and reusing. The American writer Greg Easterbrook (1996) and the UK journalist Richard North (1995) wax lyrical about all these technofixes, should anyone wish to consult them.

Both of these authors focus on the economic propulsion attached to eco-efficiency. This is an arena for new management skills as well as technology innovation that can add real wealth to an economy. For example, Carlin *et al.* (1992) believe that the US economy can grow by 0.5% along with 2 million new jobs if this path is followed. McLaren and his colleagues (1997) claim that investing in eco-efficiency can add 750 000 new jobs for Britain's unemployed. These are guesses, as jobs gained often come at the expense of jobs lost. Admittedly, the authors try to look at net job formation, but they do not take fully into account the social consequences of job restructuring. In a sustainability transition, such costs are very real and have to be sensitively handled. We are still some distance away from that.

Throughout this book we will examine both the pluses and the minuses for the Factor Four revolution. The rise of eco-auditing, life-cycle analysis, improved economic accounting and a more socially sensitive risk management process are all steps in this positive direction. So too are new approaches to ecotaxation and more proactive, voluntary, regulation. All these assist the process that is also assisted by competitiveness, and the willingness of most large corporations to take social responsibility and customer reputation very seriously indeed. But the very success of Factor Four requires regulated markets, long-term planning, paced investment and opportunistic acquisition of new corporate ventures that lock both firms and customers into a fixed embrace through which consumer choice becomes limited, and environmentalism with a flexible ecocentric face becomes progressively suppressed. One example is the rise of the genetically modified soya plant, claimed by its inventor Monsanto to be much lower in environmental burden. This is because it is engineered to be protected against Roundup, a non-systemic herbicide produced by Monsanto. Proposed regulations may well require farmers to buy new (Monsanto-provided) seed every year, in order to prevent any contamination of nearby crops. Monsanto already produces the herbicide and has acquired the seed manufacturer as part of a huge conglomerate takeover. The result may be closer to Factor Four. But in the process it has foreclosed competition and consumer choice, and may lead to the elimination of natural genetic stock. Is this really the transition to sustainability?

The *legal and institutional domain* in Figure 2.4 covers the corollary to the corporate hegemony of eco-efficiency. This means patterning regulation towards co-operative managerialism and voluntary compliance, it means seeking to integrate policy-determining processes across linked sweeps of public–private effort, and it rewards imagination and vision by bending legal rules to suit motivation and enterprise. This is the age of the trade association compacts, where groups of companies in particular industrial sectors seek a common understanding of targets, best practice and published accounts. The success or otherwise of these compacts still depends on strong regulatory surveillance, actively suspicious NGOs which are also prepared to support best practice, and full accountability in auditing and shareholders' meetings. So far, this 'new look' regulation is not proving too popular anywhere. Old-style regulators still prefer the stick and the carrot, many businesses like similar treatment with scope for competitive advantage set within a clear framework of rules and regulations, and consumer/environmental groups look for performance rather than promise (see Davies *et al.*, 1997; Beardsley *et al.*, 1997).

The domain of greatest challenge to the relatively cosy world of eco-efficiency and voluntary compliance through best practice and creative partnerships is that of *culture and civil society* (see Figure 2.4). Here sustainability enters the social and ethical realms of intrinsic rights of nature, civil rights of people and political rights of communities. We have touched on much of this in the previous chapter, though we have not yet looked very carefully at the issue of constitutional rights to sustainability.

The British philosopher John O'Neill (1993, 8–10) defines three properties of intrinsic value, namely:

- *Non-instrumental value*, i.e. a purpose of existence in itself and not as a means for another's purpose. Arguably, a flower or a bird whose existence gives people great pleasure, whether seen or not, does not fit into this pure interpretation, as there is an instrumental value in aesthetic or psychic pleasure. Yet such pleasure creates the preference to preserve and to protect the object of intrinsic value.

- *Self-possession value*, i.e. the value that is essential to the object itself and is not shared by any other object. This is the beauty or vitality or spirituality of any natural object, but again it is practically impossible not to separate its meaning from the values of the observer.

- *Objective value*, i.e. the value something possesses independently of the observer. Again this is a questionable assumption, since to bestow a value involves some act of observation, no matter how remote.

Thus, we find that the notion of intrinsic value cannot really be separated from the human condition, and that the sense of a coupled interdependency between human motive and natural meaning offers a better way forward. We examined this issue in the previous chapter and noted the excellent analysis of this particular relationship by Macnaughton and Urry (1998).

What does appear to be more significant is the character of democracy that assists or inhibits civil liberties, encourages empowerment, and generates an active and consensus-seeking civil society. Box 2.9 examines the various stages towards more comprehensive empowerment. The key issue here is the preparedness of modern society actually to uplift the disempowered into a constructive democracy. This will not be easy as local structures of micro-power can prove very tenacious. This is why procedures for face-to-face facilitation towards revelation of every other's legitimate interests become so important (see Warburton, 1998).

Figure 2.5 summarises this set of relationships. Each of the interconnecting phases complements and reinforces the others. Secure wealth creation that does not carry avoidable and uncompensatable socio-ecological burdens can only be fully achieved in

Box 2.9 Different entry points for participation, consultation, mediation and empowerment in environment and community planning

Power broking by coercion

Legitimation is a process of establishing a predetermined outcome through a manipulative process of involvement which serves to justify that a properly democratic process has been put in place. Its distinguishing characteristics are:

- a drive to get a predetermined outcome through fast track procedures;
- environmental assessments prepared after the initial decision to proceed is taken;
- selective involvement of interested parties, who are fed selective information.

Power broking by manipulation

Consultation takes place when an agency is required to do so either by law or regulatory requirement (e.g. a permitting process), or where the co-operation of stakeholders is regarded as essential to proceed with any proposed development. Its distinguishing characteristics are:

- the recognition of a legal or regulatory requirement that would be challenged if avoided;
- the pragmatic need to obtain consent from interested parties, but only selected parties;
- the deployment of facilitators to smooth the process of participation but not to reach legitimate consensus.

Sharing power

Building institutions for participation involves a proactive preparation of existing stakeholders to be enabled to be effective in technically or socially complex decision making, as well as the training of disempowered interests so that they actually identify and prosecute their particular needs. The distinguishing characteristics are:

- proactive training for empowerment, particularly for those formerly alienated from governance;
- active search for disempowered interests on an inclusive basis;
- mediation techniques geared to achieving constructive consensus.

Devolving power

Preparing communities for empowerment. Here the task is educational and capacity building so that the shift from empowerment to revelation becomes possible. This approach involves the deployment of community activators who act as a bridge between the many communities who need to be represented on any particular development or empowerment programme. The distinguishing features are:

- opportunities to participate in an atmosphere of trust and self-respect;
- ability to communicate authentically.

Community building from below. At the micro-level, people are forming self-help organisations in housing, enterprise co-operatives, health schemes, and consciousness-raising efforts. These are designed to create institutions that generate local economic self-reliance as far as is practicable, to build community confidence, and to create defensive alliances to fight crime, violence and drugs.

Source: O'Riordan (1998, 105).

a stewardship context, where ecological care and the precautionary principle are in place. For the social dimensions of the burdens to be fully valued, the principles of empowerment through revelation have to be involved, as discussed in the previous chapter. The revelatory process also helps to clarify the intrinsic values that become the precursor to ecological and civil liberties. These in turn create a society that is more likely to be self-regenerating, and hence more capable of creating secure wealth.

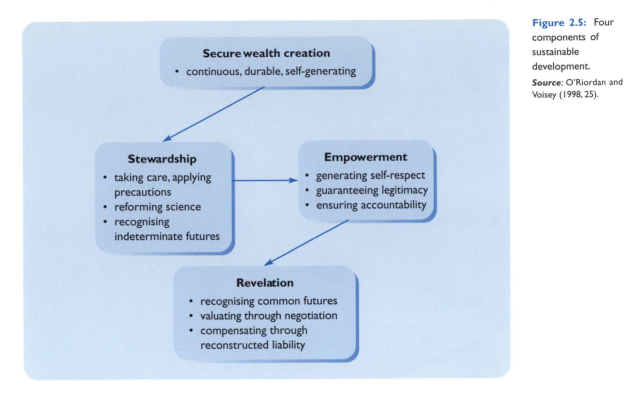

Figure 2.5: Four components of sustainable development.

Source: O'Riordan and Voisey (1998, 25).

And so on. The link to all this is a civic science of socio-ecological well-being, coupled to 'burden analysis' together with an evolving democracy where participation, revelation and social learning intertwine. The important subtext is the modulating relationship between socio-economic transformations that are taking place in an evolving culture and the particular push given to these changes by the sustainability initiative. Throughout the world, those relationships will continually vary.

Environmental space, footprints, rucksacks

The global environmental carrying capacity for a present and future humanity is a hugely subjective, relativistic and meaningless concept. It depends on a host of uncertain and interdependent assumptions that cannot be made without prejudice. An attempt to calculate it is being made through a series of approaches under the various headings of environmental space, socio-ecological footprints, ecological rucksacks or environmental burdens. *Environmental space* is the generic term, developed by the Dutch, to calculate the share of the planet and its resources that the human race can sustainably take. Connected to this concept are presumptions about the resilience of future generations and their capability to use diminishing environmental space, and huge guesses as to what really is the relationship between the natural and social fabric that underpins the multi-layered interpretations of vulnerability, resilience, compliance, resistance and adaptability. All these terms are outlined in Chapter 6.

Underlying the notion of environmental space is a fresh attempt to develop an audit of ecological–economic activity that removes the distortion of money in favour of some more generic measure. One choice is to examine materials flow, namely the total amount of energy and raw materials required to extract, transport and manufacture a product. This is called the *ecological rucksack*, as depicted in Figure 2.6. The concept

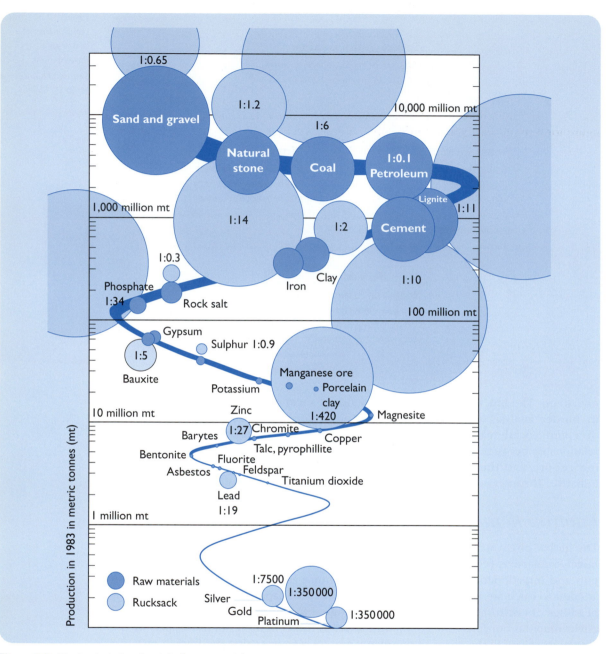

Figure 2.6: The 'ecological rucksacks' of raw materials.

Source: von Weizsäcker *et al.*, (1998, 243).

underlying this diagram is the *materials intensity per service* (MIPS). This calculation takes into account the route kilometres for any economic activity, such as the transport of food from farm to plate, or the movement of fertiliser from raw rock to field. According to von Weizsäcker and his colleagues (1997, 242), a 10g gold ring carries a rucksack of 3 tonnes. One tonne of coal carries a rucksack of 10 tonnes, and so on. The problem is less the issue of calculation as the definition of a service (is watching television a service?) and the estimation of the longevity of individual products.

Much more difficult is the calculation of the burden on society and on biogeochemical functioning. Here we enter highly contested terrain. The social impact is crucially a value matter and one for ethical judgement. One culture's utility may be the outcome of a unique combination of values. So there is no common comparator. To try to equalise or commensurate such well-being does a huge injustice to cultural identity, distinctiveness and the geography of localism. So the very act of measurement is distorting. Consider Figure 2.7 on the burden of various extractive resources. The high figure for aluminium (bauxite) in Figure 2.6 is set because of its dependence on subsidised hydroelectric power. But that burden depends on assumptions about the impact of big dams on local populations. Such impacts are hugely political and enormously variable. For instance, if the inclusive participatory consensus-driven procedures introduced in the previous chapter are sensitively adopted, the outcome would be very different from the circumstances prevailing when they were not. The debate over the 'value' of climate change reduction outlined in Chapter 7 ran full steam ahead into this cauldron of dispute. A key theme for the next generation of analysts is just how to do this measurement job more authentically.

We shall see in Chapter 6 that the application of burden analysis is enormously influential on the relationships between population, consumption, vulnerability and resilience. Right now the early stages of this work do not indicate a sufficiently holistic sensitivity to the issues raised in this debate. Sustainability, if it eventually has to have any effective meaning, will have to grapple with this extremely challenging issue for interdisciplinary environmental science.

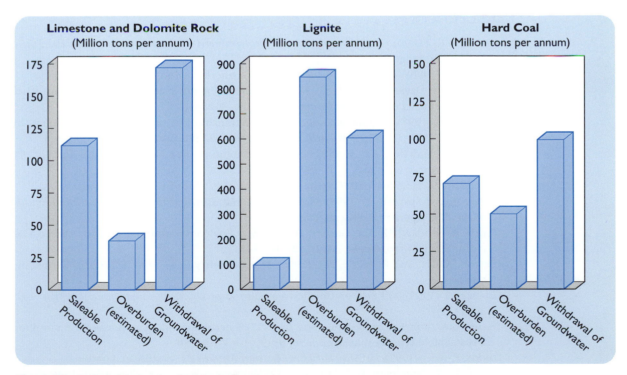

Figure 2.7: Ecological backpacks of mining in Europe.

Source: Carley and Sapens (1998, 63).

Concluding observations

The transition to sustainability has begun. We may not be able to verify it independently of other 'modernising' drives, but that is an advantage. The politics of sustainability are the politics of social and political change generally, not of environmental threat or Gaian revelation. It is better to achieve sustainability through a huge variety of channels of change, only a few of which will carry the sustainability label. In Europe generally, social and liberal democratic coalitions are now in vogue. These governments look for a so-called 'third way' blend of entrepreneurial market opportunities, social care and environmental protection. This adds up to more jobs for maintaining the social fabric, not just production, creative use of knowledge and management, and ecotax-driven technological innovation. Connected to all this is the unleashing of more participatory democracy, visioning procedures and deliberative institutions. The global economic order is under review, as are the constitutional structures of national political power and regional devolution. This is a mix that both promotes and impedes the transition to sustainability. As social justice and social responsibility rise on the international political agenda, as well as through more transparent corporate reporting procedures, so the sustainability drive will become more comprehensive. But lurking in the near shadows is the spectre of fundamental inequality, debilitating health and seemingly permanent exclusion from the human dignity of a reliable income, an education, a weatherproof shelter and a regular meal, and the respect of race, religion and gender. That growing disparity, which is occurring both globally and locally in all societies of the globe, remains the testimony to our collective failure as a species to respect our brief tenure on this unique orb.

References

Adams, W.R. (1990) *Green Development: Environment and Sustainability in the Third World*. Routledge, London.

Ayres, R.U. (1998) *Turning Point: the End of the Growth Paradigm*. Earthscan, London.

Beardsley, D., Davies, T. and Harsh, R. (1997) Improving environmental management: what works and what doesn't. *Environment*, **39**(7), 6–9, 28–36.

Bramble, B. (1997) Financing resources for the transition to sustainable development. In F. Dodds (ed.), *The Way Forward: Beyond Agenda 21*. Earthscan, London, 190–205.

Brown, L.C. and Wolf, E.C. (1988) Reclaiming the future. In L.C. Brown (ed.), *State of the World*. Norton, New York, 170–189.

Brundtland, G.H. (Chair) (1987) *Our Common Future*. Oxford University Press, Oxford.

Carley, M. and Spapens, P. (1998) *Sharing the World: Sustainable Living and Global Equity in the 21st Century*. Earthscan, London.

Carlin, P., Scodari, P.F. and Garner, D.H. (1992) Environmental investments: the cost of cleaning up. *Environment*, **34**(2), 12–20, 38–44.

Clayton, A.N.H. and Radcliffe, N.J. (1996) *Sustainability: a Systems Approach*. Earthscan, London.

Crosson, P. (1997) Will erosion threaten agricultural productivity? *Environment*, **39**(8), 4–9, 29–31.

Davies, G.A., Wilt, C.A. and Barkenbus, J. (1997) Extended product responsibility: a tool for a sustainable economy. *Environment*, **39**(7), 10–15, 36–38.

Dodds, F. (ed.) (1997) *The Way Forward: Beyond Agenda 21*. Earthscan, London.

Dooge, J.C.I., Goodman, G.T., la Riviere, J.W.M., Marton Lefevre, J., O'Riordan, T., Praderie, F. and Brennan, M. (1992) *An Agenda for Science for Environment and Development into the 21st Century*. Cambridge University Press, Cambridge.

Dunlap, R.E., Gallup, Jr., G.H. and Gallup, A.M. (1993) Of global concern: results of the health of the planet survey. *Environment*, **35**(9), 6–10, 33–39.

Easterbrook, G.E. (1996) *A Moment on the Earth: the Coming Age of Environmental Optimism*. Penguin, London.

The Ecologist (1972) Blueprint for survival. *The Ecologist*, **1**(1), whole issue.

The Ecologist (1992) Whose common future? *The Ecologist*, **22**(4), whole issue.

Environmental Data Services Ltd (1998) The spiralling agenda of agricultural biotechnology. *ENDS Magazine*, **283** (August), 18–30.

Grubb, M. (1993) *The Earth Summit Agreements: a Guide and Assessment*. Earthscan, London.

Holmberg, J. (1992) Financing sustainable development. In J. Holmberg (ed.), *Policies for a Small Planet*. Earthscan, London, 302–310.

Jordan, A.J. (1994) The politics of incremental cost financing: the evolving role of the Global Environment Facility. *Environment*, **36**(6), 12–20, 31–36.

Jordan, A.J. (1995) Designing new international organisations: a note on the structure and operation of the Global Environment Facility. *Public Administration*, **73** (Summer), 303–312.

Jordan, A. and Voisey, H. (1998) The Rio process: the politics and substantive outcomes of Earth Summit II. *Global Environmental Change*, **8**(1), 93–97.

Lafferty, W.M. and Eckerberg, K. (eds) (1998) *From the Earth Summit to Local Agenda 21*. Earthscan, London.

Lenton, T.M. (1998) Gaia and natural selection. *Nature*, **394**, 30 July, 439–447.

Macnaughton, P. and Urry, J. (1998) *Contested Natures*. Sage, London.

McCoy, D. (1997) Reflections on the double dividend debate. In T. O'Riordan (ed.), *Ecotaxation*. Earthscan, London, 201–216.

McLaren, D., Bullock, S. and Yousuf, N. (1997) *Tomorrow's World: Britain's Share in a Sustainable Future*. Earthscan, London.

Meadows, P., Meadows, D. and Randers, J. (1972) *Limits to Growth*. McGraw-Hill, New York.

Moffatt, I. (1996) *Sustainable Development: Principles, Analyses and Policies*. Parthenon, London.

Myers, N. and Kent, T. (1998) *Perverse Subsidies: Tax $s Undercutting our Economies and Environments Alike*. International Institute for Sustainable Development, Winnipeg.

North, R. (1995) *Life on a Crowded Planet*. Manchester University Press, Manchester.

Nriagu, J. O. (1990) Global metal pollution: poisoning the biosphere? *Environment*, **32**(7), 6–12, 28–33.

O'Keefe, P., Middleton, N. and Noao, S. (1993) *Tears of the Crocodile*. Pluto Press, London.

O'Neill, J. (1993) *Ecology, Policy and Politics: Well-being in a Natural World*. Routledge, London.

O'Riordan, T. (ed.) (1997) *Ecotaxation*. Earthscan, London.

O'Riordan, T. (1998) Sustainability for survival in South Africa. *Global Environmental Change*, **8**(2), 99–108.

O'Riordan, T. and Voisey, H. (eds) (1998) *The Transition to Sustainability: the Politics of Agenda 21 in Europe*. Earthscan, London.

Osborn, D. and Bigg, T. (eds) (1998) *Earth Summit II: Outcomes and Analysis*. Earthscan, London.

Sachs, W. (ed.) (1993) *Global Ecology: a New Agenda for Political Conflict*. Zed Books, London.

Sachs, W., Loske, R. and Linz, M. (1998) *Greening the North: a Post Industrial Blueprint for Ecology and Equity*. Zed Books, London.

Sandbrook, R. (1998) New Hopes for the Environment United Nations Programme. *Global Environmental Change*, **9**(2), 170–174.

Shiva, V. (1991) *The Violence of the Green Revolution*. Zed Books, London.

Toynbee, A. (1976) *Mankind and Mother Earth*. Oxford University Press, Oxford.

United Nations Environment Programme (1997) *Environmental Data Report*. Blackwell, Oxford.

United Nations Development Programme (1998) *Human Development Report 1998*. Oxford University Press, Oxford.

Von Weizsäcker, E., Lovins, A.B. and Lovins, L.H. (1997) *Factor Four: Doubling Wealth, Halving Resource Use*. Earthscan, London.

Warburton, D. (ed.) (1998) *Sustainability and Community*. Earthscan, London.

Wilson, E. O. (1989) Threats to biodiversity. *Scientific American*, **261**(3), 60–70.

World Bank (1997) *World Development Report*. Oxford University Press, Oxford.

World Resources Institute (1997) *World Resources*. Oxford University Press, Oxford.

Worldwatch Institute (1997) *State of the World*. Earthscan, London.

Further reading

There are four regular reports on the state of the globe. They provide indispensable reading for any assessment of the global predicament.

World Bank (annually) *Environment Matters: Annual Review of Environmentally and Socially Sustainable Development Projects*. World Bank, Washington, DC.

World Resources Institute (biennally) *World Resources*. Oxford University Press, Oxford.

World Watch Institute (annually) *State of the World*. Earthscan, London.

United Nations Development Programme (annually) *Environmental Data Report*. Blackwell, Oxford.

For an accessible guide to UNCED see the following.

Grubb, M. (1993) *The Earth Summit Agreements: a Guide and Assessment*. Earthscan, London.

Environment magazine (1992) Earth Summit: judging its success. *Environment*, **34**(4).

Holmberg, J. (1993) *The Road from Rio*. International Institute for Environment and Development, London.

See also the books by Dodds (1997), Osborn and Bigg (1998) and O'Riordan and Voisey (1998), cited above, for a general update.

Chapter Three

Environmental politics and policy processes

Andrew Jordan and Timothy O'Riordan

Topics covered

- Politics
- Policy processes
- Theories of power
- Policy communities
- Advocacy coalitions
- Interest groups
- Direct action

Editorial introduction

Sadly, it seems, most students of political theory do not extend their enquiry to environmental sustainability politics. Equally sadly, most students of environmental science are not fully aware of how politics works, how institutions play their evolving part in holding societies together, how political power works, how coalitions of interests form and dissolve, and why the rule of law is both observed and ignored. In short, the strengths of political enquiry and the intrigues of the sustainability transition are not fully connected.

This chapter tries to make amends for this gap. It introduces basic theoretical principles of power, democracy, coalitions, networks and interest groups. It seeks to provide the theoretical lubricant for any environmental sustainability case study. To tackle any such study, one needs to know what political institutions are, how they form, what roles and procedures they use, how they would influence and shape values, and the processes through which they learn and adapt. But institutions are not just inanimate organisations, analytical and judgemental procedures, and relationships. They are also people – individuals with personal visions and missions, social roles and career aspirations. The essence of sustainability is a personal goal of survival through sharing and caring set within the bounds of social justice and planetary tolerance. If the sustainability pathway is to be joyfully travelled, politics has to be played by people of conviction, sensitivity, compassion and adaptability.

Figures 3.1 and 3.2 show that environmental issues become politicised in two complementary ways. Figure 3.1 applies to the issue-attention cycle of public interest and organisational response promoted by the American political scientist Anthony Downs

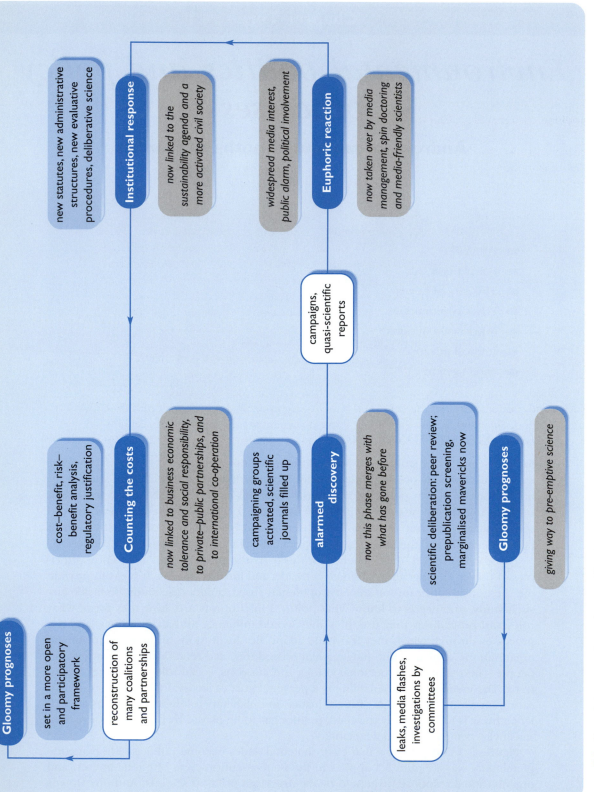

Figure 3.1: Downs's issue-attention cycle reconsidered.

in 1972. He argued that environmental issues evolve slowly in a culture of scientific enquiry and suppression, along the lines explored more fully in Chapter 1. In the past, this was usually a secretive phase during which gloomy prognoses were carefully explored behind the cacophony of day-to-day news. Nowadays, with scientists anxious to break ranks, the Internet buzzing, and the media licking their lips for any attention-grabbing story, this normally careful stage is shattered by pre-emptive science. Such, for example, is the case with endocrine-disrupting organochlorines and genetically modified food products. The politics of bias begins earlier and earlier these days.

The second phase in Downs's cycle was one of media attention, popular surprise and anxiety and a lot of political juggling. One saw this in full evidence in the early days of the BSE crisis in the UK in April and May 1996. But nowadays, these two phases have become muddled, so the distinction is decreasingly clear. One reason for this is the 'dance of influence peddling' between political minders, scientific advisory committees, investigative journalists and the non-governmental watchdogs, sometimes brokered by business.

The third phase is the reactive process, in which institutions cope and adapt, usually within predetermined parameters. But invariably this process, too, is now inseparable from the earlier two. The speed of information movement, and the hounding by campaign groups and journalists, make response more rapid than in the past. Also that response has to be more transparent, not least because of the Web, and the recognition by all players that more openness has a sporting chance of inducing credibility.

The next phase is 'counting the costs' of regulatory intervention and possible price increases because of environmental burden levies. In the past, this was a calculation based on risks and benefits, partly in party political terms, but also in the context of business competitiveness and long-term co-operation. Again the Downs model has to be modified. As noted in Chapter 1, regulatory science is becoming increasingly deliberative and participatory. So the costs and benefits are ethically loaded, with maxi-min rules in mind. Business, too, is becoming more co-operative and conciliatory, as indicated in the previous chapter. So the cost counting is taking on a different hue. But make no mistake: the costs are still calculated in terms of business and governmental tolerance, and the benefits are adjudged via political and commercial payoffs. Pure altruism in such exercises does not exist, nor does a pale variation of it flourish.

The Downs model has served environmental analysts well, but it is time to recognise its important reconfigurations for a modern age. Similarly, another model, also much used, is the effect of sudden 'crisis' in the environmental movement. In recent years, issues such as prolonged fires in Indonesia and Brazil, persistent and dangerous flooding in China and Bangladesh, whale poaching by indigenous peoples in remote Alaska and northern Norway, and wholesale 'bleaching' of coral reefs in warmed-up tropical waters, have alerted popular attention to the symbolism of cathartic events. Figure 3.2 shows the possible relationships involved. Such events excite the media because they contain all the qualities of newsworthiness. They also stimulate the NGOs, because they feed on such issues for their membership campaigns and political influence. More to the point, such events create a range of responses in institutional procedures and actual people. One reason some big companies are becoming more socially responsive is because the brightest and most able managers will not be attracted to a business with a poor public image and an uncaring approach to human and ecological rights.

Symbolic events also impinge on institutional failure – in economic analysis, in legal procedures, in evaluative mechanisms, and in democratic styles. So the merry-go-round of convulsion is beneficial in a properly functioning democracy. The key, however, is the properly functioning part. This is the basis of the present chapter.

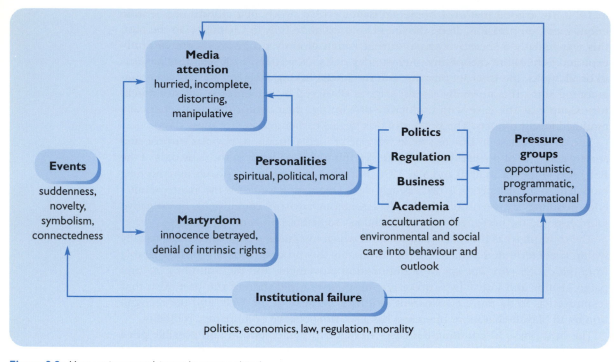

Media attention
hurried, incomplete, distorting, manipulative

Events
suddenness, novelty, symbolism, connectedness

Personalities
spiritual, political, moral

Politics
Regulation
Business
Academia
acculturation of environmental and social care into behaviour and outlook

Pressure groups
opportunistic, programmatic, transformational

Martyrdom
innocence betrayed, denial of intrinsic rights

Institutional failure
politics, economics, law, regulation, morality

Figure 3.2: How environmental issues become political.

Introduction

The success of modern environmentalism may be the loss of its special identity. Think about that. For the past 30 years, environmentalists have been trying to bully or scare us into reforming our conscience, our behaviour, our society and our economy. They have done this by seeking to make environmentalism not a separate science, or a teaching material, or a political party, or a social ethos. They want environmentalism to be our constant companion in day-to-day existence. The so-called 'greening' of society is an attempt to incorporate the full cost of living into our economy, the complete caring for the unavoidably disadvantaged into our society, and for planetary continuation into our conscience. When all that is properly achieved, we should be emancipated. Then there would be no separate environmentalism. Now, that is a political statement!

Needless to say this is not so easily achieved, as pointed out in the previous chapter. The argument present in Figure 3.1 suggests that humans relate to nature in three distinct ways:

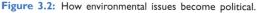 *By putting science first*, in the sense of observing, modelling, predicting, verifying and monitoring.

● *By designing with nature* in recognising that natural functions provide many services that could otherwise cost us dear (see Chapter 4), so we should recognise this inestimable benefit by adjusting to it and nurturing it. This opens up pathways to ecological economics, ecologically based ethics and eco-audits by business.

● *By giving way to Gaia.* The concept of Gaia pervades this book but is given special attention in Chapter 19. It symbolises the immersion of human life within the totality of life-maintaining creation.

These three approaches have always been inherent in the human condition. All societies model, negotiate and succumb to their natural worlds and in so doing redefine their humanness. Organisms face a contradiction of having to manipulate their surroundings in order to gain their competitive edge in the survival game. Yet they also understand the need to act as trustees for the planet and its future progeny. The two modes of thought are outlined in Box 3.1. We shall see in the chapters that follow that

Box 3.1 Ecocentrism and technocentrism

By the early 1970s, two fundamentally different attitudes towards the environment had emerged. The technocentric view is hierarchical, manipulative and managerial. The ecocentric view, by contrast, embraces community-scale natural rhythms and a morality based on ecological principles. As Figure 3.1.1 shows, however, each view has two important variants. Ecocentrists may be

Variations on the basic themes

Ecocentrists		Technocentrists	
Deep environmentalists	**Soft technologists**	**Accommodators**	**Cornucopians**
• Lack of faith in modern, large-scale technology and its need for elitist expertise, central authority and inherently undemocratic institutions • Believe that materialism for its own sake is wrong and that economic growth can be geared to provide for the basic needs of those below subsistence levels		• Believe that economic growth and resource exploitation can continue indefinitely given (a) a suitable price structure (possibly involving taxes, fees and so forth); (b) the legal right to a minimum level of environmental quality; and (c) compensation for those who experience adverse environmental or social consequences	• Believe that humans can always find a way out of difficulties, through either politics, science or technology • Believe that scientific and technological expertise is essential on matters of economic growth and public health and safety
• Recognise the intrinsic importance of nature to being fully human • Believe that ecological (and other natural) laws determine morality • Accept the right of endangered species or unique landscapes to remain unmolested	• Emphasise small scale (and hence community identity) in settlement, work and leisure • Attempt to integrate work and leisure through a process of personal and communal improvement • Stress participation in community affairs and the rights of minorities	• Accept new project appraisal techniques and decision review arrangements to allow for wider discussion and a genuine search for consensus among affected parties • Support effective environmental management agencies at the national and local level	• Accept growth as the legitimate goal of project appraisal and policy formulation • Are suspicious of attempts to widen participation in project appraisal and policy review • Believe that any impediments can be overcome given the will, ingenuity and sufficient resources (which arise from wealth)

Figure 3.1.1

Box 3.1 (continued)

divided into *deep environmentalists* and *soft technologists*. The former hold the unpopular views that other life forms have as much right to exist as humans, while the latter take ecological relationships as a guide to a more socially compatible economy and see cultural and economic diversity as the cornerstone of local well-being. Both types envisage a much more decentralised and egalitarian political and social order. Technocentrists may be divided into *accommodators* – individuals who are prepared to make some concessions to the environment but not to alter the existing political and social order of things – and *cornucopians* – confirmed optimists who believe that a properly functioning market, minimal government interference and appropriate price signals will lead to an optimal outcome for society.

Source: O'Riordan (1995, 7).

both these perspectives can be incorporated into politics, economics and ecology in a most imaginative and lively way.

All this is energised by a constructive association between environmentalism and various social movements which have their roots in other areas of social and political life. It is this infusion of vitality that gives environmental politics its encompassing and adaptive characteristics, qualities that enable it to evolve and endure. These movements are rights to basic health, to freedom of information, to eco-consumerism, to eco-feminism, to peace, justice and intrinsic civil rights, and to civic empowerment.

Figure 3.3 places all these relationships in another frame, namely the struggle between two strands of liberalism. On the one hand there is the radicalism of neo-liberalism, the freeing of economies to adapt to an ecologically and socially sustainable framework. On the other is the reduction of collectivism, the freeing of the soul from the asphyxiation of capitalism, whether state or privately run, with its subtle controls over citizens as consumers and householders. Lined up in the middle is the pluralism or emancipatory frame of humanism, creative education and citizenship responsibilities, the incorporation of environmental ethics into society and economy without going the whole hog of anarchism, yet recognising that market reformulation under capitalism is impossible.

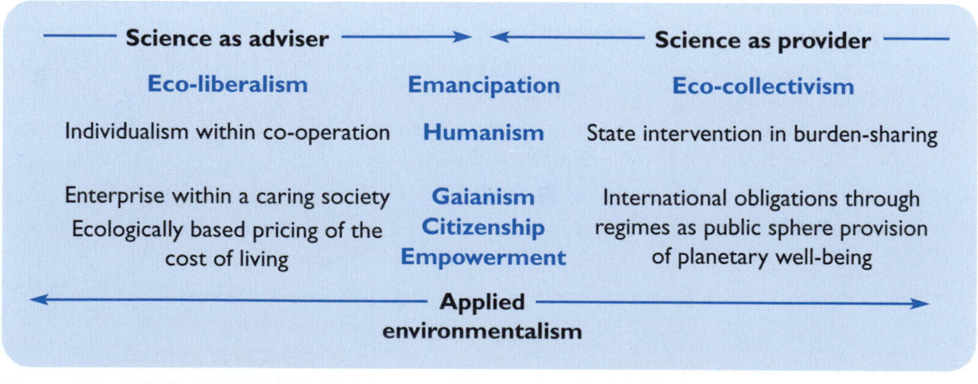

── Science as adviser ──→ ←── Science as provider ──		
Eco-liberalism	**Emancipation**	**Eco-collectivism**
Individualism within co-operation	**Humanism**	State intervention in burden-sharing
Enterprise within a caring society Ecologically based pricing of the cost of living	**Gaianism** **Citizenship** **Empowerment**	International obligations through regimes as public sphere provision of planetary well-being
←──────── Applied ────────→ environmentalism		

Figure 3.3: Different conceptions of environmental management rooted in varying degrees of radicalism. All of the ideas below are relevant for this text. The linking themes of humanism, Gaianism, global citizenship and empowerment are the 'cross-over' points in an integrated environmental science.

This is the modern sustainability politics: the extension of science into wider ways of believing and knowing, the eco-liberalisation of economics and financial accounting, and the incorporation of personal transcendentalism into the majesty of creation.

What is politics?

The Blackwell Encyclopaedia of Political Science defines politics as 'The activity by which decisions are arrived at and implemented in and for a community' (Blondel, 1991). Following this, we can distinguish five important characteristics of politics:

- *Politics is an activity.* Politics is a delicate interplay between the need to take decisions on the one hand and the superimposition of these decisions by independent alterations in society, the economy and environmental well-being on the other. There is, however, also an element of circularity in politics, with some decisions being taken in order to correct the failings of earlier decisions. It was noted in Chapters 1 and 2 that environmental politics is often triggered by failures in institutional structures: see also Figure 3.2.

- *Politics is about decision making.* Politics addresses the way in which issues emerge, how and why decisions are taken, and the extent to which they are properly implemented. Decisions are normally determined in the context of *policies*. Policies may be defined as frameworks for consistent and progressive formulation of decision conditions and implementation procedures, providing a guide for future evaluation of effectiveness. Issue formation, decision taking and implementation are stages of what is commonly termed the policy *process*.

- *Politics involves power.* Power is the ability to achieve a desired outcome. This often takes place through the action of certain interests achieved at the expense of other interests. This is why politics is often conducted by pressure groups, or lobbies, or coalitions, creating a strategy for 'getting their way'. Groups of individuals naturally differ in their view of how society should be organised and what ends it should achieve, so conflict is inherent. However, some participants in the decision-making process are more powerful than others, in that they get others to accept what they otherwise would not have tolerated. This is not just a matter of reasoned argument, physical force or threat. We shall see that power is by no means only a matter of observable manipulation or mobilisation. Much power is exercised structurally, meaning that rules, procedures and common beliefs support the interests of the powerful without the powerful having to decide on every occasion what should be allowed onto the political agenda for action. In fact some decisions are never taken, because, if taken, they would upset powerful vested interests. Other decisions are effectively 'taken' before political parties, pressure groups or individuals can have an input. Political scientists, therefore, are interested not only in how decisions are made but also in who has the *power* to make them or to avoid them. They seek to know who has power, how it is used and on what basis it is exercised.

- *Politics occurs within communities and social networks.* Politics arises because communities of people have to find a way to take joint decisions. Politics can involve balance and compromise, but in practice is often extremely divisive, creating winners and losers. The world has traditionally been divided into discrete legal communities called states. As we shall see, many environmental problems cross state borders. Policies, therefore, also have to be devised by states working together as outlined in Chapter 19.

● *Politics is everywhere.* Politics pervades every aspect of life. We noted in Chapter 1 that even supposedly objective, 'apolitical' matters like science are deeply politicised. Sport, smoking, education, abortion, transport: these are all political matters. We all participate in politics by voting at elections. But we also participate in politics when we join pressure groups, discuss the news over a drink in the pub, respond to an opinion survey, attend marches and demonstrations. Through these activities we articulate our preferences, and also adapt them at the same time. Thus, social outlooks are formed and continually reassessed. This process influences perceptions on environmental risk and institutional trust as explained in Chapter 17.

Environmental politics

Judging from the discussion in Chapter 1, Albert Einstein may well have been wrong when he famously defined the environment as 'everything that isn't me'. This is because 'me' is intrinsically part of the negotiation process that is environmental awareness. Indeed, the fact that environmental questions touch upon almost every aspect of social life may explain why decision makers have often overlooked them. 'The environment' is simply too interconnected and too amorphous for most time-pressed politicians to grasp. They tend to prefer intervening in seemingly less diffuse policy areas where policies are likely to produce rapid and observable effects. It also explains why 'the environment' cuts across conventional political lines such as social class, religion and party politics. Environmentalism is associated with both fascism on the far right of the political spectrum and anarchism on the extreme left (see, for example, Pepper, 1996, for a comprehensive discussion).

Because environmental issues are enmeshed in a huge array of policy realms, it is increasingly difficult for politicians to ignore the environmental dimension, even in formerly perceived 'aloof' policy domains such as social welfare provision, crime prevention and education. Equally problematic for environmental activists, however, is that such 'meta' policy issues are not conducive to simplistic, 'sound bite' solutions or to televised debate. The irony, therefore, is that environmental issues rarely receive a political airing, especially not in pre-election periods, simply because no politician can subsume analysis and solution into simple, tabloid dimensions. This is especially true of the sustainability debate.

The sociologist Howard Newby (1993) is not at all surprised that environmental matters excite great conflict and debate:

> *In the broadest sense they are deeply political, raising concerns about the expansion of individual choice and the satisfaction of social needs, about individual freedom versus a planned allocation of resources, about distributional justice and the defence of private property rights, and about the impact of science and technology on society. Beneath the concern for 'the environment' there is, therefore, a much deeper conflict involving fundamental issues about the kind of society we wish to create for the future.*

The relationship between 'the environment' and society is two-way. The way citizens think about and respond to 'objective' environmental resources is itself deeply political. Our mental maps are based upon the information we receive from scientists, filtered by pressure groups and the media. 'The environment' therefore is not 'out there'. It is something we ourselves construct, based on our experiences and upbringing. It mirrors the way society itself is organised, as well as the particular perceptions and beliefs of individual citizens. The psychologist Abraham Maslow pointed out that public concern

Box 3.2 Transcendentalism and environmental attitudes

The Transcendentalists were a group of American poets and philosophers who reflected the unease over the rapid changes in American society following the advent of industrialisation, trade and travel in the mid-nineteenth century. Like the European 'Romantics' such as Wordsworth and Byron, they were alarmed at a pace of change that appeared to destabilise values and undermine the commitment to stewardship and the spiritual associations with the natural world. In many ways, they were a literary version of the 'new world' communes of the modern era.

Transcendentalists believed that nature could be understood only by direct and sensitive contact. Henry Thoreau lived for two years in a wooden cabin next to Walden Pond near Boston in the United States, but carefully maintained literary links with the *Boston Globe*. Ralph Waldo Emerson was fascinated by the emerging scientific understanding of the age and realised that nature provided services for humanity in the form of ecological foundations that a spiritually disconnected humanity was incapable of recognising. 'Man' wrote Emerson in 1836, 'cannot be a naturalist until he satisfies the demands of the spirit.' Walt Whitman wrote of the danger of 'having certain portions of the people set off from the rest by a line drawn – they are not privileged as others, but degraded, humiliated, made of no account.' A fractured society would never be at ease with the planet. The founder of the Sierra Club, John Muir, wrote famously of the philistinism of crass commercialisation. On fighting the proposal to fill a Sierra valley, Hetch Hetchy, for a reservoir for thirsty San Franciscans, he raged: 'Dam Hetch Hetchy! As well dam for water tanks the people's cathedrals and churches, for no holier temple has ever been consecrated by the hand of man.' Muir was broken by the flooding of his spiritual home, but the legacy of his observation remains an enduring part of environmentalism.

for the environment tends to blossom as societies become materially richer. This has arguably been the case through the history of modern environmentalism, beginning with the nineteenth-century Romantics and conservationists (see Box 3.2).

Pulses of environmentalism do tend on balance to coincide with periods of economic prosperity – the late 1960s and the late 1980s, for example, witnessed a significant greening of party politics, the adoption of pollution control laws and the creation of environmental departments and agencies in government (see Box 3.3). In trying to understand the relationship between society and nature, political power cannot be ignored. After all, the 'most effective and insidious use of power is to prevent ... conflict from arising in the first place' (Lukes, 1974, 23). We shall see that one of the ploys used by the powerful is to distort environmental issues in a way that suits their purposes.

Box 3.3 'Greening' government

Greening government requires a mixture of 'hard' interventions to create new environmental bodies at the heart of government and 'soft' measures to improve the procedures by which they work. In 1990, the UK government put in place a number of mechanisms to 'green' the machinery of government. These included nominating a 'green' minister in each department to promote environmental concerns and prepare an annual report, creating two green Cabinet committees to address conflicts between departments, and producing new guidance documents to help civil servants to integrate environmental concerns into day-to-day decision making. There is also a

Box 3.3 (continued)

Round Table on Sustainable Development involving ministers and pressure groups, and a panel of establishment figures reporting directly to the prime minister. Now, there is a sustainable development unit in the environment department and a parliamentary audit committee to look at any policy – 'a creature to bite our own ankles' as the current Environment Secretary, John Prescott, described it. The new Labour government also has plans for a 'green book' listing the environmental consequences of its annual budgets.

How well have these institutional innovations fared? Under the Conservatives, the 'green' ministers seem to have busied themselves with 'housekeeping' matters such as waste management and not routinely considered the environmental impacts of departmental policies. Departmental reports remained patchy and few made it clear whether environmental targets

were ever met. It is difficult to gauge the success of the Cabinet committee (one was disbanded after a few years) because of the secrecy surrounding Cabinet affairs, while the policy appraisal documents were not routinely used. One looks in vain for evidence that environmental issues were a factor in large policy decisions like rail privatisation, liberalisation of the energy market or the handling of BSE. One commentator concluded that they could only be described as 'marginal in terms of public investment, political debate and public awareness' (Christie, 1994, 18). It is hard to escape the conclusion that any formal mechanism can only be as powerful as politicians want it to be. Without a strong push from the Prime Minister, departments will always find ways of 'drafting around' whatever procedures are put in place (Hill and Jordan, 1994).

Defining features of environmental politics

According to Albert Weale (1992, 5), the characteristics of an issue have an important bearing upon the kind of politics that develop around it. While the relationship is by no means deterministic, it is nonetheless possible to identify a number of core characteristics that differentiate the environment from other policy concerns.

● *'Public' goods and 'bads'.* Environmental quality is a public good. Environmental resources like the atmosphere and large areas of the ocean are common property, so, in principle, nobody can be excluded from using them. This means that while the costs of improving the environment may be concentrated in one small section of society (e.g. the owners of polluting factories), the benefits tend to be spread widely among a myriad of users. One implication is that users have an incentive to 'free ride' – that is, consume resources in the knowledge that the costs, or 'externalities', will be shared by all. We shall see below that for many years Britain benefited economically by pumping raw sewage into the North Sea because it was 'cheaper' than treating it inland. However, those who had to pay for the clean-up measures had an incentive to avoid the cost, while the losers were widely dispersed, and individually suffered too little, to mobilise into a coherent group. This is why many environmental disputes involve a conflict between a spatially concentrated and well-organised group of damagers (e.g. polluters) and a fragmented array of individuals and groups fighting for the 'public' interest or the welfare of non-human entities. Such conflicts involve a mixture of motivations and are often as much a demand for more 'democracy' (i.e. local involvement) in decision making as for the preservation of the environment. It is worth bearing in mind, however, that a new form of consumer-based environmental damage is now in vogue. Persistent chemicals, non-degradable products and antibiotics are all scattered about the landscape of consumerism. The source of the trouble is as dispersed as are the victims of the outcome, though the two are rarely the same.

● *Regulation.* Since environmental damage has its origins in otherwise socially legitimate activities like energy and food production, the role of the state is often to police the level of damage that one section of society imposes upon others. This is normally achieved using regulations, although in recent years interest has grown in market-based instruments such as taxes and charges. Economists like to think that governments act fairly in balancing costs and benefits to produce an outcome that is in some way socially 'optimal'. In practice, some groups have more of a say than others and can 'capture' the regulatory process, thereby securing the benefits of production without incurring the full social costs. The tendency for costs and benefits to be shared unequally immediately raises politically important questions of fairness and equity (see Chapters 4 and 6).

● *Complexity and uncertainty.* 'Local' problems are sometimes only the observable tip of a much more complicated and spatially dispersed set of processes. For instance, a polluted river may stem from regional agricultural practices which are in turn related to broader policy settings of subsidising food production and under-regulated international trade. Policy makers always face myriad constraints and conflicting demands, but there are good reasons for believing that scientific and procedural uncertainty is particularly acute in the environmental sector (see Chapter 1).

● *Irreversibility.* Once development exceeds the planetary capacity for self-repair, environmental assets can collapse and disappear, never to reappear. Certain environmental functions are substitutable, but many are not. Irreversibility complicates environmental politics, differentiating it from other policy areas like tax or social affairs, where problems can be more easily revisited. Irreversibility means that incremental, *ad hoc* policy responses have to be substituted for ones that foresee problems and consider them systematically. In practice, the environment often responds slowly and subtly to human intervention. Whenever the state of the environment cannot precisely be determined before or after a decision is made, the scene is set for those with vested interests to twist scientific ignorance to suit their own cause, as noted in Chapter 1. Scientists too, have their own policy objectives. Epistemic communities (Chapter 19) of experts with an authoritative claim to policy-relevant knowledge can have a decisive impact on the course of policy. The trouble here is that conventional political institutions are not terribly good at coping with genuine uncertainty. The political power game makes it difficult for policy actors to act in advance of any consensus for the need to decide definitively. In any case, politicians can find it difficult to convince their constituencies to make sacrifices when the 'payoff' may be generations away. This is why the deliberative and inclusive processes introduced in Chapter 1 become so important in the conduct of modern environmental politics.

● *Time and space.* The costs of environmental problems are often spread unevenly over time and space. Conflicts consequently emerge between those who are likely to benefit and those who are likely to lose from a particular course of action. The construction of a bypass, for example, might improve the welfare of inner city residents but at the cost of disruption to those living in the path of the new route. Balancing these different needs becomes a central question of politics. As discussed in the previous chapter, unlike many other policy areas, sustainable development forcefully insists that decision makers address the question of equity not only within but also between generations. This in turn raises difficult questions of justice and morality, to say nothing of due democratic process (see Press, 1994, for a good case study).

● *Political borders.* Problems like acidification, ozone depletion and climate change transcend political borders. Solving them requires states to co-operate in an atmosphere of trust and reciprocity. Difficulties arise when action is required from those who can legitimately say they are not to blame (see Chapter 19).

● *Policy sectors.* The environment cuts across the traditional sectors of government. When Chris Patten, as former environment secretary in the UK government, remarked that 'the most important parts of environmental policy are handled elsewhere – the levers aren't in my office', he revealed the main barrier to 'greening' the policy process. Protecting the environment often involves deep conflicts between environmental ministries and the older and often more powerful parts of government that attend to finance, agriculture, transport or energy policy. One of the great challenges of the 1990s has been to find mechanisms to integrate environmental concerns into all areas of decision making (see Box 3.3). When and if this is achieved, there will be less need for a separate environmental department.

Box 3.4 Power in environmental politics

For some political scientists, politics *is* the study of political power. Power is the ability to get one's way, if necessary at the expense of others. However, political scientists fundamentally disagree about the nature and exercise of power. In a landmark text, Stephen Lukes (1974) considers the methodology of testing for the exercise of power between those who exercise it and those who are manipulated by it. For Lukes, the issue was how far the process was observable. His concern was partly methodological, namely, how far could the exercise of power actually be studied and proved. But he was also at pains to point out that certain dimensions of power are designed to be unobservable, hence their potency. By contrast, Ham and Hill (1993) are fairly typical of mainstream policy analysts, who characterise power through patterns of democratic structures and processes, including the formal machinery of the state.

First-dimension power (Lukes) or pluralism (Ham and Hill) deals with the observable exercise of power by vested interests with political, economic or social clout (elites). They do so by recourse to the law, or to superior information, or because of their centrality to formal arenas of

decision making, or if thwarted, through their ability to cause trouble or to be disobedient. Power is exercised by moral or political authority, by force or by manipulation. In elite democratic structures, small groups of interests control the show and usually have the skills or the influence to ward off adversaries. Such groups gain particular solidarity by forming policy coalitions around key advisers and decision makers in certain well-defined arenas. Pluralism, however, is still an important feature of a democracy which allows interests to form, be invigorated or create alliances in the face of new policy issues and agendas. It is this capacity to mobilise by articulating particular bias that is the hallmark of pluralism.

Second-dimension power (Lukes) or elitism/neo-pluralism (Ham and Hill) is less observable. This is partly because elites can manoeuvre to avoid contentious issues being discussed, or because governments are persuaded to kick awkward issues into 'touch' by setting up a long-winded review or a committee of inquiry, or because scientific uncertainty means that no quick decision can be taken. The chemical producers relied upon second-dimension power

Box 3.4 (continued)

to delay the imposition of controls on CFC production (see Box 19.7). Elitists and neo-pluralists like this interpretation because it allows them to point out that interests can be influential even when they are not physically present in decision forums. The very 'popularity' of a position or a lobby group gives it political weight. Politicians are always 'looking over their shoulders' when issues relevant to particular lobbies, such as bird protection and the role of the Royal Society for the Protection of Birds (RSPB), are at stake.

Third-dimension power (Lukes) is exercised when dominant interests manipulate matters so that subordinate interests are persuaded that their 'free will' is actually being served. This is more ubiquitous than is usually thought, and is associated with cultural envelopment by advertising, popular icons in sport, and the creation of supportive social networks that link status identification with certain patterns of behaviour, value outlooks or consumer purchasing. The common desire to own a large, powerful motor car is a very good example of this (see Box 3.7)

● *Non-human entities.* Alone among the different domains of policy, environmental politics directly addresses the relationship between humans and the non-human world. If pushed, most people express support for the environment. But how far should we ascribe moral significance to plants and animals, especially if it involves self-sacrifice and higher economic costs? As noted in Box 3.1, those who adopt an *anthropocentric* view regard nature as primarily a source of resources, material and amenity satisfaction. Those who feel that the non-human world has interests and moral significance independent of human valuation – what economists term 'existence value' – are more *bioethical* in their perspective. The clash of these two perspectives has been at the root of environmental conflict through the ages.

Political power

There is no single agreed concept of political power (Box 3.4). At its simplest, it is the ability to get someone to do what they otherwise would not have done. There are essentially three main models of politics: *pluralism, elitism* and *structuralism (Marxism)*. None is necessarily more 'right' than the other two, because they tend to focus upon different aspects of society and have distinct views about how power is exercised. These differences are best investigated via detailed case studies of real-life conflicts (see Box 3.5).

Pluralists believe that political power is widely, though by no means evenly, spread throughout society. Since democratic elections are too infrequent and partial in scope to provide a check on the accumulation of political power, the job of democratising the policy process falls upon pressure groups. There are pressure groups in each and every sector of policy making, representing interests as diverse as animal rights and single-parent families. Groups glow and fade according to public opinion, but pluralists argue that there will always be groups to represent most of the important sides to a dispute. It is a matter of considerable significance to the politics of power, however, that the interests of those least able to mobilise into groups, or whose demands are least effective in political terms, are not always represented by pressure groups or political lobbyists.This point is emphasised in Box 2.9 in the previous chapter.

Box 3.5 Bedfordshire bricks: a case study of corporate power

In the late 1970s, the London Brick Company submitted plans to extend production in Bedfordshire, a traditional site of brick making in the Midlands of England. A number of groups opposed the move on the grounds that it would pollute the local environment. Following discussions with the environmental regulator, the company said it would fit the best practicable pollution control equipment. The dispute looked to be heading in favour of the developers when the local government planning authority tried to impose strict air quality controls on the proposed works. London Brick responded by threatening to move production to an alternative site, putting in jeopardy hundreds of local jobs at a time of national recession. Eventually, the local authority backed down and the licence was granted without any additional pollution controls.

In applying the three main theories of power to the saga, Andrew Blowers (1984) found evidence of pluralism at several important stages. However, London Brick was a privileged participant in the policy process and enjoyed close links with government. Economic factors were dominant. Despite the strength of environmental protest and some delay, the company achieved the outcomes it had originally wanted. It received strong backing from the supposedly neutral environmental regulatory inspectorate and used its significant financial resources and political connections to outflank other groups. When events seemed to be running against it, the company deployed its ultimate weapon – an 'investment strike' – to very good effect. Blowers concludes that while material interests eventually triumphed, structuralist–Marxist theories are too crude to explain the precise distribution of outcomes which were also influenced by more pluralistic forces. Significantly, although such theories worked best in this particular case, had the economy not worsened towards the end of the study period, pluralist claims 'would have been upheld'. This suggests that pluralism in environmental politics is most clearly evident under conditions of relative prosperity and is less evident during periods of recession.

While there undoubtedly are very powerful groups in particular policy sectors, no single group is continuously successful or capable of distorting the whole political system. According to pluralists, government's nominal task is to act as an impartial referee. By integrating the various demands placed upon it into a set of workable policies that suit the public interest, it is said to work on behalf of society as a whole. The fact that it is internally fragmented into different client interests circulating around individual departments of state prevents it acting solely in the interests of one group. The competition between government departments provides another check on the accumulation of power.

Pluralists have been criticised for paying too much attention to pressure groups and ignoring the influence of underlying structures, ideology and ideas, which tilt the political playing field in favour of certain interests. They also suggest that the state is receptive to all groups when some are granted privileged access to decision makers and important committees ('insiders') while others are not ('outsiders') (see Box 3.6).

Neo-pluralism developed in the USA with the recognition that business was often in a 'privileged' position compared with other groups. According to Lindblom (1977), the fact that all governments, whether Left or Right, need a healthy economy to provide jobs, prosperity and growth for welfare support requires them to adopt measures which are in the interests of business without business

Box 3.6 Pressure groups: 'insiders' and 'outsiders'

According to Wyn Grant (1989), pressure groups fall into two main groups. Insiders are consulted regularly because their views are regarded as legitimate and important. Outsiders, in contrast, are excluded from decision-making processes or choose not to participate. Farmers have often been granted access to decision makers because of the key function they play in society, whereas groups like Greenpeace and Friends of the Earth hold views that mark them out as outsiders. The question, then, is how do groups pass through the gate of influence? How do they get to sit upon the important committees and panels? Martin Smith (1995) argues that access is determined by ideology and a group's willingness to play by the 'rules of the game' – in other words, by accepting the government's terms of entry. Insider status brings with it responsibilities such as preserving confidentiality, or supporting a particular world view or policy position. This partly explains why business groups and developers find it easier to gain an audience with ministers on a routine basis than green pressure groups. Groups favouring more ecocentric approaches, such as Greenpeace, have often found themselves ideologically opposed to government in promoting policies that place limits on economic growth. But access works both ways. It is in Greenpeace's interest not to be too closely associated with government; it gains support from some people precisely because it is an 'outsider'. Hence, the interests of FoE and Greenpeace become relevant for all environmental policies and decisions, even if these groups are not physically represented in the lobbying process. It is this so-called second dimension of power that gives neo-pluralists a particular line of argument in the modern political stage (see Box 3.4). The politics of pressure groups generally are now well covered by Rawcliffe (1998).

having to take any observable action. Instruments of 'circularity' in modern democracies, like the mass media, help to structure politics, removing 'grand majority' issues concerning the fundamentals of the political order (e.g. the right to own property) from the agenda and leaving citizens to debate an endless range of 'secondary' concerns.

According to neo-pluralists, power operates through a 'second' dimension (see Box 3.4) when issues are removed from the agenda by pre-emption and exclusion. 'Non-decision' making, as it is termed, can take a number of forms. Whereas pluralists assume that most grievances are brought fully into the open, neo-pluralists argue that many are not. Some groups lack the resources needed to make themselves heard, find it difficult to gain entry to decision makers or, more fatalistically, conclude that the chance of bringing about change is so slim as to be not worth fighting for. There is an endemic bias in society favouring the powerful and prejudicing others, which is contained within the very structure and institutions of society. Some issues are organised into politics, whereas others are systematically excluded by what political theorists call a 'mobilisation of bias'. This bias operates in the interests of *status quo* defenders', namely privileged or elite groups.

Some of the more extreme environmentalist demands are undoubtedly marginalised by those in power, but other, less radical demands are suppressed simply because the *status quo* is seen by many to be 'natural and unchangeable ... divinely ordained and beneficial' (Lukes, 1974, 24). The sociologist Michael Redclift (1992, 40), for example, suggests that:

Environmental problems ... are the outcome of a series of choices, many of which we make collectively as a society. The epicentre of these choices is the developed world, and most of these choices are so culturally grounded that few people recognise them as choices at all; they are routinely depicted as 'needs' rather than wants.

Road-building lobbies, for instance, argue that new roads are 'needed' to cater for some projected level of use, without ever having to justify the assumptions on which such a calculation is based. That lack of justification is not an accident. Political institutions dictate that to be the case.

Strong vested interests in the form of pressure groups and established procedures serve to protect and reinforce the *status quo*. Former pluralist American political scientists like Dahl and Lindblom (1976, xxxix) accepted a much more radical exercise of power in modern societies which creates an 'indoctrinated complacency', secured through the mass media and the education system. Vance Packard (1954) described the ability to manipulate human behaviour by the creation of unwanted 'needs' in his classic study of the power of corporate advertising, *The Hidden Persuaders*. This has much in common with deep green claims that the social consensus about the materialist basis of modern societies is manufactured by clever marketing and cultural indoctrination. For Redclift (1992, 38), then:

> *The underlying assumptions about our relationship with the environment, that support the idea of 'progress' in advanced industrial societies, have tended to become normative impositions ... . The implications of examining the underlying social commitments of our society, rather than alternative refinements in environmental policy options, may be very radical indeed.*

Individuals continue to act in ways which they know are damaging to the environment, not because they are irrational or unconcerned about the environment, but because they are socially and institutionally 'locked in' to unsustainable patterns of daily activity from which they find it hard to break out. There is no better example of this than the transport sector (see Box 3.7).

Box 3.7 Human agency and social structure: the transport dilemma

One of the enduring puzzles in politics is the exact relationship between conscious human agents and broader social structures, which include habits, customs and social rules to bureaucratic organisations and economic systems. Political scientists disagree over whether explanations should be based upon the deliberate actions of individuals or whether they derive from wider social structures. We see this in the debate about power, with pluralists emphasising the role of individual pressure groups and Marxists the economic basis of society. This pluralist–structuralist perspective is nowadays very evident in the transport–environment debate (Department of Environment, Transport and the Regions, 1998). On the one hand, the anti-car coalition is increasing because of the recognition that too many vehicles on the road creates costly congestion and lowers the competitiveness of business. In addition, the National Health Trusts are worried about the longer-term consequences for air pollution-related ill health. Yet the car is both a structural economic force and a powerful status symbol. No wonder it is difficult to control, for all the variants of power are in evidence in this highly contested policy domain.

It is widely agreed that current rates of traffic growth in Britain are manifestly unsustainable. Forecasts made by the UK government in 1989 suggested an overall growth of 83–143% by 2025. The policy in the past has been to provide the necessary road space to meet these sorts of predictions, but the truth is sinking in that such an approach is self-defeating and cannot be accommodated without serious environmental damage and economic dislocation.

Box 3.7 (continued)

Cars produce global warming gases and local air pollutants. The roads they travel upon and the associated development that comes with them are a major source of habitat loss and development on 'greenfield' sites. But what is to be done about it? We live in what Mrs Thatcher once called a 'great car economy', and the institutional barriers to changing it are high. The roads lobby is strong and well represented in government. Cars are an icon of the modern age and the freedom and choice they provide are highly cherished. Road traffic and economic growth are regarded as inextricably intertwined. The environmental movement, on the other hand, has to rely upon a mixture of reasoned argument and direct action. Individuals would be willing to get out of their cars if reasonable alternatives were available, but often they are not. The difficult task facing the government is how to find a balance between these competing demands over a time frame that prevents irreversible environmental damage and semi-permanent gridlock in parts of Britain. The conventional power analysis, coupled with the possibility that many of the electorate are structurally, psychologically and, in social status terms, esteem-dependent on car ownership for their identity suggests that the metamorphosis towards a non-car-using society will have to be very painful indeed.

While there undoubtedly are powerful institutional mechanisms supporting the *status quo,* several commentators have noted the rise of a 'new' politics around issues such as gender relations, the environment and technology that are considerably broader in scope than the 'old' politics of employment rights and wages. We see this in areas such as food production, nuclear power, pollution and other 'lifestyle' concerns. Once dominated by cohesive policy communities of scientists and technical experts, these areas are characterised by more pluralistic forms of politics in which exclusion mechanisms no longer function as effectively as they once did. We will return to the question of how established policies become destabilised later in this chapter.

Structuralists accept there is an identifiable elite in society which exercises political control over decision making, but argue that it arises from the basic economic structure of society. Simply put, society is divided into those that own resources and those that do not. For Marxists, those that own resources are well represented in business, government, universities and scientific establishments, whereas those that do not are forced to sell their labour. The relationship between classes is essentially exploitative. Like neo-pluralists, Marxists believe that the state does not hold the ring between different interests but represents the interests of the capital-owning class. The state intervenes to ensure economic growth continues and class relationships are preserved. The capitalist system survives by denying its class basis. According to Marxists, the welfare state and environmental controls are put in place to pacify critics and keep the conflict between classes to manageable levels.

Wherever possible, sensitive problems that reveal the true economic relationship between classes are 'technocratised' – that is, passed to scientists and technical experts to deal with in an 'apolitical' manner. Tools like environmental appraisal, as well as supposedly democratic procedures such as the planning system, risk management and EIA, are mechanisms to convert 'political and normative issues ... into bogus technical ones' (Sandbach, 1980, 104). For Marxists, then, environmental problems reflect the unequal distribution of resources. There can be no final solution to such problems when the social system from which they arise is structured so unequally (Sachs, 1993).

Structuralists are also fascinated by positional power, where the levers of policy are handled by a few specialist or highly influential groups in the interests of ensuring peaceable compliance. Thus the very patterning of organisational relationships in policy communities and in the manoeuvring of lobbyists creates and reinforces structures of special bias. Intriguingly, many well-established environmental groups, including Friends of the Earth (FoE) and Greenpeace, do now make use of structural influence.

Structuralism is also akin to second- and third-dimension power as outlined in Box 3.4. Structures of organised influence shape attitudes and create their own self-perpetuating biases. This is why the tendency in even the best-developed democracies is towards neo-pluralism and structuralism in a reinvigorating materialism. Pluralism and, latterly, deliberative and inclusive processes are the best means of protection against this tendency.

The policy process

Most political battles centre on the way policies are developed by governments on behalf of interests embedded in society. Explaining how policies are made and implemented offers revealing insight into the way society as a whole operates. Policy itself is a slippery term (see Box 3.8) and there are numerous theories and models available to understand it. Models of decision making can be divided into those that treat actors as self-interested, utility-maximising individuals and those that regard them as satisficers, whose rationality is bounded, responding to what is socially appropriate as opposed to what is optimal. Accordingly, some models treat governments as unified actors, whereas others present policy as the outcome of battles between different departments. Moving on a policy stage, some theories of implementation take the perspective of decision makers and are essentially top-down in their view of implementing agents. Others view implementation through the eyes of implementing officials and are more bottom-up.

Box 3.8 What is 'policy'?

Defining what we mean by 'policy' is not easy. A dictionary definition is 'a course of action or principle adopted or proposed by a government'. However, some policies comprise a whole series of decisions, layered on top of one another. Policies are often put in place, then immediately fine-tuned to fit a new problem or altered in the light of a sudden discovery of unexpected information. Policies might be sanctioned by governments, but the detailed aspects of implementation are often left to lower levels of government and quangos, i.e. non-departmental, official, though not directly accountable, public bodies. When a local pollution control officer negotiates with the owner of a polluting factory over the details of a site licence or emissions permit, (s)he is arguably remaking policy. We must not, of course, also forget the question of power. Many policies are only ever statements of intent, which governments have no expectation of enacting. Many important environmental laws are arguably adopted to appease environmental groups and remove issues from the agenda. For example, many of the regulations under the Control of Pollution Act 1974 were never implemented. For neo-pluralists, much policy making is a charade designed to maintain the *status quo* and deflect challenges to the existing allocation of resources. Policy, therefore, must also be thought of in terms of courses of inaction as much as courses of action. One of the most influential voices in policy analysis, Aaron Wildavsky (1979, 387), reminds us that 'policy is a process as well as a product. It is used to refer to a process of decision making and also to the product of that process'.

Box 3.9 The 'textbook' policy process

Traditionally, political scientists have tended to study individual institutions such as parliament or the courts in isolation. But in the 1960s the policy process began to be seen in terms of a series of sequential parts or stages. These are (1) problem emergence; (2) agenda setting; (3) consideration of policy options; (4) adoption of policy options; (5) implementation; and (6) evaluation. There are advantages in adopting such an approach. For instance, it emphasises that policy is a process, involving many different parts of the government. It is simple and intuitively appealing. However, a 'stagist' approach simplifies what in practice is an interconnected process. According to more modern accounts, it fails to explain what drives issues from one stage to the next, overlooks the fact that stages are not always sequential (for example, policies are 'made' as they are implemented) and is too rigid and top-down in its view of non-governmental actors (Sabatier and Jenkins-Smith, 1993). Despite these criticisms, the 'stagist' view persists because political scientists need a metaphor to help to reduce the complexity of the policy process to a more manageable form.

According to McCormick (1991, 7), environmental policy is 'public policy concerned with governing the relationship between people and the environment'. According to the 'textbook' view of policy (see Box 3.9), the first task facing environmental groups is to get a particular problem on the agenda for discussion and, if possible, consideration by policy makers. Policy makers choose among the possible course of actions the most appropriate set of policy responses, which are eventually adopted and handed on to administrators to be implemented. Our view of this process depends upon which theoretical lens we choose to look through.

Power and the policy process

Pluralists believe that the agenda-setting stage is open and competitive. The political system as a whole is seen to be open and accessible to groups, although which particular coalition triumphs varies from sector to sector and from policy to policy. Once adopted, though, policies must still be steered through the reefs and shoals of the implementation process. If policies do not gain the support of implementing agents – and often they do not – they are unlikely to result in concrete action on the ground. The notion of implementation 'failure' has been a theme of public policy since the 1970s. While some analysts bemoan the inability of central government to get its way, others regard performance deficits as another powerful check on the power of central government. For pluralists, then, outcomes are likely to be unpredictable because of the volatility and competitiveness of the policy process. The continuous conflict between groups for resources across the range of policy sectors prevents one particular class or elite from dominating the policy process.

For neo-pluralists, institutional factors mean that the agenda-setting process is fundamentally biased towards corporate power. Non-decision making removes many potentially controversial issues from the agenda. The state, meanwhile, is forced to cultivate close relationships with business and professional interests because of the key functions they perform in society. Policy is said to be made and implemented within fairly small groups of actors, clustered around particular government departments. Consequently, policy outcomes are likely to suit the preferences of business rather than environmental groups, whose influence is rather more short-lived.

Finally, Marxists believe the state is simply the instrument of the capital-owning class. Whatever pluralism exists is merely a sham to legitimate the process of capital accumulation. For Sandbach (1980, 125), pressure group demands for greater consultation and public participation reveal how far they have been 'duped into the dominant political ideology of pluralism' when the pressures to develop 'are closely related to the nature and demands of modern capitalism'. In the longer term, policy outcomes are likely to fit the needs of business, although there may be short-run triumphs for environmental interests (see Box 3.5). Hence, how business in general accommodates to the environmental and sustainability agenda is a process of considerable interest.

Policy making in policy networks

In recent years, policy networks have emerged as the preferred tool for examining the policy process, both nationally and increasingly in the EU (see Jordan, 1998b). Network analysis recognises that the modern policy process tends to be segmented into highly specialised sectors. Networks of specialist interest and interested specialists form around particular government departments to trade information and share the burden of decision making. They comprise a relatively small number of professionals from government and pressure groups who have sufficient interest and technical ability to provide an input into policy development. Policy networks describe the closed and consensual nature of the policy process. They indicate the fact that much policy is made in informal institutions outside the reach of formal democratic controls such as parliamentary scrutiny or ministerial responsibility. Networks are mechanisms for exchanging information, communicating and exchanging resources. They create order by funnelling policy in certain directions; they simplify by limiting the number of problems to be addressed and options to be considered. Networks are themselves shaped by pressure groups whose activities are nowadays highly influential (Box 3.10).

Box 3.10 Pressure groups and environmental politics

Pressure groups form to promote common interests which their members feel are not being given sufficient articulation by the existing arrangement of bids. There are two major kinds: *public interest* (or *cause*) groups (which act in the name of the public good generally) and *private interest* or *NIMBY* (not in my back yard) groups. *Civic organisations* often straddle the two but serve community interests in a defined locality, usually a city or town.

Pressure groups are usually non-party-political, run by professional or active amateurs, have variable accountability to membership, mobilise to recruit support and membership income, and provide special services for members (newsletters, information campaigns, political updates and low-cost access to favoured sites). Groups aim to embarrass governments, to provide counter-establishment interpretation of key issues, to use the law to promote a particular right, or to correct ill-managed regulation, or simply to educate, raise awareness and provide various channels for articulated protest.

In times of large parliamentary majorities, pressure groups act as powerful lobbies to influence backbench MP attention, and are especially active around the policy select committees of both the Commons and the Lords. Pressure groups find it difficult to form coalitions, especially across the environmental and social/poverty/justice arenas. This is partly a matter of inertia, partly a response to member interests and partly a result of declining member income. In essence, then, environmental groups are good on the old

Box 3.10 (continued)

environmental agendas but really weak and somewhat ill-prepared for the multiple coalitions needed for sustainable development.

There are three theories of pressure group formation, which form a start to any further analysis.

● *Resource mobilisation.* This is the concentration of expertise, coalition support and organised campaigning that provides momentum, attention and unavoidable presence. The key is the generation of fresh outlooks, counteracting viewpoints and co-ordinated criticism, together with constructive alternatives. All of these require the support of individuals and groups, who lend political weight through legitimacy.

● *Social movements.* Any society undergoes changes in values and concerns, sometimes convulsively but more often at a less observable pace. Feminism, environmentalism and iconism around fashion models are examples of a genre that is universal. Environmental groups create and capture social movements so are part of the changes that make societies evolve. Green consumerism, ecological architecture and women's environmental networks are all examples of environmental capturing of consumerism, professional development and feminism.

● *Political opportunism.* The policy process, as described in the main text, allows for entry into the structures of power. This may be due to a hiatus in the policy community structure, as, for example, with the recent rise of interest in organic agriculture. Or it may be due to a novel issue, such as the trials of genetically modified crops, which creates the scope for new networks of activism to form. The use of the Internet as both a bonding agent and a campaigning channel has assisted opportunistic coalitions of direct action or 'value-added' information transfer to be far more effective than otherwise would be the case.

Marsh and Rhodes (1992) distinguish between different types of network, ranging along a continuum from highly integrated *policy communities* at one end to loosely integrated *issue networks* at the other. Policy communities tend to have a fairly stable membership and are closed to 'outsiders' (see Box 3.7), while issue networks have a much larger membership, multiple decision-making centres and loci of power, and a relatively open structure. The best example of a policy community is to be found in the agricultural sector. Since the Second World War, a policy community of farmers and government officials from the Ministry of Agriculture has supported a policy of subsidised production (see Smith, 1995, 81–92). Groups which called attention to the environmental damage of intensive farming were effectively marginalised. Since 1985, however, pluralism has been at work. The financial cost of the Common Agricultural Policy, the futility and injustice of paying wealthy farmers to produce for subsidised surplus at the expense of increasingly valued wildlife and scenic amenity, and the rise of co-ordinated environmental and rural well-being lobbies, transformed UK agricultural policy in a decade (see Cobb *et al.,* 1998). Blending elements of pluralism, neo-pluralism and Marxism, Rhodes (1995, 11) believes the main challenge facing UK central government is to find ways to steer these networks which 'have a significant degree of autonomy from the state and are not accountable to it'. So, although neo-pluralists are correct in saying that there are strong links between business and government at the sectoral level, it remains difficult to co-ordinate policy across the various policy networks – what has been termed 'departmental pluralism'.

Policy change over time

The problem with network theory is that it seems to explain only *stability* rather than change. This is partly because it was developed at a time of close relations between government, business and the trade unions – what political scientists refer to as the 'old' politics of corporatism. The 1980s witnessed a number of important political, economic and social changes which destabilised policy communities, creating a 'new' set of political alliances. Smith (1993) suggests that several factors bring about policy change when there is a strongly institutionalised policy community. These include:

- *changing international relations* (e.g. interventions by the EU);
- *economic and social changes* (e.g. the rise of 'new social movements' such as environmentalism);
- *the emergence of new problems to which the community has no immediate solution* (e.g. climate change) – these may trigger internal divisions as participants disagree over policy responses;
- *autonomous actions by the state or political leaders* – politicians may suddenly decide to consult new groups or sever links with existing members of the network;
- *new technologies which create new problems and new solutions* (e.g. biotechnology).

In recent years, several attempts have been made to develop comprehensive theories of the policy process. One of the most successful is Paul Sabatier's model of *social learning* (Sabatier and Jenkins-Smith, 1993). It is worth discussing in detail because it tries to draw together insights from more stage-based models and provides a convincing explanation of why policies change over time.

Like policy network theorists, Sabatier's advocacy coalition framework divides the policy process into discrete *subsystems* of actors such as academics, bureaucrats and journalists, who play a part in the generation, dissemination and evaluation of policy ideas. Within each subsystem there are likely to be a small number of *advocacy coalitions* comprising actors with different institutional affiliations who nonetheless share a set of policy beliefs towards policy formulation and implementation. Each coalition seek to realise its beliefs by adopting a strategy containing various policy interventions. Normally, there are between two and four coalitions, but in politically tranquil subsystems there may be only one. It is the struggle between these coalitions that drives the policy process. Policy brokers in the form of civil servants, the courts and elected officials try to adjudicate between rival claims and strategies, achieve feasible compromises, and generally keep conflict down to a reasonable level. Coalitions seek to turn their beliefs into policies by outlearning their adversaries in the wider struggle to keep abreast of events and in tune with changes to the parameters of the subsystem. These include the basic attributes of the problem in question, the distribution of natural resources, and cultural values in society and the constitutional framework, and are relatively stable over time.

Sabatier suggests that the beliefs of the different coalitions are organised into three distinct layers:

- *deep core beliefs and fundamental ideas* define a person's basic philosophy;
- *near (policy) core beliefs* determine a person's fundamental values and strategies;
- *secondary elements* relate to the way in which policies should be implemented.

This hierarchy is arranged in order of decreasing resistance to change, with secondary elements being the most fluid.

At any one time there is likely to be a dominant coalition, which sets the intellectual framework within which minor or individual policy decisions are made, together with a series of minority coalitions. While changes at the secondary level emerge incrementally as different coalitions engage in a process of policy-oriented learning, changes in the core aspects of policy require an exogenous 'shock' from outside the subsystem. These include changes in public opinion, impacts from other subsystems, economic and social developments, and so on. 'Shocks' challenge all the advocacy coalition's understanding of reality, triggering the search for new ideas. The dominant advocacy coalition must respond to, or better still anticipate, them in a manner consistent with its beliefs to retain its pre-eminent position. While minority coalitions can hope to have some of their beliefs incorporated in policy by outlearning their adversaries, achieving substantial change depends upon their playing a waiting game until an exogenous shock significantly increases their political resources. At this point 'luck' becomes a significant factor: 'The process must be frustrating at times, as actors who have worked for years to gain an advantage over their competitors ... suddenly find their plans knocked awry by [external] events' (Sabatier, 1993, 22).

The *rational actor model* is based on a consistent ordering of priorities, on predictable underlying values. The fundamental premise is optimising utility, or totality of satisfactions of efficiency-seeking actors. In the pattern outlined in Figure 3.4, this efficiency preference is tempered by other considerations, notably equity, social justice and environmental tolerance. There is no reason in principle why rational actors cannot cope with such matters, though the model retains its reliance on consistency and priority.

The *sequential decision model* depicted in Figure 3.5 is altogether different. It begins with the exercise of elite power, through which a predetermined pattern of decisions occurs, based on standard operating procedures. These may result in resistance and protest. If the criticism is based on strong advocacy coalitions, it may be sufficient to trigger a more pluralistic and deliberate approach to decision making, with inclusionary, power-sharing principles being invoked. This would only run its course to a second, strategic, decision if the old elite order breaks down. In the arenas of nuclear power, intensive agriculture and road transport, a degree of that fragmentation has begun to show. But beware, the old order has not really gone away. Appearances can be deceptive. The full sequence of Figure 3.5 is rarer than one might imagine.

A case study of policy change over decades: British coastal water quality

In Britain the question of coastal water quality has traditionally been left to a professionalised policy community comprising engineers and local authority interests. The preferred policy for dealing with sewage was to dispose of it along pipes into the sea. This fitted with the general philosophy of environmental policy, which was to dilute and disperse wastes by emitting them along long pipes and tall chimneys into the environment, where they were thought to dissipate harmlessly (Weale, 1998). A marine treatment advocacy coalition, which was well represented in the water industry and in Whitehall, regarded sea disposal as a better environmental solution than treating it on land in sewage works, which is still costly and creates environmental impacts. In practice, many of the pipes were old and very poorly located, and sewage collected on

Figure 3.4:

Rational actor
decision model.

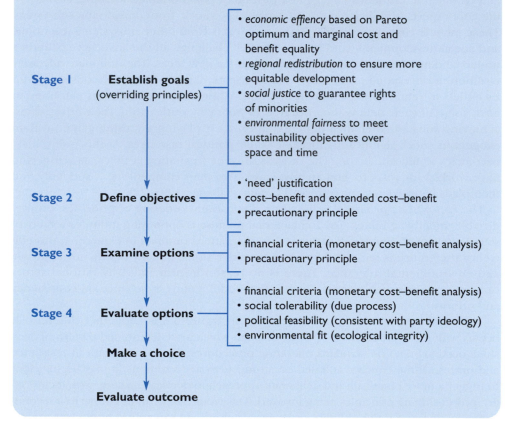

- When confronted with a choice a decision is made.
- That decision is based on preferential ranking of alternatives.
- Preferential ranking is both consistent and predictable.
- The highest ranking alternative is always selected.
- Given the same range of choice, the same decision is always made.

Stage 1   **Establish goals**
(overriding principles)

- *economic effiency* based on Pareto optimum and marginal cost and benefit equality
- *regional redistribution* to ensure more equitable development
- *social justice* to guarantee rights of minorities
- *environmental fairness* to meet sustainability objectives over space and time

Stage 2   **Define objectives**

- 'need' justification
- cost–benefit and extended cost–benefit
- precautionary principle

Stage 3   **Examine options**

- financial criteria (monetary cost–benefit analysis)
- precautionary principle

Stage 4   **Evaluate options**

- financial criteria (monetary cost–benefit analysis)
- social tolerability (due process)
- political feasibility (consistent with party ideology)
- environmental fit (ecological integrity)

Make a choice

Evaluate outcome

beaches and bathing areas. Amenity groups campaigned from the 1950s for the construction of longer and better-sited pipes. The Treasury, however, pointed out that public money was short and, having received evidence from scientific experts that sewage was not a cause of serious diseases like typhoid, the DoE decided to leave matters as they were. Until the 1980s, non-decision-making devices served to organise dirty beaches out of politics via a mobilisation of vested interests (Jordan, 1998a).

Nowadays, the question of coastal water quality is part of the political mainstream and is widely debated in parliament and the media. Stories about dirty beaches coated with sewage and sanitary products regularly make front-page news. Gone is the old policy community, replaced by a much broader issue network of actors including environmental groups, water companies, national regulatory bodies, tourist interests, local chambers of commerce and European authorities. What was, from the government's point of view, once a relatively well- 'managed' and tranquil policy area has become much more unstable as new ideas and domestic regulatory structures have forced the government to justify principles and practices that were implicit or simply rhetorical.

Why do relatively stable policies in areas or subsystems dominated by well-established policy communities undergo sudden and convulsive periods of change? Why do

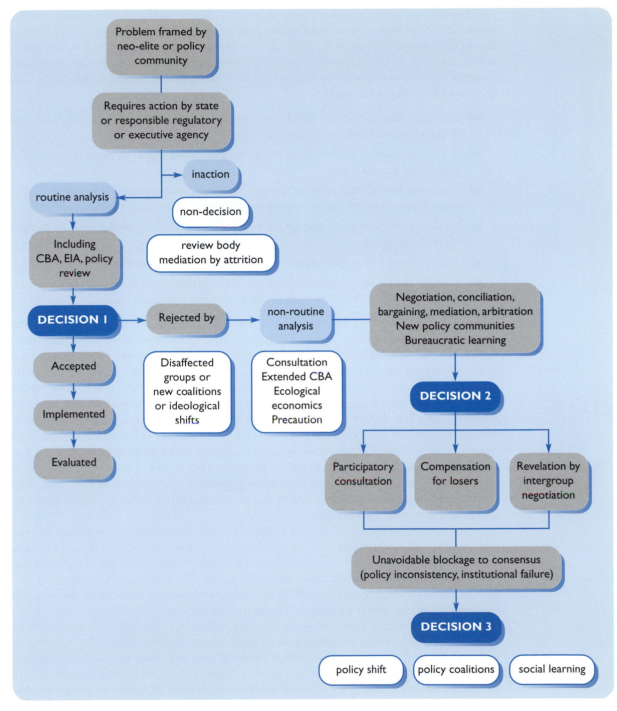

Figure 3.5: Multiple-sequence decision model.

policies that appear to have firm support and a solid ideological underpinning exhibit wholesale shifts which then stabilise around a new equilibrium position? We look at this in the context of Figures 3.1 and 3.2. The changes can be traced back to the 1980s and a number of developments which culminated in a significant shift in policy towards the improvement of recreational waters using inland treatment methods such as disinfection. These included:

- *interventions by the EU,* which demanded higher environmental standards and greater inland treatment of sewage;
- the *privatisation of the water industry* and the creation of new, independent regulatory bodies;
- *changing public attitudes* to sewage pollution and greater *demands for year-round recreation* in coastal waters (e.g. surfing, sailing and sub-aqua sports);
- higher-profile *campaigning by environmental pressure groups* such as Surfers Against Sewage;
- *improvements in inland treatment technology* such as disinfection;
- *improvements in epidemiological understanding* coupled with a greater interest in minor health effects such as diarrhoea and conjunctivitis.

How do we explain these changes? In the 1950s and 1960s, the nature of the coastal environment (strong tides and turbulent waters), conventional wisdom, scientific understanding and legal structures strongly supported sea disposal, which functioned as the dominant policy framework. Members of the marine treatment advocacy coalition believed in the efficacy and cost-effectiveness of dilution and dispersion, with a secondary-level preference for long disposal pipes over expensive treatment on land. Underpinning this coalition was a deep core belief in the ability of science and technology to predict the behaviour of bacteria and pathogens in water and thus to optimise the amount of pollution. In addition, there was a presumption that marine currents functioned as a self-cleansing process. This is a very technocratic view to adopt, but it was buttressed by established scientific judgement at the time. The marine treatment coalition dominated the professionalised water policy community and policy was made consensually.

In the 1980s, an inland treatment coalition began to gain a hold in policy-making circles. While the British scientific establishment, the government, the House of Lords and, interestingly, the older amenity groups advocated marine treatment, environmental and new amenity groups, other member states, the European Commission and backbench MPs were concerned about the waste-assimilative capacity of the sea and counselled greater inland treatment. The normative difference in the policy core of the two coalitions reflects a deep dispute over the extent to which pollution should be reduced, as well as the appropriate role of scientific uncertainty in decision making. In essence, the marine treatment coalition believed that pollution should be optimised (i.e. waste should be emitted up to a certain damage threshold), whereas the inland treatment advocacy coalition regarded waste minimisation at source as the preferable long-term objective, with financial costs being a less pressing consideration. In addition, the precautionary principle was invoked regarding the putative health effects, pushed by the recreation and tourist lobbies.

Matters were eventually brought to a head by the privatisation of the water industry. This policy, determined outside the coastal treatment dispute, involved the private water supply and waste disposal companies, which, unlike their publicly run but government-controlled predecessors, could recoup the costs of extra investment from the consumer. This 'cost pass-through' strategy of the consumer paying was helped by a legally much tougher European Commission threat to take the UK to the European Court over its violation of the Bathing Waters Directive. Today, both coalitions are now at work within the more open context of the modern water policy network, and each leaves a discernible imprint upon policy outputs. Policy is now based upon a mixture of long pipes and inland treatment, with expensive disinfection in major resort areas such as Jersey. However, the cost of these investments has had to be met by water charge payers, who have seen their bills soar since privatisation. The new water companies, which have enjoyed large profits since the sale, have become extremely unpopular with the public, demonstrating the recursive nature of policy making (see Box 3.10). In recent years, an economic efficiency advocacy coalition has formed around the problem of how to finance the coastal improvement programme, which runs into many billions of pounds. This new coalition is dominated by the domestic price regulator, the Office of Water Services (OFWAT). OFWAT is primarily concerned with achieving a better deal for water charge payers by minimising compliance costs, and is much less preoccupied with the scientifically rooted dispute between the two other advocacy coalitions. It is not statutorily responsible for the stewardship of the water environment, a matter for the Environment Agency. This division of official duties between two regulatory bodies in the public domain illustrates yet again how sustainable development faces a demanding uphill struggle with regard to institutional inertia and policy contradiction.

So how did this secular shift in thinking come about? It is possible to identify the influence of changes in the basic parameters of the subsystem:

- *The problem became more severe.* More sewage was created as tourism at coastal resorts boomed and its non-biodegradable content increased, making it harder to disperse in sea water. Evidence of eutrophication in the North Sea came to light and fears grew on the spurious claim advanced by mischievous and maverick members of a local FoE group that AIDS could be transferred in polluted sea water.

- *Social attitudes to pollution changed.* People became more intolerant of even fairly low levels of contamination, while demand grew from sailors and surfers for year-round recreation in nearshore waters.

- *Emergence of powerful environmental groups.* Groups such as Greenpeace had the legal and financial resources to campaign much more effectively than their forerunners.

- *Technological developments.* Pipe-laying technology and disinfection improved, offering viable alternatives to short pipes.

These forced proponents of marine treatment to defend their beliefs. The questioning of existing solutions in the face of 'new' problems triggered some learning within the marine treatment coalition as water engineers, following practice in the USA, began to construct longer sea outfalls in the 1960s, but they were held back by financial controls.

Nonetheless, the core of British policy remained substantially intact. New information about the more minor health risks associated with sewage was rejected by the marine treatment coalition because it was not consistent with its policy core values, whereas it found a receptive audience in the pressure group community since it appeared to demonstrate the fallibility of marine disposal. In fact, groups such as Greenpeace – a group excluded from the formal policy community – tried to alter perceptions and beliefs by commissioning its own epidemiological research in a bid to outlearn adversaries.

As recently as 1989, the marine treatment coalition held a tight hold over the direction of policy; the best hope for opponents such as Greenpeace lay in waiting for one or more external shocks to alter the resources and opportunities of all actors (see Figure 3.2). These came from two directions:

- *External policy demands.* The pressure from the EC and other North Sea states for the UK to treat its sewage at source, based on the EC principles of shared responsibility and the precautionary principle, was introduced in the 1980s as an effective policy tool.

- *Impacts from other domestic subsystems.* Privatisation introduced management techniques and opened the doors holding back investment.

However, these external 'events' did not translate automatically into policy change; they had to be skilfully exploited by members of the minority coalition. EC policies, for instance, provided new campaigning opportunities for environmental groups. Similarly, privatisation helped to open up the hitherto closed policy community to greater external scrutiny.

As yet, however, the system-wide learning process has not created a new dominant coalition and new alignments are still in a state of flux. Although there has been significant change at the level of policy core beliefs, the guiding philosophy of the investment programme is still marine treatment. We may now be witnessing the beginning of another period of stability in the water sector as a hybrid advocacy coalition emerges as a result of the interactive combination of beliefs and ideas.

References

Bennet, G. (1992) *Dilemmas: Coping with Environmental Problems.* Earthscan, London.

Blondel, J. (1991) Politics. In V. Bogdanor (ed.), *The Blackwell Encyclopaedia of Political Science.* Blackwell, Oxford.

Blowers, A. (1984) *Something in the Air.* Harper & Row, London.

Carley, M. and Christie, I. (1997) *Managing Sustainable Development.* Earthscan, London.

Carter, A. (1999) *A Radical Green Political Theory.* Routledge, London.

Christie, I. (1994) Britain's sustainable development strategy. *Policy Studies,* **15**(3), 4–20.

Cobb, R.D., Dolman, P. and O'Riordan, T. (1999) The politics of sustainable agriculture. *Progress in Human Geography,* **16**(2).

Dahl, R. and Lindblom, C. (1976) *Politics, Economics and Welfare.* Harper & Row, New York.

Department of Environment, Transport and the Regions (1998) *A New Deal for Transport: Better for Everyone.* The government's White Paper on Transport. The Stationery Office, London.

Downs, A. (1972) Up and down with ecology: the issue-attention cycle. *Public Interest,* **28**, 38–50.

Grant, W. (1989) *Pressure Groups, Politics and Democracy in Britain* (2nd edn). Philip Allen, London.

Greenaway, J., Street, J. and Smith, S. (1992) *Deciding Factors in British Politics: a Case Studies Approach.* Routledge, London.

Ham, C. and Hill, M. (1993) *The Policy Process in the Modern Capitalist State.* Harvester Wheatsheaf, Hemel Hempstead.

Hill, J. and Jordan, A. (1994) The greening of government: lessons from the White Paper process. *ECOS,* **24**(3/4), 3–9.

Jordan, A. (1998a) The construction of a multi-level environmental governance system. *Environment and Planning* C, **17** (in press).

Jordan, A. (1998b) Step change or stasis? EC environmental policy after the Amsterdam summit. *Environmental Politics,* **7**(1), 227–235.

Lee, K. (1993) *Compass and Gyroscope: Integrity Science for Policy and the Environment.* Island Press, New York.

Lindblom, C. (1977) *Politics and Markets.* Basic Books, New York.

Lukes, S. (1974) *Power: a Radical View.* Macmillan, Reading.

Marsh, D. and Rhodes, R. (eds) (1992) *Policy Networks in Britain.* Clarendon Press, Oxford.

McCormick, J. (1991) *British Politics and the Environment.* Earthscan, London.

Newby, H. (1993) One world, two cultures: sociology and the environment. *Network,* **50** (May), 1–8.

Nugent, N. (1993) *The Government and Politics of the EU.* Macmillan, Basingstoke.

O'Riordan, T. (1995) Core beliefs and the environment. *Environment,* **37**(8), 6–10, 23–27.

Packard, V. (1954) *The Hidden Persuaders.* Random House, New York.

Parsons, W. (1995) *Public Policy.* Edward Elgar, London.

Pepper, D. (1996) *Modern Environmentalism: an Introduction.* Routledge, London.

Press, D. (1994) *Democratic Dilemmas in the Age of Ecology: Trees and Toxics in the American West.* Duke University Press, Durham, NC.

Rawcliffe, P. (1998) *Environmental Pressure Groups in Transition.* Manchester University Press, Manchester.

Redclift, M. (1992) Sustainable development and global environmental change. *Global Environmental Change,* **2**, 32–42.

Rhodes, R.A.W. (1995) *The New Governance: Governing without Government.* ESRC, Swindon.

Sabatier, P.A. (1993) Policy change over a decade or more. In P.A. Sabatier and H.C. Jenkins-Smith (eds), *Policy Change and Learning.* Westview Press, Boulder, CO.

Sabatier, P.A. and Jenkins-Smith, H.C. (1993) *Policy Change and Learning: an Advocacy Coalition Approach.* Westview Press, Boulder, CO.

Sachs, W. (ed.) (1993) *Global Ecology.* Zed Books, London.

Sandbach, F. (1980) *Environment, Ideology and Policy.* Basil Blackwell, Oxford.

Smith, M. (1993) *Pressure, Power and Policy.* Harvester Wheatsheaf, Hemel Hempstead.

Smith, M. (1995) *Pressure Politics.* Baseline Books, Manchester.

Wallace, H. and Wallace, W. (1996) *Policymaking in the European Union.* Oxford Univrsity Press, Oxford.

Weale, A. (1992) *The New Politics of Pollution.* Manchester University Press, Manchester.

Weale, A. (1998) Environmental policy. In I. Budge, I. Crewe, D. McKay and K. Newton (eds), *The New British Politics.* Longman, Harlow.

Wildavsky, A. (1979) *Speaking Truth to Power.* Macmillan, Reading.

Young, S. (1993) *The Politics of the Environment.* Baseline Books, Manchester.

Further reading

There is a huge literature on politics and policy. You might like to start with Christopher Ham and Michael Hill's book entitled *The Policy Process in the Modern Capitalist State* (1993) or *Deciding Factors in British Politics* by John Greenaway and colleagues (1992). More in-depth coverage is provided by Wayne Parsons (1995). Good textbooks on environmental politics are surprisingly thin on the ground, but try Stephen Young (1993) or *Managing Sustainable Development* by Mike Carley and Ian Christie (1997). Moving up to the European level, currently the best bets are Helen Wallace and William Wallace (1996) or Neill Nugent (1993), although neither book says much about the environment. A good introductory text on EC environmental policy processes is still to be written. In the meantime, Nigel Haigh's (1998) *Manual of EC Environmental Policy* describes every item of EU environmental legislation and its impact upon Britain. Jordan (1998b) provides an introductory analysis of how EU environmental policy is made and implemented. Finally, political theories only really come to life when they are applied to case studies of environmental problems in 'real life'. Good examples can be found in the books by Graham Bennett (1992), Andy Blowers (1984) and Peter Rawcliffe (1998). A recent book by Alan Carter (1999) summarises various philosophies underlying green political thought.

Chapter Four

Environmental and ecological economics

W. Neil Adger

Topics covered

- Growth and development
- Externalities
- Monetary valuation
- Cost–benefit analysis
- Equity and efficiency
- Ecological economies

Editorial introduction

Economists are often seen as the villain of the piece by non-economists. They give as good as they get, so the argument is often fiery and protracted. What is less recognised is that economists themselves are divided as a disciplinary species. The tribalists of growth, game theory, international trade and development, and labour markets find much of the work on environmental economics rather trivial and unimaginative. After all, much of that work is based on conventional neo-classical economic theory. To deviate too far from that would invoke wrath, not just derision. So there is often a defensiveness amongst environmental economists as to how to proceed when faced with the sustainability edifice. Sustainability is holistic and it involves social values, equity justice, fairness and ethical considerations over how to care and be precautionary. The discussion in Chapter 2 is some distance away from conventional economic theory. While equity and distributional matters certainly are central to that theory, what is not so easily handled are the cultural and political settings through which groups have, or are denied, power, and some control over their destiny. So it is not enough to indicate that a loser can be compensated by gainers and be no worse off. The kind of compensation on offer, how that is discovered and through what means it is monitored in terms of the values of the compensated individual or group go far beyond the realm of neo-classical economics. This is why the economics profession sometimes has difficulty in justifying its theory and method. It is the holistic approach to valuation, power, democracy and deliberation that still eludes much of the economics profession.

   In this chapter, Neil Adger is fully aware of this context and alive to the difficult issues of relating equity, ethics and ecology into economic analysis. This is partly why his chapter follows the chapters on deliberative science, sustainability politics, and political and policy processes, and why the ecology and biodiversity chapter follows on. Economists are not immune to the criticism outlined above. Hence the interest in ecological economics, the lively debate over ecosystem function valuations, and the emergence of parallel accounts of social well-being in association with the more familiar gross domestic product calculations. Politicians and economists do want comparative indices, if for no other reason than to see how well an economy is doing. GDP, for all its many faults, is still a measurable and replicable indicator of a particular realm of economic activity. It measures the volume of goods and services provided in a given time period, and, for the most part, it is comparable year on year. The UN Development Programme (1998, 16–37) has spent the last decade extending GDP to a *human development index* (HDI). This covers several measures of well-being, including life expectancy, income, adult literacy and enrolment in primary, secondary and tertiary education. Figure 4.1 summarises the position of major countries on both measures of GDP and HDI.

The HDI is important since it provides a useful indicator of change in many aspects of fundamental human development. Of interest is the finding that of 174 countries studied, 98 rank higher on HDI than on GDP *per capita*, measured on a purchasing power parity basis. This suggests that development is moving to support conditions favourable to human well-being, and not just goods and services *per capita*. The link between the two is by no means obvious, and much depends on politics, personality and history.

A more ambitious index has been advanced by Robert Prescott-Allen (1999) in co-operation with a large number of organisations. This is the *well-being of nations* assessment method. The aim is to provide an accessible and straightforward way of identifying diagnostic features and selecting and combining indicators. The key is to determine a nation's goals, set these in specific indicators, identify performance criteria, and then measure and combine the findings into a series of relationships between ecosystems and human conditions. The result is complex but very comprehensive. It is not strictly comparable but shows more clearly how far a development path is moving towards or away from sustainability.

All such approaches to composite measures of well-being depend on the acceptance of the methodology by official agencies, and on the national capability to generate appropriate data. The last factor is a justifiable one, for it allows the fundamental premises of economics, namely the political economy of human and planetary well-being, to be better analysed and observed than any other means devised so far. 'Green' accounting is fine as a tool, but it tries always to convert to a monetary measure. It is useful as a means of indicating the depletion of resources, and for measuring how far care and social maintenance are real economic assets. But in terms of sustainability economics, it is only a partial stage in the evolution of a more comprehensive and comparative account of the true 'commonwealth of nations'. At a local level, the deliberative and inclusive search for *sustainability indicators* is the complementary approach to determining the true 'common good' (see Farrell and Hart, 1998).

Efficiency, scarcity and scale in environmental economics

Economics is a social science. Economic analysis is primarily concerned with the well-being of people, through the allocation and consumption of the factors of economic production. Economics interacts with the environment because the natural resources which form one major factor of production are inherently scarce. The other two fac-

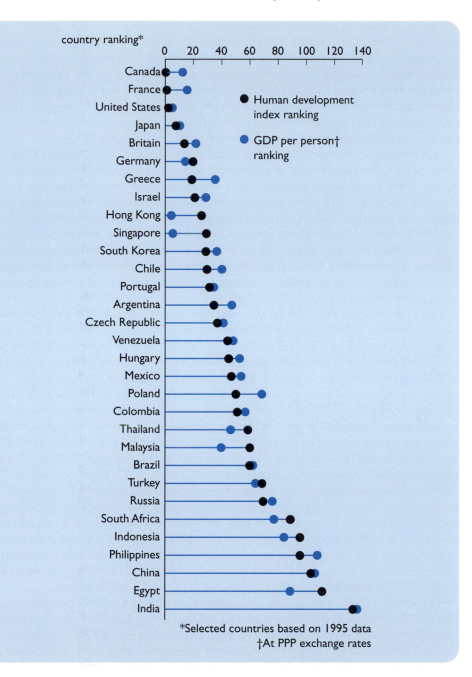

Figure 4.1:
Human development index.

Source: *The Economist*, 12 September 1998, 138 based on data from the UN Development programme.

country ranking*

Human development index ranking

GDP per person† ranking

Canada
France
United States
Japan
Britain
Germany
Greece
Israel
Hong Kong
Singapore
South Korea
Chile
Portugal
Argentina
Czech Republic
Venezuela
Hungary
Mexico
Poland
Colombia
Thailand
Malaysia
Brazil
Turkey
Russia
South Africa
Indonesia
Philippines
China
Egypt
India

*Selected countries based on 1995 data
†At PPP exchange rates

tors are labour, and capital in the form of technology and knowhow. In principle, the application of capital and labour should increase the availability of natural resources, through extraction and recycling. In addition, a socially valued and habitable environment is also a scarce resource, the allocation, ownership and management of which is a major focus of environmental economics. These principles, which place economics within the realm of the social interest and the natural environment, are often overlooked in conventional economics. Many of the global environmental problems detailed in other chapters are often argued to be the outcome of economic 'growth mania' or a belief in progress through the ever-expanding use of the world's natural resources, often concentrated in the hands of a relative minority (see Chapter 2). Environmental and ecological economics addresses this fundamental issue by trying to show where economics and environment are in conflict and where they are complementary. This chapter outlines how economics tools have been used to address the questions of sustainable development. These approaches are not, however, designed to be used in isolation from the social and political as well as environmental consequences of all individual and collective actions on the environment.

It is sometimes assumed that 'economics' is the cause of much negative and undesirable environmental change and degradation. Indeed, the scale of economic activity is one of the major issues addressed by economists. Since economic growth is one of the major objectives of government policy, the consequences of growth have been shown, for example, to be associated with the loss of biodiversity globally; the pollution of marine and fresh water; and other global-scale environmental problems. In addition, the reliance on fossil fuels within the world's industrial economies has increased atmospheric concentrations of greenhouse gases in the period since the Industrial Revolution, potentially causing large-scale impacts of climate change and irreversible impacts on many ecosystems.

Can growth of the economy be decoupled from its environmental impact? There are two approaches to this issue within environmental economics. The first is to characterise the economy as a system in which the desirable goal is a *steady state*, where input and throughput are minimised but human well-being, or welfare, is maximised (following Daly, 1992, for example). The second is to argue that the combination of economic growth and scarcity of good environmental quality will eventually raise the effective demand for environmental quality to such an extent that pollution and other environmental degradation become no longer politically or socially tolerable. Collective action on society's behalf by government or international bodies will then take place to meet this effective demand for environmental quality. This has come to be known as the *environmental Kuznets curve* hypothesis (see Barbier, 1997; Rothman and de Bruyn, 1998). The original Kuznets curve was a hypothesised link between income levels in a country and the distribution of income. This stated that rising income levels were correlated with rising inequality up to a threshold where at higher levels of income, inequality would begin to decline again (Kuznets, 1955). In the same manner, it is hypothesised that since good environmental quality is in effect a luxury good, then higher environmental quality is demanded only when incomes reach a threshold level – hence economic growth is in effect eventually 'good for the environment'.

The relationships between various ambient pollutants and the level of *per capita* income is demonstrated in Figure 4.2 (see Shafik, 1994). Rising economic activity can cause environmental problems but in certain circumstances and with appropriate institutional arrangements can lead to enhanced environmental quality. The upper panel representing issues such as potable and safe drinking water, would appear to improve with increasing incomes. In these cases, increasing incomes provide the resources for investment in public services such as sanitation or water and soil conservation. But for other impacts on the environment, such as *per capita* carbon dioxide emissions, it is clear that richer is more polluting. A key issue is whether there are many aspects of environmental quality where the middle panels are more appropriate. The examples in Figure 4.2 are of concentrations of SO_2 and smoke in urban areas, where there would appear to be some evidence of a 'turning point': concentrations worsen during the early stages of industrialisation, but as incomes rise they turn downwards with increased effective demand for enhanced environmental quality.

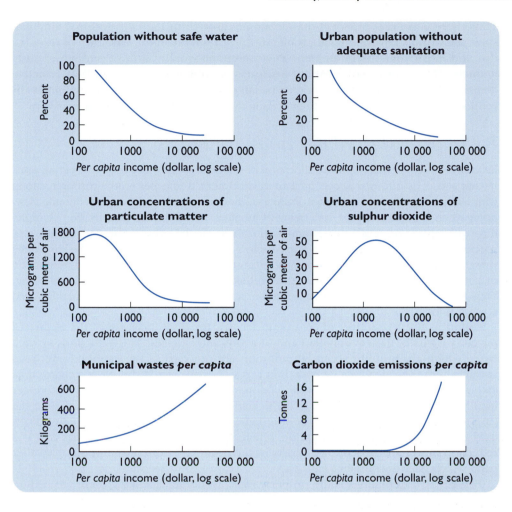

Figure 4.2:
Relationships
between ambient
pollutants and
income.
Source: World Bank
(1992, 11).

If many aspects of the environment and change conform to this model, the implications of this analysis are enormous. Economic growth *per se* will lead eventually to environmental conservation. But there are a number of environmental problems, particularly on the global scale, which it is not possible to grow out of. These include climate change, where the absolute stock of greenhouse gases in the atmosphere as well as the rate of emissions are important; and the preservation of biological diversity, where much habitat or species loss is effectively irreversible. The critiques of the environmental Kuznets curve can therefore be summarised as follows.

- Only certain parts of environmental quality are open to this analysis – primarily those such as urban air quality where data are available. This limits the overall consideration of whether turning points exist. Neither are there considerations of technological change and the role it can play in changing the impact of economic activity.

- There is no guarantee that environmental quality standards will rise, since the incidence of pollution often falls on those at the bottom end of the income spectrum, who are also least politically influential.

- Some environmental change is effectively irreversible. Relying on growth to bring about desirable use and conservation of scarce environmental resources is fundamentally flawed and goes against the precautionary principle.

Thus there is no real prospect that economic growth of the conventional kind can solve the world's environmental problems. Improvements in the efficiency of resource use and improvements in ambient environmental quality do not happen spontaneously. They require appropriate institutional arrangements as well as effective demand from interest groups acting on behalf of society. Nevertheless, economic intervention does have a role to play in managing the consequences of economic activity.

Pollution, externalities and economics

Economics is concerned with the allocation of scarce resources between competing uses. In that sense, as outlined above, a clean environment is another scarce resource, along with labour, capital and knowledge. Pollution, then, is important from an economic viewpoint when it impacts on the well-being of humans. It can do this directly (for example, through impacts on health) or indirectly (for example, by causing the decline in a rare species, the loss of which causes some people a loss of well-being). Under this economic view, it is not enough for a human activity to cause some emission as a result of production or consumption. Pollution requires this emission to cause some change in well-being. Thus pollution, under an economic definition, requires both a physical impact and related economic loss.

Pollution is a by-product of economic activity. It occurs because of the nature of property rights of the environment and of goods and services. The reductions in well-being which are brought about by pollution are not part of the economic decision about whether to produce or consume a good. Pollution is therefore regarded as external to decisions made by private economic agents. But if, for example, a producer uses land to site their waste, and pays the landowner to provide that service, then the externality is being internalised. The externality (still an economic part of pollution) is only that side-element of production which harms other people. External cost is a reduction in welfare caused to one party by the production or consumption of another party which goes uncompensated.

The key to externalities is the property right. Generally speaking there are private property rights for economic goods, but less often for economic bads. I have my transistor radio, but you can help yourself to the noise. The purpose of regulation by government is to curtail the use of private property right. Because external effects are not always calculable or even known, so externalities persist and become even more ubiquitous. It is the accumulation of such externalities that make up global environmental change. The asymmetry between gain and loss allows externalities to persist. This general outcome is usually called *market failure*.

A further issue in economic definitions of pollution is that the external costs sometimes occur only above a certain threshold level of emission of the pollutant. Various examples of this include low levels of lead in the air, or 'clean beaches' which have low levels of bacteria present as a result of sewage discharge. The assimilative capacity of the environment is different for each type of pollution. There are, however, examples where the environment has zero assimilative capacity. This is notably the case for persistent, toxic and bioaccumulative substances such as organochlorines, and volatile organic compounds.

The standard theory of the economics of the environment shows that following this set of principles and definitions, there is indeed an optimal level of pollution (see for example Baumol and Oates, 1988). This level occurs when the total marginal external costs are equal to the total marginal private benefits. However, if the market for goods and services does not take the pollution into account, there will always be the incentive by private agents to produce more than the optimal level of pollution. There are various means employed by governments, acting on behalf of all those who experience the external costs, to reduce the pollution. These include regulations on emissions, regulations on technologies, taxes and charges. Often there is a mixture of all of these mechanisms (or instruments), as in the cases of carbon taxes and waste management in the UK, as shown in Box 4.1.

Box 4.1 Two examples of pollution taxes

UK tax on solid waste in landfills

The actual tax rates were announced in the 1995 UK Budget:

Normal waste	£7 per tonne
Inactive waste (non-organic)	£2 per tonne
'Optimal tax' as calculated in theory	£3–8 per tonne

How did the tax rate come about? The government set a rate which was politically feasible. Any similarity between actual and theoretical 'optimal' tax is purely coincidental.

How are the monetary costs of noise pollution estimated? Or of the loss of 'green field' sites? It is almost impossible to estimate an optimal tax rate. It is therefore easier to calculate a minimum or 'lower bound' tax rate, and raise it subsequently when key interests have got used to the principle.

Who pays the tax and who spends the revenue from pollution taxes? Households pay through indirect local taxes to local authorities, which pay the tax into environmental trusts. But the UK government traditionally argues against giving revenue to local authorities – the 'hypothecation' issue. It argues that central government prioritises tax revenue for all citizens across all possible public expenditure. This makes the landfill tax novel in the UK case. This point is also covered in Chapter 18.

Taxes on carbon in fossil fuels

Carbon dioxide is a pollutant because:

● it contributes to the greenhouse effect and hence climate change;

● it is jointly produced with other non-global pollutants in fossil fuel consumption.

Chapter 7 illustrates how complex the relationship is between carbon dioxide and other greenhouse gas emissions in the present and the impacts of climatic changes some decades in the future. The 'optimal carbon tax' would reduce the level of emissions on a global scale to the level where the marginal external costs of carbon dioxide emissions (the impacts of climate change) were equal to the marginal private benefits of using fossil fuels.

Sources: Powell and Brisson (1994); Powell and Craighill (1997); OECD (1994).

Table 4.1.1: Actual carbon taxes in Europe.

	Tax (Ecu per tonne CO_2)	Existing taxes (Ecu per tonne CO_2)	As percentage increase
Denmark	5.5–11.1	26.3	21–42
Finland	1.1	19.1	6
Italy	1.7	39.9	4
Netherlands	0.4	15.9	3
Norway			
coal	13.8	32.5	42
fuel oils	15.7		
gasoline	40.6		
natural gas	40.6		
Sweden	37.9	38.2	99

Source: OECD (1994).

The most efficient tax, given the optimal level of pollution, is that which brings about this optimal level. This is illustrated diagrammatically in Figure 4.3. The optimal tax rate is calculated as:

Tax rate = Marginal external cost at the optimal pollution level

Its derivation is explained with the aid of Figure 4.3. Without the presence of a tax any firm is concerned only with its profit level, or marginal private benefit (MPB). When this marginal private benefit is positive there is an incentive for the firm to expand production. In the absence of any external regulation it will produce its output up to the level Q_1. However, this results in a high environmental impact, or a high marginal external cost (MEC). Reducing the level of pollution will have a cost to the firm and a benefit to those currently suffering from the pollution. The optimal level of pollution is that point which minimises the sum of the cost to the firm of reducing output (abatement costs) and the damage costs of the pollution. Thus it can be shown that this solution occurs where the MPB and the MEC cross and the level of output is Q_2. In order for a tax to make the firm want to produce only Q_2 rather than Q_1, it must make the profit (MPB) of the firm zero at Q_2. In other words, it must move the MPB line downward to become the dotted line (MPB-T). The level of tax required to do this is the vertical distance between these lines, T. The impact of taxes to bring about environmental objectives is complex, since firms may move out of producing present output into something more damaging. Similarly, it is argued that reducing output also affects employment and has other knock-on effects, hence unilateral taxes involve many trade-offs. Certainly the costs of reducing the emissions of different types of pollutant vary widely across different sectors of the economy, as shown for the USA in Figure 4.4. But economists have also shown that the costs of reducing emissions are minimised through market mechanisms such as taxes precisely because of this variation.

The advantages of market based approaches to government intervention in the free market are that market-based instruments bring about an efficient resource allocation. Firms with higher costs of production will be those whose production is reduced. Economists argue that:

Figure 4.3:
Calculation of optimal pollution tax rate.

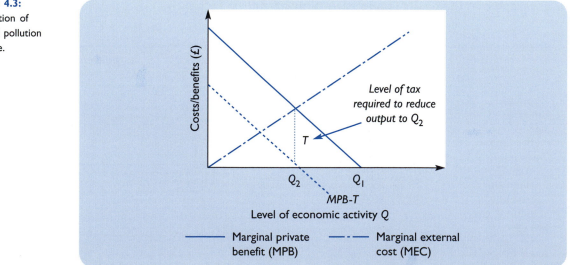

Marginal private benefit (MPB)   ―·― Marginal external cost (MEC)

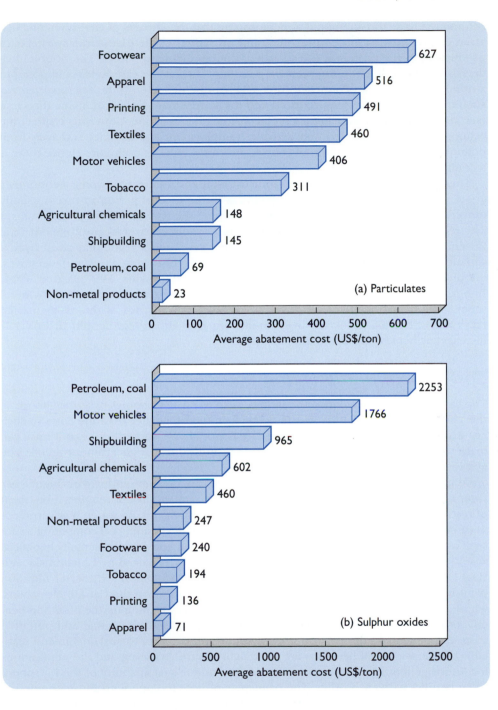

Figure 4.4:
Costs of reducing levels of (a) particulates and (b) sulphur oxides in sectors of the US economy.
Source: Hartman *et al.*, (1997).

- across-the-board regulation by licences is inefficient as regulation places the cost of emission reduction equally on all firms, even when the optimal levels may be very different; and

- institutional capture occurs in regulatory structures (see Hahn, 1989; Hanley *et al.*, 1990) this refers to the likelihood that the discharger is able to influence the regulator as to what standards to apply.

Economists have also attempted to demonstrate that the impacts of environmental taxation can be positive on employment and on reducing the distortionary effects of production in the economy, the so-called 'double dividend' effect. The double dividend arises because imposing taxes on pollution, such as the carbon taxes imposed in a number of European countries (Box 4.1), can be designed to be *revenue neutral* such that other taxes can be reduced and the government budget remains the same (see also Chapter 2). Reducing tax on income or on employment can thus redirect the economy towards more desirable activities (employment and income) and away from polluting activities. In addition, of course, such a tax reduces pollution. Hence there is a double dividend.

These arguments are, however, controversial and contested. Some economists argue that shifting towards indirect taxes (as in the case of pollution taxes) usually exacerbates pre-existing tax distortions even where the revenues from pollution taxes are 'recycled' and are hence neutral, leading to reduced overall employment (e.g. Bovenberg and de Mooij, 1994). However, other economists argue that given *typical* situations of both levels of unemployment and heavily 'distorting' taxes and incentives, the dividends from environmental taxation are real (e.g. Ekins, 1997; Pearce, 1991; Bohm, 1997; Komen and Peerlings, 1997). The arguments for the double dividend seem to be persuasive to many government finance ministries, with environmental taxes and other market-based instruments, such as those highlighted in Box 4.1, increasingly being devised and implemented.

Cost–benefit analysis and valuation

The basis of public intervention in markets, such as those examples described above, is that the benefits to society as widely defined should be greater than the costs to society. Although this type of decision making is not always formalised, economics can make explicit what decisions are being made by individuals or government agencies by employing the tools of cost–benefit analysis.

To give an example, the increase in water quality brought about with higher *per capita* incomes (Figure 4.2) does not come about automatically but requires investment of resources. Faced with the demand for clean and safe drinking water, a government must decide how much to spend on this compared with spending on competing issues such as health or education. Investment in conservation and environmentally beneficial activities can in fact save costs in the future. In a study reported by Chichilnisky and Heal (1998), New York city avoided paying more than $6 billion for a water filtration plant by investing $1 billion in restoring the soil ecosystems in the Catskill Mountains, whose watersheds supply New York with its water. Thus weighing up the costs and benefits of investment and other decisions, while keeping both the actual and potential costs and benefits to the environment into account, can lead to sustainable and equitable decisions. Not all decisions are so straightforward, however, and the implications for future generations, for present-day equity and for sustainability of the biosphere often conflict with each other. The principles of how to cope with these competing issues are examined in the section below. First there are some principles which need to be applied in carrying out such cost–benefit calculations.

An analysis of the impacts of decisions made on behalf of society requires a clear distinction and delineation of the limits of appraisal. Such appraisal often has economic dimensions. Where all factors, both economic and non-economic, are brought into an economic framework, this is known as extended cost–benefit analysis, a standard economic tool which has been applied to many public sector decisions (see Brent, 1996).

In general the steps within this analysis are:

● Step 1: definition of proposed reallocation or change in resource utilisation.

● Step 2: delineation of the relevant costs and benefits.

● Step 3: quantification of the environmental and physical impacts of the proposed or observed change.

● Step 4: valuation of the relevant effects.

● Step 5: discounting of the temporal cost and benefit flows.

● Step 6: deciding on the desirability of the proposed change and policy recommendations.

● Step 7: sensitivity analysis and scaling activities.

This can be summarised as a set of calculations which leads to step 6, the decision rule. A common form of this is to appraise desirability on the basis of a positive net present value (NPV). In other words, an action or a project is desirable from an economic perspective if the present value of the benefits outweigh the costs, or if its NPV is greater than zero:

$$NPV = \sum_t \frac{B_t - C_t}{(1+r)^t}$$

where B_t is benefit at time t, C is cost at time t, and r is the discount rate.

These steps are carried through in the example presented in Box 4.2. There the example of examining the desirability of mangrove replanting in a tropical coastal zone demonstrates that much of cost–benefit analysis of environmental decisions requires appraisal of issues which have real and direct economic impacts. It further illustrates that many activities which are in some ways environmentally beneficial, in a similar manner to the pollution tax issue discussed above, can in fact be justified on economic grounds. In addition, the benefit of the environmental functions, in this case the function that mangroves play in protecting coastal defences, can add weight to such arguments when converted into monetary terms.

Box 4.2 To plant or not to plant? Applying cost–benefit analysis

The first step in carrying out a cost–benefit analysis is to delineate the resource issue and potential environmental change. In this case, the resource efficiency of using land, labour and capital resources to rehabilitate or restore mangroves in the coastal areas of Vietnam is examined. Reversing the trend of the previous decades, restoration of mangrove areas has occurred along the coast of Vietnam where mangroves previously had been converted to agricultural or other uses, or had been degraded because of herbicides used during wartime. Mangroves are important habitats in terms of species and functional diversity and provide many local resources (see Field et al., 1998; Ewel et al., 1998 for example). Identification of the functions and services and the incorporation of these values into policy and the encouragement of appropriate property rights are therefore necessary first steps in promoting sustainable utilisation of such resources.

The second step in this appraisal is the identification of costs and benefits, in this case the environmental costs of conversion of mangroves, or their rehabilitation. When the issue to be investigated has been identified (conversion of mangroves, or rehabilitation of mangroves), the costs and benefits, which occur at different times, are assessed together.

Box 4.2 (continued)

Table 4.2.1: Illustrative table of cost–benefit calculations for mangrove rehabilitation over 20-year time horizon.

Year	Extractive benefits (000 VND per ha)	Sea dike protection (000 VND per ha)	Total costs (000 VND per ha)	Total benefits (000 VND per ha)	5% discounting factors	Discounted net benefits
0	0	205	521	205	1.000	–315 000
1	633	209	83	842	0.952	723 000
2	633	213	83	845	0.907	692 000
3	633	217	83	849	0.864	662 000
4	633	220	83	853	0.823	634 000
5	639	224	83	862	0.784	611 000
6–20	22 150	3 536	3 723	25 686	0.746–0.377	10 469 000
Total NPV =						13 476 000

Notes: Extractive benefits include timber, honey from beekeeping and fish. They do not include valuation of biodiversity or of the links to offshore fisheries (see Ruitenbeek, 1994; Barbier and Strand, 1998).
Costs include maintenance and initial planting costs.
VND 11 000 = US$1 approximately.

Source: based on Tri *et al.*, (1998)

The data on costs and benefits come from three districts in northern Vietnam where yields of timber, fish and honey extracted from such sites over time and the costs of planting have been estimated. The timing and scale of these costs and benefits are shown in Table 4.2.1, where the next stages of calculating the discounted net present value are presented (steps 3 and 4), following the mathematical notation in the main text.

The major indirect benefit, and the principal reason for rejuvenating the stands, is in the role of stands in protecting the extensive sea-dike systems present along much of the low–lying deltaic coast of northern Vietnam. This indirect benefit is estimated through a model where the major parameters determining the value of the protection are the width and age of the stand, and the local hydrological features (see Tri *et al.*, 1998), giving the benefits in Table 4.2.1. This table demonstrates how the cost–benefit calculations are made. The discount factors for a 5% discount rate, calculated as $1/(1+r)^t$, are shown: the present value at each year represents the net benefits (total benefits – total costs × discount factor). Net present value represents the comparable value at the present time and is the sum of these discounted net benefits.

Is mangrove planting desirable? This is step 6 of the cost–benefit analysis and is taken by reference to net present value or the benefit to cost ratio. A benefit to cost ratio is the ratio of the present value of benefits to the present value of costs, and is an alternative and equivalent indicator to NPV. A project is desirable if the B–C ratio is greater than 1. In this case the benefits outweigh costs in a ratio of 4:1 to 5:1, which means mangrove rehabilitation can be justified on economic grounds.

The final step (7) is sensitivity analysis and scaling, where appropriate. This is undertaken in this example by examining sensitivity of the results to discount rates and to the inclusion or exclusion of the value of particular functions. Figure 4.2.1 illustrates that the direct benefits from mangrove rehabilitation are more significant in economic terms than the indirect benefits associated with sea-dike protection. It is clear from Figure 4.2.1 that, even if no indirect benefits were available, the direct benefits from mangrove rehabilitation justify this activity as economically desirable. This is shown by the positive net present values at all discount rates considered, thus

Box 4.2 (continued)

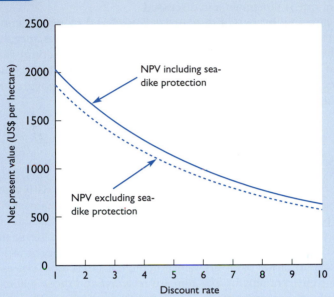

Figure 4.2.1

also demonstrating that the economic desirability of mangrove planting in these circumstances is also sensitive to the discount rate chosen.

The case study therefore illustrates each of the steps in this appraisal of costs and benefits. In the mangrove case presented here, investment in mangrove planting in Vietnam's coastal zone has various impacts on coastal residents, with the benefits outweighing the costs. The estimation of these requires both socio-economic surveys and modelling of the sea-dike maintenance function. It would appear that planting is desirable and this result is robust to various scenarios of discount rates and changes in the benefits stream.

A critical issue in all such cost–benefit analysis is that highlighted in the mangrove example, of how to treat uncertain costs and benefits over time. If costs and benefits which occur in the future are not counted as highly as those closer to the present, then this has been argued to bias such analysis towards actions with environmental impacts. Typically, many environmental problems caused by economic activity, such as the pollution issues highlighted in Figure 4.2, become apparent only at some time well into the future. For example the impacts of nuclear power are the costs of decommissioning or of storage of nuclear waste, and of health effects linked to certain processes which may accrue only over period of a century or more. Similarly, using up exhaustible natural resources such as oil may have environmental consequences in terms of the scarcity of the resources, which may be apparent only when this scarcity actually occurs in the future.

However, discounting has a sound intuitive basis. Discounting future costs and benefits to the present day is carried out to reflect observed economic behaviour. The future is discounted in capital markets and in investment and consumption decisions more generally. Discounting also reflects the necessity to appraise decisions on behalf of society. Society may wish to give greater weight to the future, particularly where the environment is concerned, and hence would adopt a low positive discount rate. A simple discussion on the rationale of discounting and its application is given in Hanley and Spash (1993) and Pearce (1993) (see also Price, 1993; Markandya and Pearce, 1991). In the example in Box 4.2 the sensitivity of the results to changes in the discount rate is illustrated and discussed.

Choosing the rate of discount is considered by many economists to be somewhat arbitrary and dependent on whether the project to be appraised is being undertaken in the public or private domain. A range of real discount rates from 1 to 20% have been used in many circumstances, but rates at the lower end of this range tend to reflect the time preferences implicitly applied by governments in investments on behalf of society (see Markandya and Pearce, 1991).

Monetary valuation methods and techniques

Alternative and appropriate methods

The major constraint on applying cost–benefit analysis is that not all environmental changes are amenable to economic valuation. Clearly many environmental changes have an economic dimension, such as productivity changes, health effects, amenity gains and losses, and the existence value of conservation or loss. Diverse methods have been developed by environmental economists to capture the economic aspect of these changes. It is possible to divide monetary valuation methods into those where preferences of individuals for environmental protection are *revealed* and those where preferences are elicited directly (the *stated preference* techniques) as shown in Figure 4.5. Revealed preference techniques are most applicable when values for use of the environment are important, where stated preferences allow probing of non-use elements of economic value.

In addition to these techniques shown in Figure 4.5, the economic cost of environmental harm is often characterised by the cost of replacing the asset. So, for example, the replacement cost of the loss of soil due to erosion from agricultural land may involve the market cost of fertilisers to replace the soil fertility so that productivity is maintained. But this replacement cost estimate does not reflect the demand for soil fertility by the farmer. The cost to the farmer is best represented as the actual value of the loss of production of agricultural outputs from the less fertile land, rather than the hypothetical cost of replacement of soil. The following sections concentrate on meth-

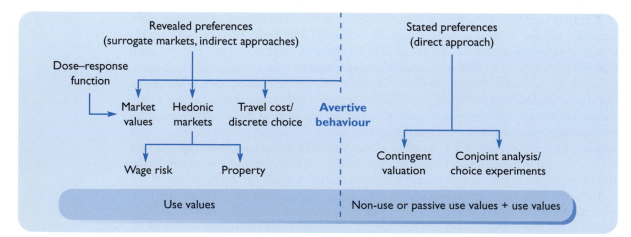

Figure 4.5: Revealed and stated preferences for environmental protection.

Note: Each of these research techniques has limitations and advantages. The vast numbers of studies allows comparison of the determinants of values, through meta-analysis (e.g. Smith and Huang, 1995; Smith and Kaoru, 1990; Brouwer *et al.*, 1997).

ods of revealed and stated preference approaches to valuation, namely the travel cost method, the hedonic pricing method and the contingent valuation method. Further information can be found in Bateman (1993), Bateman and Turner (1993), Turner *et al*. (1994) and other texts.

Travel cost method

The travel cost method evaluates the recreational use value of resources, hence measures one aspect of use values which tend to be non-consumptive. In other words, recreational use of beaches, forests or other areas tends not to directly extract or use the resources on which they depend. The travel cost method is a survey technique, whereby visitors to a site are asked a series of questions to ascertain their place of residence, necessary socio-economic information, frequency of visits to the particular and other similar substitute sites, means of travel, and cost information about the trip. From these data, visit costs can be calculated and related to visit frequency so that a demand function can then be used to estimate the recreation value of the whole site.

The method was developed in the 1960s in the USA for estimating the value of outdoor recreation, particularly as information for management of national parks and other assets. The method is somewhat restricted in the range of impacts and changes in which it can provide economic values, but it is of use in the estimation of value in coastal environments, where recreational use of beaches and other resources represents a significant demand. Examples of travel cost estimates include those for beach resorts, where the quality of the beach affects demand, hence the environmental quality has a marginal value (Bell and Leeworthy, 1990). Various estimates of recreational value of forests and other resources exist, including those by Tobias and Mendelsohn (1991), Maille and Mendelsohn (1993) and Mercer *et al*. (1995). These latter studies investigate the recreational value of forest resources in the tropics.

The value for a specific recreation site is estimated under the travel cost method by relating demand for that site (measured as site visits) to its price (measured as the costs of a visit). A simple travel cost method model can be defined by a trip-generation function, where visits to a recreational site are a function of costs and of a set of other socio-economic variables. This method leads to an estimate of consumer surplus as that element of the value of the recreational experience which is not directly paid for through markets. There are various caveats to this value being used directly for any resource. These caveats include whether there are substitute sites, whether the visitors are valuing particular attributes of the site, and whether the visitation rate and distance can be taken as an indication of recreational value (see Bateman, 1993). Given these caveats, the method is useful for determining recreational value of coastal resources. It is, however, ultimately limited in the scope of environmental impacts which can be addressed.

Hedonic pricing method

The hedonic pricing method relies upon the assumption that the local environmental quality will determine the price of capital assets such as property, and that differences in these prices can be taken as an indicator of marginal value of environmental change. The environmental factors, however, are only a subset of property price determinants, which for residential houses may include, amongst other factors, the number of rooms and accessibility to shops and workplaces. The general specification of a hedonic price model is therefore:

Property price = f (size and characteristics of property, access, environment)

The equation states that house price is a function of the property characteristics such as number of rooms, distance to local facilities in the case of a residential property, and some measure of local environmental quality. Thus, given the availability of data on both property prices and the other non-environmental characteristics of the property, the influence of environmental quality on the price of the property can be determined. The presence of woodland, for example, often enhances the value of property – hence woodland has this value which can be observed through the proxy of residential house prices. Garrod and Willis (1992) show that the impact of changes in forest cover towards broadleaved forest increases the price of a house by £43 per unit change, while increasing the proportion of coniferous forest cover decreases adjacent house prices by £141 per unit. This is based on a survey of over 1000 house sale transactions across the UK in the late 1980s. Similarly, location of some polluting source or of local nuisance hazardous waste sites will lower property prices. Ketkar (1992) demonstrates that in New Jersey the cleaning up of any of that state's hazardous waste sites will increase the median house price for adjacent areas by $1300 to $2000 based on prices in the early 1980s.

This method, as with the travel cost method, is limited in its applicability to valuing the impacts of environmental change in coastal zones. The method is data-dependent and can reasonably be applied only where the environmental asset under consideration is well understood within the purchasing decisions of house or property owners. However, it can give some estimates of the availability of recreational assets as well as the impact of risk of inundation or flooding or other environmental risks.

Contingent valuation method

The contingent valuation method (CVM) is a method for placing monetary values upon assets and impacts which do not have market prices. It achieves this by constructing a hypothetical market and asking individuals, for example, what they are willing to pay (WTP) towards preservation of a particular environmental good. The contingent valuation method therefore relies upon individuals' expressed preferences, rather than simply observing their preferences for environmental quality by how they actually behave in property or other markets.

The advantages include the fact that the method allows the researcher to ask questions about and estimate both use and non-use values. The disadvantages of the method are that respondents may not believe in the credibility of the hypothetical markets, and that without an actual marketplace, stated WTP may not equate to what would actually be paid. The flexibility of these techniques in valuing many aspects of environmental quality has led to a voluminous literature on this subject (e.g. reviewed in Mitchell and Carson, 1989; Cropper and Oates, 1992), as well as voluminous critiques of the method and economic valuation more generally (e.g. Sagoff, 1998). Part of the controversy stems from the influence that this technique now has in determining liability for damage assessment in the US legal system, to the extent that the estimation of damages from the oil spill of the *Exxon Valdez* in Prince William Sound, Alaska, in the early 1990s was partially determined by using a CVM survey.

Applying the contingent valuation method is a resource-intensive research technique, as extensive testing and focusing of the question in hand is required, and large representative samples of users and non-users of the particular environmental resource are usually required. The survey techniques are outlined in Mitchell and Carson (1989) and a major set of case studies and guidance on theoretical developments in Garrod and Willis (1999) and Bateman and Willis (1999). Most effort in such studies is required to determine what the hypothetical market actually is: if respondents are being asked whether they are willing to pay to preserve the environment, are they consistent in their understanding of the asset being valued? This can be overcome by questioning on specific aspects of the asset. For example, a survey in Tobago to elicit willingness to pay to preserve a marine park found that respondents to the survey were actually responding in terms of specific characteristics of the coral reefs that they were enjoying, namely the fish diversity and the coral diversity and that they were willing to pay more for non-congested use of the reefs.

Contingent valuation surveys result in estimates of mean willingness to pay for the defined environmental assets, which can be converted to aggregate willingness to pay through multiplying by the relevant population. As the name of the technique implies, these values are contingent on the information provided, the type of question asked and the so-called payment vehicle. In other words, respondents can be provided with little or much information about what they are valuing, and this affects their responses. Respondents can be asked either 'how much are you willing to pay?' (the so-called open-ended question) or 'are you willing to pay £10?' (a dichotomous choice), resulting in different estimates of value. In terms of the payment vehicle, respondents are placed in the hypothetical situation that they are paying into a trust fund for protection of the relevant environmental asset; or that they pay an entrance fee to enjoy the resource; or that they would pay through extra dedicated taxation. Again research has shown that responses to different hypothetical payment devices, as well as the other information and formats, determine the estimates of willingness to pay.

The results from a contingent valuation study presented in Table 4.1 illustrate these changes in mean willingness to pay from different so-called elicitation methods. In the study valuing landscape preservation in the Norfolk and Suffolk Broads in eastern England (Bateman *et al.*, 1995), there were three methods of eliciting WTP responses:

● open-ended: 'what are you willing to pay?'

● dichotomous choice: 'would you pay £X?' (where X is different for different sub-samples)

● iterative bidding: asking a series of yes or no questions to arrive at a willingness to pay.

The authors of this study deduce from the results in Table 4.1 that the dichotomous choice question leads to upward anchoring. In other words, the first amount asked

Table 4.1: Estimates of willingness to pay for recreation and amenity for Norfolk and Suffolk Broads, UK.

	Sample size	Mean WTP[1]	Median WTP	Std devn	S.E. mean	Min. bid (£)	Max. bid (£)
Open-ended WTP study	846	67.19	30.0	113.58	3.91	0.0	1250.0
Iterative bidding WTP study	2051	74.91	25.0	130.1	2.87	0.0	2500.0
Dichotomous choice WTP study	2070	140	139	n/a	n/a	n/a	n/a

Note:
1 Includes as zeros those who refused to pay anything at all.

Source: adapted from Bateman *et al.* (1995).

would be responded to positively by respondents even if their actual willingness to pay was somewhat lower. Open-ended elicitation approaches, however, lead to the phenomenon of free riding in responses. If a respondent is intensely and passionately interested in preservation of the Norfolk and Suffolk Broads, they are able with this question to overstate their hypothetical willingness to pay to ensure that the mean is raised. Thus different elicitation methods lead to different responses.

All of these effects in contingent valuation are important to consider both in carrying out such a valuation exercise and in interpreting studies. The limitations of the contingent valuation technique remain, however, so the validity and robustness of the estimates are dependent on the acceptance of a hypothetical market by respondents. Critics of the technique argue that decision makers are led to create the markets which have been suggested as hypothetical in such studies, and that decisions on environmental preservation are separated in many respondents' perceptions from market transactions (see, for example, Sagoff, 1998; Burgess *et al.*, 1998). Furthermore, it is argued that some respondents may feel powerless to 'get out' of the hypothetical survey situation, even though they have no faith in their own estimates (see Burgess *et al.*, 1998).

The arguments concerning economic valuation are often put in terms of whether some estimate, how ever flawed, of monetary values of the environment is better than an absence of estimates. Proponents of valuation highlight the vital role of economics in *demonstrating* the value of ecosystem services and functions. This can however lead to real environmental problems being assessed with a partial view of economic dimensions of value, which are but one element: 'because ecosystem services are not fully captured in commercial markets or adequately quantified in terms comparable with economic services and manufactured capital, they are often given too little weight in policy decisions' (Costanza *et al.*, 1997, 253).

Contingent valuation techniques stir up much passion between economists and non-economists. Many of the ethical and interdisciplinary perspectives introduced in Chapter 1 are of relevance here. There is, as yet, no resolution to this matter. Most commentators look to a mix of political, ethical and economic techniques, but there is no methodology yet devised for combining all these approaches.

Sustainable development: meaning and criteria

It is now commonly asserted that sustainable development is development which does not jeopardise future well-being through reduction of the capacity of the environment to meet the legitimate needs of all future generations (following the World Commission on Environment and Development, 1987). The central aspects of such definitions appear to be their focus on the environmental basis for human activity and the time dimensions of development and well-being. The Brundtland definition emphasises in particular the long-term aspects of sustainable development, and equity for future resource users. An ecological economics approach to sustainability which incorporates sensitivity to ecological processes requires that resources be allocated in such a way that they do not threaten the stability either of the system as a whole or of key components of the system. To quote from a well-known definition from this sub-discipline of economics:

an ecological economics of sustainability implies an approach that privileges the requirements of the system above those of the individual. Since the valuation of resources deriving from ecologically unsustainable preferences is itself unsustain-

able, there is no advantage in giving special weight and special privilege to such valuations. What is important in the approach is the ability of the system to retain the resilience to cope with random shocks, and this is not served by operating as if the present structure of private preferences is the sole criterion against which to judge system performance. (Common and Perrings, 1992)

This approach incorporates the ecological notions of stability and resilience, which require diversity and other ecological criteria. These matters are discussed in Box 4.3 and Chapters 5 and 6. So, for the economist, sustainable development requires consideration of :

- efficiency in resource allocation;
- equity in the present and future;
- resilience of ecological and social systems.

This conceptual framework for sustainable development, as set out in Table 4.2, is one of many ways by which these necessary criteria can be developed. Efficiency in resource use simply means maximising the economic welfare or well-being derived from the use of finite resources. The means by which resource allocation can be judged to be sustainable in welfare economic terms is through Pareto improvements in aggregate well-being over time. The notion of Pareto improvements means ensuring allocation so no-one is demonstrably and avoidably worse off, following compensation where necessary. An economy should seek to maximise the economic well-being derived from resource use, taking into account the temporal dimensions of such use, as outlined in the section on cost–benefit analysis above.

There are two necessary aspects to equity, namely a temporal dimension based on the necessity to incorporate resource users in the future for whom unsustainable resource use in the present would foreclose options, and present-day equity. Each of these aspects can again be incorporated into resource use criteria but requires explicit acknowledgement of the underlying ethical position or definition of fairness.

Box 4.3 Why is ecological resilience important for human welfare?

Ecological resilience is defined and elaborated in Chapter 5. But it is argued by ecological economists that it is an important component of sustainable development, since all human welfare is ultimately dependent on the biosphere, and the resilience of ecosystems is an important characteristic in this sustainability equation. These arguments have been summarised by Kenneth Arrow and colleagues (1995), who argue that ecological resilience is important for human welfare for three related reasons:

- Discontinuous change is often observed in eco-system functions, including human life support functions, and these changes are associated with loss of productivity.

- loss of resilience may lead to irreversible change (e.g. individual species loss) and hence loss of options for future use.

- decreased resilience and change to unfamiliar states increases the chance of uncertain negative welfare impacts of environmental change.

The relationship between ecological resilience and sustainable development is discussed in the text. Is the concept of social resilience related? This is discussed in Chapter 6.

Table 4.2: Economic perspectives on modelling sustainable development.

Criteria for sustainable development	Economic interpretation	Measured through
Efficiency	Present economic welfare maintained and maximised	max [benefits (B) – costs (C)]
Equity	Welfare non-declining over time (intergenerational equity) Extremes in equity rules: (i) Resource allocation increasing total welfare (Pareto) (ii) Resource allocation benefiting poorest (Rawlsian)	max $[a_i . (B_i – C_i)]$ where a_i = distributive weight for stakeholder group
Ecosystem resilience	Maximise ability to withstand shocks and uncertain impacts of change	*Ecological resilience*: proxies of diversity and functional integrity

Source: Adger (1997).

From Chapter 2 we see that in practice, it is not possible to operate a Pareto-style optimal allocation of environmental goods. Invariably and increasingly, the poor and vulnerable lose out. Indeed, it seems that the weakest are often more and more adversely affected by losses of environmental well-being. The modern economic growth pattern is simply incapable of bringing equity into practice.

The definition of equity, or fairness, is always relative. In effect, sustainable development can be judged against any equity rule: making explicit such rules allows comparison of the distribution of resource use against implicit rules taken by government policies or social responses to existing situations. In Table 4.2, two examples of equity rules are presented.

One common assumption is that any change which increases the well-being of the total population is equitable. This implies that who receives the benefits is unimportant and resource allocation simply needs to be Pareto-improving. In other words, it does not matter who within that population loses or gains as long as the net result is positive. An alternative fairness rule is that any change in resource allocation, to be sustainable, should benefit only the least well-off individuals within a population. This principle of maxi-min suggests that welfare improvement is obtained if and only if the welfare of the poorest individuals improves (Rawls, 1971).

Resource allocations can be judged against such rules, and if they do not meet the criterion then such allocations are not sustainable. In reality there is a continuum of equity rules between Pareto and Rawlsian equity rules. Resource allocation decisions by individuals and by collective action have implicit positions on equity almost always lying somewhere on the spectrum between these two extremes. As most definitions of sustainable development urge consideration of equity, some decision rule on equity is required: the default position is the acceptance of the implied equity in the *status quo* position in whatever is the relevant population. This may not be 'fair' but it is the implicit judgement if equity is not explicitly considered.

Equity therefore depends on the incidence of costs and benefits to stakeholder groups. One way in which this is incorporated is to weight the impacts of changes on different stakeholder groups either through adjusted cost–benefit procedures or through non-economic, poverty alleviation criteria (e.g. Brent, 1996). The intuition here is that costs and benefits to lower income groups should be weighted more heavily than those to higher income groups. The efficiency criterion is therefore

$$\max \left[a_i \cdot (B_i - C_i) \right]$$

where a_i = distributive weights for each income group (i). Distributive weights can be estimated as the proportional difference in income in the ith group to the mean income:

$$a_i = \left(\frac{\overline{Y}}{Y_i} \right)^b$$

where $\overline{Y}$ = mean income of the total population, and Y_i = income of ith income group. Thus resource and environmental costs to groups with income lower than the mean are weighted higher than costs to groups with income greater than the mean. b is a power factor representing implicit inequality aversion, which can be represented in a number of ways. The more weight is given to equity, the greater the aversion to inequality and the higher the power factor.

There are various observable measures which act as proxies for the weighting factor (b). First, it could be related to the observation that at higher levels of income: further income brings decreasing marginal well-being or utility. If the Rawls maxi-min rule were adopted then this would approve a development decision only where the benefits accrued to the worse-off individual and everyone else was a loser (see Brent, 1996).

Ecological economics: the value of ecosystem services

For sustainability, the efficiency requirement is to maximise the benefits of development over the costs of development, where these incorporate all benefits and costs, including marketed economic activity as well as non-marketed activity. As discussed above, equity rules can be operationalised for economic values. In essence this means determining who are the gainers and who are the losers and which stakeholder groups are affected by environmental changes. The benefits and costs associated with the various ecosystem functions and services can be weighted to reflect underlying aversion to inequality, with the weights used depending on the equity rule adopted. The issue of social resilience at the community level can also be investigated. The issue here is to determine whether institutional changes, property rights and access to resources, and demographic changes make a social system more or less sustainable.

In estimating the benefits of ecosystem services, Costanza et al. (1997) demonstrate that the efficiency of resource use requires consideration of the ecosystem functions of ecological resources. The results of their analysis suggest that the annual value of services from the world's ecosystems is nearly double the world's marketed GNP. The current economic value of the non-marketed services and functions of ecosystems is $33 trillion per year (global GNP = $18 trillion per year). The calculation is undertaken by a two-stage process:

- the study derives the value per hectare of the world's ecosystems from published studies;
- the values are scaled up by multiplying by the area of the ecosystems to arrive at a global figure for their functional values.

The results for selected ecosystems show that the non-marketed services of coastal and oceanic ecosystems have higher values than those of many terrestrial ecosystems, by many orders of magnitude. It should be noted, as explicitly stated in Costanza *et al.* (1997), that these are non-marketed values. Cropland, which features as contributing only $92 per ha of ecosystem services and functions, provides the human world, along with rangeland and aquatic systems, with all of its food. But because this activity is to a large extent traded through markets, it is not included in ecosystem functions. So clearly the life support functions (to humans) of these and other ecosystems are captured only in an ephemeral fashion.

To examine how these estimates of value are derived, the example of two coastal ecosystems is presented in Figure 4.6, that of mangrove and coral reef ecosystems. In Figure 4.6, it is demonstrated that the value of these ecosystems is in the order of $6075 per ha for coral reefs and $9990 per ha for mangroves, based on the published estimates in Costanza *et al.* (1997). In both cases, the major ecosystem functions which are valued are coastal protection, nutrient cycling, food production and recreation. For coral reefs almost half their value is derived from recreational use, based on the mean of estimates of a range of economic indicators, from direct revenues spent by tourists in visiting coral reefs in Florida to consumer surplus estimates of recreational use in Australia, to expenditures in the Caribbean.

For mangroves, the coastal protection function value is based on the substitution cost for other protection; nutrient cycling is based on replacement cost for waste treatment plant; food production is based on market price from fishing studies; and the recreational value of mangroves is based on estimates from Trinidad and Tobago and Puerto Rico.

The derivation of particular function value estimates in the Costanza *et al.* (1997) study can be criticised. However, the authors of this study demonstrate that a range of

Figure 4.6: Non-marketed values of coral reefs and mangroves.

Source: following Costanza et al. (1997)

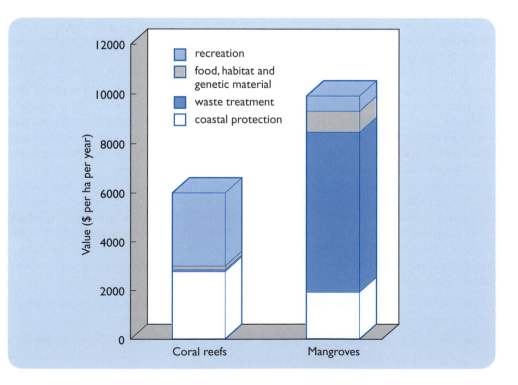

constraints exist on determining relative economic values, and particularly on scaling these values in a manner which makes them meaningful for appraising the sustainability of resource use.

As discussed above, economic values as presented in many studies are criticised because they capture only part of ecosystem value, as they relate to human well-being. Hence they ignore all ethical reasons for conserving the world's ecosystems. From an economics perspective there are a number of criticisms of the approach to deriving values, as alluded to in the discussion of mangroves above. The values are partial in the Costanza *et al.* (1997) study in that they ignore many other aspects of economic well-being from the ecosystems chosen, even within the economic approach, and are based on data which is often not comparable across studies of the same function, let alone across the different functions.

In scaling up such values there appear to be two major constraints. The first is that an area basis of the world's ecosystems may not be appropriate: ecosystems overlap and quality of ecosystem is ignored. But the most fundamental criticism of the approach from an economic perspective is that such scaling up ignores relative scarcity. If a single environmental function of an ecosystem, converted into its impact on human welfare, is independent of location, such as with the carbon retention function, then scaling up economic values can be partially valid.

Conclusions

This chapter provides a brief overview of some highly topical and controversial debates in the environmental field from the perspective of economics. Economic techniques can be used to examine the motivations and rationale behind many aspects of society's interaction with its environment, on both an individual and a collective basis. Economics can also define those aspects of sustainability which are economic in dimension and which impinge on equity and resilience within the environment on which sustainability ultimately depends. Thus economics can be analytical and descriptive of the society–environment interactions observed in the world around us. But since the environment is systematically overlooked in many economic decisions, there is a clear case for intervention in the workings of society through public policy. This is another vigorous area of research for economists – the incorporation of the environment in decision making requires comprehensive understanding of environmental change, including its uncertainties and dynamics, and the impact of this on human well-being. The scope and limitations of economics in this area are ultimately determined by the political and social context within which it operates.

References

Adger, W. N. (1997) *Sustainability and social resilience in coastal resource use*. GEC Working Paper 97–23, CSERGE, University of East Anglia, Norwich and University College London.

Arrow, K., Bolin, B., Costanza, R., Dasgupta, P., Folke, C., Holling, C. S., Jansson, B. O., Levin, S., Mäler, K. G., Perrings, C. and Pimentel, D. (1995) Economic growth, carrying capacity and the environment. *Science*, **268**, 520–521.

Barbier, E. B. (ed.) (1997) Environmental Kuznets curve: special issue. *Environment and Development Economics*, **2**(3).

Barbier, E. B. and Strand, I. (1998) Valuing mangrove fishery linkages: a case study of Campeche, Mexico. *Environmental and Resource Economics*, **12**, 151–166.

Bateman, I. (1993) Valuation of the environment, methods and techniques: revealed preference methods. In R. K. Turner (ed.), *Sustainable Environmental Economics and Management: Principles and Practice*. Belhaven, London, 192–265.

Bateman, I. J. and Turner, R. K. (1993) Valuation of the environment, methods and techniques: the contingent valuation method. In R. K. Turner (ed.), *Sustainable Environmental Economics and Management: Principles and Practice*. Belhaven, London, 120–191.

Bateman, I. J. and Willis, K. G. (eds) (1999) *Valuing Environmental Preferences: Theory and Practice of the Contingent Valuation Method in the US, EU, and Developing Countries*. Oxford University Press, Oxford.

Bateman, I. J., Langford, I. H., Turner, R. K., Willis, K. G. and Garrod, G. D. (1995) Elicitation and truncation effects in contingent valuation studies. *Ecological Economics*, **12**, 161–179.

Baumol, W. J. and Oates, W. E. (1988) *The Theory of Environmental Policy*, 2nd edn. Cambridge University Press, Cambridge.

Bell, F. and Leeworthy, V. (1990) Recreational demand by tourists for saltwater beach days. *Journal of Environmental Economics and Management*, **18**, 189–205.

Bohm, P. (1997) Environmental taxation and the double dividend: fact or fallacy? In T. O'Riordan (ed.), *Ecotaxation*. Earthscan, London, 106–124.

Bovenburg, A. L. and de Mooij, R. A. (1994) Environmental levies and distortionary taxation. *American Economic Review*, **84**, 1085–1089.

Brent, R. J. (1996) *Applied Cost Benefit Analysis*. Edward Elgar, Cheltenham.

Brouwer, R., Langford, I., Bateman, I., Crowards, T. and Turner, R. K. (1997) *A meta–analysis of wetland contingent valuation studies*. GEC Working Paper 97–20, GSERGE, University of East Anglia, Norwich and University College London.

Burgess, J., Clark, J., and Harrison, C. M. (1998) Respondents' evaluations of a contingent valuation survey: a case study based on an economic valuation of the wildlife enhancement scheme, Pevensey levels in East Sussex. *Area*, **30**, 19–27.

Chichilnisky, G. and Heal, G. (1998) Economic returns from the biosphere. *Nature*, **391**, 629–630.

Common, M. (1995) *Sustainability and Policy: Limits to Economics*. Cambridge University Press, Cambridge.

Common, M. S. and Perrings, C. (1992) Towards an ecological economics of sustainability. *Ecological Economics*, **6**, 7–34.

Costanza, R., d'Arge, R., de Groot, R., Farber, S., Grasso, M., Hannon, B., Limburg, K., Naeem, S., O'Neill, R. V., Paruelo, J., Raskin, R. G., Sutton, P. and van den Belt, M. (1997) The value of the world's ecosystem services and natural capital. *Nature*, **387**, 253–260.

Cropper, M. L. and Oates, W. E. (1992) Environmental economics: a survey. *Journal of Economic Literature*, **30**, 675–740.

Daly, H. E. (1992) *Steady State Economics*, 2nd edn. Earthscan, London.

Ekins, P. (1997) On the dividends from environmental taxation. In T. O'Riordan (ed.), *Ecotaxation*. Earthscan, London, 125–162.

Ewel, K. C., Twilley, R. R. and Ong, J. E. (1998) Different kinds of mangrove forests provide different goods and services. *Global Ecology and Biogeography Letters*, **7**, 83–94.

Farrell, A. and Hart, M. (1998) What does sustainability really mean? The search for useful indicators. *Environment*, **40**(9), 4–9, 26–31.

Field, C. B., Osborn, J. G., Hoffman, L. L., Polsenbourg, J. F., Ackerly, D. D., Berry, J. A., Bjorkman, O., Held, A., Matson, P. A. and Mooney, H. A. (1998) Mangrove biodiversity and ecosystem function. *Global Ecology and Biogeography Letters*, **7**, 3–14.

Freeman, A. M. (1979) Hedonic prices, property values and measuring environmental benefits: a survey of the issues. *Scandinavian Journal of Economics*, **81**, 154–173.

Garrod, G. and Willis, K. G. (1992) The environmental economic impact of woodland: a two stage hedonic price model of the amenity value of forestry in Britain. *Applied Economics*, **24**, 715–728.

Garrod, G. and Willis, K. G. (1999) *Economic Valuation of the Environment: Methods and Case Studies*. Edward Elgar, Cheltenham.

Hahn, R. W. (1989) Economic prescriptions for environmental problems: how the patient followed the doctor's orders. *Journal of Economic Perspectives*, **3**(2), 95–114.

Hanley, N. and Spash, C. L. (1993) *Cost Benefit Analysis and the Environment*. Edward Elgar, Cheltenham.

Hanley, N., Hallett, S. and Moffat, I. (1990) Why is more notice not taken of economists' prescriptions for the control of pollution? *Environment and Planning A*, **22**, 1421–1439.

Hartman, R. S., Wheeler, D. and Singh, M. (1997) The cost of air pollution abatement. *Applied Economics*, **29**, 759–774.

Ketkar, K. (1992) Hazardous waste sites and property values in the state of New Jersey. *Applied Economics*, **24**, 647–659.

Komen, M. H. C. and Peerlings, J. H. M. (1997) The effects of the Dutch 1996 energy tax on agriculture. In W. N. Adger, D. Pettenella and M. C. Whitby (eds), *Climate Change Mitigation and European Land Use Policies*. CAB International, Wallingford, 171–185.

Kuznets, S. (1955) Economic growth and income inequality. *American Economic Review*, **45**, 1–28.

Maille, P. and Mendelsohn, R. (1993) Valuing ecotourism in Madagascar. *Journal of Environmental Management*, **38**, 213–218.

Markandya, A. and Pearce, D. W. (1991) Development, the environment and the social rate of discount. *World Bank Research Observer*, **6**, 137–152.

Mercer, E., Kramer, R. and Sharma, N. (1995) Impacts on tourism. In R. Kramer, N. Sharma and M. Munasinghe (eds), *Valuing tropical forests: methodology and case study of Madagascar*. World Bank Environment Paper No. 13, World Bank, Washington DC.

Mitchell, R. C. and Carson, R. (1989) *Using Surveys to Value Public Goods: the Contingent Valuation Method*. Resources for the Future, Washington DC.

OECD (1994) *Managing the Environment: the Role of Economic Instruments*. OECD, Paris.

O'Riordan, and Voisey, H. (1998) The political economy of the sustainability transition. In T. O'Riordan, and H. Voisey (eds), *The Transition to Sustainability: the Politics of Agenda 21 in Europe*. Earthscan, London, 3–30.

Pearce, D.W. (1991) The role of carbon taxes in adjusting to global warming. *Economic Journal*, **101**, 935–948.

Pearce, D. W. (1993) *Economic Values and the Natural World*. Earthscan, London.

Powell, J. C. and Brisson, I. (1994) *The assessment of social costs and benefits of waste disposal*. WM Paper 94–06, CSERGE, University of East Anglia, Norwich and University College London.

Powell, J. C. and Craighill, A. (1997) The landfill tax. In T. O'Riordan. (ed.), *Ecotaxation*. Earthscan, London, 304–320.

Prescott–Allen, R. (1999) *The Wellbeing of Nations: Reviewing the Wellbeing and Sustainability of 180 Nations*. Earthscan, London.

Price, C. (1993) *Time, Discounting and Value*. Blackwell, Oxford.

Rawls, J. (1971) *A Theory of Justice*. Harvard University Press, Cambridge, MA.

Rothman, D. S. and de Bruyn, S. (eds) (1998) The Environmental Kuznets curve: special issue. *Ecological Economics*, **25**(2).

Ruitenbeek, H. J. (1994) Modelling economy–ecology linkages in mangroves: economic evidence for promoting conservation in Bintuni Bay, Indonesia, *Ecological Economics*, **10**, 233–247.

Sagoff, M. (1998) Aggregation and deliberation in valuing environment public goods: a look beyond contingent pricing. *Ecological Economics*, **24**, 213–230.

Shafik, N. (1994) Economic development and environmental quality: an econometric analysis. *Oxford Economic Papers*, **46**, 757–773.

Smith, V. K. and Huang, J. C. (1995) Can markets value air quality? A meta-analysis of hedonic property value models. *Journal of Political Economy*, **103**, 209–227.

Smith V. K. and Kaoru, Y. (1990) Signals or noise? Explaining the variation in recreation benefit estimates. *American Journal of Agricultural Economics,* **72**, 419–437.

Tobias, D. and Mendelsohn, R. (1991) Valuing ecotourism in a tropical rainforest reserve. *Ambio*, **20**, 91–93.

Tri, N. H., Adger, W. N. and Kelly, P. M. (1998) Natural resource management in mitigating climate impacts: mangrove restoration in Vietnam. *Global Environmental Change*, **8**, 49–61.

Turner, R. K., Pearce, D. W. and Bateman, I. J. (1994) *Environmental Economics: an Elementary Introduction*. Harvester Wheatsheaf, Hemel Hempstead.

United Nations Development Programme (1998) *Human Development Report 1998*. Oxford University Press, Oxford.

World Bank (1992) *World Development Report 1992: Development and the Environment*. Oxford University Press, New York.

World Commission on Environment and Development (1987) *Our Common Future*. Oxford University Press, Oxford.

Chapter Five

Biodiversity and ethics

Paul Dolman

Topics covered

- What is biodiversity?
- Human attrition of biodiversity
- Predictions of a global extinction spasm
- Why conserve biodiversity? Biodiversity ethics
- Approaches to biodiversity conservation
- Integrating conservation and sustainable development

Editorial introduction

It is very easy for humans to harm the Earth: they may profess otherwise, but they still do so. Word and deed rarely match. Some of the most atrocious acts are made in the name of peace and divine love. So it should not be surprising to learn that from the time of entry, humans have used simple but voracious tools, namely fire and the axe, to establish settlement, clear away enemies, frighten game or produce charcoal. The chainsaw and the bulldozer are admittedly quite a step up on the technological ladder, but the net result is the same, albeit on a different scale.

The distinctions in the positioning of humanity in the divine and Gaian ordering, as suggested in Box 1.7, underpin the divergence in outlooks towards the natural world illustrated in Figure 3.1. The text in Chapter 1, however, stressed that the natural world reveals itself only through human intent. So it is hardly surprising that the ethics of biodiversity management cross over scenic values, property rights, Gaian conscience and the emergence of an equity justice dimension to sustainability, as further illustrated in Box 3.2. What follows, therefore, is a delicate mixture of the themes carried in Chapters 1, 2, 3, 4 and 19. With an appreciation of the changing character of science, and the politics of sustainability, it is now possible to begin the long overdue odyssey of marrying ecology to economics. This is why Chapters 4 and 5 should be seen as a coupled pair.

Science and biodiversity

Nowadays many are obsessed with the potential loss of species and habitats – so much so that the UN Convention on Biodiversity was signed in Rio and duly ratified in 1994 to try to put a legal stop to unrestrained damage. The term 'biodiversity' applies to the variety and variability of living organisms from genes to elephants. As this chapter shows, in terms of evaluating the role of biodiversity in maintaining ecosystem services, it is not so much a matter of numbers as of *functional redundancy*. So a major task for scientific enquiry is not just to record but also to establish the functional significance of species in a variety of ecosystems. It is therefore not a matter of how many species can the Earth afford to lose but how many critical species simply must be retained.

Why worry about biodiversity losses?

The consequences of the loss of up to a third of all living species within a stretched generation sober the mind. This is a scientific, ethical, political and economic issue that is of profound significance to the future of humanity. Plants and organisms provide the essential basis for many medicines, genetic strains of food crops, and industrial products. The Chinese utilise some 5000 of their estimated 30 000 plants for medicinal purposes and some 2500 plants were similarly used in the former Soviet Union (Reid, 1995, 15). Although only one in 10 000 potential plant 'hits' yields a commercially profitable drug, modern screening techniques make this 'hit rate' viable. Already 25% of prescription drugs use active ingredients from plants, worth $45 billion in 1985 (Reid, 1995, 15). Yet today only about 110 plants provide virtually all the world's food. Clearly, loss of potential scientific knowledge which can subsequently be put to the good of the world is one of the many reasons why we should mourn.

The other reasons are just as powerful. *Moral concern* and *aesthetic pleasure* are two more. We ought to be worried if we are caring humans, one of many millions of different genetic constellations. The American biologist David Erenfeld (1978) argues that species and their habitats should be preserved

> *because they exist and because this existence is itself but the present expression of a continuing historic process of immense antiquity and majesty. Long-standing existence in Nature carries with it the unimpeachable right to continual existence.*

This point is taken up in the present chapter. To care either comes internally from a personal conscience that is directed by cultural norms, or it may be imposed externally by legal duties or by granting rights of existence to living organisms – rights that provide statutory backing to protection. At present neither structure is well developed in modern society.

This is, as yet, promissory money. It awaits the institutional design hinted at in Chapter 19 for its delivery. Meanwhile, the 'real' money comes through poaching and ecocriminality, namely the removal of endangered species for highly profitable smuggled trade. Barber and Pratt (1998, 4–10) survey the illegal removal of coral reef and associated fish delicacies in the Indo-Pacific region. To make matters worse, the fishermen use cyanide to stun the aquarium-destined species. Cyanide is lethal in low doses, though it is still unknown just how many fishermen and their families suffer as a result. The illegal trade is worth $1.2 billion annually, $1 billion of which consists of exports to the dining tables of Hong Kong and $200 million to the aquaria of the USA. We do not know the cumulative effects of this atrocity: by definition, its scale and local impacts are unknown.

Possible solutions

What can be done? The answer is plenty, apart from improving our science.

Debt for nature swaps appear to be on the wane but still have merit. The idea is to trade the writing down of an unpayable Third World nation debt for local cash, bonded by a rich country organisation, so as to fund the conservation and protection of genetic reserves. To date only a small amount of debt has been written off – about $1 billion for some 1 million hectares of reserves. The main problem is the guarantee of sound cash flow and of good local management. Many impoverished nations resent the whole idea of ecological conditionality to a system of indebtedness which they feel is the product of a greedy capitalist world (Patterson, 1991).

Biodiversity prospecting is a variant of North–South technology transfer. Here the idea is to finance a local conservation agency, with the help of local folk knowledge of ethnobiology, to prospect for plants or organisms that could be of value for genetic modification into pharmaceuticals or disease-resistant strains, or biological pesticides. In one famous example, the US pharmaceutical giant Merck pays a Costa Rican organisation INBio both to look for species of potential value and to conserve precious habitat (Blum, 1993; Reid, 1997). Any profit from a successful formulation would be shared with INBio. Such an arrangement is fine so long as at least one jackpot drug is hit and there is an endless supply of cash to finance ever more sophisticated medical technology.

Transferable property rights is an ingenious idea that has not yet been properly tried out. The purpose is to protect a critical ecosystem by transferring the loss of the right to develop, say for minerals or hydropower, or for forest products, to another site, possibly in another country, where such development is environmentally more tolerable. Clearly there would have to be some compensation for the transfer of development rights, but the conserver economy should not be any worse off. The conserved area might still benefit from controlled 'green tourism' by charging an entry permit for visitors to film or simply experience the retained biological diversity.

Paying for intellectual property rights. Intellectual property rights (IPRs) are the rights of knowledge capable of being privatised and sold commercially or catalogued in the public interest. IPRs operate in two ways. A commercial company that fashions a drug out of a plant holds an IPR on the patent. So too should the native population who successfully practise ethnobiology. Their knowledge is also an IPR and is worth a varying amount to a scientist, a tourist or a biochemical corporation. But capturing IPRs is not easy. It requires conscientious legal innovation, essentially to enclose the intellectual commons. So far little legal progress has been made, but the field is ripe for the creative union of the ecologist, the lawyer, the anthropologist and the economist. Very early studies suggest that the value of medicinal plants could be of the order of $10 to $60 per ha, while a low estimate of the pharmaceutical value of a rich tropical forest could be $20 per ha. The total value of tropical forest, both in terms of its role in absorbing carbon dioxide and in its bequest status for future generations, could lie between $550 and $2200 per ha – four times nominal commercial value (Brown and Moran, 1993; Brown and Adger, 1993).

This translates into an interesting calculation. Land prices in tropically forested countries are now about $300 per ha. The conservation value of the trees storing carbon could be as much as $1000 per ha. Then it would pay the rich world to buy the development rights for the most prized of forests for any sum between these two figures. To do so would mean a transfer in property rights – both developable and intellectual – to some global trusteeship agent. At present, there is no such arrangement. But it should come now that the real 'global' price for a tropical forest is in our sights (Swanson, 1992; Pearce, 1993).

The UN Convention on Biodiversity

This important convention establishes the principle of sustained management or protection of biological reserves, seeks adequate financial measures, guarantees access to IPRs, and encourages appropriate technology transfers. Every nation has to prepare a national plan for conserving and sustaining biodiversity, and to monitor its genetic stock. Also every nation has a responsibility to safeguard key ecosystems and the indigenous knowledge of ethnobiology, and to be protected from the entry of biotechnology products that might undermine or endanger the raw genetic stock. Crucially, the convention protects the rights of nations to 'enclose' their biogenetic commons but also to share their IPRs – at a price. This is why the INBio–Merck agreement is viewed as such an important precedent.

Raustiala and Victor (1996, 20) summarise the current state of play within the UN Convention on Biological Diversity.

- It is a framework convention with international scope for subsequent agreements via protocols and other binding commitments. This is in part because of the inadequate basis of scientific knowledge on habitat functioning and species relationships.

- Each nation must develop national programmes for the conservation and sustainable use of biodiversity 'in accordance with its particular conditions and capabilities'. This means monitoring potentially deleterious activities, establishing an adequately protected scheme of sites and management plans, and rehabilitating degraded areas.

- Every strategic decision in policy should formally take biodiversity into account, in accordance with this programme. This could be done by transferring property rights so that a levy on any development would be used to finance biodiversity enhancement nearby, but not actually on the development site.

The Convention is by no means settled. Disagreements remain over intellectual property rights, biodiversity prospecting and safeguards to truly wild areas. Even with various protocols in place, a process expected to take at least a decade, actually verifying and implementing on the ground is likely to prove enormously problematical, at least until better institutional safeguards are in place.

A more troublesome issue is the safeguarding of the local knowledge of indigenous peoples in such a manner as to maintain their evolutionary stock of biological understanding while converting this asset into a genuine economic good. The obvious issues are how the intellectual property right of these people is to be defined, and what price should be paid to protect and utilise that right. Darryl Posey (1997), a trenchant critic of the current moves to manipulate local people into underselling their knowledge, believes that independent and international safeguards should be put in place as part of the protocol package of the Convention on Biological Diversity.

Local economic benefits. The scope for incorporating biodiversity gains into local agriculture and forest harvesting arrangements is immense (see Temple, 1998, for a good review). This can be achieved by removing perverse subsidies and by encouraging supportive practices. Such arrangements have to be made on a habitat-wide basis, so would benefit from sufficient flexibility as to enable whole landscapes and associated communities to come on board. The kinds of deliberative and inclusive processes outlined in Chapter 1 are relevant here.

As can be seen from this chapter, biological diversity is a classic case for interdisciplinarity. Any natural system is disturbed by the human hand, at least to some extent. Climate change and toxicological damage are trigger forces whose long-term effects cannot yet be known. Linking people to the land is now a cardinal component of adaptive management. Arguably, every private right carries a biodiversity tag: so far we have neither recognised that nor put it into operation.

Prospect

Degradation and loss of natural habitats, ecosystems and species, and disruption of the climate regulation system are escalating and threaten to reduce global biodiversity severely. This has important consequences for human welfare and for the carrying capacity of the planet that will affect many generations to come. There is currently no agreement on the scale of the problem or on appropriate solutions. The scientific disciplines of ecology, conservation biology and climate modelling have much to offer; they can help in understanding the scale and implications of biodiversity impacts, and guide the development of policies for the use of natural resources and mitigating human impacts on the biosphere. However, formulating and implementing solutions also involves value judgements concerning politics, ethics and spirituality. Different groups, all claiming to be seeking the sustainable use of natural resources, remain locked in unresolved conflict as an inevitable consequence of their opposing value and belief systems.

What is biodiversity?

The term 'biodiversity' is now widely used but is rarely defined and can mean many different things to different people. Following the official definition in Article 2 of the Convention on Biological Diversity, Harper and Hawksworth (1995) propose that biodiversity be considered at three ascending levels: *genetic diversity, organismal diversity* and *ecological diversity,* which are explained in Box 5.1. Ideally, a full appraisal of biodiversity would consider each of these. Such assessments, however, are hindered by a lack of scientific knowledge. It is currently not possible to assess genetic and organismal biodiversity at a global scale.

Biodiversity is most often discussed in terms of species richness, and a measure of global species richness would be helpful in assessing the function of biodiversity and the consequences of its loss. However, even such apparently simple estimates are fraught with problems and uncertainty. Only a small proportion of the Earth's species have been identified to date (some 1.4–1.6 million species), with major biases towards a few intensively studied groups such as birds, mammals and beetles. At current rates of taxonomy, it would be anything from one to six centuries before all the Earth's species are identified. Measurement of 'species richness' is made even more difficult as we lack a single unifying definition of what actually constitutes a species! Traditional taxonomy is often based on external structure or appearance. However, molecular techniques have shown that such 'morpho-species' often include several genetically distinct species. In addition, species concepts developed in the study of vertebrates and insects have almost no relevance to the evolutionary diversity of fungi and other groups of micro-organisms; microbiologists cannot yet quantify microbial diversity. From current knowledge, Stork (in Reaka-Kudla *et al.*, 1996) estimates that the total number of insect species in the world may be 5 million, and for all plant, animal and microbe species the global total is likely to be 5–15 million.

Box 5.1 Definitions of biodiversity

Genetic diversity refers to the variability within and between populations of a particular species. Current genetic diversity has accumulated over long time-spans and reflects past climates, population structure, ecological interactions and natural adaptation, as well as random events. Genetic diversity may allow an evolutionary response to environmental change. Diversity in wild populations of cultivated crops provides a genetic reservoir that can be used in engineering or breeding new strains, for example that tolerate different climatic or soil conditions, or confer resistance to pathogens and pests. Once lost, genetic diversity can be replaced only by the slow accumulation of new mutations.

Organismal diversity is a measure of taxonomic diversity. Although commonly equated with species richness, the term 'organismal diversity' is preferable. It is not yet possible to define 'species'

in groups such as micro-organisms and, in addition, it is important to consider the extent to which taxa differ, rather than just counting the number of forms – conserving two closely related species of beetle will protect less biodiversity than conserving two radically different species such as a beetle and a seahorse (see Figure 5.1.1).

Ecological diversity considers spatial variation in the composition of species assemblages and in ecosystem processes. Ecological diversity may be measured at ascending scales: from local *assemblages,* through widespread *communities* or *ecosystems,* to biogeographical regions and *biomes.* Although such ecological diversity is of fundamental importance in a holistic evaluation of biodiversity, it cannot be quantified in a purely objective manner. At each scale of spatial or ecological resolution, assessments rest on subjective 'classification' of continuous variation between notional categories.

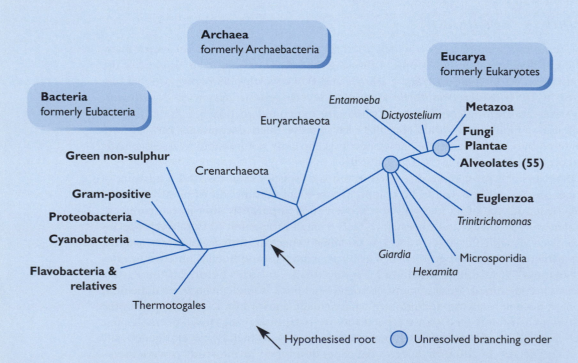

Figure 5.1.1: When we consider the overall diversity of life on this planet, we see that the three fundamental, deeply rooted divisions are between the Bacteria, the Archaea (formerly Archaebacteria) and the Eucarya – a division that includes all animals, plants, fungi and protists. We are but one species among millions within the Animalia, which is just one division of the Eurcarya: from a holistic perspective we are closely related to fungi and plants (redrawn from Embley *et al.,* 1995).

As introduced in Chapter 1, biodiversity cannot be fully understood or valued solely in terms of rationalist science. The values that people attribute to biodiversity, 'nature' and wilderness, their perceptions of biodiversity loss and their attitudes towards conservation, are subjective, culturally conditioned and experiential. Tropical deforestation is now a powerful cultural and political metaphor, due to a combination of astounding biological diversity and media-accessible beauty, juxtaposed with potent emotive images of destruction. Most people who reject the idea of 'sustainable' whaling quotas are not influenced by a rational evaluation of the evidence regarding population size and harvest levels but by deeply held feelings and spiritual beliefs. This tension between the Cartesian tradition of rational objective science and subjective cultural, ethical values will be elaborated throughout this chapter.

Human attrition of biodiversity

Human impacts on biodiversity have not been confined to the modern age. The arrival of humans in the Americas, Oceania and Madagascar was associated with the loss of over 100 genera of large mammals. Before becoming extinct, mega-herbivores such as the moas of New Zealand, giant ground sloth and mastodont of North America and temperate rhinoceros and elephant of Europe would have had important impacts on ecosystem structure and function. Thus many ecosystems were no longer truly 'natural' long before their first clearance for agriculture or the arrival of European settlers. The spread of Polynesians across the Pacific islands during the last 1000–4000 years resulted in the extinction of approximately 2000 bird species, 15% of the world's avian diversity (Chapin et al., 1998). Pre-agricultural human cultures also transformed their environment through the use of fire, primarily to create open grasslands and boost populations of herbivore prey. Australian vegetation was transformed by Aboriginal firestick farming, and the first human settlers in North America expanded the prairies at the expense of forests. Similarly, in Africa human use of fire caused expansion of grasslands and savanna and a reduction in evergreen forests (Bond and van Wilgen, 1996).

Human impacts on terrestrial biodiversity escalated with the development and spread of agriculture. This resulted in wholesale replacement of forest and other natural habitats by simplified ecosystems, of lower biodiversity but much increased human carrying capacity. Most of the natural habitats that once cloaked the land surface of the planet are now reduced to small isolated fragments, within a landscape controlled by agricultural, industrial and urban land uses (see Table 5.1). Losses currently continue at high rates, greatest in the biodiversity-rich tropics. Habitat loss reduces the genetic diversity remaining within species and will cause escalating species loss from the isolated fragments that remain (through impeded dispersal, reduced population size and vulnerability to chance events).

In addition to such direct destruction, many other human activities are degrading the ecological integrity and biodiversity of those habitats that remain. By 1983, desertification of arid lands through overgrazing, cultivation and fuelwood harvest amounted to 9 million km^2 (Primack, 1995), while unsustainable agriculture has accelerated soil erosion throughout the tropics (see Box 2.5 and Chapter 11). In many rangelands of New Zealand, Australia and the former North American prairies, native vegetation has been largely replaced by alien plant species that are more tolerant of grazing by introduced sheep or cattle. In the USA, half of the total length of rivers and streams are significantly polluted and only 2% is free-flowing; of the US freshwater fauna 20% of fish, 36% of crayfish and 55% of mussels are already extinct (Reaka-Kudla et al., 1996).

Table 5.1: Loss of terrestrial wildlife habitats (% loss, expressed as a proportion of original area of habitat).

Globally

Primary forest (by late 1980s)	76%	Tropical forest (by late 1980s)	50%
Coral reefs	10%		

Europe

Primary forest	99%	Coastal temperate rainforest	99%
Mediterranean region forest	90%		

North America

Old-growth forest, Douglas fir region of Oregon and Washington	83–90%	Prairies in many US states	>99.9%
Coastal temperate rainforest logged:		Native grassland/oak savannas, Williamette Valley, Oregon	99.5%
Washington	75%	Riparian ecosystems, Arizona and New Mexico	90%
Oregon	96%	Primary eastern deciduous forest (US)	99%
Coastal redwood forest, California	85%		

Central and South America

Rainforest, S. Mexico	>90%	Atlantic rainforest, Brazil	>95%
Tropical dry forest, Pacific Central America	>98%		

Africa

All original wildlife habitat, S. of Sahara	65%		

Primary tropical forests:

Gambia	89%	Rwanda	80%
Ghana	82%	Zaire	57%
Kenya	71%	Zimbabwe	56%
Madagascar	75%		

Asia

All original tropical wildlife habitat	67%		

Primary tropical forests:

Bangladesh	96%	Philippines	97%
China	99%	Sri Lanka	86%
India	78%	Thailand	73%
Indonesia	51%	Vietnam	76%
Malaysia	42%		

Mangrove forests:

India, Pakistan and Thailand	>75%

Australia

Primary forest	95%

New Zealand

Primary forest	76%	Coastal temperate rainforest	72%

Sources: Primack (1995); Noss *et al.* (1995).

Goats and pigs released onto oceanic islands by European seafarers to provide food have devastated endemic vegetation, threatening the habitat on which many endemic species depend. Acid rain impacts temperate and boreal forests and lakes: acidification is apparent in 39% of lakes in Sweden and 34% of lakes in Norway (Primack, 1995). The list of such impacts is potentially endless.

Some agricultural landscapes initially retained substantial residual biodiversity in semi-natural habitats and landscapes (see Box 5.5 later in this chapter). Major land-use changes have resulted in the direct loss of semi-natural habitat and degraded the quality of that which remains (see Table 5.2). Agricultural intensification has had further severe effects on biodiversity within the farming ecosystem itself. Birds are good indicators of such change as they tend to be well monitored, have complex feeding and habitat requirements and are often near the top of the food chain (Dodds *et al.*, 1995). In Britain, long-term monitoring has revealed large population declines in most bird species associated with farmland (see Table 5.3). Although a few of these species are 'rare' (historically restricted in numbers and/or range: corncrake, stone curlew, cirl bunting), the others were once common but have undergone unprecedented declines. The concern is therefore not only with the survival of a handful of rarities but also involves a suite of species that have been taken for granted as an integral part of the countryside. Such declines result from the conversion of pasture to arable, loss of mixed farming, regional specialisation, loss of spring-sown cereals, spring tillage and winter stubble, increased use of herbicides, insecticides and inorganic fertilisers, field drainage, and a switch from traditional haymaking to intensive silage production (Dodds *et al.*, 1995). Similar patterns are emerging at a European scale. Most bird species with an unfavourable conservation status in Europe (SPECs: species of European conservation concern) were once common and widely distributed but have undergone substantial population declines. Farmland holds more of these species than

Table 5.2: Loss of semi-natural habitats through land-use change.

Lowland heathland

England	75% loss 1800–1985, 60% loss 1949–1983
Sweden and Denmark	60–70% loss 1860–1960, further losses since
Plateau des Tailles, Ardennes, France	99.6% 1770–1970 (from 28 000 ha to <100 ha)
Belgo-Dutch Campine	90% loss c.1770–1980s (1500 ha remaining)
The Netherlands	95% loss by 1980s (from 800 000 ha to 42 000 ha)

Lowland neutral grasslands (species-rich hay meadow and pasture)

Britain	95% loss 1947–1983

Wet grasslands (grazing marshes)

England and Wales	Of a historic 1 200 000 ha, 220 000 ha remains of which only 20 000 ha is agriculturally unimproved wet grassland of high conservation value

Lowland calcareous grassland

Britain	80% loss 1949–1983
Dorset, England	89% early 1800s–1980s (from 28 000 ha to 3000 ha)
Champagne, France	99% loss 1948–1981

Hedgerows

Britain	23% net loss 1984–1990

Sources: Noirfalise and Vanesse (1976); NCC (1984); Tubbs (1985); Farrell (1989); RSNC (1991); DoE (1993); RSPB/EN/ITE (1997).

Table 5.3:
Population changes
in farmland bird
species in Britain.

Bird species	Habitat	Status in Britain
Grey partridge	arable	82% decline in 25 years
Corncrake	hay meadows	now globally threatened
Stone curlew	arable, grassland	50% decline in 25 years
Lapwing	arable, pasture	62% decline
Turtle dove	arable	77% decline in 25 years
Barn owl	rough grass, hay	50% decline in 50 years
Skylark	grassland, arable	58% decline in 25 years
Swallow	farmland, villages	43% decline in 25 years
Song thrush	farmland, woods	73% decline in 25 years
Tree sparrow	farmland	89% decline in 25 years
Linnet	arable	52% decline in 25 years
Yellowhammer	farmland	small decline
Reed bunting	wetlands, farmland	61% decline in 25 years
Cirl bunting	mixed farming	steep decline in 25 years
Corn bunting	arable	80% decline in 25 years

Source: Dodds *et al.* (1995).

any other habitat, and agricultural intensification has been identified as the most common threat, affecting 42% of avian SPECs (Tucker and Heath, 1994).

As discussed in Chapters 7 and 8, human impacts on biodiversity now operate at the scale of global biogeochemical climate regulation. Some species assemblages may already have been impacted by global climate change. In the cloudforests of Costa Rica, warming has raised the height at which cloudbanks form, drying forests: many species of bird have extended their ranges up the mountain slopes, highland lizard populations have declined and disappeared, and amphibians have suffered a massive die-back – in one 30 km² study area 40% of the frog and toad species became extinct during the 1990s (Pounds *et al.,* 1997). Bleaching of coral reefs (loss of symbiotic pigmented algae, which can lead to coral death) has become more frequent, widespread and severe since the early 1980s. A study of Florida's coral reefs in 1997 revealed extensive damage and a dramatic increase in diseased coral, while in 1998 coral bleaching affected over 1000 km of inshore reefs in Australia's Great Barrier Reef (Henderson, 1998). The most disturbing thing about these events is that they merely herald the severe disruption to species and ecosystems that is to come (see Box 5.2).

In previous periods of Quaternary climate change, species range shifts were able to occur within continent-wide natural landscapes. Despite this, the Quaternary is characterised by extinctions of large vertebrate herbivores and their predators and by the loss of the park-tundra biome and its associated biota (Huntley *et al.*, 1997). This time, the overwhelming extent of habitat loss and fragmentation has trapped many populations in vulnerable sites, separated by a sea of inhospitable degraded land. Major species loss is inevitable.

Predictions of a 'global extinction spasm'

The combined impacts of habitat loss and fragmentation, ecosystem disruption and global warming are a vicious pincer that will destroy much biodiversity. Current extinction rates are 100 to 1000 times higher than the average seen throughout the fossil

Box 5.2 Consequences of global climate change for biodiversity

At a genetic level, significant morphological and/or physiological evolution will be rare over timescales relevant to forecasting future climate change. Species will simply not have time to 'adapt' to the new climates by evolution, placing many in a perilous position (Huntley *et al.,* 1997). The major adaptive response available is that of range shift.

Smith *et al.* (1992) compared the current distribution of terrestrial 'life zones' or biomes (such as rainforest or tundra) with that predicted under 2 × CO_2 climate change scenarios from four global circulation models (GCMs):

- all scenarios predicted a significant shift in the distribution of life zones, ranging from 39 to 55% of terrestrial land area;
- northern life zones will be especially vulnerable, due to the greater amplitude of climate change and smaller relative land area at northern latitudes;
- all scenarios investigated resulted in a severe decline in tundra due to expansion of boreal forest. For boreal conifer forest, potential expansion of the northern limits (by some 2.6 million km^2) is likely to be offset by large range contractions to the south (6.4 million km^2), giving an absolute decline of 66% (Cramer and Leemans, 1993);
- similar modelling by Prentice *et al.* (1993) suggests that in Africa, the two biomes that are currently most rare (semi-desert and broadleaved evergreen forest) are predicted to decrease by 81% and 69% respectively in response to a tripling of atmospheric carbon dioxide.

These bioclimatic models relate life-zone distributions to mean climate variables. Predictions consider potential distributions only once biomes have reached equilibrium with the changed climate. They lack ecological resolution and do not consider temporal dynamics. In contrast, forest gap models consider the fate of individual trees in a small forest plot. They simulate the demographic processes of mortality ('gap formation'), establishment and growth and thus predict the dynamic temporal trajectory of climate responses. Gap models incorporate reciprocal feedback between tree growth and stand composition and shade, soil moisture and soil nutrient dynamics, giving a high degree of ecological resolution. However, their fine spatial scale does not incorporate species range shift within regional landscapes: the nested integration of scales remains an important challenge.

Despite incomplete knowledge, a number of predictions may be made from existing models, paleo-ecological evidence and our understanding of ecology.

Paleo-ecological evidence from previous episodes of Quaternary warming shows that:

- many invertebrates (particularly land snails and beetles) and other taxa with short-generation times, such as grass and herbs, were relatively mobile, but
- many tree species moved slowly and will lag significantly behind shifting climatic zones, and
- many taxa are expected to suffer marked population declines, reducing genetic diversity and increasing chances of stochastic extinction.

Changes in the frequency of severe climatic events (drought, killing frost, storms) can have profound impacts on vegetation composition (Prentice *et al.,* 1993; Cramer and Leemans, 1993):

- rapid, large-scale vegetation change may occur, especially at forest–grassland boundaries impacted by increased frequencies of drought and associated forest fires;
- models that deal better with rainfall variability, seasonality and extreme events are required in order to assess ecosystem vulnerability (Markham, 1996).

Where one forest type is expected to be replaced by another (such as the invasion by temperate deciduous species of boreal conifer forest and shifts between tropical forest zones), responses may be case-specific:

- some forests may have high inertia and resist invasion by species favoured by the new climate (Smith *et al.,* 1992; Cramer and Leemans, 1993),

Box 5.2 (continued)

● but where climates exceed physiological requirements of current species, rapid dieback may occur long before replacement tree species enter and establish themselves (Cramer and Steffen, 1997; Bugman, 1997). Thus, even where the forest area will be stable or greater at bioclimatic equilibrium, over the next few centuries we can expect a major pulse of carbon release.

The Quaternary fossil record contains many examples of non-analogue assemblages that differ from any extensive modern community. From this, and by modelling the climate space occupied by current species' distributions, we know that species have highly individualistic responses to climate (Sykes, 1997; Huntley *et al.*, 1997). This has major consequences:

● 'communities' will not migrate as intact units and many familiar assemblages will dissociate;

● species linked by tight mutualisms may be constrained to migrate together, but many loose (facultative) mutualistic associations (such as pollination and seed dispersal) will uncouple;

● range shifts of competitors, predators and invertebrate 'pests' may have unexpected consequences for assemblages which they invade.

In some cases, subtle changes in key species may have major knock-on effects. In tropical forests, disruption of flowering or fruiting of key plants that provide food during seasons of resource scarcity may impact on numerous birds, mammals and insects (Markham, 1996)

record since the dawn of the Cambrian. This is just the beginning: the expected extinction of currently threatened species could increase this rate by a factor of 10 (Pimm *et al.*, 1995). It is hard to predict accurately the rates of species loss that will result. To date, many of the species that have been driven extinct by humans were rare species endemic to small oceanic islands or other isolated habitats. Understanding such extinctions is of little help in predicting future impacts on continental and global biodiversity (Chapin *et al.,* 1998). For a few well-studied groups such as birds and large mammals we have a good understanding of the distribution, biology and status of most species. However, for the vast majority of named species we know virtually nothing other than what they look like, and lack any real understanding of their ecology, of whether they are widespread or extremely localised, beyond risk or in imminent danger of extinction. An even larger part of biodiversity is completely unknown. The likely scale of extinctions has been estimated from rates of tropical deforestation and species richness/geographical area relationships, from numbers of local restricted-range species, or from extinction functions and current threat categories. Eleven such estimates published during 1979–1994 ranged from 0.6% to 30% of global species being lost per decade. Over half of these estimates predict rates greater than 5% per decade for the foreseeable future. Such uncertainty is inevitable, but the scientific consensus is that many hundreds of thousands, possibly millions, of species are threatened with extinction (Reaka-Kudla *et al.*, 1996).

At least six previous episodes of mass extinction have been identified from the fossil record; so is the current extinction spasm just one more infrequent but natural event that should not concern us greatly? The most severe previous mass extinction took place at the end of the Permian epoch, about 250 million years ago, when approximately half of all animal families then existing on the planet disappeared. However, it is believed that we are now witnessing a mass extinction spasm as great as, or perhaps even greater than, any known from the geological record, while the current *rate* at which species are being lost is without precedent (Lawton and May, 1995). Such species loss is effectively irreversible (Chapin *et al.*, 1998): it took the process of evolution 50 million years to regain the number of families lost during the Permian transition. Perhaps the most fundamental distinction between this and previous mass extinctions is that those were driven by extrinsic factors, such as pulses of volcanic activity or meteor impact, while the course of the current extinction episode can be influenced by human choice (Chapin *et al.*, 1998). So should we conserve biodiversity? And if so, why?

Why conserve biodiversity? Biodiversity ethics

Our material existence is dependent on the use of natural resources. The material values of biodiversity may be classified between *consumptive use, productive use, option* or *opportunity* value, and *non-productive use* (see Box 5.3). As explained in the previous chapter, these values may be expressed in economic terms.

Box 5.3 'Values' of biodiversity?

Anthropocentric material values (empiricist/rationalist)

Consumptive use (life support): Local use and consumption of natural resources, such as food, fish, meat, building materials, timber, fuel, medicines

● Cultural value system: Socially regulated commons

● Resource use paradigm: Sustainable management, resource conservation

Productive use: Material resources (food, timber, etc.) *Option value* opportunities, e.g. potential medical drugs, genetic resources

● Cultural value system: Market economics

● Resource use paradigm: Exploitation, genetic engineering

● Cultural value system: 'Utilitarian' societal ethic

● Resource use paradigm: Interventionist resource management regulated for the common good; emphasis on 'progress' and 'improvement'; increasing use of ecological management

Non-productive use, indirect values: Ecosystem services, e.g. pollination, regulation of pest species through predator complexes, nursery habitats for fisheries (e.g. mangrove swamps), soil fertility, nutrient cycling, aquifers, flood control, waste assimilation capacity, climate regulation

● Cultural value system: Societal utility, stewardship, intergenerational equity

● Resource use paradigm: Ecologically sensitive management, sustainable resource use, bioremediation, ecological engineering

Anthropocentric cultural values (subjective)

Value of biodiversity: Societal and individual well-being through contact with nature; human-oriented spirituality

● Cultural value system: Aesthetics, philanthropic ethics, spiritual beliefs

● Resource use paradigm: Nature and landscape conservation, interventionist management, ecological education

Ecocentric values

● Cultural value system: Belief that the natural world has unquantifiable intrinsic value; respect for non-human life

● Resource use paradigm: Fully inclusive biodiversity conservation; emphasis on natural dynamics; active restoration of ecological processes, or alternatively, non-intervention 'wilderness' areas

Many traditional societies in the developing world still rely on natural or near-natural ecosystems to supply fuelwood, meat, fruit, fish, medicine, fibres and building materials (*consumptive use*). Consumptive use is important but often substitutable. A consequence of the shift from consumptive use of biodiversity for sustaining life to reliance on market economics is the separation of culture and communities from the land and from nature and a loss of human cultural diversity. The next chapter suggests that the resilience of societies and their ability to absorb disruption and change may be related to the diversity and resilience of the ecological systems on which they depend.

In addition to consumptive uses, further material resources are extracted or harvested commercially, often as the basis for industrial production such as fisheries, food and timber. Although this *productive use* of natural resources is important in global and local economies, only a small part of biodiversity is involved. Most of the global diversity of plant species currently has negligible commercial value in terms of *productive use* (see Table 5.4). The conservation and sustainable management of economically key species and resources may be important, but productive use does not provide an inclusive basis for biodiversity conservation.

The pre-eminence of materialist values means that biodiversity is often regarded as a suite of natural resources to be exploited and developed. This mindset reflects values and beliefs developed from the 16th-century Enlightenment: the goal is progress, by applying rational thought, reductionist science and technology to harness and control nature, resulting in interventionist and engineering approaches to natural resource management (Shiva, 1989). The consequences ramify through every degraded farmscape, stream, forest and rangeland on the planet. This approach is well illustrated by the introduction of Nile perch to lakes of the African Rift Valley. Protein production and thus the *productive value* of the ecosystem were enhanced, but both 'naturalness' and biodiversity were severely reduced. Many species of cichlid fish were lost. These were an astonishingly diverse group that had evolved in these lakes over millions of years. In addition, the ecological resilience of the system was reduced, with increased variability in catches (Jennings and Kaiser, 1998).

Much material value of natural resources and biodiversity is now defined by market economics. Purist capitalist economists prioritise the pursuit of profit through the free market: when applied to natural resources this often has dire consequences for both biodiversity and human welfare (Shiva, 1989; Hildyard, 1991). Market forces have driven continued harvesting of large whale species, rhinos and tigers, which has severely reduced numbers, range and genetic variation. In many tropical countries, deforestation is being driven by the harvesting and export of timber to finance industrialisation and service foreign debt. In countries such as

Table 5.4: The negligible importance of global plant diversity to productive use.

	Number of species	*Per cent of global total*
Global plant species richness	*c.* 250 000	
Plants used in agriculture (often just local use)	6000	2.4 %
Plants with economic use (includes agriculture)	1000–2000	0.4–0.8 %
Plants important in world trade	100–200	0.04–0.08 %
Plants providing the bulk of the world's food crops	15	0.00006 %

Source: Heywood (1993).

Malaysia and Indonesia, timber products have been among the top export earners, accounting for billions of dollars per year (Primack, 1995). Market economics continues to drive the destruction of diverse semi-natural farmed landscapes and their associated biodiversity (see Tables 5.2 and 5.3).

When considering the utilitarian, societal benefits arising from resource use, it is important to consider to whom those benefits flow. For example in Thailand, companies have replaced natural forest by fast-growing Australian eucalyptus; but eucalyptus cannot provide the wide range of foods, medicines, fodders and building materials which many villages on the edges of the market economy need to fill out their subsistences and which can be supplied by patches of secondary forest (Lohman, 1991). Similarly, tiger prawn ponds have replaced huge areas of Thailand's rich mangrove forests in order to grow a single species, *Penaeus monodon,* for luxury markets in Japan, the USA and Europe (Lohman, 1991; Gujja and Finger-Stich, 1996). In 1993, Thailand earned $2 billion from shrimp exports, while in Bangladesh shrimp exports were more important than any other agricultural product. The scale of these returns is impressive, but intensive production has transformed the enterprise from one dominated by small independent farmers to one controlled by large multinational conglomerates. At the local level, displacement of other land uses causes unemployment, and villagers are excluded from resources once held in common. The use of industrial quantities of fish to feed the shrimps has devastated fisheries in the Gulf of Thailand; farms have depleted freshwater supplies and contribute to eutrophication and red tides. Ponds have a life span of only 5–10 years: around 15 000 hectares of degraded land was abandoned in Thailand, Ecuador, Indonesia and India in 1994 alone (Gujja and Finger-Stich, 1996).

Currently unexploited biodiversity and unexplored genetic resources may form the basis for new pharmaceutical drugs or crops (see Box 5.1). This *option* or *opportunity value* may provide a strong economic argument for conserving biodiversity that may have potential economic value or material benefits for society. Severe limitations in knowledge of organismal and genetic diversity provide a strong precautionary basis for retention of biodiversity that may have potential unexplored value. But where genetic resources are exploited by the market, fully screening the myriad of potentially useful compounds is so prohibitive in scale that it is not feasible in the short to medium term. Companies find it more profitable to 'mine' the accumulated ethnobotanical knowledge of indigenous peoples. Once candidate species have been identified, product development may shift to pharmaceutical laboratories. Where the industrial product is still to be extracted from the natural plant, rather than synthesised, this may be most efficiently achieved using intensive plantations. Either way, there is little short-term value in retaining the biodiversity of the forest, or the cultural diversity of forest peoples. For example in Thailand, Sankyo's plantations of *Croton sublyratus,* a medicinal herb used by the Japanese to treat ulcers, has impoverished the very ecosystem in which the genetic stock used on the plantation evolved (Lohman, 1991).

Many governments and international bodies attempt to regulate the free action of the market, adopting a utilitarian ethic: that resource use should be controlled by society for the common good (see Box 5.3 and Chapter 8). As detailed in Chapters 1 and 4, utilitarian ethics have been extended to incorporate equity and obligations to future human generations, which are fundamental to principles of sustainability in the Convention on Biodiversity and to the concept of global citizenship. This gives an emphasis on ecosystem services that sustain human material welfare.

Global utility: ecosystem services

Biodiversity may provide indirect but vitally important ecosystem services (*non-productive use*: see Box 5.3). Examples include the regulation of pests through predator complexes, pollination of commercial crops, regulation of soil fertility, nutrient cycling and flood control, the capacity of the environment to assimilate waste, and the regulation of global climate through complex biogeochemical feedback mechanisms (Daily, 1997). The discussion that follows complements the previous chapter, in which valuation of ecosystem services is explored.

Although further large-scale habitat loss, environmental degradation and mass extinctions are expected, scientific ecology does not currently predict a *failure* of global life support mechanisms. In this sense the indispensability of terrestrial ecosystems such as tropical forests and wetlands has often been overstated. In fact, there is great scientific uncertainty regarding the future contribution of terrestrial ecosystems to climate regulation. Increased temperature and fertilisation by carbon dioxide and other atmospheric pollutants (such as oxides of nitrogen) may shift many ecosystems from being net carbon sinks to net carbon sources. In the future, highly productive managed secondary forests, or even grasslands, may have greater carbon sequestration value than some old-growth forests or other 'natural' ecosystems (Walker and Steffen, 1997).

At a local or regional scale, non-substitutable ecosystem support services are critical in sustaining human carrying capacity (Cohen, 1997; Daily, 1997). Many of the poorest people in the world live in ecologically fragile areas. Intensive exploitation of natural resources is damaging biodiversity and has reduced the quality of human carrying capacity over much of the Earth's land surface. Given the perilous state of much of the global human population in terms of nutrition, health, water supplies and economic and material security, further disruption to ecosystem services will have profound consequences for human material welfare.

Is biological diversity important in maintaining such ecosystem services? A number of plausible hypotheses regarding the functional role of species diversity within ecological communities have been proposed (see review in Johnson *et al.*, 1996). If there is tight coupling, then reductions in biodiversity will lead to reduced ecosystem stability and resilience (the ability to absorb perturbations; see also Chapter 4). Alternatively, if there is significant *redundancy* or *functional similarity* between species, it may be possible to lose much organismal diversity before serious disruption occurs. The urgent need for answers has forced a synthesis of the historically distinct disciplines of holistic ecosystem ecology and reductionist autecological science. Experimental studies suggest that species diversity could affect many ecosystem processes, including photosynthetic carbon gain, productivity, nutrient cycling and ecosystem stability. However, the nature and magnitude of effects vary between ecosystems and with the functional response measured: in some cases diversity has little effect at all (Johnson *et al.*, 1996; Bengtsson *et al.*, 1997; Chapin *et al.*, 1998). At present too few studies have been conducted to allow convincing generalisations (Chapin *et al.*, 1998). Studies to date have considered relatively simple systems. In these, the gain or loss of a few species may be more likely to give detectable effects than would occur in complex systems. Alternatively, disruption of complex webs of biotic interactions in species-rich systems may initiate cascade effects not seen in simple systems (Chapin *et al.*, 1998). Additional uncertainty comes from the fact that we currently know very little about the extent and functional significance of microbial diversity in soils: few free-living microbes can be cultured and definitions of 'species' are unclear (Bengtsson *et al.*, 1997). The uncertainty imposed by global climate change further emphasises the importance of a precautionary approach. Current 'communities' are expected to dissociate and reform in novel assemblages (see Box 5.2) and species that are functionally similar at present may perform significantly different roles. Even if species richness does not play an important role in maintaining ecosystem processes now, it may be important to ensure that a wide range of species are available to fulfil new roles when conditions change (Bengtsson *et al.*, 1997; Chapin *et al.*, 1998), and thus provide the raw material for interventionist ecosystem manipulation. To achieve this, we will need to conserve the natural systems, ecological processes and dynamic landscapes within which these species occur.

However, in many cases it is likely that ecosystem services may be served by simplified, managed ecosystem components. For example, wastewater treatment and flood control can be achieved using heavily managed or even artificial wetlands, and in agricultural catchments watercourses may be protected from siltation by woodland or grassland buffer strips with little requirement for a full complement of organismal diversity or 'naturalness'. In heavily modified landscapes such as are found in the UK, many rare plant and invertebrate species occur as just a small number of individuals at only a handful of localities. It is hard to argue that their loss would have any impact on ecosystem function. Their value lies elsewhere, in the socio-economic domain.

Cultural values, intrinsic value and ethics of biodiversity

> *The position that nothing in the natural world has intrinsic value, that the whole conservation movement is motivated only by narrow utilitarian aims centered on human health and prosperity, corrodes in the long run the public image of the movement. Highly dedicated persons who cannot help but work for conservation and for whom it is a vital need to live with nature are confused by what they take to be the utter cynicism of scientists and experts who use purely utilitarian, flat language in their assessment of environmental risk, 'genetic resources', and extinction. These experts are often seen as traitors.*
> (Naess, 1986, 505)

Something as simple as the skylark's song has inspired poets, from Theocritus in the third century BC through to Chaucer, Shakespeare, Wordsworth, Keats, Pope, Browning and Shelley, while paintings by Frida Khalo and Georgia O'Keefe evoke spiritual feelings in many people. Such cultural values of biodiversity may be expressed in terms of direct economic value (see Chapter 4). *Amenity value* of wildlife and landscapes for recreation and tourism may be substantial, providing a materialist, market justification for conservation. More intangible and abstract values have also been evaluated: contingent valuation methodologies have shown that many people are *willing to pay* significant amounts of money to ensure the continued existence of species, biomes or landscapes. Richard North (1995) sought to demonstrate the grossly subjective basis from which people value existence. The Flow Country of north-east Scotland, the most extensive area of blanket bog in the world (initially covering more than 400,000 ha) and one of the few near-natural areas remaining in Britain, was threatened by widespread commercial afforestation during the 1970s and 1980s (Ratcliffe and Oswald, 1987). North points out that the public outcry that prevented this destruction came from influencing and mobilising the opinions of people who had never even visited the area and had little understanding of its ecology. North then feels justified in dismissing such subjective motivations for biodiversity conservation as irrational and, by implication, invalid. Many rationalists, empiricist scientists and economists share North's disdain for subjective judgements. Moral, cultural and spiritual motivations for biodiversity conservation are often trivialised as mere 'aesthetics' or romanticism. However, this ignores fundamental issues, such as the subjectivity of ethics (Soule, 1986).

Many people would agree with the supposition that rape, murder and torture are somehow wrong, or that society should attempt to prevent them occurring. But why? There can be no rational, objective proof of 'wrongness' that does not ultimately rest on underlying subjective precepts. In fact, deterministic understanding of evolution shows that, in certain circumstances, such behaviour may increase individual fitness (the contribution to the gene pool of the next generation). Biology has been used as a rational justification for competition, selfishness, disregard for the importance or suffering of others, the glorification of strength and a capacity for violence. Such values are implicit in capitalist economics of the unregulated free market and, for some, are seen to provide a workable basis for allocating scarce natural resources. Many societies have endorsed such behaviour, with cultural codes that proscribe when, and in what circumstances, it is and is not appropriate to kill, or subjugate. Biology has been used to justify eugenics and racial cleansing. Despite this, many choose to transcend biological determinism with more 'humanitarian' ethics that adopt precepts such as a presumption against violence and causing unnecessary suffering, in favour of compassion and altruism. The application of such ethics has now been extended to the concept of global citizenship, both now and for future generations (see further dicussions in Chapters 1, 4 and 19).

The decision to limit ethics and compassion to human well-being is subjective. Arne Naess introduced the idea that non-human life and its habitat has *intrinsic value* that should be respected by human beings, independently of the usefulness of the non-human world for human purposes. In Chapter 1 it is suggested that 'nature' exists only through human perception and cognition. But the diversity and complexity of forms of life and their myriad interactions may exist independently of human cognition or perception, and thus have *intrinsic value*. Evolution is not directional, with an ascending pecking order that places *Homo sapiens* at the apex of 'progress', but a radiating divergence of myriad lineages from a cluster of starting points (see Figure 5.1). One divergent tendril has no more intrinsic value than any other successful strategy.

> *A clearcut or even a mile-wide strip-mine will heal in geological time. The extinction of a species, each one a pilgrim of four billion years of evolution, is an irreversible loss. The ending of the lines of so many creatures with whom we have traveled this far is an occasion of profound sorrow and grief. Death can be accepted and to some degree transformed. But the loss of lineages and all their future young is not something to accept. It must be rigorously and intelligently resisted.*
> (Snyder, 1990, 176)

Nature may have intrinsic value, but our perceptions of this are filtered by conditioning and merge with spirituality. Many people's experiences of being in nature engender calmness, peace, wonder, exhilaration, veneration, awe or resilience. The recognition of this was instrumental to the development of a nature conservation movement in Britain (Lowe, 1983). During the 19th century, humanitarian philanthropists experienced a strong moral and aesthetic revulsion to the social degradation of industrial cities, with their slums, poverty, crime and perceived breakdown of moral values. In addition to establishing schools and legislating against child labour and slavery, philanthropists tried to preserve places not yet 'corrupted' by urban and industrial expansion, regarding the 'countryside' as a place that people could go to for clean air and to rejuvenate their spiritual health. Similarly in North America, a passionate belief in the beneficial and spiritual value of experiences in nature was fundamental to Thoreau, Leopold and Muir and the development of the wilderness preservation movement (see Box 3.2).

Culture and values are in a state of rapid flux and their future is uncertain. To achieve a sense of global citizenship we need to construct, educate and develop an altruistic, compassionate society: ecological education and access to nature may contribute to this. Ecology shows that we are part of an interconnected system and in this sense no aspect of life exists independently of our influence. Ecology is the study of endless chains of cause and effect, of interactions. The value of nature as a metaphor is that outer truths may also serve when we turn inwards. Through nature it is apparent that we are but part of an interconnected whole, that our actions cannot fail to have consequences, and that we cannot avoid responsibility for these. Snyder (1990) and others have drawn parallels between the philosophies of intrinsic value and Buddhism. The philosophy and ethics of Buddhism teach an understanding of cause and condition, of the arising of volition, and of the consequences of action. Humans may use such wisdom to make informed decisions so that they may act with compassion, in such a way as to cause less suffering to arise, in recognition of the intrinsic value and interdependence of all life.

Approaches to biodiversity conservation

To address fully the problems of biodiversity loss requires profound economic and political change. In the short to medium term, biodiversity conservation attempts to slow rates of loss and to retain distinctive components, giving the opportunity of future restoration.

An important step is that of defining priorities and targeting resources for effective conservation. Priorities for vulnerable species have been defined at national and international scales, as in the IUCN lists of globally threatened species, the US Endangered Species Act (ESA) or UK Red Lists. Although such priorities may be defined by objective evaluation of data for population size, range, changes or threats, on occasion resources are targeted preferentially to distinctive species that have important aesthetic or cultural associations. For example in England, the statutory conservation agency English Nature included photogenic and popular appeal among biological and scientific criteria in selecting species for its Recovery Programme (see Figure 5.1). Such an approach may be taken to extremes: in the US during the 1980s, over half of the funds available for developing and implementing recovery plans under the ESA were spent on just a handful of species with high public appeal, some of which were not even highly threatened (Rohlf, 1991). The US Act illustrates problems arising from exclusive focus on species conservation. The most recent assessment of the status of endangered species by the US Fish and Wildlife Service, in 1995, shows improvement for fewer than 10% and declines in nearly 40%, while the status of about a third of the species was still unknown (Bean, 1998). The belief that a focus on charismatic species will additionally safeguard ecosystems and the vast numbers of other less well-known or attractive species has worn thin, and ecosystem- or habitat-based approaches are increasingly favoured (Markham, 1996; Bibby, 1998). Compared with the inefficiencies of a species-by-species approach, ecosystem conservation directly addresses the primary cause of many species' declines (destruction or modification of habitats), offers a meaningful surrogate to surveying every species, and provides a cost-effective means for simultaneous conservation and recovery of groups of species (Noss *et al.*, 1995). Despite repeated calls for reinterpretation or revision of the US ESA, safeguarding and

Figure 5.1: Public support for conservation may be enhanced by publicising action for species such as the dormouse described in *Natural World*, the magazine of the UK Wildlife Trusts Partnership (1996), in the following terms: 'With its golden-brown fur, baby pink feet and classic mammalian good looks, dormice have always been popular.' In England, schoolchildren and the general public have been enlisted in surveys of the declining dormouse population. Conservation targets for the dormouse are now detailed in the UK Biodiversity Action Plan, with the aim of maintaining and increasing numbers where it still occurs and re-establishing populations in at least five counties where it is no longer found. The future of many of the remaining populations depends on careful management of the woodlands in which they occur. Rotational coppicing, a traditional form of management carried out since prehistoric times, ensures a complex aerial architecture that favours the dormouse, allowing it to forage for its food of fruit, nuts, flowers and buds. **Source**: Stephen Dalton/NHPA.

management of habitats, ecosystems and metapopulations remains inadequate, particularly on private lands (Rohlf, 1991; Bean, 1998). In contrast, the European Habitat and Species Directive lists habitats that must be protected by the member states, while for species protected by the directive the emphasis is on ensuring favourable conservation status through active habitat management and restoration. In the UK, biodiversity action plans are developed for important habitats, or for individual species whose conservation can often be achieved through habitat conservation and wider landscape measures that will also benefit a suite of associated biodiversity.

Most conservation effort is geographically targeted. By 1993, 5.9% of the Earth's land surface had been designated in protected areas, and only 3.5% as strictly protected scientific reserves or national parks (Primack, 1995). Many existing national parks may be too small to protect 'viable' populations in the long term, and many other biodiversity-rich regions or habitats currently receive inadaquate protection. An important challenge is the development of objective yet practical techniques for the optimal selection of further geographic areas for conservation in a way that maximises biodiversity conservation benefits (Bibby, 1998). At a global scale, Balmford and Long (1995) identified some potential priority nations for further action, based on discrepancies between the area protected (or financial investment) and that expected from a regression of these measures on a biodiversity index. However, political borders often have little relationship to biologically meaningful boundaries: biogeographical regions provide a more rational framework for analysis of resource allocation (Bibby, 1998). Objective biogeographical assessment is made practicable by nested hierarchical classification of ecoregions and by analysing the current extent and configuration of habitats, current rates of conversion and degree of protection (Noss *et al.*, 1995; Bibby, 1998). Alternatively, conservation effort may be prioritised in 'hotspots' of outstanding species richness or in centres of endemism. For example, 20% of the world's bird species are confined to just 2% of the land surface (ICBP, 1992): concentrating conservation effort in such areas will give disproportionate benefits (see Box 5.4). However,

Box 5.4 Priority areas for global conservation: endemic bird areas

Figure 5.4.1: Locations of the 221 endemic bird areas (EBAs) defined by ICBP (1992).

In a global study of endemism in birds, the International Council for Bird Preservation (1992) identified priority regions for conservation action.

● Bird species whose geographical range was less than 50 000 km² were defined as 'restricted range species': these numbered 2609, or 27% of the world's birds.

● 51 000 individually located records of restricted-range species were mapped by GIS and used to define 221 endemic bird areas (EBAs), which embrace 95% of all restricted-range birds.

Box 5.4 (continued)

- 76% of EBAs are in the tropics.
- Habitat destruction is the greatest threat to restricted-range birds, affecting 62% of those currently threatened with extinction under IUCN criteria.
- Comparision with the distribution of protected areas revealed that, in more than half (133 of 221) of the EBAs, less than 5% of the area was protected.
- Priorities for action were defined by combining scores for the biological importance of each EBA and scores for the level of threat.

Threat score	Biological importance score			
	★★★	★★	★	Total
★★★	12	21	28	61
★★	46	28	25	99
★	31	19	11	61
Total	89	68	64	221

Score for importance and threat

★★★ = 79 EBAs ☐
★★ = 87 EBAs ☐
★ = 55 EBAs ☐

Figure 5.4.2: Numbers of endemic bird areas falling in different categories of biological importance and threat, and the method for combining them.

knowledge of species distributions needed to identify such areas is available for only a few well-known taxa, particularly birds and plants. The degree to which hotspots defined from one taxonomic group will effectively encompass less well-known taxa varies with the scale of resolution and with the taxa considered (Reid, 1998).

Successful ecosystem conservation must emphasise protection of large, interconnected landscapes (Noss et al., 1995). Huntley et al. (1997) suggest that conservation strategies should reflect important regional differences in the response of biodiversity to previous episodes of climate change. Paleo-ecological evidence shows that northwest Europe and eastern North America have had low-diversity floras and range shifts of thousands of kilometres. Here, biodiversity loss may be reduced by a network of habitats and corridors to facilitate migration. In contrast, tropical regions and lower-latitude mountainous areas have been characterised by a high diversity of specialist, range-restricted species and a qualitatively different response to climate change, dominated by changes in relative population size and local range shifts within topographically diverse landscapes. Huntley et al. (1997) suggest that, in such regions, the preferred strategy would be to set aside numerous and diverse protected areas that individually and collectively hold the widest possible variety of habitats. To minimise the hazards of migration as populations move in response to changing climate, new areas that have a high altitudinal diversity should be protected. Such areas would allow organisms to migrate altitudinally over short distances within the same park in response to climate change (Chapin et al., 1998). In Costa Rica, 7700 hectares of forest was set aside to form a corridor several kilometres wide providing an elevational link between two large conservation areas (Primack, 1995).

In some protected areas biodiversity may best be conserved by non-intervention. However, in many cases, 'leaving nature to take its course' will be less effective than informed ecological management. At one extreme, North American conservationists are often appalled, or downright derisory, when European colleagues attempt to explain the curious gardening they carry out in the name of nature conservation, whereby tiny patches of habitat, often of artificial origin, are carefully managed with the aim of retaining ecologically impoverished, early successional habitats and preventing woodland resurgence. But, in the short term, such approaches are the only possible means of conserving biodiversity in semi-natural habitats and landscapes that result from the interplay of ecology and human land use over millennia (see Box 5.5). However, it is increasingly recognised that this approach may not guarantee the long-term viability of specialist species trapped in isolated fragments of habitat in a warming world. Even among the well-studied fauna of Europe there is uncertainty in the precise habitat requirements of many species, particularly for invertebrates: a precautionary approach to biodiversity conservation would be to re-establish near-natural disturbance dynamics at large spatial and temporal scales. Habitat re-creation should be targeted at areas that will consolidate and link remnant fragments (Kirby, 1995). Attention is being given to reinstating dynamic processes of disturbance and regeneration that once operated on a landscape scale, such as catchment processes and free-ranging large herbivores. There is growing momentum for re-establishing populations of key vertebrates, a policy supported by the EU 1992 Habitat and Species Directive, which encourages member states to restore former indigenous species. For example, successful reintroduction programmes of Eurasian beaver have occurred in Norway, the Netherlands and Austria and re-establishment may soon follow in Scotland, and conservationists are now discussing the possibility of reintroducing the European bison to heathland landscapes in lowland England.

Box 5.5 Conservation in semi-natural habitats and agricultural landscapes

Extensive natural habitats and natural ecological processes are now virtually absent from many terrestrial regions. For example, in Britain 75% of the land surface is farmed, while the wildwood has been reduced to fragments of ancient woodland that occupy just 1.3% of the land (Ratcliffe, 1984). But components of the original biodiversity survived in 'semi-natural' habitats: modified vegetation structures controlled by human management but derived from the original species pool.

Traditional land use happened to create some of the *micro-habitats* and structures that once resulted from the dynamic ecological processes of disturbance and regeneration that had taken place in the natural landscape (flooding, trampling and grazing by large herbivores, forest fires, storms). In this way, human land use fortuitously provided conditions that many specialist species need for their survival and regeneration (see Figure 5.5.1). Valuable semi-natural landscapes are characterised by subsistence or low-intensity agriculture, with high ecological diversity at landscape scales and considerable species diversity. Although not natural, such landscapes often have high conservation value. Examples include:

- agro-forest systems found in Mediterranean *dehesa* and many tropical areas;

- extensive grazing and transhumance systems found in montane regions of Europe and elsewhere;

- high landscape diversity with fine-scale mosaics of semi-natural habitats created by subsistence economies; in Europe these include
  - stock grazing of lowland heathland, chalk grassland, neutral pastures and wet marshland;
  - extraction of resources from reed beds (to provide roofing materials), hay meadows (to provide fodder) and semi-natural woodlands (to provide timber and firewood).

Box 5.5 (continued)

Figure 5.5.1: In Eastern England, ecological succession taking place within shallow pools created by digging of peat for fuel in oligotrophic calcareous fens supports many plant species, including the threatened fen orchid *Liparis loeselii*. Such habitat is now being re-created using large machinery. In the past, ephemeral pools created by beaver dams and the activity of moose and European bison would once have provided a similar habitat. (Copyright: Richard Hobbs.)

The progressive shift from a subsistence rural economy to the market has led to

● abandonment of some habitats and systems of land use that are no longer viable;

● conversion by intensive agricultural practices that reduce landscape diversity and are highly damaging to the residual biodiversity (see Tables 5.2 and 5.3).

Conservation management

In Western Europe, most surviving fragments of semi-natural habitat are now protected by site-safeguard designations. But where such sites are left unmanaged, habitat structure changes through succession and many specialist and rare species are lost, usually to be replaced by common and widespread species gained from the surrounding impoverished species pool. Initial management during a *restoration phase* may involve severe measures, for example woodland clearance. Subsequent management aims to sustain characteristic habitat structures, by reinstating or mimicking traditional systems of land management, for example harvesting of reed, coppicing woodland, or stock grazing on grasslands and heathlands. Here, a number of conservation paradigms may be identified:

● Rotational management creates a dynamic mosaic of patches at different stages of succession and caters for species with different micro-habitat and structural requirements.

● Small-scale management patches facilitate dispersal of insects and reptiles.

● Structurally diverse interfaces cater for species with complex requirements.

This approach is epitomised by the management of European lowland heathland – an impoverished dwarf-shrub habitat of nutrient-poor, sandy soils (Dolman and Land, 1995). Such heaths support many specialist species, mostly invertebrate, plus a few vertebrates such as the sand lizard and smooth snake. They were created by forest clearance, ephemeral agriculture and extensive stock grazing. For thousands of years, access to the natural resources of heathlands was vitally important to local rural populations, who grazed their animals on the commons and cut gorse for fodder or firewood, turves for fuel, and heather and bracken for livestock-bedding and thatching. This land use set back succession, removed nutrients, created disturbance and perpetuated the habitat. Unproductive heathland is no longer viable in the modern agricultural economy, and catastrophic habitat loss and fragmentation have occurred through conversion to farmland, afforestation, urban development, or abandonment and ecological succession (see Table 5.2). Conservation

Box 5.5 (continued)

management is resource-intensive and can be costly. *Restoration management* involves clearing woodland, scrub and bracken by herbicide or cutting, and in some instances stripping soil and vegetation to remove accumulated nutrients. *Maintenance* can be conducted by rotational burning of heather on larger sites, or by cutting, but restoring grazing is the ideal. Rather than relying on commercial graziers, managers are increasingly using traditional breeds of cattle, sheep or ponies: these hardy animals are less productive than modern farm breeds but can survive on the poor forage and tackle regenerating trees. Creating or maintaining bare ground, sandy banks and vertical sand faces is important for habitat diversity, particularly for invertebrates, reptiles and early successional plants.

Despite their cultural importance and rich biodiversity, semi-natural habitats are a severely filtered and reassembled remnant of the original species pool of natural forested landscapes, with a bias to species of early successional habitats and a loss of species dependent on old growth. In Britain, recognition of the unique value of ancient wood pastures (that have had a continuity of woodland containing very old, decaying trees and still maintain assemblages of saproxylic invertebrates and fungi from the original primary forest) has awakened interest in restoring natural woodland dynamics.

Not surprisingly, European conservationists are often envious of the vast 'wilderness' areas remaining in the national parks of North America. But in reality, most of these are far from natural and are still reeling from earlier disruptive management and ecological degradation. The history of the US National Park Service reveals tension between materialist 'utilitarian conservationists' such as Pinchot, who sought sustainable consumptive use of the national resource, and 'aesthetic conservationists' such as Mather, who commodified scenic beauty, game and fisheries (Sellars, 1997). The combination of their anthropocentric values (see Box 5.3) led to spraying forests with insecticide, control of predators, manipulation and artificial stocking of fish populations, severe overgrazing and proliferation of introduced exotic species. Until the 1970s, the US National Park service attempted to suppress forest fires (a policy also pursued in Canada), greatly modifying forest structure and regeneration dynamics (Keiter and Boyce, 1991). National Park Service policies have now been redesigned to permit ecological processes, including fires and predators, to prevail. The grey wolf, eradicated from the Greater Yellowstone ecosystem through aggressive predator control, was finally reintroduced in 1995. But new tensions have arisen between proponents of 'ecosystem management' and passionate advocates of non-intervention (Keiter and Boyce, 1991; Grumbine, 1994). One of the most controversial issues in recent decades has been the use of management-ignited fires to reduce hazardous fuels and mimic natural fire dynamics. Implementing ecosystem management (or natural process management) requires the recognition of ecosystems that defy conventional political boundaries, necessitating inter-agency co-ordination of management policy, and the accommodation of ecological imperatives with long-standing economic interests (Keiter and Boyce, 1991).

Management of public lands in North America is beset by controversies, for example those concerning rangeland degradation in the western USA, logging in Oregon, Washington and British Columbia, or the débacle over the 'Lands for Life' proposals for Ontario. Each of these involves unresolved conflict between economic interests and extractive industries and cultural values imposed by a wider society (Keiter and Boyce, 1991). Such problems are magnified in the protected area systems of many developing countries, where external Western values often set the conservation agenda. Here, the first national parks were created within the mindset of the colonial annexation of land and its natural resources (Pearce, 1991): in some instances neo-colonialism is now reinforced by the economics of wildlife tourism. Exclusion of local people from decision making and from the benefit streams flowing from use of natural resources in protected areas favours unregulated 'illegal' extractive uses, such as game harvest, stock grazing, logging and 'poaching'. This threatens the integrity of many national parks. Where park authorities attempt to enforce regulations, conflict may arise, as graphically illustrated by the history of the Kenyan National Park Service and the killing of park wardens in India.

> *Wildlife conservation programmes in the third world have all too often been premised on an antipathy to human beings. In many countries, farmers, herders, swiddeners and hunters have been evicted from lands and forests which they have long occupied to make way for parks, sanctuaries and wildlife reserves. This prejudice against people is leading to new forms of oppression and conflict. Biologists, who seek to preserve wilderness for the sake of 'science', have been a major force in fomenting such prejudice.*
> (Guha, 1997, 14)

Biosphere reserves, created by the UNESCO Man and Biosphere Programme, are intended to serve as models demonstrating the compatibility of conservation and sustainable development of benefit to the local people. By 1994, 312 reserves had been created, covering approximately 1.7 million km^2 (Primack, 1995). But commentators such as Neumann (1997) remain highly critical of many integrated conservation–development projects (ICDPs), particularly those of the World Bank and WWF, arguing that they are often coercive, extend state authority into remote areas, fuel conflict over land rights, and reflect outdated First World attitudes concerning 'primitive peoples'. An example of an apparently successful biosphere reserve is the Kuna Yala Indigenous Reserve, Panama (Primack, 1995). The protected area comprises 60 000 ha of tropical forest, but also 60 Kuna villages with a population totalling 30 000. The Kuna practise traditional medicine, farming and forestry and also regulate the activities of external scientists who visit to study both the forest and the people, collecting research fees and insisting on the employment of local guides. The Kuna also control the rate and type of economic development that takes place within the reserve. This appears to be a perfect model for the empowerment of local peoples and their potential to take control of their own destiny as well as the environment. However, the exposure to outside material influences is inevitably changing the cultural values and beliefs of the younger generation, who are beginning to question the need to rigidly protect the reserve.

Conclusions

We as a species are systematically transforming and impoverishing the biological functioning of the planet. Chapter 19 examines why this is the case, and what may be done to avert this avoidable disaster, suggesting that a 'nature's conscience' may be triggered in a new form of citizenship education. But, as shown in Chapter 2, poverty and desperation are powerful forces for transformation, as much as those of technology and wealth. Environmental and ecological problems cannot be resolved in isolation: we cannot hope to reduce the loss of ecosystems and species unless we also address fundamental questions of human well-being, poverty and disempowerment (Hildyard, 1991).

References

Balmford, A. and Long, A. (1995) Across-country analysis of biodiversity congruence and current conservation effort in the tropics. *Conservation Biology*, **9**, 1539–1547.

Barber, C.V. and Pratt, V.R. (1998) Poison and profits: cyanide fishing in the Indo-Pacific. *Environment*, **40**(8), 4–9, 28–34.

Bean, M.J. (1998) Endangered species, endangered act? *Environment,* **41**(1), 12–18, 34–38.

Bengtsson, J., Jones, H. and Setälä, H. (1997) The value of biodiversity. *Trends in Ecology and Evolution*, **12**, 334–336.

Bibby, C.J. (1998) Selecting areas for conservation. In W.J. Sutherland (ed.), *Conservation Science and Action*. Blackwell Science, Oxford, 176–201.

Blum, E. (1993) Making biodiversity conservation profitable: a case study of the Merck INBio agreement. *Environment*, **35**(4), 16–20, 39–43.

Bond, W.J. and van Wiglen, B.W. (1996) *Fire and Plants*. Chapman & Hall, London.

Brown, K.R. and Adger, W.N. (1993) *Forests for international offsets: economic and political issues of carbon sequestration*. Working Paper GEC 93–15, CSERGE, University of East Anglia, Norwich.

Brown, K.R. and Moran, D. (1993) *Valuing biodiversity: the scope and limitations of economic analysis*. Working Paper 93–09, CSERGE, University of East Anglia, Norwich.

Bugman, H. (1997) Gap models, forest dynamics and the response of vegetation to climate change. In B. Huntley, W. Cramer, A.V. Morgan, H.C. Prentice and J.R.M. Allen (eds), *Past and Future Rapid Environmental Changes: the Spatial and Evolutionary Responses of Terrestrial Biota*. NATO ASI Series, Vol. I 47. Springer-Verlag, Berlin, 441–453.

Chapin III, F.S., Sala, O.E., Burke, I.C., Grime, J.P., Hooper, D.U., Lauenroth, W.K., Lombard, A., Mooney, H.A., Mosier, A.R., Naeem, S., Pacala, S.W., Roy, J., Steffen, W.L. and Tilman, D. (1998) Ecosystem consequences of changing biodiversity. *Bioscience*, **48**, 45–52.

Cohen, J.E. (1997) Population, economics, environment and culture: an introduction to human carrying capacity. *Journal of Applied Ecology*, **34**, 1325–1333.

Cramer, W.P. and Leemans, R. (1993) Assessing impacts of climate change on vegetation using climate classification systems. In A.M. Solomon and H.H. Shugart (eds), *Vegetation Dynamics and Global Change*. Chapman & Hall, New York, 190–217.

Cramer, W.P. and Steffen, W. (1997) Forecast changes in the global environment: what they mean in terms of ecosystem responses on different time scales. In B. Huntley, W. Cramer, A.V. Morgan, H.C. Prentice and J.R.M. Allen (eds), *Past and Future Rapid Environmental Changes: the Spatial and Evolutionary Responses of Terrestrial Biota*. NATO ASI Series, Vol. I 47. Springer-Verlag, Berlin, 415–426.

Daily, G.C. (1997). *Nature's Services: Societal Dependence on Natural Ecosystems*. Island Press, Washington DC.

Dodds, G.W., Appleby, M.J. and Evans, A.D. (1995) *A Management Guide to Birds of Lowland Farmland*. Royal Society for the Protection of Birds, Sandy.

DoE (1993) *Countryside Survey 1990: Summary Report*. Department of the Environment, London.

Dolman, P.M. and Land, R. (1995) Lowland heathland. In W.J. Sutherland and D.A. Hill (eds), *Managing Habitats for Conservation*. Cambridge University Press, Cambridge, 267–291.

Embley, T.M., Hirt, R.P. and Williams, D.M. (1995) Biodiversity at the molecular level: the domains, kingdoms and phyla of life. In D.L. Hawksworth (ed.), *Biodiversity: Measurement and Estimation*. Chapman & Hall, London, 21–33.

Erenfeld, D. (1978) *The Arrogance of Humanism*. Oxford University Press, Oxford.

Farrell, L. (1989) The different types and importance of British heaths. *Botanical Journal of the Linnean Society*, **101**, 291–299.

Grumbine, R.E. (1994) What is ecosystem management? *Conservation Biology*, **8**, 27–38.

Guha, R. (1997) The authoritarian biologist and the arrogance of anti-humanism: wildlife conservation in the Third World. *The Ecologist*, **27**, 14–19.

Gujja, B. and Finger-Stich, A. (1996) What price prawn? Shrimp aquaculture's impact in Asia. *Environment*, **38**(7), 12–15, 33–39.

Harper, J.L. and Hawksworth, D.L. (1995) Preface. In D.L. Hawksworth (ed.), *Biodiversity: Measurement and Estimation.* Chapman & Hall, London, 5–12.

Henderson, S. (1998) Climate change may destroy coral reefs. *Marine Pollution Bulletin*, **36**, 320.

Heywood, V.H. (ed.) (1993) *Flowering Plants of the World.* Batsford, London.

Hildyard, N. (1991) Liberation ecology. *The Ecologist*, **21**, 2–3.

Huntley, B., Cramer, W., Morgan, A.V., Prentice, H.C. and Allen, J.R.M. (1997) Predicting the response of terrestrial biota to future environmental changes. In B. Huntley, W. Cramer, A.V. Morgan, H.C. Prentice and J.R.M. Allen (eds), *Past and Future Rapid Environmental Changes: the Spatial and Evolutionary Responses of Terrestrial Biota.* NATO ASI Series, Vol. I 47. Springer-Verlag, Berlin, 487–504.

ICBP (1992) *Putting Biodiversity on the Map: Priority Areas for Global Conservation.* International Council for Bird Preservation, Cambridge.

Jennings, S. and Kaiser, M.J. (1998) The effects of fishing on marine ecosystems. *Advances in Marine Biology*, **34**, 201–351.

Johnson, K.H., Vogt, K.A., Clark, H.J., Schmitz, O.J. and Vogt, D.J. (1996) Biodiversity and productivity and stability of ecosystems. *Trends in Ecology and Evolution*, **11**, 372–377.

Keiter, R.B. and Boyce, M.S. (eds) (1991) *The Greater Yellowstone Ecosystem: Redefining America's Wilderness Heritage.* Yale University Press, New Haven, CT.

Kirby, K. (1995) *Rebuilding the English Countryside: Habitat Fragmentation and Wildlife Corridors as Issues in Practical Conservation.* English Nature Science, No. 10. English Nature, Peterborough.

Lawton, J.H. and May, R.M. (eds) (1995) *Extinction Rates.* Oxford University Press, Oxford.

Lohman, L. (1991) Who defends biological diversity? Conservation strategies and the case of Thailand. *The Ecologist*, **21**, 5–13.

Lowe, P.D. (1983) Values and institutions in the history of British nature conservation. In A. Warren and F.B. Goldsmith (eds), *Conservation in Perspective.* Wiley, Chichester, 329–352.

Markham, A. (1996). Potential impacts of climate change on ecosystems: a review of implications for policymakers and conservation biologists. *Climate Research*, **6**, 179–191.

Naess, A. (1986) Intrinsic value: will the defenders of nature please rise? In M.E. Soule (ed.), *Conservation Biology.* Sinauer, Sunderland, MA, 504–515.

NCC (1984) *Nature Conservation in Great Britain.* Nature Conservancy Council, Peterborough.

Neumann, R.P. (1997) Primitive ideas: protected area buffer zones and the politics of land in Africa. *Development and Change*, **28**, 559–582.

Noirfalise, A. and Vanesse, R. (1976) *Heathlands of Western Europe.* European Committee for the Conservation of Nature and Natural Resources, Council of Europe, Strasbourg.

North, R.D. (1995) *Life on a Modern Planet: a Manifesto for Progress.* Manchester University Press, Manchester.

Noss, R.F., LaRoe, E.T. and Scott, J.M. (1995) *Endangered Ecosystems of the United States: a Preliminary Assessment of Loss and Degradation. Biological Report 28.* US Department of the Interior, National Biological Service, Washington DC.

Patterson, A. (1991) Debt for nature swaps and the need for alternatives. *Environment*, **32**(10), 4–13, 31–32.

Pearce, D.W. (1993) Saving the world's biodiversity. *Environment and Planning A*, **25**(6), 755–760.

Pearce, F. (1991) *Green Warriors: the People and the Politics behind the Environmental Revolution.* Bodley Head, London.

Pimm, S.I., Russell, G.J., Gittleman, J.L. and Brooks, T.M. (1995) The future of biodiversity. *Science*, **269**, 347–350.

Posey, D.A. (1997) Protecting indigenous peoples' rights to biodiversity. *Environment*, **38**(8), 6–9, 37–42.

Pounds, J.A., Fogden, M.P.L., Savage, J.M. and Gorman, G.C. (1997) Tests of null models for amphibian declines on a tropical mountain. *Conservation Biology*, **11**, 1307–1322.

Prentice, I.C., Monserud, R.A., Smith, T.M. and Emanuel, W.R. (1993) Modeling large-scale vegetation dynamics. In A.M. Solomon and H.H. Shugart (eds), *Vegetation Dynamics and Global Change.* Chapman & Hall, New York, 235–250.

Primack, R.B. (1995) *A Primer of Conservation Biology.* Sinauer Associates, Sunderland, MA.

Ratcliffe, D.A. (1984) Post-medieval and recent changes in British vegetation: the culmination of human influence. *New Phytologist*, **98**, 73–100.

Ratcliffe, D.A. and Oswald, P.H. (eds) (1987) *Birds, Bogs and Forestry: the Peatlands of Caithness and Sutherland.* Nature Conservancy Council, Peterborough.

Raustiala, K. and Victor, D.C. (1996) Biodiversity since Rio: the future of the Convention on Biodiversity. *Environment*, **38**(4), 16–20, 37–45.

Reaka-Kudla, M.L., Wilson, D.E. and Wilson, E.O. (eds) (1996) *Biodiversity II. Understanding and Protecting our Biological Resources.* Joseph Henry Press, Washington DC.

Reid, W. (1995) Biodiversity and health: prescription for progress. *Environment*, **37**(6), 12–15, 35–45.

Reid, W. (1997) Strategies for conserving biodiversity. *Environment*, **39**(7), 16–20, 39–43.

Reid, W.V. (1998) Biodiversity hotspots. *Trends in Ecology and Evolution*, **13**, 275–280.

Rohlf, D.J. (1991) Six biological reasons why the Endangered Species Act doesn't work – and what to do about it. *Conservation Biology*, **5**, 273–282.

RSNC (1991) *Losing Ground: Vanishing Meadows.* Royal Society for Nature Conservation, Lincoln.

RSPB, EN, ITE (1997) *The Wet Grassland Guide.* Royal Society for the Protection of Birds, Sandy.

Sellars, R.W. (1997) *Preserving Nature in the National Parks. A History*. Yale University Press, New Haven, CT.

Shiva, V. (1989) *Staying Alive: Women, Ecology and Development.* Zed Books, London.

Smith, T.M., Shugart, H.H., Bonan, G.B. and Smith, J.B. (1992). Modelling the potential response of vegetation to global climate change. *Advances in Ecological Research*, **22**, 93–116.

Snyder, G. (1990) *The Practice of the Wild.* North Point Press, New York.

Soule, M.E. (1986) Conservation biology and the real world. In M.E. Soule (ed.), *Conservation Biology*. Sinauer Press, Sunderland, MA, 1–12.

Sykes, M.T. (1997) The biogeographic consequences of forecast changes in the global environment: individual species' potential range changes. In B. Huntley, W. Cramer, A.V. Morgan, H.C. Prentice and J.R.M. Allen (eds), *Past and Future Rapid Environmental Changes: the Spatial and Evolutionary Responses of Terrestrial Biota.* NATO ASI Series, Vol. I 47. Springer-Verlag, Berlin, 427–440.

Taylor, D. (1997) Saving the forests from trees: alternative products from woodlands. *Environment*, **39**(1), 6–11, 33–36.

Temple, S.A. (1998) Easing the travails of migratory birds. *Environment*, **40**(1), 6–9, 28–32.

Tubbs, C. (1985) *The Decline and Present Status of the English Lowland Heaths and their Vertebrates.* Nature Conservancy Council, Peterborough.

Tucker, G.M. and Heath, M.F. (1994) *Birds in Europe: their Conservation Status.* Birdlife International, Cambridge.

Walker, B. and Steffen, W. (1997) An overview of the implications of global change for natural and managed terrestrial ecosystems. *Conservation Ecology* [online] **1**(2), 2. URL: http://www.consecol.org/vol1/iss2/art2

Chapter Six

Population, adaptation and resilience

W. Neil Adger and Timothy O'Riordan

Editorial introduction

There is a vexed and endless argument as to what precisely causes environmental degradation on a global scale. Some argue that the main culprit is population growth, others overconsumption by the rich, others still blame ill-advised transfers of technology, and yet others look to distortional pricing and ill-defined property rights leading to misuse of the commons and excessive consumption of privately owned resources. An extension of these rights into formal law would, they argue, force nations, corporations and individuals the world over to recognise such rights in all aspects of development. Critics counter that such formal legal arrangements would stifle creativity and initiative, and encourage legalistic bureaucratic arrangements. Others see a malaise of misapplied ethics corroding the soul into materialism and self-centred acquisition.

All these forces are at work. A truly sustainable society has characteristics that would be regarded as intolerable to most modern populations, except, possibly, to those wholly untouched by Western 'civilisation'. There would be no innovation in technology that would lead to non-replenishable resource use or to excessive acquisition of wealth. Where accumulation did take place this would be shared with those less fortunate on the presumption that the flow of support would be reversed if fortunes also reversed. Such a society would also operate only within the limits of self renewal, or it would work to create new resources and habitats out of the profits of removing existing non-renewable or non-substitutable resources.

The ethical nature of the concept of sustainable development, based on sustainable communities, presumes that a part of all property rights is for the common good, inextricably bound up with the principles of precautionary stewardship, as outlined in

Chapter 1. Any environmental impact incurred when altering the land, or the adaptability of individuals or groups to tolerable change, would have to be compensated for by some form of equivalent sharing of that common good.

Underlying this discussion is the crucial issue of how many people, following what consumption patterns, and with what moral outlooks, can the Earth handle. Because science cannot tell us, we have to work this out for ourselves. In Chapter 1 we noted that this process will have to be localised and thoroughly participatory, and conducted in such a way as to incorporate equity considerations and social learning. Constructing appropriate institutions for the process is one of the major challenges in all nations and communities.

More vexing is the matter of increasing divergence of income and opportunity between the rich and the poor, both globally and in most nations. The share of global income accruing to the poorest 20% fell from 2.3% to 1.4% between 1980 and 1996, while the richest fifth rose in their wealth share from 70% to 88% in the same period (Carley and Spapens, 1998, 161). In every OECD country, the gap between the top and bottom 20% in terms of asset ownership has widened over the same period. According to the UNDP World Development Report 1998 (29–30), in Brazil the poorest 50% of the population received 18% of the national income in 1960, falling to 11.6% in 1995. The richest 10% rose in their proportion from 54% to 63% during the same period. In Russia, the poorest 20% of the population require two or three sources of income just to survive at all, and are 11 times poorer than the top fifth of the population. So underlying the population–consumption issue is that of equity and access to the global commons, linked to redistribution of wealth and increases in economic opportunity. Whether redistribution can, or indeed should, be done at the expense of further wealth creation for the rich remains unanswered: so far, the sustainability debate has not addressed this matter at the highest political levels.

This is why the consumption and political contexts of resource use issues raised in Chapter 2 are central questions. One cannot address the population question without a much clearer idea about the role for science, for participatory democracy, and for politically framed valuation. Hence the positioning of this chapter at this point in the book as a whole. Only when the context of human population, adaptation and resilience is accounted for is it possible to address the climate change issue, hence its location in the chapter that follows.

The ideology and polarisation of population issues

The size and structure of the global human population is one of the critical factors in determining the sustainability of environment and society into the 21st century and remains one of the most polarised and ideological of debates in the latter half of the 20th century. In particular, some aspects of global environmental change have resurrected debates on population 'crises' and the spectre of Malthus, who in the 18th century foresaw only the misery of famine and societal collapse as a result of population pressure. There is no doubt that the global human population is growing at an unprecedented rate. However, the headline figures are not as important as the changing population structure. Admittedly, a steep population distribution pyramid with a high proportion of young people (Figure 6.1(a)) for some countries which tend to have low *per capita* incomes, guarantees population growth and associated detrimental impacts on demand for land for agriculture and settlement. A more even pyramid (as in Figure 6.1(b)) signifies an ageing population which, as average life expectancy creeps upwards towards 85 years, also presents major economic and social problems to those societies experiencing them. By 2020, around a fifth of the Western European and Japanese populations will be over 70 years old. This will place a potentially enor-

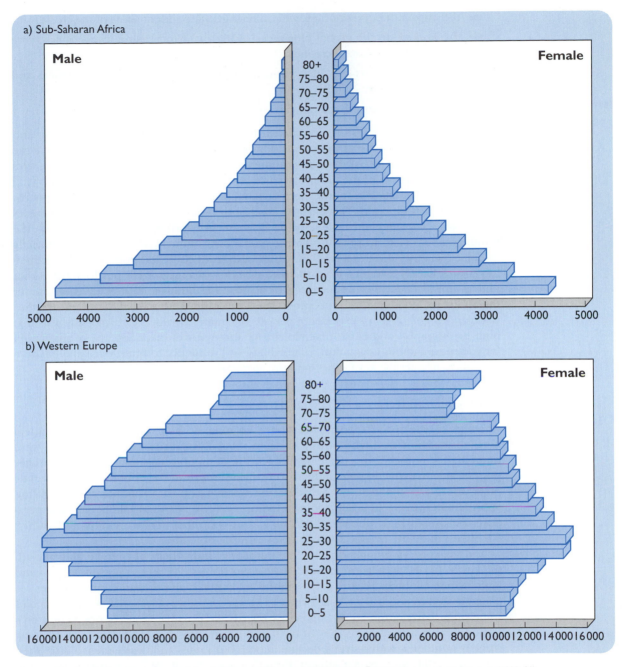

Figure 6.1: Population structures in (a) sub-Saharan Africa, and (b) Western Europe. *Source*: MacKellar *et al.* (1998, 99).

mous burden on private savings, or on publicly supported care provision, and on the nature of the productive economy in general. This particular time bomb still has to be diffused as Western nations grapple with a competitive global economy, the need for high levels of public and private investment, and the potentially crippling effects on the public sector in financing provision of caring for the elderly and the unproductive members of the population. So the major population issue is not only the absolute size of the global human population, but rather:

● the impact of population growth on resource-dependent areas of the world;

● the impact on the global biosphere of high and increasing levels of resource consumption and expectations;

● the societal means by which populations adapt to changing circumstances and resource pressures; and

● the opportunities for change in the overall population trends.

This chapter reviews these issues. First it looks at the so-called 'numbers' issue and assesses the current state of the population transition. The population transition is the reduction in overall population growth rates as a result of falling birth and death rates witnessed over the centuries in many parts of the industrialised world. The following sections review political responses to the global population challenges and the underlying issue of consumption. This includes an examination of what it is in modern societies that produces an unquenchable appetite for material consumption that, along with the numbers issue, leads to pressure on the global environment. The chapter then briefly examines whether there is a crisis in world food production and population by examining contradictory evidence from the past decades. The argument put forward is that human population is not undermined by an absolute limit on food production; rather the access or 'entitlement' of people to food and other resources determines whether populations go hungry or experience periods of famine. Finally, the chapter suggests that both demographic and environmental change form part of the landscape of adaptation for societies, and this adaptation, vulnerability and resilience to both social and environmental change are the key factors in future sustainable societies.

The population transition

Changes in the world's population have an elegant biological and mathematical inevitability associated with birth and death rates. Even if the world's birth rate were immediately reduced to replacement level, the demographic momentum of younger populations in many regions of the world will still result in the doubling of the world's population over the next few decades. The world's population is currently doubling approximately every 40 years, adding a billion every 10 years, and the younger generations are already born who should create a future population of 8.5 billion by 2020 (*cf.* 6.0 billion in 1998), irrespective of any foreseeable outcome.

The structure of populations and the number of individuals of child-bearing age are the key to future population growth. This proportion is different at present in low- and high-income countries (Figure 6.1), with the result that high population growth rates are concentrated in countries whose profile is closer to Figure 6.1(a), mainly in Africa and some other parts of the South. Figure 6.2 reveals the possible range of population projections on various assumptions of birth and death rates for 2100. The most realistic curve is number 2, with a strong decline in both birth rates and death rates to 2025 then constant fertility, or zero net growth. Even that puts the global population at twice the present number. Yet it has been calculated that for a sustainable economy at present Western levels of consumption, a population of 2.5 billion would be the maximum for an average standard of living equivalent to that found in Spain today (Arizipe *et al.*, 1992, 64). Such calculations depend enormously on assumptions about consumption, technology, resilience and vulnerability, by culture, location and economy. It is difficult to talk of carrying capacity or, indeed, population tolerance if these concepts are ignored.

Carrying capacity underlies many discussions of human population and its limits at the global scale, yet it is a term which is widely used and misunderstood. As Zaba and Scoones (1994) have shown, carrying capacity 'in the popular imagination is generally

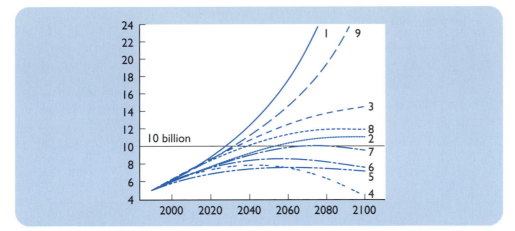

Figure 6.2: Population projections for the year 2100 based on various assumptions about the timing and rates of fertility decline and mortality. The very high rate (1) is simply an extension of existing trends for 1985–1990, and the next highest (9) assumes constant birth rates in Africa and Asia, but a 10% increase in mortality. The lowest projection (4) assumes rapid fertility decline to 1.4 by 2025, and the next two up (5 and 6) relate to the rapid onset of replacement fertility at 2.1, beginning in 1990 (5) with medium mortality, or in 2025, (6) with constant mortality rates. (2) assumes strong fertility and mortality decline until 2025, then constant; (3), UN fertility decline 25 years delayed, UN medium mortality; (7) UN mortality decline 25 years delayed, fertility rate of 2.1 in 2025; (8) life expectancy of 80/85 years and fertility rate of 2.1 in 2025). The chances of having a population lower than twice that of today in 100 years time are less rather than more likely, depending, of course, on the success of Agenda 21, and many other socio-economic and environmental factors.

Source: Arizipe *et al.* (1992, 63).

associated with vague notions of faraway places which are unable to produce enough food for their growing populations' (p. 197). They show that carrying capacity is simply the maximum population that can be sustained indefinitely in a given environment, but that this has been translated in many ways by those commentators arguing that the Earth will reach its carrying capacity for humans. In reality, a single number for carrying capacity is not a useful concept, dependent as humans are on their environments in complex ways. Kenneth Arrow and colleagues (1995, 521) go further and argue that the term is 'meaningless because the consequences of both human innovation and biological evolution are inherently unknowable'. They argue instead that a focus on the resilience of the ecosystems on which humans depend is more appropriate.

The latter part of the 20th century has witnessed a unique take-off in human population numbers (MacKellar *et al.*, 1998, 95). It took all of human history to 1800 to create the first billion: the sixth billion was created in just a decade, 1987–1997. This trend is the consequence of medical advance, greater food availability, improved health care generally and some significant changes in the educational, cultural and economic status of women. The demographic transition for a country such as Sweden is shown in Figure 6.3, common to much of Europe in the period from the late 18th to the mid-20th centuries. The transition is not automatic and has been triggered by different mechanisms for many countries. It is described in three stages. The initial stage (associated with the pre-industrialisation era in Europe) is characterised by both high birth rates and high death rates. The second stage of reduction in death rates, particularly of infant mortality, typically occurs in response to economic and social gains, with the third stage of reduced birth rates coming about as a result of complex social change including universal literacy and changing social aspirations.

McMichael (1993) argues that the currently observed stalling of many poorer countries between the second and third stages (the so-called 'demographic trap'), which normally

Figure 6.3: The demographic transition for Sweden and for a typical poor country.

Source: adapted from McMichael (1993, 115).

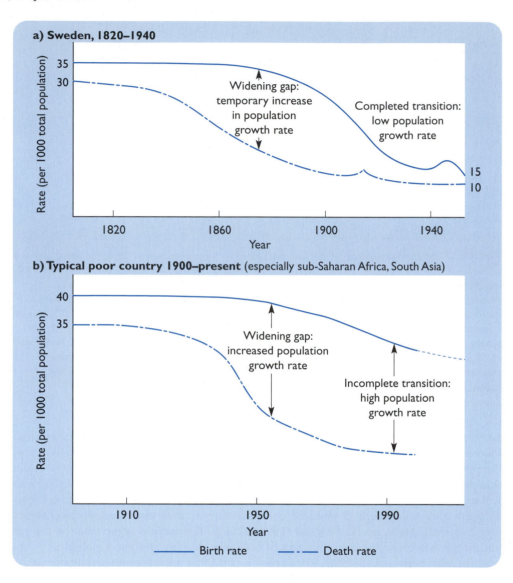

a) Sweden, 1820–1940

Widening gap: temporary increase in population growth rate

Completed transition: low population growth rate

Rate (per 1000 total population)

Year

b) Typical poor country 1900–present (especially sub-Saharan Africa, South Asia)

Widening gap: increased population growth rate

Incomplete transition: high population growth rate

Rate (per 1000 total population)

Year

——— Birth rate   — - — Death rate

propels population growth, is a consequence of poverty itself. This is because the parental motivations for large family size are primarily economic and social, and the economic opportunities necessary to bring about the social conditions for reduction of average family size are thus absent. In some poor countries such as Bangladesh and Mozambique, children work in the domestic economy from an early age, and schooling is frequently interrupted by demands from the household to earn some additional income by 10 years old. The completion of the transition in Europe in the 19th and 20th centuries was fortuitous in that there was a strong base of improved nutrition, education and public infrastructure concurrent with economic growth, partly based on colonial expansion. The latter of these factors also provided an outlet for emigration from the countries of Europe, as well as the impacts of war, resulting in the completion of this transition (McMichael, 1993, 114–116). The challenges facing the world's poor countries, where death rates have been tackled to an extent but where birth rates are still high today, are much greater.

Table 6.1 shows that the pattern of birth rates has generally slipped behind that of death rates, especially in Africa. Over the period 1950–55 to 1990–95 life expectancy as a

Table 6.1: Global demographic trends since 1950. The key is the decline in reproductive rates in the past decade or so. This is due partly to better health care for women and children, and partly to more successfully integrated family planning programmes. The fertility rates in the developing world remain the focus of attention, and a routeway to more serious effort at improving equity and justice for disadvantaged groups.

	Total population (millions)			Growth rate			Life expectancy (both sexes)			Total fertility rate		
	1950	1970	1990	1950-55	1970-75	1990-95	1950-55	1970-75	1990-95	1950-55	1970-75	1990-95
World	2520	3697	5285	1.78	1.96	1.57	46.4	57.9	64.4	4.97	4.88	3.10
More industrialised	809	1003	1143	1.20	0.81	0.40	66.5	71.2	74.4	2.77	2.36	1.70
Less industrialised	1711	2695	4141	2.05	2.37	1.88	40.9	54.6	62.3	6.13	5.96	3.48
Africa	224	364	633	2.23	2.56	2.81	37.8	46.0	53.0	6.64	6.67	5.80
Asia	1403	2148	3186	1.90	2.27	1.64	41.3	56.3	64.5	5.86	5.06	3.03
Europe	549	656	721	0.96	0.60	0.15	66.1	70.8	72.9	2.56	2.14	1.58
Latin America and Caribbean	165	283	440	2.68	2.44	1.84	51.4	61.1	68.5	5.87	4.98	3.09
North America	166	226	278	1.80	1.10	1.05	69.0	71.5	76.1	3.47	2.01	2.06
Oceania	13	19	26	2.21	2.09	1.54	60.8	66.6	72.8	3.84	3.21	2.51

Source: Rahman and Huq (1998, 94).

whole increased from 40.9 to 62.3 years, while total fertility rate (numbers of children per child-bearing woman) fell from 6.0 to 3.5. The critical point here is the rapidity of the improvement in life expectancy of women and children in some poor countries. This has led to the classic and steep pyramid-style demographic structure in poorer countries, as depicted in Figure 6.1(b), with ten times the number of children under the age of 5 compared with adults over 60. Because such children are already born, it is relatively easy to estimate future populations over the next 20 years. The forecasts, as evident in Figure 6.2, diverge after that because of the huge uncertainties in socio-economic assumptions, in health patterns, and in the underlying social position of women.

The emergence of novel disease patterns, and the re-emergence of diseases which were thought to have been checked in the past, present enormous potential challenges for some countries. These may include AIDS and HIV, 'diseases of poverty' even in relatively rich countries, and changes in vector-borne diseases as a result of climate change (see Cliff and Haggett, 1995; McMichael *et al.*, 1997). Higher mortality rates from malaria or AIDS may check total population size in the short run, but the impact of the reduction in health and well-being of the adult population, particularly the workforce, increases the number of 'dependants' in a population and makes people feel less secure. This is illustrated in Box 6.1 for the phenomenon of AIDS and HIV infection in the southern Africa region.

Box 6.1 AIDS and HIV in southern Africa

One of the greatest social and economic threats to southern Africa is a continuation of the uncontrolled spread of the immune virus HIV, coupled to continued high ratios of tuberculosis. Both trends are likely to have very serious economic consequences for the size and effectiveness of the workforce, and for the cost of human suffering, often borne by families already impoverished and by hapless orphans with no chance of economic security.

According to a report by Claire Bisseker (1997, 1), more than 2.5 million South Africans are infected with the HIV virus, and in many areas over one in four women below 20 suffer from the generic disease of HIV infection. This is a particularly troublesome issue, for it falls within the zones of poverty, historical racism, gender relations in different South African cultures, and the under-capacity of local health services to treat, pre-treat and educate all those involved. The scale of the effort required is simply enormous.

Table 6.1.1 summarises the incidence of AIDS cases in the Southern African Development Commission (SADC) region for the period 1995–1996. There is obviously some variation

in accuracy of the reporting data, and indeed there is some evidence that only a quarter of all cases are actually recorded. Loewenson and Whiteside (1997, 10–11) caution against drawing hasty conclusions from these figures. They conclude:

● The number of HIV-positive people among sexually active populations should not exceed 20–40% at its peak.

● The spread of HIV becomes exponential when about 2% of the sexually active population is infected.

● The period of exponential spread to the peak of infection is comparatively short, about 10 years.

● The epidemic may then become endemic with about 10% infected at any one time.

● Current prevention programmes may not have stopped the spread of the epidemic, but they have saved many thousands of lives, and hence are cost-effective for the present.

The overall conclusion of those studying the problem is that AIDS will reduce future popula-

Box 6.1 (continue)

Table 6.1.1: The incidence of AIDS in southern Africa.

Country	Reported AIDS cases (as at 1995/1996)	AIDS rate per 10 000 of population	Population size estimate
Angola	895[1]	0.78	11 404 445
Botswana	3 451	25.75	1 340 300
Lesotho	936	4.53	2 065 608
Malawi	44 775	42.56	10 519 867
Mauritius	132	1.05	1 253 677
Mozambique	3 118	1 74	17 878 125
Namibia	5 101	33.08	1 542 000
South Africa	10 351	2.49	41 551 800
Swaziland	590	5.23	1 127 500
Tanzania	82 174	27.51	29 870 000
Zambia	34 000	35.52	9 570 760
Zimbabwe	54 744[2]	46.23	11 840 544

Note: Computation of AIDS rates was estimated by employing growth rates to calculate estimates of the population at last date of reporting AIDS cases for each specific country.

1 UNAIDS Data, December 1996.
2 HIV, STI and AIDS Surveillance, Zimbabwe Ministry of Health and Child Welfare, 1st Quarter Report 1996.

Source: *Weekly Epidemiological Record* No. 27, World Health Organisation, 5 July 1996.

tion growth in poor African countries almost to the point of stability. This is serious enough in demographic terms because of possible transmissibility from one generation to another. But much more serious is the prognosis that the remaining population will be less economically active with greater real cost to the local health services. On the economic front, forecasts are hazardous because of the nature of weak parts of the African economy: even with a healthy population it is difficult to forecast growth trends. Add the unknowns of both illness and loss of productive capacity, plus the even more unknown costs of likely health care, and one can readily see that the HIV and AIDS issue has severe social as well as economic consequences. Loewenson and Whiteside (1997, 20–22) make a best guess at 25% depression of economic output over the next 20 years. This significant impact falls on economies with low *per capita* earnings, low levels of investment and savings, and insufficient capacity to provide public services. The number of working-age people may fall by as much as 20% over the next 20 years in southern Africa. With a concomitant rise in the costs of worker training and further education generally, the costs of maintaining an employable population may rise by as much as 500% over the same period.

MacKellar and his colleagues (1998, 110) conclude that the best-guess estimate for 2050 is a population total between 8.1 billion and 12.0 billion, and for 2100 a range of 5.7 billion to 17.3 billion. Summarising their position:

● Virtually every scenario forecasts an additional 2 billion people over the next three decades.

● Over 90% of this growth will be in the developing world.

● Seven countries will account for over 80% of this increase, namely China, India, Nigeria, Indonesia, Bangladesh, Pakistan and Brazil. Social and economic change in those nations will have a huge influence on global population as the growth rate diminishes.

- Smaller numerical increases will take place in the poorest nations, but the environmental and social burden in such countries will be much greater, especially if present trade patterns continue to discriminate against them.

- Ageing populations are a major issue. Even under pessimistic assumptions, over a fifth of the whole population will be over 60 by 2100, with a more likely estimate at 26%. In the developed world, the proportion of people over 80 will increase to nearly 15% by 2100, with potentially enormous implications for pensions, health care costs, individual family caring and housing needs.

- Changing lifestyles are proving to be the greatest imponderable in forecasting future population numbers and consumption habits. Diet, participation in the labour force and reducing gender inequality all have huge implications for numbers and timing of children. One variable which is a major determinant of demographic change is educational opportunities for women (see also Box 6.2).

Box 6.2 The Cairo Conference

The third UN Population Conference . . . was a watershed global event. It succeeded in both shifting concern about world demographics into gender-sensitive, people-centred sustainable human development and propelling sensitive and ideologically charged population issues into the public domain. (Chen *et al.*, 1995, 5)

The American population analyst Lincoln Chen charted the changing themes of each of the three UN International Conferences on Population and Development:

- 1974 Bucharest: the developing world wanted development as the 'best contraceptive'.

- 1984 Mexico City: the developing world sought health and social uplift as the solution.

- 1994 Cairo: the developing world was pushed to accept women's social and economic liberation.

The difference lay primarily in the changing demographics and the new participatory role of a wide range of non-governmental organisations. The developing world faces huge population pressure with the glint of a downturn in fertility rates. Health, education and women's rights are the focal concerns of over 2000 NGOs which were active at Cairo, assisted by prominent women politicians.

The result was a World Programme of Action that placed the central focus on sustainable development, safe reproductive rights (health care and adequate childbirth facilities), women's empowerment through gender equity and economic uplift (including micro-credit schemes), and greater individual freedom to choose sexual intercourse and pregnancy. These were hard-won agreements. Cultural and religious fundamentalists disputed the scope for independent choice, and gender rights were often played down in favour of male and female responsibilities.

The programme is estimated to cost $17 billion over the period 1993–2000, rising to $27.5 billion by 2015 (Sen, 1995, 36). Of this, $10 billion is earmarked for family planning, $5 billion for reproductive health and $1.3 billion for HIV/AIDS treatment. But these resources have not been allocated in the period since the conference, nor is there significant co-ordination between child welfare programmes, primary health care, primary education attainment and community family planning clinics, nor between the UN agencies, developing nation states and NGOs.

The subsequent UN Conference on Women, held in Beijing in 1996, did not significantly push these agendas along. The two intervening years saw a regrouping of interests opposed to

Box 6.2 (continued)

the rapid emancipation of women, and the restriction of Northern aid, for ideological purposes, for family planning and abortion clinics.

In the light of trends highlighted at the Cairo Conference, the aim of campaigning groups and agencies was to ensure adequate aid, capacity building and economic freedom for women's rights in a world still struggling to come to terms with sustainable development. These issues are patently important for human development and more holistic for sustainability as well as merely for the issue of population.

Education raises women's status within the household, enabling them to follow their own, usually lower, desired family size rather than that of their partner. Parents' decisions to endow girls with even a modicum of education, as opposed to none at all, reflects a fundamental shift in the fairness with which women are treated. It sets in motion an irreversible shift in women's perceptions, ideals and aspirations. (MacKellar *et al.*, 1998, 118–119)

Population, consumption, needs and wants

A common approach for estimating the global impact of human population on the environment is the equation

Impact (I) = Population $(P) \times$ Affluence $(A) \times$ Technology (T)

Such approaches have been used to appraise carrying capacity, arguing that inevitable consumption increases (A) mean that population (P) should be limited. Although it has already been highlighted that carrying capacity is not a useful concept in this area, nevertheless some interesting estimates do emerge from such analysis. Vitousek *et al.* (1986), for example, show that the biophysical limits to economic activity are indeed real. They hypothesise that the Earth's net primary production (NPP) is the total food resource of the biosphere (being the basis of all maintenance growth and reproduction of living organisms), and hence is one absolute limit on human activity. They show that approximately 40% of NPP is currently being appropriated directly or indirectly for human consumption, and argue that this appropriation is detrimental to the survival of other species which depend on these ecological processes. Consumption and technology are obvious leverage points for minimising the impact of human populations and also interact in determining present and future fertility patterns. This is the message of the UN Human Development Report (UNDP, 1998, 81–84), namely that with the application of 'clean technology' and the 'knowledge economy' it should be possible to create wealth and lessen the environmental footprint.

Figure 6.4 outlines a visual variant of this equation, comparing the total materials flows of a German citizen and an inhabitant of a poor developing nation. The figure demonstrates that the former consumes natural assets at an average rate five times the equivalent demand from the impoverished and excluded. The challenge remains as to how future populations can increase the security of their livelihoods while at the same time not overwhelming the resilience of the environment on which they depend. This is why the consumption issue is so problematic environmentally and politically. At its very essence lie four crucial bones of contention:

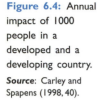

Figure 6.4: Annual impact of 1000 people in a developed and a developing country.

Source: Carley and Spapens (1998, 40).

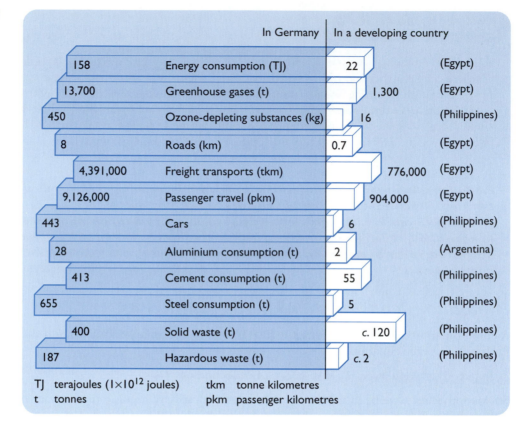

	In Germany	In a developing country	
Energy consumption (TJ)	158	22	(Egypt)
Greenhouse gases (t)	13,700	1,300	(Egypt)
Ozone-depleting substances (kg)	450	16	(Philippines)
Roads (km)	8	0.7	(Egypt)
Freight transports (tkm)	4,391,000	776,000	(Egypt)
Passenger travel (pkm)	9,126,000	904,000	(Egypt)
Cars	443	6	(Philippines)
Aluminium consumption (t)	28	2	(Argentina)
Cement consumption (t)	413	55	(Philippines)
Steel consumption (t)	655	5	(Philippines)
Solid waste (t)	400	c. 120	(Philippines)
Hazardous waste (t)	187	c. 2	(Philippines)

TJ terajoules (1×10^{12} joules)   tkm tonne kilometres
t   tonnes   pkm passenger kilometres

● How to develop technology for the economies of the developing countries that does not further increase South dependency on North technology and corporate hegemony.

● How to share the 'intellectual property rights' of technological progress so that the appropriate mix of technology and social institutions for the South is 'owned' by the South (see, for example, Posey, 1996; Wirth, 1997 for contrasting views).

● How to organise a trading regime that respects the different environmental and social capabilities of producer nations as responsible elements of process and production methods, so that sustainability considerations enter the global trading pattern. This is a particularly difficult issue as, currently, the World Trade Organisation does not properly acknowledge environmental and social rights in its guidelines and legalities of international trade (Brack, 1998). Thus it is quite possible for trading agreements to undermine sustainability objectives and to worsen the capability of a trading nation to create a consumption pattern suitable for its culture and population structure.

● How to couple aid to sustainable capacity building in such a way that future populations are empowered in ways that meet the needs of reproductive health and community enterprise.

The anthropologist Mary Douglas and her colleagues (1998, 197) define 'wants' as an essential construct of neo-classical economics, and the reflection of consumer rationality and diminishing preferences as more and more of the same goods and services become available:

Wants are the desires of a rational being: they are ordered logically in a hierarchy of claims on resources. The ordering has no guaranteed connection with needs. The economist does not have the official doctrine as to how wants arise.

(Douglas *et al.*, 1998, 197–198)

Yet, as Douglas argues, wants are both economically and ideologically defined. Poor people subsist: they may have needs but they cannot afford to purchase or pursue them. In sociological and psychological terms, however, such needs are vital for the wholeness and the stability of the person. Wants are not a function of personal survival: they are determined by external prescription; commodities do not satisfy desire, they are only the tools for satisfying it. The economics of consumption is the pattern of authority and social alliances made manifest and distinctive by the circulation of goods. So goods become emblems of social relations and social norms.

Tackling patterns of consumption according to these arguments is not simply a matter of resource-use efficiency but involves recognition that consumption is a social construct. The consumption element of the analysis is inextricably part of the power relations and social networks of complex societies experiencing the biophysical limits of economic activity. This discussion raises the relationship between opportunities, expectations, globalism generally, and the advertising culture. Wants and needs cannot be dissociated in the modern age of global consumer reach. Self-esteem is as much a factor in determining both the ability to consume and the personal demand for 'basic needs', as is the advertising blitz that often demeans individual self-respect. It is not possible to conceive basic needs and personal well-being outside the complicated patterns of informal and formal social support and personality.

Population and food availability

One of the major potential threats associated with rising population is the possibility of large-scale food shortage at the global level. A number of concerned commentators have pointed out that the aggregate agricultural production of the world per head of population has not been sustained since the Green Revolution of the 1960s: in other words, population is outstripping food production (e.g. Brown, 1993). However, a contribution to this debate by Tim Dyson (1994, 1996) demonstrates that this Malthusian fear may be ill-founded on a global scale, though it certainly requires careful analysis in different regions of the world.

Consider the trends in *per capita* cereal production for the major regions of the world in Figure 6.5, which are based on data compiled by the FAO and reported in Dyson (1994). There does appear to be a decline in *per capita* cereal production in many regions of the world, even when based on 5-year moving averages (lines in these figures), which smooth out the annual variation. The volatility of the annual points is due to changes in regional cereal harvest, since population changes are relatively smooth. The variation in production is due to the impact of climate variability. It is also argued that the variation in world cereal production has been rising over time because of the reliance on fewer varieties (see Hazell, 1985). The greater that reliance, the greater the probability that adverse climate conditions or a virulent pest will wipe out a crop over a larger area. Thus the increased homogeneity associated with the Green Revolution and with loss of agro-diversity generally may have negative consequences for world food security. But the global trend in *per capita* cereal production, accounting for about half of total human calorific intake, is down between the early 1980s and early 1990s by about 4% at around 358 kg *per capita* in 1991 (Dyson, 1994).

Figure 6.5:

Per capita cereal production.

Sources: FAO, Dyson (1994).

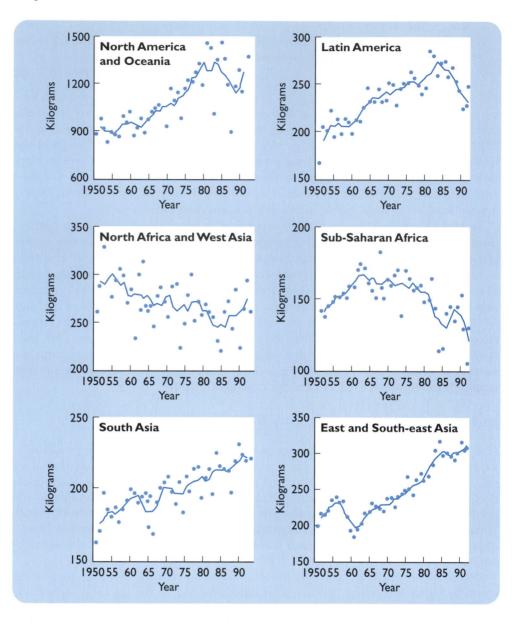

The explanation for this trend depends, however, on a number of complex political-economy factors rather than changes in population *per se* (summarised in Table 6.1). In short, the 1980s decline in production in North America and Oceania of 2.2% per annum across the decade is the major factor in this trend. The countries in this region contain 6% of the world's population and produce over 20% of its cereals. The high levels of domestic subsidy for cereals production in the European Community and the USA in that decade kept production higher than domestic needs in those countries. This subsequently depressed world market prices for cereals due to the subsidies paid for exports of cereals from Europe and the USA (see Ritson and Harvey, 1996, for an account). Part of the response to this crisis in these regions was to take land out of agricultural production, thus reducing the area harvested *per capita* by over 3% (Table 6.2). The impact of this market distortion is seen in the other cereal-exporting coun-

Table 6.2: Regional measures of population and cereal production, 1980–92. Although *per capita* production has been decreasing, this is in part accounted for by reduction in harvested area over the decade. Yield improvements continue in all regions.

Region	Population, 1990 (millions)	Population, growth rate, 1980–90 (% per year)	Total cereal production, 1992 (m tonnes)	Average per capita cereal production, 1987–92 (kg)	Measures of cereal production annual % change 1981–86 to 1987–92		
					Per capita production	Area harvested per capita	Yield
North America and Oceania	303	1.0	428.3	1185	-2.2	-3.1	0.9
Latin America	441	2.1	114.1	245	-1.8	-2.8	1.1
North Africa and West Asia	272	2.8	70.0	262	0.8	-1.4	2.2
Sub-Saharan Africa	502	3.0	54.6	133	-0.5	-2.6	2.0
South Asia	1191	2.3	276.3	220	0.5	-2.6	3.0
East and South-east Asia	1794	1.6	567.2	305	0.5	-1.4	1.9
Europe and USSR	790	0.5	441.8	601	0.1	-1.6	1.7
World	5295	1.7	1952.2	356	-0.7	-2.3	1.6

Note: Figures cover 98% of total population representing all countries with populations over 5 million in 1990. The same figures are represented on the six graphs in Figure 6.5 except for Europe and the former USSR, which follow the same trend as North America and Oceania due to the predominance of domestic market intervention in determining cereal production.

Source: adapted from Dyson (1994).

tries such as in Latin America, where the observed trend in reduced *per capita* food production, as seen in Figure 6.5, can be explained by the impact of US and EC policies on production in Argentina (Dyson, 1994). Similarly, part of the decline in regions such as Western Asia are a result of increased cheap imports in this period. No doubt there are bleaker spots: the decline in sub-Saharan Africa in Figure 6.5 is exacerbated by the impact of the mid-1980s and early 1990s droughts in that region.

Overall, the Malthusian spectre of reduced food availability cannot be concluded from these data, because the areas of the world where the majority of people live still experience rising *per capita* food production. Part of the explanation of overall global decline is the reduction of land area associated with the impacts of the agricultural policies of industrialised nations, and the reduction in land area under cereals (in the data presented) caused by switching to other food crops. Diets are becoming more diverse and hence cereals are not such an important part of the world's calorific intake.

Do these data suggest that at the global scale the world is likely to go hungry or to run out of resources to meet its basic needs? This is the threat which has been implicit since the work of Malthus in the late 1700s. He suggested that population growth, being exponential, would eventually outstrip any possible rise in food production, which tends to rise in a linear fashion. In the intervening centuries, neo-Malthusians have taken this message and updated it to encompass all resources (see Findlay, 1995, for a review).

Thus the population–food scarcity issue requires greater elaboration. The population–poverty relationship remains critical for many parts of the world. This is less to do with simple demographics than with the distribution of consumption and wealth. The arguments concerning famine as an outcome of population growth have long been recognised by social scientists to be spurious. The causes of famine are numerous but are as often associated in this century with civil strife and conflict as with absolute food availability (e.g. Devereux, 1993; Chambers, 1989).

Resilience and vulnerability

Access to food, in contrast to its production, is the most important explanatory variable in food security and the resilience of populations. This has been developed into a theory of *entitlements* by Amartya Sen (1981). He argues that entitlements are actual or potential bundles of commodities which individuals can access and that most famines are caused by circumstances of entitlements failure caused by human political action. Thus he explains how all the major famines of the past century, such as the Bengal famine beginning in 1943 and the famine in China in 1958–60 were a result of lack of access to food, not of absolute scarcity. Drawing on these insights, the underlying vulnerability of societies to the poverty and resources issue in the context of population pressure becomes important. Under this approach, it is social groups and individuals who are vulnerable to changes in their socio-economic and environmental circumstances, while adaptation to such changes provides opportunities through diversification or migration (see Adger, 1999, for an example of the application of this approach). An alternative geographical perspective hypothesises that regions and areas can be defined as vulnerable or critical based on the environmental and socio-economic pressures on them, stemming from the work of Jeanne Kasperson and colleagues (1995). They argue that identifying *criticality* brings about opportunities to overcome adversity or stress through the focused application of technology and new institutional devices. The onset of criticality is a manifestation of ill-adapted institutional arrangements, acting without precaution or care for the vulnerable.

How societies adapt when there are external pressures on them leads to consideration of the ecological terms *resilience* and *fragility*. Resilience in ecology as outlined in Chapter 4 is defined in one of two ways: the ability to withstand change, or the capacity to restore and replenish following some externally imposed shock. Fragility, or susceptibility, is the degree of change associated with human-induced stress. The key to this concept is the coupled notion of ecological buffering and social learning. Buffering is the in-built, or managed, capability of the natural fabric to be restored, either by its own capability or through human intervention. We have seen in Chapter 5 how sensitive management of ecosystems can reduce their fragility. Thus the coupling of both human resilience and ecological sensitivity into a single, interactive totality might help to increase the carrying capacity of the planet. Improvements in civic science, participatory democracy and understanding of ecological economic relationships would allow this broader framing of resilience to be placed in a sustainability framework.

Vulnerability is a measure of the enforced exposure to critical stress, or hazard, combined with the restricted capacity to cope. Note that vulnerability is a function of

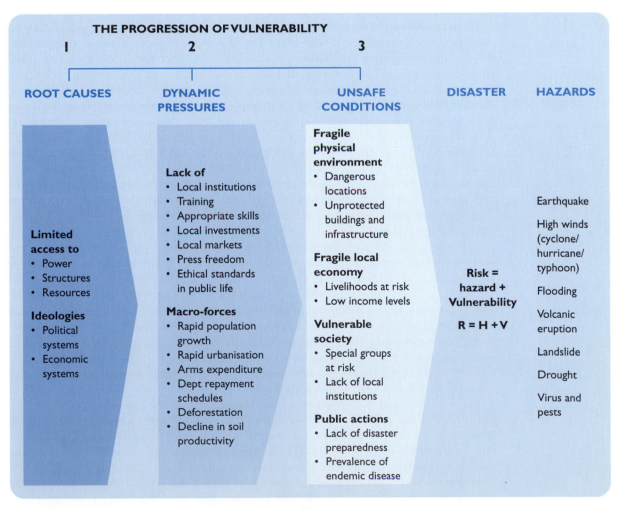

Figure 6.6: Links between vulnerability and exposure to natural hazards.
Source: Blaikie *et al.* (1994, 109).

powerlessness: it is created as people face phenomena beyond their control, and, at times, their comprehension. Vulnerability can arise out of second-dimension power through which the exposed are deliberately kept ignorant of their danger or their positioning. Many risks are the result of people being wilfully misinformed for a variety of political and commercial reasons (see Chapter 17). Figure 6.6 shows how vulnerability and exposure to natural hazard intertwine. Piers Blaikie and his colleagues (1994, 20–42) document in some detail the links between population growth, urban concentration, natural resource degradation and vulnerability. They seek to show that vulnerability is a political and ethical issue, not a natural state. Global economies, corporate interests, property rights abuses and distorted information flows combine to create conditions where environments become critical, and people vulnerable.

Criticality is the coupling of a propensity of an ecosystem to degrade with vulnerability, that is, a human circumstance that reduces the capacity to cope with institutional reform as an ecosystem degrades. Like vulnerability, criticality is the coupling of a natural process with a social process. Because criticality is a function of vulnerability, so vulnerability is a prior condition. The aim of sustainability is to reduce, and hopefully eliminate, vulnerability through equity and justice considerations, so that the conditions that give rise to criticality are lessened.

As vulnerability is lowered and criticality reduced, so resilience increases. But in an ecological sense, resilience relates to the functioning of the system rather than the stability of the component populations. Economists and other social scientists also argue that resilience is the key to sustainability in the wider sense (Common, 1995; see Chapter 4). Certainly resilience is related to stability, but it is not clear whether these characteristics are always desirable, for example in evolutionary terms. Social resilience is connected to ecological resilience (see Adger, 1997, for a review) in a synergistic way. The resiliences of social systems are themselves related in some way, though still undefined, to the resilience of the ecological systems on which they depend. This is most clearly exhibited within social systems which are dependent on a single ecosystem or single resource. Taking the concept of resilience from the ecological sciences and applying it to social systems is controversial. One point of legitimate argument is that there are indeed essential differences in socialised institutions compared with ecosystem behaviour.

One way forward is to examine the relationship between ecosystem functioning and socio-economic dependence. We note this in Chapter 13 with regard to agro-ecosystems. Another way, identified in Chapter 1 and further developed in Chapters 3 and 17, is that of institutional persistence, based on trust, legitimacy and accountability. This is the basis of one definition of social fabric, namely the capability to observe and learn. This is why we take such an interest in property rights and the inclusion of commons responsibilities. The social fabric holds together in a pattern of norms and behaviours that knits the household to the community, and the community to the state. If these bonds are fragmented, social resilience breaks down and vulnerability crops up in exposed and disadvantaged areas.

There is the issue of dependency of social systems on the environment itself. The point here is the degree to which communities and institutions, which are directly dependent on natural resources, are themselves linked to the resilience of the ecosystem. In short, are there direct linkages between ecological and social resilience? In this sphere, research in the areas of human ecology are relevant. In determining the parallels between social and ecological resilience, potential indicators for the concept focus on the links between social stability (of populations within social systems) and resource dependency. Resource dependency is defined by the reliance on a narrow range of resources leading to social and economic stresses within livelihood systems.

An example of the resilience of institutions can be found in the ability of communities to deal with the forces of globalisation which impinge upon them. Social capital, ecological resilience and social resilience are all tested when upheaval and stress are placed on institutions: an example here is the management of traditional collective resources. Commonly managed resources are being degraded throughout the world through the breakdown of property rights or inappropriate privatisation (Berkes and Folke, 1998). Nowhere is this more clear than in tropical coastal resources such as fisheries, mangroves in the intertidal zone and other assets, as discussed in the example in Box 6.3.

The dependency of individuals within a resource system does not necessarily depend on reliance on a single crop or fish stock, but in some circumstances on dependence on an integrated ecosystem. This is particularly the case with coastal resources, as argued by Bailey and Pomeroy (1996) in the context of coastal regions of Asia:

fishing communities are best understood as dependent not on a single resource but on a whole ecosystem. This expanded understanding of tropical coastal resources is the key to stability for households and communities in South East Asia's coastal zones.   (Bailey and Pomeroy, 1996, 195)

Box 6.3 Vulnerability and resilience and collective resources in coastal Vietnam

This box illustrates how changing property rights and institutions affect social resilience. In Vietnam, the system of government changed during the 1990s from central planning towards a market-oriented economic policy. Part of this trend has been the privatisation of assets which were previously managed either by the state through co-operatives or traditionally under common property rights, particularly in lowland and coastal Vietnam. Mangroves are one of the coastal resources which have been privatised to a large extent. They provide a variety of functions from protecting coastlines against a rising sea level to offering food and sanctuary for young fish populations, to cleansing sediment and nutrient from polluted coastal rivers. Traditionally, mangrove areas were commonly owned and managed, and because their value was collectively recognised, they were cared for.

As the mangrove is converted to food production or to commercial aquaculture, so changes in indicators such as property rights and income distribution highlight changes in resilience and vulnerability for coastal communities. Studies have shown that the poorer households rely on the mangrove more for their livelihood than the wealthy, who are more engaged in the private commercial activity utilising the coastal resources. The ability of the community to maintain sustainable common property management of the remaining mangrove and fishing areas is undermined by changes in the property rights and changes in inequality brought about by externally driven enclosure and conversion. Some 11% of all income generated in a case study area is dependent on the mangrove resources. At the household level, the poor, dependent families suffer disproportionately with the loss of the habitat functions. But at the community level there is also a loss of resilience as families compete for the remaining resources, leading to non-co-operation in use of the ecosystem. As some families gain in the new commercialism, others disproportionately lose since they have no fallback when the formerly commonly managed resource is denied them. By this means, vulnerability is created and resilience is undermined.

Source: based on Adger *et al.* (1997).

Box 6.3 examines the relationship between resilience and changing property rights in an example from coastal Vietnam. When an economy moves from collective to private through incorporation of commonly managed resources, the implications for livelihoods are very mixed for different groups in the community. Population numbers, therefore, become only part of the equation: institutional arrangements that determine vulnerability and resilience are actually more relevant for the transition to sustainability. These institutions are intensely location and culture-specific.

Migration is an important factor in resilience but is a phenomenon whose presence, confusingly, is cited both as evidence for instability and as a component of enhanced stability and resilience, depending on the type of migration examined. This confusion arises because migration occurs for a plethora of reasons, which are often classified as 'push' and 'pull' factors. Push migration is that movement of people caused by a deleterious state of affairs in the home locality (such as loss of assets), while pull migration is the demand to move caused by attractive circumstances elsewhere (often in urban areas). However, the important aspect of migration in the context of stability and resource dependency is that migration allows people to cope with increasing resource pressures by spreading risk in their livelihood strategies. Evidence of the impact of this urbanisation trend in both developed and developing countries demonstrates that the links between rural and urban economies remain strong. This may be facilitated by decreasing transport costs. However, the urbanisation processes themselves can create enormous social and environmental problems, particularly where the population growth is rapid (reviewed in Drakakis-Smith, 1995, 1996, 1997).

In summary, migration and population movement is one important aspect of the ability of societies to cope with demographic and other social and environmental stresses. Resilience, in both its social and ecological manifestations, is an important criterion for the sustainability of development and resource use. Each of these social and ecological aspects has several empirical indicators, but no single indicator captures the totality of resilience. Social resilience can be examined, for example, by reference to economic, demographic and institutional variables in both temporal and spatial fashions. The example in Box 6.3 also demonstrates the analysis of social resilience characteristics, acknowledging the difficulty in analysing resilience at different scales.

Population and adaptation

There is an inescapable link between patterns of democracy and ownership of development rights in defining and explaining vulnerability, adaptation and resilience. The valuation of ecological and social functions in a rigorous interdisciplinary fashion and their application to finding sustainable development paths is a major challenge for social and environmental sciences. We shall see in Chapter 19 that this also represents a major challenge to environmental education and, if interdisciplinarity is to be transferred from the academic world, to society as a whole.

This chapter has also shown the need to examine the ultimate adaptability of the most vulnerable humans in conditions of stress. Nowadays, 90% of those killed in modern conflict are innocent civilians, compared with 5% at the turn of the century (UNDP, 1998, 35–36). Over the past decade, armed conflict has killed 2 million children, disabled 4–5 million and left 12 million homeless. In addition, there are 8 million orphans associated with the AIDS/HIV pandemic, many of whom also live in conflict zones. By 2010, over 10% of children in areas most heavily affected by AIDS/HIV will have lost at least one parent. The UNDP's Human Development Report (1998, 35–37) highlights the impact of civil strife, violence and crime and its interactions with poverty and resilience. Armed conflicts have a habit of persisting: 20 years in Afghanistan, 15 in Sudan, 14 in Sri Lanka and 10 years in Somalia. It is in these poorest areas that military expenditure is soaring, and 100 million active landmines remain to be found. In South Africa, for example, the police are coping with over 380 known criminal networks, many linked to global networks. Well over 90% of criminal acts are not brought to justice in South Africa, so it is not surprising that much serious violence goes unreported:

> *Conflict destroys years of progress in building social infrastructure, establishing functioning government institutions, fostering community led solidarity and social cohesion, and promoting economic development. When conflicts finally peter out and the death tolls are tallied up, countries face formidable challenges of reconstruction and reconciliation requiring resources far beyond their reach as they emerge from years of destruction.* (UN Development Programme, 1998, 36)

All these factors undermine social resilience. Co-operation at all levels of governance (formal and informal) is required if this cycle of despair among the most desperately vulnerable is to be broken. This chapter has shown that simply focusing on the total number of the global human population does not answer the fundamental questions of where sustainability may lie and what the challenges are for adaptation to the populous planet we will experience in the next century. These are both ethical and scientific issues.

References

Adger, W. N. (1997) *Sustainability and social resilience in coastal resource use*. GEC Working Paper 97-23, CSERGE, University of East Anglia and University College London.

Adger, W. N. (1999) Social vulnerability to climate change and extremes in coastal Vietnam. *World Development*, **27**(2), 249–269.

Adger, W. N., Kelly, P. M., Ninh, N. H. and Thanh, N. C. (1997) *Property rights and the social incidence of mangrove conversion in Vietnam*. GEC Working Paper 97-21, CSERGE, University of East Anglia and University College London.

Arizipe, L.R., Costanza, R. and Lutz, W. (1992) Population and natural resource use. In J. Dooge, G.T. Goodman, T. O'Riordan, M. de la Riviere and M. Lefevre (eds), *An Agenda for Science for Environment and Development into the 21st Century*. Cambridge University Press, Cambridge, 61–78.

Arrow, K., Bolin, B., Costanza, R., Dasgupta, P., Folke, C., Holling, C. S., Jansson, B. O., Levin, S., Mäler, K. G., Perrings, C. and Pimentel, D. (1995) Economic growth, carrying capacity and the environment. *Science*, **268**, 520–521.

Bailey, C. and Pomeroy, C. (1996) Resource dependency and development options in coastal south-east Asia. *Society and Natural Resources*, **9**, 191–199.

Berkes, F. and Folke, C. (eds) (1998) *Linking Social and Ecological Systems*. Cambridge University Press, Cambridge.

Bisseker, C. (1997) South Africa's national AIDS review. *AIDS Analysis Africa*, **8**(3), 10–11.

Blaikie, P., Cannon, T., Davis, I. and Wisner, B. (1994) *At Risk: Natural Hazards, People's Vulnerability and Disasters*. Routledge, London.

Brack, D. (ed.) (1998) *Trade and the Environment: Conflict or Compatibility?* Earthscan, London.

Brown, L. (1993) *State of the World 1993*. Norton, New York.

Carley, M. and Spapens, P. (1998) *Sharing the World: Sustainable Living and Global Equity in the 21st Century*. Earthscan, London.

Chambers, R. (ed.) (1989) Vulnerability, coping and policy. Special issue of *IDS Bulletin*, **20**(2).

Chen, L.C., Fitzgerald, W.M. and Bates, L. (1995) Women, politics and global management. *Environment*, **37**(1), 4–9, 31–33.

Cliff, A. and Haggett, P. (1995) Disease implications of global change. In R. J. Johnston, P. J. Taylor and M. J. Watts, (eds) *Geographies of Global Change: Remapping the World in the Late Twentieth Century*. Blackwell, Oxford, 206–223.

Common, M. (1995) *Sustainability and Policy: Limits to Economics*. Cambridge University Press, Cambridge.

Devereux, S. (1993) *Theories of Famine*. Harvester Wheatsheaf, Hemel Hempstead.

Douglas, M., Gasper, D., Ney, S. and Thompson, M. (1998) Human needs and wants. In S. Rayner and E. Malone (eds), *Human Choice and Climate Change. Vol 1: The Societal Framework*. Battelle Press, Washington DC, 195–264.

Drakakis-Smith, D. (1995, 1996, 1997) Third World cities: sustainable urban development. Parts I, II and III. *Urban Studies*, **32**, 659–677; **33**, 673–701; **34**, 797–823.

Dyson, T. (1994) Population growth and food production: recent global and regional trends. *Population and Development Review*, **20**, 397–411.

Dyson, T. (1996) *Population and Food: Global Trends and Future Prospects*. Routledge, London.

Findlay, A. (1995) Population crises: the Malthusian specter? In R. J. Johnston, P. J. Taylor and M. J. Watts (eds), *Geographies of Global Change: Remapping the World in the Late Twentieth Century*. Blackwell, Oxford, 152–174.

Hazell, P. B. R. (1985) Sources of increased variability in world cereal production since the 1960s. *Journal of Agricultural Economics*, **36**, 145–159.

Kasperson, J. X., Kasperson, R. E. and Turner, B. L. (eds) (1995) *Regions at Risk: Comparisons of Threatened Environments*. United Nations University Press, Tokyo.

Loewenson, R. and Whiteside, A. (1997) *Social and Economic Issues of HIV/AIDS in Southern Africa: a Review of Current Research*. Southern Africa AIDS Information Dissemination Services, Harare, Zimbabwe.

MacKellar, L., Lutz, W., McMichael, A. J. and Subrke, A. (1998) Population and climate change. In S. Rayner and E. Malone (eds), *Human Choice and Climate Change. Vol. 1: The Societal Framework*. Battelle Press, Washington DC, 89–193.

McMichael, A. J. (1993) *Planetary Overload: Global Environmental Change and the Health of the Human Species*. Cambridge University Press, Cambridge.

McMichael, A. J., Haines, A., Slooff, R. and Kovats, S. (eds) (1997) *Climate Change and Human Health*. World Health Organisation, World Meteorological Organisation and United Nations Environment Programme, Geneva.

Posey, D. (1996) Protecting indigenous peoples' rights to biodiversity. *Environment*, **38**(8), 6–9, 37–45.

Rahman, A. and Huq, S. (1998) Coastal zones and oceans. In S. Rayner and E. Malone (eds), *Human Choice and Climate Change*, Vol. 2. Battelle Press, Columbus, Ohio, 145–202.

Ritson, C. and Harvey, D. (eds) (1996) *The CAP and the World Economy* (2nd edn). CAB International, Wallingford.

Sen, A. K. (1981) *Poverty and Famines: an Essay on Entitlement and Deprivation*. Clarendon Press, Oxford.

Sen, G. (1995) The World Programme for Action: a new paradigm for population policy. *Environment*, **37**(1), 10–15, 34–37.

United Nations Development Programme (UNDP) (1998) *Human Development Report*. Oxford University Press, Oxford.

Vitousek, P. M., Ehrlich, P. R., Ehrlich, A. H. and Matson, P. A. (1986) Human appropriation of the products of photosynthesis. *Bioscience*, **36**, 368–373.

Wirth, D. A. (1997) International trade agreements: vehicles for regulatory reform? *University of Chicago Legal Forum*, **1977**, 221–273.

Zaba, B. and Scoones, I. (1994) Is carrying capacity a useful concept to apply to human populations? In B. Zaba and J. Clarke (eds), *Environment and Population Change*. Ordina, Liège, 197–219.

Chapter Seven

Climate change

Timothy O'Riordan

Topics covered

- Climate change, metaphors of nature and integrated science
- The science of climate change, greenhouse effect and human interference
- Predicting future climate change
- Evaluating the impacts of future climate change to adjust the rate of change
- The politics of climate change and the Kyoto Protocol
- Equity, contraction and convergence of CO_2 emissions, and post-Kyoto politics
- Integrated assessments and community empowerment

Editorial introduction

Dealing with climate change is the ultimate test for environmental science in the round. Climate change is a truly global issue: one molecule of 'greenhouse gas' carries with it the same package of impacts irrespective of where or how it is emitted. And each additional molecule emitted today is not just everybody's concern. It creates a cumulative headache for four generations to come. The likely effects of human-generated climate change will, however, be experienced almost in inverse proportion to innocence and blame. Those peoples and countries which contribute most to the emissions of radiative forcing gases are, for the most part, least likely to be most inconvenienced, impoverished or physically vulnerable to the consequences of their behaviour. The inhabitants of small island nation states and impoverished coastal areas of larger nations living barely a metre above present sea level will, on the other hand, suffer progressive deterioration of their freshwater resources, their coastal tourism industry and, ultimately, their physical existence. Yet their contribution to the cause of their plight is minuscule.

Climate change therefore will test science, science–policy relationships, global environmental agreements, the economics of response, the politics of coalition building across interests and generations, the morality of individual 'lifestyle choices' in the light of the innocence–blame contradiction, and the collective ethics of responding, coping, adapting and sharing in a world that is not yet one. The coming decade will bring all these connections into a common strand of analysis and moral judgement.

In an interesting analysis of how the science of acid rain, ozone depletion and global warming evolved, Michael Kowolak (1993) reached the following conclusions.

- *The discoveries accumulated over a number of years,* in the case of acid rain for almost a century, climate change for 75 years and ozone depletion for over a decade. For the most part, this was a process of hypothesis testing, careful observation, painstaking theory building and measured publication. These conditions do not always pertain to the 'pressure-cooked', commercially funded and media-inquisitive science of today.

- *The breakthroughs came almost by accident,* but always in the form of a pool of data based on historical trends and supported by boring but vital monitoring and recording. The bias is somewhat different today. Research councils hard pressed for funding and assailed by clamouring science teams find it difficult to support long-term monitoring. Luckily, the advances in telemetry and automatic recording make it possible to receive reliable data from isolated places, including the surface of the oceans, at much lower running cost. Overall, the science of data collection has been transformed (see Chapter 2).

- *Scientists learn through networks of communication and experimentation.* None of these great global problems was diagnosed by a single individual, though in the history of discovery one or two names stand out. In all three cases, it was multidisciplinary workshops and communities of scientists that unravelled the mystery. Nowadays, the test is to integrate rather than multiplicate, for the modern mystery is achieving a socially just consensus in which science and policy and human choice couple inextricably.

- *Most of the science was only partially funded by governments.* A substantial part was privately paid for, and some was even undertaken at the scientists' own expense. Megabuck science tends to follow rather than precede great discoveries.

This suggests that, for science at least, truly unexpected discoveries are rare, though always possible. Bob Kates and Bill Clark (1996) examine the history of scientific surprises over the past 25 years. They conclude that surprise is necessary if science is to influence policy, and that surprise is an inevitable consequence of the interactions between humans and their environment. They quote the Canadian scientist Buzz Holling (1995): 'surprises occur when causes turn out to be sharply different than was conceived, when behaviours are profoundly unexpected, and when action produces a result opposite to that intended – in short when perceived reality departs quantitatively from expectation.'

Like Jasanoff and Wynne (1998) in probably the most comprehensive review of science and knowledge formation (see Chapter 1), Holling believes that these discontinuities between expectations and revelation are shaped by underlying metaphors, models and belief systems that are almost 'culturally accepted paradigms' in their own right. This means that they are rarely examined or even taken into consideration in the formulation and testing of theories. Holling proposes five such metaphors. These are important because, in one form or another, they underlie almost all predictive models of integrated analytical assessments of future states of global economies and climate change. They also form the basis of many 'scenarios' of future interpretations of economics, societies and democracies as described by Gallopin and Raskin (1998). In addition, they reflect part of the cultural theoretic approach to risk perception discussed in detail in Chapter 17. This relationship between science and scenario is now gripping the climate debate, so is given much attention in this chapter.

Holling's five metaphors are as follows.

- *Nature cornucopia*, namely a belief in smooth exponential growth where nature unfolds her riches in response to human imagination, innovation and competitive advantage.

● *Nature anarchic*, where nature is skittish, unpredictable and always likely to thump humanity on the head, and where close, small-scale, localised human–nature relationships can create creative equilibria.

● *Nature balanced*, where the name of the game is logistical and rational approaches to understanding and adaptation by seeking to 'know' how a turbulent future might evolve and be managed.

● *Nature resilient*, where nature is both brittle and malleable, predictable and discontinuous, but ultimately indestructible, though always subject to violent and at times catastrophic alterations.

● *Nature evolving*, which is a set of beliefs around co-evolution of humans and nature along the lines suggested in Chapter 1, namely the openness of revelation to create genuine partnership between humans and nature through coupling of evolutionary destiny.

This last metaphor is by far the most interesting for climate change policy. It presumes a systems coupling of behaviour, culturally interpreted expectations of future states of society and economy, and a physical complex of ocean–atmosphere–ice–biota interconnections, of multiple causation and diverse outcome. The causation comes from a slow build-up of a sequence of non-linear states that are full of surprises and discontinuities. This accumulation of historical interaction over centuries of co-evolution can produce short-term and catastrophic outcomes for, say, health or water availability that strain the vitality and peaceability of societies. Above all, the solutions have to be adaptive, evolutionary, learned and shared. According to Holling (1995, 34):

> *The problems are not amenable to solutions based on knowledge of small parts of the whole, or on assumptions of constancy or stability of fundamental relationships – ecological, economic or social. [Assumptions of constancy] produce policies and science that contribute to a pathology of rigid and unseeing institutions, increasingly vulnerable natural systems and [social] dependants.*

We shall see in this chapter that the 'concentration effect' of climate change politics has begun to integrate environmental science in the evolutionary ways outlined in Chapter 1 and summarised by Holling. The huge challenge of reaching 'contraction' of greenhouse gas emissions by a process of equitable 'convergence' of responsibility and accountability will dominate this science–policy–equity debate for the decade to come. In the process, a truly interdisciplinary environmental science has to come of age if this troublesome issue is successfully to be faced. And even if we collectively fail to deliver on the 'Kyoto promises' by 2010 (of which more below), then the monitoring of the subsequent effects on vulnerable societies and economies, as well as the most appropriate measures to increase their coping responses, will inevitably become part of the climate science family. So whichever way we turn, interdisciplinarity is on the move, prodded by outcomes we cannot escape.

The science of climate change

Climate change science is essentially the product of an amazingly inventive period of interdisciplinary natural systems science over the past 20 years. The technology of easy and inexpensive electronic communication, notably with the advent of email, plus the well-funded international air travel attached to research budgets, has revolutionised the character of scientific collaboration. One meeting creates a friendship which is intellectually sealed by subsequent email. Huge teams of motivated and skilled people can generate models, assessments, scenarios and peer review yet hardly move a muscle. This is why climate change science has such credibility, why it is so intensely politicised, and why major international groupings of recognised specialists continue to be well funded – at least for the time being.

The basic science of the 'greenhouse effect' is well known (see Box 7.1). So, too, is the global carbon budget (Box 7.2). The Earth is in long-term heat balance, receiving radiation from the Sun and losing the same amount of heat to space. If the Earth lacked an atmosphere the average surface temperature would be around –18°C. The current average global temperature is +15°C. This 33°C difference is caused by so-

Box 7.1 The greenhouse effect and the concept of climatic forcing

Figure 7.1.1 shows that there are two immediate controls on the temperature of the Earth, the incoming solar radiation and the insulating effect of the gaseous atmosphere and its clouds. Although not invariable, the energy from the Sun is almost constant and the main climatic changes today are the result of changes in the composition of the atmosphere. The diagram shows the incoming short-wave solar radiation set at 100 units, and 30% of this is reflected by clouds or the Earth's surface. Half the incoming radiation warms the Earth's surface, and this then radiates long-wave radiation (infared), which passes less readily through the atmosphere, leading to the warming of the lower atmosphere, so that there is a difference in temperature shown here as the 'greenhouse' effect of 33°C. If the absorption by the greenhouse gases increases, this differential will increase. Thus we can build up a warmer surface (including the oceans) and lower atmosphere, which can only occur if the outgoing energy is less than that arriving from the Sun. Thus currently the balance shown here is not present, and in energy units the Earth in space receives 240 watts for each square metre of its surface (W/m²) and radiates about 236 W/m². The surplus of about 4 W/m² of absorbed over emitted radiation (the calculated value for a sudden doubling of CO_2) is called radiative forcing. Over time, the warmer surface will cause a sufficient increase in emitted infared energy to bring the Earth back into balance with the incoming solar radiation – but with a warmer surface than before.

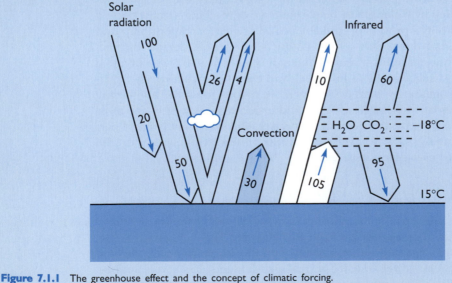

Figure 7.1.1 The greenhouse effect and the concept of climatic forcing.
Source: Jäger and Ferguson (1991, 53).

Box 7.2 The global carbon budget

The numbers in Figure 7.2.1 show the fluxes between reservoirs and the estimated amount of carbon within each reservoir in gigatonnes (10^9 tonnes). It will be noted that the fluxes balance – this is, however, not easy to achieve and requires a fairly high sink to the oceans and an uptake of CO_2 by increased photosynthesis by land vegetation to balance the estimated loss due to deforestation. Many estimates show a net loss from land vegetation and soils due to forest clearance and desertification. Although it may seem surprising that we cannot accurately account for the fate of the huge amount of fossil carbon burned each year, it will be seen that the quantity is a small proportion of the total flux and a very small proportion of the carbon tied up in the major reservoirs. In addition, fluxes across the ocean and the land surfaces are extremely difficult to measure on a global basis.

Source: Houghton *et al.* (1990).

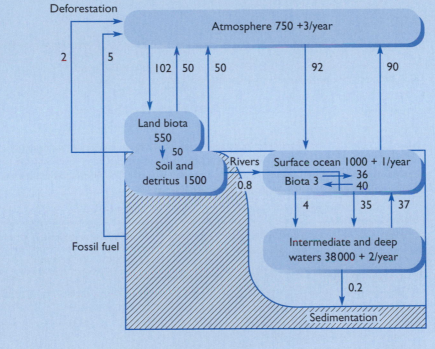

Figure 7.2.1 The global carbon budget.

called 'greenhouse' gases, which absorb the re-radiated long-wave radiation and warm the lower atmosphere as a result. The character of the effect is more of a thermal blanket than a greenhouse, but 'greenhouse gas' is a metaphor that is here to stay. The principal 'natural' greenhouse gases are water vapour, carbon dioxide, methane, oxygen and nitrous oxide. These five gases have been perturbed by human activity in the form of industrialisation and land-use change, and have been added to by a group of synthetic chemicals, notably chlorofluorocarbons, hydrofluorocarbons, perfluorocarbons and sulphurhexafluoride. Table 7.1 lists the evolution of the main greenhouse gases over the past 200 years, and Table 7.2 summarises the relative global warming potential of the main greenhouse gases.

A major theme for the scientific effort was to try to distinguish the 'signal' for any purely human-induced effect on climate change from the 'noise' of natural variability.

Table 7.1: The evolution of the principal greenhouse gases. This table provides the basis for both climate sensitivity calculations and for the equity analysis underlying the contract and converge scenarios. It is worth noting that these calculations are regarded as fairly robust in an arena where uncertainty abounds.

	CO_2	CH_4	N_2O	CFC-11	HCFC-22 (a CFC substitute	CF_4 (a perfluoro-carbon)
Pre-industrial concn	~280 ppmv	~700 ppbv	~275 ppbv	zero	zero	zero
Concentration in 1994	358 ppmv	1720 ppbv	312 ppbv[1]	268 pptv[12]	110 pptv	72 pptv[1]
Annual concn change[3]	+1.5 ppmv	10 ppbv	+0.8 ppbv	zero	+5 pptv	+1.2 pptv
Annual rate of change[3]	+0.4%	+0.6%	+0.25%	zero	+5%	+2%
Atmospheric lifetime	50–200[4]	12[5]	120	50	12	50 000

Notes:

1 Estimated from 1992–93 data.
2 pptv = parts per trillion (10^{12}) by volume.
3 The growth rates of CO_2, NH_4 and N_2O are averaged over the decade beginning in 1984; halocarbon growth rates are based on recent years (1990s).
4 No single lifetime for CO_2 can be defined because of the different rates of uptake by different sink processes.
5 This has been defined as an adjustment time which takes into account the indirect effect of methane on its own lifetime.

Source: IPCC (1997, 19).

Table 7.2: Global warming potential of selected greenhouse gases. The global warming potential is a measure of radiative forcing power set against a CO_2 baseline. Though somewhat controversial in subsequent meaning, these calculations are generally well regarded. The figures formed the basis of the agreements made over the 'six pack' of greenhouse gases at Kyoto.

Gas	Chemical formula	Lifetime (years)	Global warming potential		
			20 years	100 years	500 years
Carbon dioxide	CO_2	70–200	1	1	1
Methane	CH_4	12±3	56	21	6.5
Nitrous oxide	N_2O	120	280	310	170
HFC group	C-H-F	1.5–50	5000	3000	500
Sulphur hexafluoride	SF_6	3200	16 300	23 900	34 900
Perfluorocarbons[1]	C-F	3–10 000	6 000	8 000	14 000

Note:

1 Averaged over a number of formulations.

Source: IPCC (1997, 26).

This is not at all easy. The history of climate change can be studied in the paleo-botanical record of changing vegetation and, by inference, through the concentrations of CO_2 trapped in ancient ice sheets. Figure 7.1 provides the evidence of significant variations of global temperature over the ice ages, and the close correlation with CO_2 and methane levels. Subsequent evidence has found that variations in the uptake of CO_2 by plants, and by shallow seas in particular (see Chapter 8), have a profound interactive effect on the rates of CO_2 generation and sequestration, as part of a natural process of accommodation.

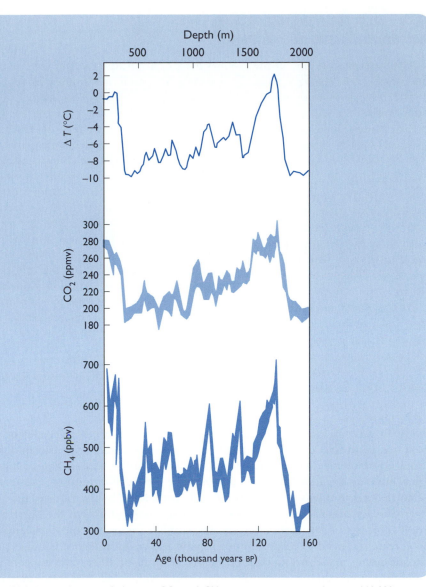

Figure 7.1: Global ice volumes and changing CO_2 and CH_4 concentrations over the past 160 000 years. The values for atmospheric carbon dioxide and methane were measured from bubbles of air trapped in the ice. The ice volumes are derived from measurements of ocean oxygen isotope values but can be broadly checked from measurements of past sea level. It will be seen that they move in sequence, but complex feedbacks are involved (larger areas of ice in glacial periods correspond with reductions in the area of tropical rainforest; cold periods mean less active vegetation; for example), so it would be unwise to regard the correlation between atmospheric carbon dioxide levels and temperatures as one of simple cause and effect. Thus it may not be used to predict the future global temperatures.

Source: Houghton *et al*. (1990).

Other natural influences affect climate. In Chapter 8 we note the huge effects of ocean currents, and especially the El Niño Southern Oscillation, in altering climate. More recently there has been a flurry of excitement over variations in sunspot activity and associated rates of solar radiation (see Box 7.3). Volcanic activity, and alterations in erosive processes affecting aerosol and dust concentrations, also affect the natural vari-

Box 7.3 Sunspots and global warming

Sunspots are patterns of variable energy on the solar surface, some 'dark' and relatively cool, some 'bright' and more radiative. Both atmospheric and oceanographic records show a variation in air and sea temperature in accord with the 10–12-year sunspot cycle, during which the Sun can get 0.1% brighter. The subsequent rise in sea surface temperatures of up to two to three times the norm may be due to sympathetic weather patterns adding further to the solar radiative effect. Other research indicates that increased sunspot activity leads to a higher level of stratospheric ozone formation, which in turn transfers heat towards the lower levels of the atmosphere. A third area of interest is the variation in magnetic flux, which improves the formation of ice particles and hence more rainfall, fewer lingering clouds and greater radiative loss. All this is still in its infancy, and the null hypothesis holds. But the point remains that separating human signals from natural noise in climate will continue to influence the debate, depending on what metaphors of nature are wheeled in to explain the relationship.

Another row is brewing over the accuracy of 'ground-truth' temperatures of the planet, recorded by satellite. Via microwave audio technology, the satellites gauge the radiation given out by oxygen molecules in the atmosphere. Over time, the satellites lose altitude due to friction in the upper atmosphere. This means that a correction has to be inserted into the measurement. Because of the enormous sensitivity about accuracy, the precise formula for this correction is a matter of dispute. After all, the calibration of the general computer models depends on extremely precise temperature measurements, as there is a possibility that a warming surface and a cooling troposphere may be decoupled.

Sources: Solar rash and earthly fever, *The Economist*, 21 February 1998, 111–112; Solar influx., *The Economist*, 11 April 1998, 97–98; A heated controversy., *The Economist*, 15 August 1998, 77–78.

ability. In general, these variations cause short-term (i.e. up to a decade) or very long-term (i.e. centuries) climate fluctuations. They are episodic, unpredictable, synergistic and accommodative within the natural resilience of the planet. The problem arrives when humans are added to the equation. The presence of humans causes additional perturbations, not least 'time-warped' CO_2 emissions. By the phrase 'time warp' we mean the accumulation of fossil CO_2, long ago sequestered in the Earth by ancient tropical forests of the Carboniferous age. Humans figure in climate science, for they also create the need to respond in economic and political terms. Humans also set up discontinuities and distributional problems, because of differential impacts on health, economics, international relations and personal well-being. It is this complicated coupling that was referred to in the editorial introduction.

The major scientific report covering all these questions is that of the Intergovernmental Panel on Climate Change (IPCC). The IPCC was established in 1988 under the auspices of the World Meteorological Organisation (WMO), the UN Environment Programme (UNEP) and the International Council on Scientific Unions (ICSU). The original task of the IPCC was to look at the natural science, the possible consequences and the options for response. This led to the creation of three working groups, one on the science, one on adaptation and one on socio–economic dimensions. For the time being, we will concentrate on the Second Assessment Report of Working Group I (IPCC, 1997). Readers looking for a longer summary of the history of climate change science should read Jäger and O'Riordan (1996).

Figure 7.2 summarises the recorded change in CO_2 levels over the last 1000 years. The notable feature is the spurt in concentration since 1850. Records from Mauna Loa Observatory in Hawaii show a steady rise (37%) in CO_2 since 1957. Overall, the

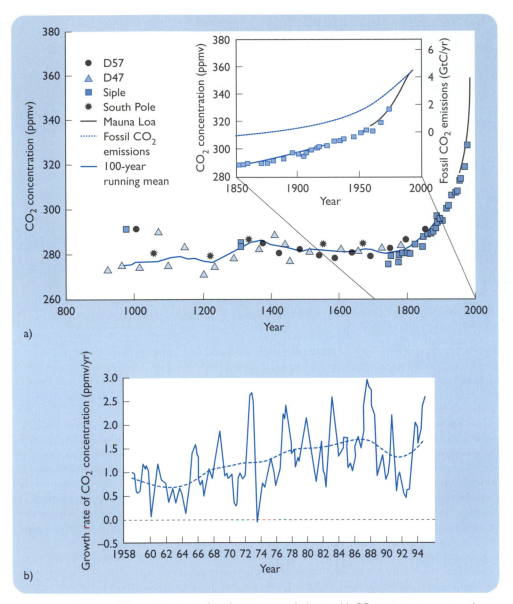

Figure 7.2: Changing CO_2 concentrations based on various calculations. (a) CO_2 concentrations over the past 1000 years from ice core records (D47, D57, Siple and South Pole) and (since 1958) from Mauna Loa, Hawaii, measurement site. All ice core measurements were taken in Antarctica. The smooth curve is based on a 100-year running mean. The rapid increase in CO_2 concentration since the onset of industrialisation is evident and has followed closely the increase in CO_2 emissions from fossil fuels (see inset of period from 1850 onwards). (b) Growth rate of CO_2 concentration since 1958 in ppmv/yr at Mauna Loa. The smooth curve shows the same data but filtered to suppress variations on timescales less than approximately 10 years. *Source*: IPCC (1997, 18).

increase is from about 280 ppmv in pre-industrial times to about 380 ppmv today. Add to this changes in methane of around 1000 ppbv (140%) and in N_2O of some 40 ppbv (13%), and the total increase in radiative forcing seems to be about 2.45 watts per square metre. From Figure 7.3 it will be seen that this increase is offset by changes in

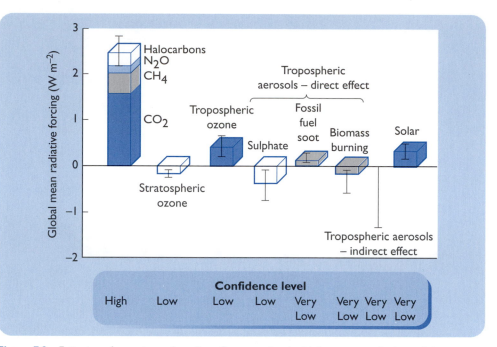

Figure 7.3: Estimates of warming and cooling effects associated with human perturbations of the atmosphere. The figure shows estimates of the globally and annually averaged anthropogenic radiative forcing (in W m^{-2}) due to changes in concentrations of greenhouse gases and aerosols from pre-industrial times to the present (1992) and to natural changes in solar output from 1850 to the present. The height of the rectangular bar indicates a mid-range estimate of the forcing, while the error bars show an estimate of the uncertainty range, based largely on the spread of published values; the 'confidence level' indicates the author's confidence that the actual forcing lies within this error bar. The contributions of individual gases to the direct greenhouse forcing are indicated on the first bar. The indirect greenhouse forcings associated with the depletion of stratospheric ozone and the increased concentration of tropospheric ozone are shown in the second and third bars, respectively. The direct contributions of individual tropospheric aerosol components are grouped into the next set of three bars. The indirect aerosol effect, arising from the induced change in cloud properties, is shown next; quantitative understanding of this process is very limited at present and hence no bar representing a mid-range estimate is shown. The final bar shows the estimate of the changes in radiative forcing due to variations in solar output. The forcing associated with stratospheric aerosols resulting from volcanic eruptions is not shown, as it is very variable over this time period. Note that there are substantial differences in the geographical distribution of the forcing due to the well-mixed greenhouse gases (mainly CO_2, N_2O, CH_4 and the halocarbons) and that due to ozone and aerosols, which could lead to significant differences in their respective global and regional climate responses. For this reason, the negative radiative forcing due to aerosols should not necessarily be regarded as an offset against the greenhouse gas forcing.
Source: IPCC (1997, 19).

the stratospheric ozone layer, increases in sulphate aerosols (due to SO_2 emissions), biomass burning and an unknown effect of tropospheric aerosols, some of which may be stimulated by fluctuations in cosmic rays. The diagram is heavily influenced by low or very low confidence levels in these latter variables. Nevertheless, the evidence of increased global warming is strong, as indicated in the temperature sequence depicted in Figure 7.4.

The IPCC reports indicate a huge transfer of CO_2 in and out of organic life in the sea and on land. Table 7.3 suggests that total human-caused emissions of CO_2 amount to

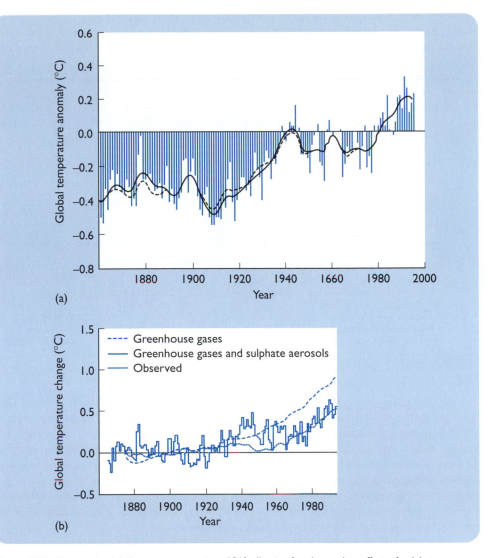

Figure 7.4: Changes in global temperatures since 1860 allowing for the cooling effect of sulphate aerosols. These data have strengthened the reliability of the models and guided policy makers into accepting the principles of human-induced effects in global warming.

(a) Combined land surface-air and sea surface temperatures (°C) 1861 to 1994, relative to 1961 to 1990. The solid curve represents smoothing of the annual values shown by the bars to suppress sub-decadal timescale variations. The dashed smoothed curve is the corresponding result from IPCC (1992).
Source: IPCC (1997, 39).

(b) Simulated global annual warming from 1860 to 1990 allowing for increases in greenhouse gases only (dashed curve) and greenhouse gases and sulphate aerosols (solid curve), compared with observed changes over the same period.
Source: IPCC (1997, 39).

Table 7.3: Annual average anthropocentric carbon budget for 1980–89 (GtC per year). This forms the basis of the 'forcing' scenarios that are discussed in the text. Note the considerable uncertainty over the data, due to natural variability and inevitably insufficient time series and spatially representative monitoring.

CO$_2$ sources
(1) Emissions from fossil fuel combustion and cement production 5.5 ± 0.5[1]
(2) Net emissions from changes in tropical land use 1.6 ± 1.0[2]
(3) Total anthropogenic emissions = (1)+(2) 7.1 ± 1.1

Partitioning amongst reservoirs

(4) Storage in the atmosphere 3.3 ± 0.2
(5) Ocean uptake 2.0 ± 0.8
(6) Uptake by Northern Hemisphere forest regrowth 0.5 ± 0.5[3]
(7) Inferred sink: 3–(4+5+6) 1.3 ± 1.5[4]

Notes:
1 For comparison, emissions in 1994 were 6.1 GtC/yr.
2 Consistent with Chapter 24 of IPCC WGII (1995).
3 This number is consistent with the independent estimate, given in IPCC WGII (1995), of 0.7 ± 0.2 GtC/yr for the mid- and high-latitude forest sink.
4 This inferred sink is consistent with the independent estimates, given in Chapter 9 of IPCC WGI (1995), of carbon uptake due to nitrogen fertilisation (0.5 ± 1.0 GtC/yr), plus the range of other uptakes (0–2 GtC/yr) due to CO$_2$ fertilisation and climatic effects.

Source: IPCC (1997, 20).

7.1±1.1, gigatonnes carbon (GtC) annually. Of this, the oceans absorb about 2 GtC, forests around 0.5 GtC and the biota generally some 1.3 GtC. The error bars around these sequestration figures are large, because the long time series of robust monitoring are simply not available, and the variability of the fluxes, activity by activity and year by year, is enormous. It will never be possible to be accurate in such data: the aim is to narrow the error to acceptable zones for true interaction with equally variable policy options and economic forecasting models. Climate change science is a confidence-building 'game' in more ways than one, as we shall see below in the debate over the 'flexible instruments' attached to the Kyoto Agreement.

One aspect of the science that was new in the 1995 Second Assessment Report was the cooling effect of aerosols of sulphates, caused by the emissions of SO$_x$ from industrial processes. These are short-lived particles, although, if continuously produced, they have longevity by virtue of a failed pollution control strategy. The science of this particular feature was very cleverly done, with a calculated cooling effect of –0.8 watts per square metre as a central estimate. This may be sufficient to lower the observed enhanced greenhouse gas effect by 0.4°C between 1920 and 1990, closer to observed changes (see Figure 7.4). This was an important policy-related breakthrough. The better calibration with recorded temperature changes and micro-scale fluctuations (ocean current movements, volcanic eruptions) led to greater confidence in the models. This in turn helped to rout the doubters, or those whose 'contrarian' views buttressed the coal, oil and chemical lobbies (see Gelbspan, 1997, and Box 7.4). Arguably it was this discovery, with its unequivocal radiation of human-induced temperature rise, which led the IPCC to conclude that:

the observed warming trend is unlikely to be natural in origin – the balance of evidence suggests that there is a discernible human influence on global climate.

(IPCC, 1997, 11)

Box 7.4 Climate politics

In Chapter 1, the possible role of sceptical, and ideologically motivated, scientists in challenging 'multistream' environmental science was noted. Ross Gelbspan (1997), an American journalist, documents the work of the *contrarians,* a group of scientists partly funded by the Global Climate Coalition, a 54-member group of oil, gas and coal interests, who claim that global warming simply has not been proven, and that there is time to delay while the science is improved. Gelbspan (1997, 40–43) documents that one of the leading contrarians, Pat Michaels of the University of West Virginia, testified under oath that he had received $165 000 in research income from oil and coal interests as well as being financed by industry to publish his own climate science magazine. Other sceptics, Robert Balling and Fred Singer, are also openly financed by industrial sponsors. The issue here is not so much the sponsorship but the propensity for lobbying groups to use the findings of these scientists for their own ends. According to Gelbspan, two major US lobby firms are supplied by information from this group of scientists and are associated with the delegations from a number of Gulf states. The Global Climate Coalition admits to spending over $1.9 million on lobbying in 1993 alone. The total expenditure of the leading five US environmental lobbies is $2.1 million annually.

This conclusion formed the basis of the sequence of political climate change negotiations that will be reported on below. Without this enormous background of work, involving a report 572 pages long, written by 350 scientists and peer-reviewed by another 500 scientists from all over the globe, the subsequent policy issues would have been even more contested.

This evidence allowed the IPCC working group to come to the following conclusions.

- The basic models are well validated. Even short-term episodic fluctuations such as the eruption of Mount Pinatubo in the Philippines in 1993 could be factored in with a high degree of confidence.

- The new evidence since 1990 is basically that of the halogenated substances and the sulphate aerosol data. The actual 'science' is a combination of past extrapolation, statistical analysis of global and regional mean temperatures, observed variations due to short-term fluctuations, paleodata, and highly sophisticated statistical modelling. The fundamentals of the models are fairly secure and stable. It is on this basis that climate science is now credible. It appears to have earned its spurs.

- The most likely mean global temperature rise lies between 1.5 and 4.5°C with the best guess at 2.5°C. In 1990, this forecast figure was 3°C, and in 1994 it was 2.8°C. More significant are the regional temperature variations, over which there is much less confidence but a recognition of great variability in the lower latitudes.

- Variations in water vapour, cloudiness, ocean circulation, ice and snow reflection, and interactive changes in soil moisture and vegetative cover seriously complicate the systems being studied. These are a classic example of Holling's evolutionary and adaptive science.

- Projected global sea level rise is estimated to be 49 cm by 2100, down from 66 cm in the 1990 projection. Thermal inertia of the huge oceans means that this rise will be delayed but will continue to accelerate thereafter unless radiative forcing emissions are reduced.

● The global hydrological cycle will be activated, with increased precipitation in high latitudes in winter. The complications of the aerosol additions make it very difficult to predict changes in regional rainfall patterns, such as the scale of the Asian monsoon and the pattern of drought. Arguably, human response to alterations in the timing and intensity of rainfall may alter water availability far more than climate change.

Bill Clark and Jill Jäger (1997, 23–27) assess the significance of the Working Group I report. They point out that this was the most inclusive scientific exercise of its kind with the result that it had global authority. But it also suffered from 'group think', almost obsessional peer review, neglect of maverick positions, and possibly too much consensus.

Speculative issues such as sudden ice melts, breakdown of ocean circulations and wholesale methane release from suddenly melted permafrost are given short shrift. For example, John Ezard (1998, 7) reported satellite observations of a large crack in the Larsen B ice shelf in Western Antarctica. Within only two years, 15 000 km^2 of ice could break away into warmer waters. At risk are the larger shelves of Ronne and Ross (see Box 7.5), though no scientist is willing to go on record to estimate their disintegration in less than 200–500 years. Such speculation is wonderful for the media, of idle interest to nature cornucopians and grist to the mill of nature anarchists. At the very least it funnels more money into sea-ice research.

Clark and Jäger also note the broadening of the frame of the assessments from global warming to climate change, from CO_2 to a 'six pack' of greenhouse gases, from simulation to statistically interpreted observation, from linearity to discontinuity, and from climate science to interactive human–nature science. This has enabled the science to pinpoint three key variables for future scientific analysis:

● the radiative forcing effects of particular economic processes and pollutants;
● the capacity of biogeochemical sequestration in oceans and terrestrial biota as altered by global warming;
● the sensitivity of the climate system to both these processes, and to human-induced responses that may accelerate or dampen such changes (see Box 7.4).

Impacts, adaptations and mitigation

The Working Group II report looked at the effects of climate change on nine natural ecosystems and ten managed systems such as agriculture and financial services, including insurance, that provide the goods and services for human well-being. This is a much more hazardous business, because regional climate change projections remain tenuous, and the effects of human activity can be very large and even more unpredictable. For example, a study of the effects of climate change on water, soils, vegetation and health in southern Africa (Hulme and Downing, 1996) concluded that there would be discernible shifts in water availability and soil structure and fertility, but that these would be masked by responses caused by changes in land ownership, water rights, food markets, uptake of technology, and alterations in the social status of women, among many other factors.

What is of interest here is the significance of the metaphors underlying views on the issues of prevention, adaptation and mitigation. The preventionist sees vulnerability everywhere and argues for the precautionary principle. The adaptationist follows some of the line taken by Michael Stocking in Chapter 11 and by Neil Adger in Chapter 6 and looks to culturally acquired responsiveness to environmental perturbations.

Box 7.5 The catastrophic scenario

Paleogeological research has suggested that, from time to time, huge rises of sea level have swamped coasts in relatively short periods (up to a century). It is possible that human intervention could do the same things. But it is only possible: the best guess is that any significant Antarctic ice melt would occur only in 200+ years (Figure 7.5.1). Those responsible for these guesses will not be around to be proved right or wrong.

Antarctic meltdown?

If the collapse of the Larsen ice shelf is due to global warming then further melting may follow, with disastrous results

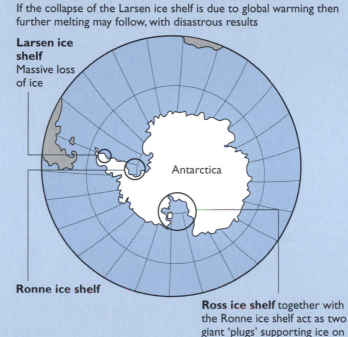

Larsen ice shelf
Massive loss of ice

Antarctica

Ronne ice shelf

Ross ice shelf together with the Ronne ice shelf act as two giant 'plugs' supporting ice on the Antarctic land mass

Further melting

If the Ronne and Ross ice shelves both melt (**1**) it will be as if two plugs are pulled on the massive layer of ice sitting on the top of the land, allowing some of it to slip off (**2**); once in the relatively warmer water, melting will be accelerated.

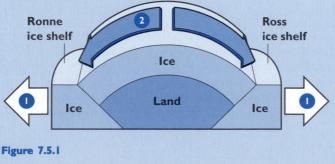

Ronne ice shelf

2

Ross ice shelf

Ice

Land

1 Ice

Ice **1**

Figure 7.5.1

The interpretation of adaptation is heavily biased by these orientations and the pre-dominance of Northern scientists on the review panels. Kates (1997, 32) points out that the agriculture scenarios were heavily influenced by perceptions of changes in biotechnology, management responses and market pricing. Yet the models also indi-cate that cereal production will decline in the South, with as many as 2 billion additional people being malnourished. There are real social costs in adaptation, and at the household levels these can be severe and oppressive – for example, the denial of schooling for starving children forced to work in distant fields to maintain minimal calorie intake.

The mitigation scenarios depend hugely on policy developments, international agreements, technological advance, pricing, regulation and changes in human outlook. There is technical scope for energy efficiency to reduce 60% of current energy usage (Von Weizsäcker et al., 1998). A wholesale switch to non-carbon renewables, extensive technological efficiency measures and changing human choices could reduce CO_2 emission by 67% by 2100, and put off the doubling of CO_2 emissions from pre-industrial levels by almost another century, especially if sink enhancement could absorb another 10–20% CO_2.

What this particular assessment shows is that climate change cannot realistically be separated from a host of other factors affecting the life chances and behaviours and outlooks of the human family. More to the point, the nature metaphors do influence how people and pressure groups interpret this evidence, with the cornucopians sin-cerely seeing market opportunities in adaptation and mitigation, and the evolutionists playing for integrated assessment approaches and participatory adaptive strategies. The egalitarians will worry far more about the widening disparity of flexibility for sur-vival between the adaptive and the vulnerable, particularly because of the huge asymmetry of blame.

All this helps to explain the poor analysis of adaptation in the report. To begin with, there is no set interpretation of an adaptation strategy. The most common framework is drawn from the hazard literature:

- reduce vulnerability by various anticipatory actions;
- absorb losses that are tolerable and diversify household income;
- spread losses widely among the vulnerable;
- modify activities to reduce sensitivity to climate change, or to take advantage of new opportunities;
- relocate activities to less vulnerable areas; and
- restore affected areas to their original condition.

This is not conducive to a strategy that is place- and people-sensitive. The difficulty with adaptation is that it is so vocationally specific and culture-determined. Here is where sound research using local people partnered by geographers or anthropologists can create an understanding and a response package that combines national economic and social factors to village-level patterns. The bias of research remains with hugely expensive models, integrated assessments and international meetings, while at the vil-lage level the potential for supportive adaptation research remains unfulfilled. Even in research effort, inequity is magnified. Yet, as we note next, climate change is a reality, prevention a troubled prospect, and adaptation essential. Yet we hardly know how to begin to increase adaptive capacities in the face of multiple drivers that bear so very differently on the myriad localities where culture and land interact. This must be the priority for future climate change science.

Climate change: economics, politics, ethics

Climate change is the ultimate interdisciplinary issue. The physical system is coupled, and it is also inextricably connected to human response. Indeed, we may be witnessing a convulsive change in our perception of nature. Here is the American writer Bill McGibbon (1990, 54) in poetic anxiety:

> *The idea of nature will not survive the new global pollution . . . We have changed the atmosphere and we are changing the weather. By changing the weather, we make every spot on earth man-made and artificial. We have deprived nature of its independence, and that is fatal to its meaning.*

Nature's independence *is* its meaning; without it there is nothing but us.

We have noted in the introduction to Chapter 5 that there are other views on the interpretation of nature. McGibbon is nevertheless indicating that we live in a tango dance of human-discourse physical process, and that there can be no separate nature. We are in it together, and we have to get out of it together.

The third working group report of the IPCC was by far the most troublesome to write, and the most revealing of the inability of social scientists to reach accord (see O'Riordan, 1997, 34–39 for a longer version of what follows).

In Chapter 1, we noted the inability of social scientists to share a common outlook or a common methodology. This is evident in their own judgement of their efforts, as well as in the perceptions by informed outsiders. In Table 7.4 it can be seen that WGI members of the First Assessment Report process of the IPCC in 1990–92 (Skea, 1996, 7–9) felt comfortable with Group I's competence on scientific inclusiveness. But outsiders took a very jaundiced view of the socio-economic assessment, both as a process and as an all-important executive summary. It seems that this arena is more evidently regarded as subject to political bias. Yet as Jasanoff and Wynne (1998, 76) state, science of any kind is infused with culturally and politically impregnated norms, expectations and self-justifying reinforcement. It is precisely because 'the science' is seen as 'objective' and 'reliable' that the social science is regarded as 'chaotic' and 'manipulated'.

But the Working Group III report, and a much more extensive analysis in four impressive volumes edited by Steve Rayner and Liz Malone (1998a), reveals just how *Continued p. 190*

Table 7.4: Views on the adequacy of the IPCC working groups in terms of representativeness of the science and accuracy of the policy-makers' summary reports. The data suggest that group members themselves had more confidence in their work in Group I compared with Group II and especially Group III. The image of the socio-economic element was of bias and muddle, with much more intervention by lobbies in the latter process. This fact also coloured the view that the 'social science' was infiltrated.

In your opinion, did the IPCC working group policy makers' summaries adequately summarise the full reports?

	Group members			Non-members		
	Yes	*No*	*Can't judge*	*Yes*	*No*	*Can't judge*
WG-I (Science)	76%	22%	2%	79%	8%	13%
WG-I (1992) Update	82%	15%	3%	76%	8%	16%
WG-II (Impacts)	60%	25%	15%	23%	8%	69%
WG-III (Responses)	46%	17%	38%	16%	9%	76%

Table 7.4: (continued)

In your view, did any of the following groups play a significant role in influencing the policy-makers' summaries?

| | WG-I | | WG-II (Impacts) | | WG-III (Responses) | |
	All responses	Group members	All responses	Group members	All responses	Group members
Governments	42%	45%	47%	78%	67%	91%
International sponsoring organisations	52%	54%	26%	35%	22%	39%
Environmental lobbies	23%	24%	29%	29%	26%	26%
Industrial lobbies	10%	13%	7%	—	43%	52%

In your opinion, did the policy-makers' summaries send a suitably clear message to policy-makers?

| | Group members | | | Non-members | | |
	Yes	No	Can't judge	Yes	No	Can't judge
WG-I (Science)	78%	17%	5%	92%	4%	4%
WG-II (Impacts)	55%	40%	5%	18%	20%	62%
WG-III (Responses)	51%	46%	4%	13%	20%	68%

Source: Skea (1996, 7–9).

Box 7.6 The value-of-life dispute

This dispute centres on the role orthodox economic theory may play in providing a cost estimate of future deaths directly attributable to future climate change. David Pearce and his colleagues (1996), responsible for writing Chapter 6 of the Working Group III report for the IPCC, argued that the cost of a statistical life (COSL) lost could be calculated by estimating the willingness to pay for a change in the threat of death caused by mitigation of radiation-forcing gases. The authors admitted that there was always a danger that people could not interpret the dangers they, or their offspring, might face, and hence queried just how representative are any willingness-to-pay figures. They also grappled with differences in income, in normative considerations of cause and blame, and in the role of ethics within an economic formulation based on efficiency considerations. The authors admitted that any deviation from conventional economic theory would be a political or normative act that would be based on a 'moral imperative'. They claimed (page 196) that the 'theoretical economic basis of this approach is weak'. Nevertheless, under pressure from critics, and also through the review process generally, they did produce weighted willingness-to-pay figures for future cost estimates. Eventually the team adopted a combined approach based on surrogates for willingness to pay, derived from contingent value studies (see Chapter 4), estimates of the costs of various safety measures, and compiled changes of net losses of human capital.

Calculations such as these ran head-on into a dispute over the role of 'root causes' of poverty,

Box 7.6 (continued)

low rates of capital formation and differential future growth rates. Economists tend to take these calculations as matters of fact, rather than the outcome of particular political power relationships and military histories. More to the point is the serious difficulty of incorporating culturally loaded values into such calculations, as well as of applying very different ethical rules as to the treatment of 'burden analysis' on a variety of populations. In the end, the working group had to fudge the issue by opening up a debate on the merits of descriptive ('fact-based') versus prescriptive ('morally based') approaches to these calculations, while leaving the original economic theory intact. The result was a range in calculations for COSL for the non-OECD countries from $81 billion (1.6% of GDP) to $115 billion (3.4% of GDP). Such variations obviously make a difference to the cost–benefit analysis of adaptation and mitigation, an issue that remains to be resolved. A fuller treatment of this complex issue is provided by Demeritt and Ruthman (1999).

Table 7.5: The broad findings of IPCC Working Group III on socio-economic dimensions.

Chapter	Title/subject	Main points
2	Decision-making frameworks	Quantitative indices cannot apply. Subject probabilities will be controversial. It would be better to deploy stakeholder round tables using scenario techniques. International risk sharing should yield substantial benefits.
3	Equity and social considerations	Because impacts vary from causes, equity issues are paramount. Countries may be unwilling to accept compensation for possible damages. Feasible, fair criteria will have to be sequentially negotiated. Developing countries urgently need capacity building to strengthen their negotiating hand.
4	Inter-temporal equity	The prescriptive (ethically loaded and precautionary-based) discount rate should be around 2%. The descriptive (allowing for adaptation and innovation) discount rate should be around 7%.
5	Applying cost–benefit analysis	It is all but impossible to estimate the marginal costs and benefits of abatement on a global basis. Decision analysis can help to cope with uncertainties over the timing of implementation strategies. Only when institutions can cope with the ethical aspects of trade-offs can efficiency and equity be separately analysed. The most important aspect of cost–benefit analysis lies in the discipline of its calculations.
6	The social costs of climate change	There is no agreed way to measure social costs, but willingness-to-pay techniques offer the best hope. Damage estimates range from 2% of gross domestic product (GDP) to 6% of GDP (depending on the valuation of a statistical life) to 2100 and up to 35% of GDP for a 10°C rise in temperature by 2300. The cost of a tonne of emitted carbon ranges from $5 to $125, depending on assumptions about social cost and discount rates.

▶

Table 7.5: (continued)

Chapter	Title/subject	Main points
7	A generic assessment of response options	The best hope lies in economic and technological development resulting in efficiency of energy intensity, amounting to 10 to 40%. Renewables offer best promise at the local scale in developing countries. Deforestation should be slowed, and reforestation is worth $30 to $60 per tonne of carbon sequestered. To reach a reduction of 2 to 5 gigatonnes (30% of present emissions) would entail a cost of between $50 and $700 for each tonne of carbon, depending on the strategy adopted, the relative price of renewables, and the success of reforestation efforts.
8	Estimating the costs of mitigation	Baselines determine all mitigation cost assumptions. Infrastructure decisions will be critical in calculating costs. Models still do not capture perfectly the full range of mitigation costs. No-regrets gains depend enormously on assumptions of how much social cost is or is not captured by the market.
9	A review of mitigation cost studies	10 to 30% of efficiency improvements could be achieved at little or no cost. The costs of mitigation are almost unknown in developing countries. If carbon tax revenues are used to remove other distortionary taxes, there should be real economic gains. In the absence of a highly favourable allocation of carbon emission rights, the likely magnitude of emissions reduction will be particularly costly for developing countries.
10	Integrated assessment of climate change	Models may be policy-optimising or policy-evaluative. The initial costs of abatement depend on the economy's dynamics, technological adaptation damage estimates and discount rates. The guiding principle of cost–benefit analysis – optimisation – should be replaced with an emphasis on precautionary targets, risk aversion and other physical criteria.
11	An economic assessment of policy changes	The final mix of strategies will have to combine mitigation and adaptation. A harmonised international carbon tax may offset national subsidies, so the final emissions reduction is indeterminate. Tradable quotas make the marginal cost of reduction uncertain. A great deal depends on how the tax or permit revenue is recycled. This in turn will depend on compliance, levels of assumption and social welfare estimates of payoff.

Source: O'Riordan (1997, 37).

difficult it is to obtain a social science perspective on a coupled human–nature phenomenon. Social scientists squabble among themselves even about the purpose, meaning and relevance of their labours. In the WG-III report, the authors make a distinction between a *prescriptive approach*, namely where judgements and ethical interpretations are made explicit for policy makers to recognise, and a *descriptive interpretation*, based on analytical nationality, where judgements are suppressed. This style was favoured by IPCC's top brass, for the panel itself was established by a political process to inform it, not to persuade or embarrass it. The consequence was an uneasy compromise in which both interpretative styles were placed side by side, without proper integration. This made the controversial conclusions vulnerable to casual dismissal at best, or to mud-slinging at worst (see Box 7.6). Honestly won and very sincere attempts to produce one of the most comprehensive assessments of the socio-economic implications of climate change and of mitigation and adaptation strategies were lost in the ensuing crossfire.

Table 7.5 presents the principal findings of the socio-economic report. The policy makers' summaries are the all-important elements. The introductory text makes it clear that climate change is an equity issue as much as an efficiency matter, and that good understanding of the possible consequences, no matter how inadequate at present, at least provides a guide as to the scale of measures required to respond. But the sheer complexity of the topics, and the coupled uncertainties between process, interpretation and response, made it impossible to come up with clear recommendations. And because of the ambivalence over the prescriptive role, it fell on the non-governmental organisations and the policy advisers to complete the job.

Rayner and Malone (1998b, 5) heroically try to summarise the findings of the huge, three-year Battelle study, *Human Choice and Climate Change*. Here are their conclusions, with an additional gloss.

- View the issue of climate change holistically, not just as a problem of emission reductions. We shall see that this is the task facing the post-Kyoto agreement, namely to create a 'culture change' in outlook, in education, in social discourse and in techno-economies, in order to convert climate change 'beneath the skin' of modern society.

- Recognise that, for climate policy making, institutional limits to global sustainability are at least as important as environmental limits. This point is covered below. What is significant here is that the current institutional structures that are being created to deal with climate change depend on that climate change for their legitimacy and survival. This is why climate change is such a tough political nut to crack.

- Prepare for the likelihood that social, economic and technological change will be more rapid and will have greater direct impacts on human populations than climate change. This may be the salvation. But unless these changes are integrated into climate change strategies, they could act at cross-purposes. For example, tax reform and fiscal incentives for long-term technological shift have to be politically buttressed if they are not to succumb to destruction by competitive global markets.

- Recognise the limits of rational planning. Indeed, this is the point about incorporating the prescriptive element into integrated science, for its own survival in a globalised and politicised world.

- Employ the full range of analytical perspectives and decision aids from the natural and social sciences, and the humanities, in climate change policy making. There is much room for art, for storytelling, for civic demonstration, and for community mobilisation. These are all tools that the humanities can deploy in association with scientists and community leaders. Why not, for example, enable school pupils to be their own energy managers, gain from the proceeds saved by generating a 'sustainability fund', and being assessed and evaluated for future job potential on their active, command contribution?

● Design policy instruments for real-world conditions, rather than try to make the world conform to a particular policy model. Climate change is a global problem that needs a global commitment through local action. Global–local cannot be separated. It is not so easy to respond to the glib phrase 'think globally – act locally': it is actually necessary to think *and* act *both* globally *and* locally.

● Incorporate climate change concerns into other more immediate issues, such as employment, defence, economic development and public health. Climate change is a long-haul policy track. To push the policy wagon down that track needs the political engine of immediacy. Here is where reforms to health, to technological change and to changing civil rights, which in turn modify consumer behaviour, come in sequentially.

● Take a regional and local approach to climate policy making and implementation. LA21 may well become a valuable routeway for comprehensive regional sustainability, within which climate change adaptation and mitigation responses become legitimised.

● Direct resources into identifying vulnerability and into promoting resilience, especially where the impacts will be largest. This is covered in detail in Chapters 1, 2, 4 and 6.

● Use a pluralistic approach to decision making. Here we see a call for inclusive participation based on the kinds of deliberative processes outlined in Chapters 1 and 3.

Here is the thoughtful conclusion by Rayner and Malone (1998b, 32), echoing the points made by Holling in the introduction.

> *To commit oneself, one's family, firm, community or nation to just one viewpoint is to gamble that it will turn out to be right and the others wrong. It is far more likely that all will be partly right and all will be partly wrong. Recognizing this, and stewarding the land of intellectual pluralism necessary to maintain multiple viewpoints and a rich repertoire of policy strategies from which to choose, is what promoting social resilience, sustainable development, and climate change governance is all about.*

Now, how far is the 'Kyoto process' going to listen to that?

Beyond the Kyoto Protocol

The Rio Conference gave birth to the UN Framework Convention on Climate Change (UNFCCC). This point was noted in Chapter 2, and will be analysed in a complementary way in Chapter 19. Jäger and O'Riordan (1996, 12–26) cover most of the main points, summarised in Figure 7.5. The UNFCCC was bedevilled by the unwillingness of the major emitters to accept any legally binding and similar targets, on the grounds that this would distort their competitive advantage, fail to recognise the differential responsibilities, and ignore the inevitably wide range of costs associated with equal percentages of greenhouse gas emission reductions. For example, Norway is primarily a hydro-power nation and cannot cut CO_2 emissions without adding to the cost of offshore oil and gas production (by reducing or eliminating flaring) or by politically intolerable increases in the cost of transport. Since road vehicles, ferries and planes knit that nation together,

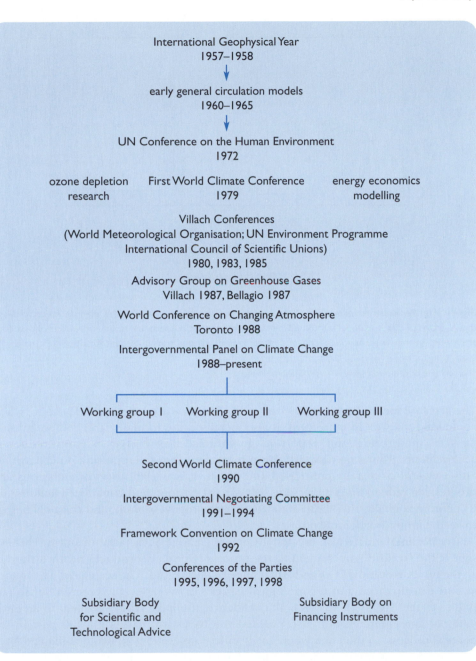

Figure 7.5: The history of climate change negotiations since 1970.

Source: Jäger and O'Riordan (1996, 13).

there is no political coalition that could deliver such a package with any guarantee of being re-elected. Similarly, Canada and Australia have huge *per capita* CO_2 emissions for a relatively small population base (Figure 7.6), because of energy-intensive resource extraction industry, much of which maintains jobs in peripheral economies. To eliminate those would also be political suicide in the short term. So equal *per capita* reductions are not on for the moment: they have to be converged over a century. UNFCCC tried to ignore this reality and paid the price. Kyoto faced up to it.

The UNFCCC created a Conference of the Parties (CoP). This is the signatories to the convention, who are, in effect, the political governing council to a legal process.

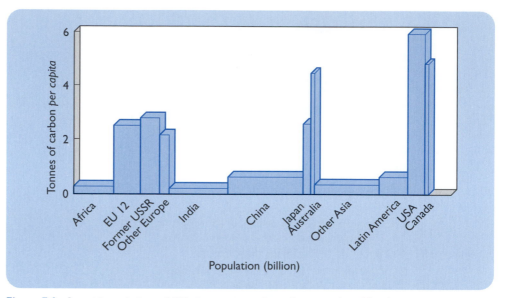

Figure 7.6: *Per capita* emissions of CO_2 by country and population numbers. This diagram makes it clear that the United States, Europe and the former USSR are the main contributors, but that China and India are waiting in the wings. Any indication of a move towards serious trading of emissions between these blocs would transform future climate change negotiations.

Source: Jäger and O'Riordan, (1996, 11) from Grubb (1995, 42).

The first CoP met in Berlin in 1995 as CoP I. This produced the Berlin Mandate, which committed the CoP to a protocol. It also increased the formality and comprehensiveness of monitoring and reporting greenhouse gas emissions, and reduction achievements. This is no easy task, for in many countries the reporting procedures are either absent or heavily politicised and insecure. So verification of emissions and progress on reductions is a crucial aspect of trust creation and confidence building. To its enormous credit, the Berlin Mandate set the process in train, and certainly helped to lay the groundwork for Kyoto.

The Berlin Mandate also created a Climate Change Secretariat in Bonn. This is a sign of the German commitment to the process (Germany produces nearly a third of all the EU emissions of CO_2) and also a huge budgetary aid. The secretariat also helped to sort out financial and technological issues via two specially created Subsidiary Bodies to the UNFCCC. All this built confidence, the most vital ingredient to an international agreement. Berlin also enabled the NGOs to maintain a full part in the political process, to the point where many official delegations to Kyoto contained NGO representation. Bearing in mind that the all-important policymakers' summaries to the IPCC Second Assessment Report were shaped by NGOs, these organisations are effectively legitimised. They are also locked into the outcome, no matter how much they may subsequently protest from the touchline in the post-agreement 'interpretative' phases. John Lanchberry (1998) provides an authoritative account of the pre-Kyoto negotiations, while the reader interested in the post-UNFCCC process should consult Victor and Salt (1994). Kyoto itself is best covered by Ott (1998).

The Kyoto Protocol, if fully ratified, will be legally binding on all parties. At the Third CoP, held in Kyoto 8–11 December 1997, the 113 countries present completed a 48-hour, almost continuous negotiation to reach the following outcome.

Ratification

This will require the signatures of 55% of all Annex 1 (24 OECD, six former Soviet Union and five Eastern and Central European) nations, as well as the signatures of those nations whose 1990 CO_2 emissions amounted to 55% of all CO_2 emissions in that year. Since the USA accounts for some 23% of such emissions, this second 'trigger' gives it a *de facto* veto. On such matters, the USA is normally in the hands of the senate (as this is a foreign affairs issue). But, in all probability, the constitutional lawyers will find a formula through which the administration can nudge the senate into an agreement, subject to a series of 'deals', of which more below.

Differentiated targets

The big breakthrough in Kyoto was to allow different countries to be subject to variable targets, and for the target date to be a four-year 'window', 2008–12, rather than a specific year. The overall commitments are laid out in Table 7.6. We shall see below that this list reflects intense political lobbying by Russia, Australia and Canada, with Japan playing a discreet but equally cantankerous role. These figures are the result of 'blood and guts' bargaining of the old kind of horse-trading. They take into account a whole host of factors:

- Likely increases of greenhouse gases under business-as-usual scenarios. The USA, for example, may increase CO_2 emissions by up to 35% before falling back to −7% by 2008–12.

- The cost of industrial restructuring and technological transfer, notably for the emerging European economies on whose reliable and crime-reduced future the West depends for its security.

- The likely removal of uneconomic but highly CO_2-intensive plant, caused by economic transformation or economic collapse. This is particularly the case for Russia, where as much as 50% of its 1990 emissions may have disappeared by 2008, just because of the collapse of the old economic order.

- Purely political calculations of what is tolerable to an electorate, who for the most part are mildly concerned but driven by short-term economic lifestyle pressures, and for many of whom climate change is a 'grandchildren' problem.

The 'six pack' of greenhouse gases

Kyoto did not stop at CO_2, CH_4 and N_2O: it added the three industrial synthetics of perfluorocarbons, hydrofluorocarbons and sulphur hexafluoride, which have a high capacity for global warming. At the insistence of Japan, these last three will be reduced from 1995 levels, not 1990 levels as is the case for the top three. But the fact that these gases are included at all was a tribute to the IPCC, which pushed for their incorporation.

Sinks

One of the most contested issues following Rio was the failure to recognise that some countries had very large 'mop-up potential' for spare carbon, by virtue of active soils and tree growth. This was a particular theme of the fierce critique by Agarwal and Narain (1991), who accused the USA of 'environmental colonialism' in its treatment of the South. The trouble with sinks is that the science of their operations, and the political economy of their longevity, are simply too uncertain. This is where equity strikes

Table 7.6: The negotiated targets at Kyoto: six pack emissions reduction for Annex I countries for the period 2008–12, based on 1990 levels.

	Emission reduction (%)
Annex I countries as a whole	–5.2
European Union	–8.0
USA	–7.0
Canada, Japan, Hungary, Poland	–6.0
Croatia	–5.0
Russia, Ukraine, New Zealand	0.0
Norway	+1.0
Australia	+8.0
Iceland	+10.0

These figures show evidence of special pleading and horse-trading. The Norwegians argued for their hydropower, the Australians for their mining, the Icelanders for thermal power, and Russia and Ukraine for peaceful co-existence in a time of economic turmoil.

The EU is currently in tough negotiations over its 8% reduction allocation. It is pressing for as much to be achieved domestically by specific policy measures, rather than by flexible instruments of offsetting.

Currently proposed targets for 2010, based on 1990 CO_2 emission levels (pre-Kyoto in brackets)

Portugal	+27%	(+40)	Netherlands	–6%	(–10)
Greece	+25%	(+30)	Italy	–6.5%	(–7)
Spain	+15%	(+17)	Belgium	–7.5%	(–10)
Ireland	+13%	(+15)	United Kingdom	–12.5%	(–10)
Sweden	+4%	(+5)	Austria	–13%	(–25)
Finland	0	(0)	Denmark	–21%	(–25)
France	0	(0)	Germany	–21%	(–25)
			Luxembourg	–28%	(–30)

CO_2 emissions from fossil fuel combustion (10^9 tonnes C)

	1996	% change from 1980
North America	1.75	8.2
Latin America	0.33	13.2
EU	0.96	0.8
CIS, Central and Eastern Europe	0.90	–37.0[1]
Middle East	0.25	41.0
Africa	0.22	19.0
Asia/Pacific	2.00	31.0
Total OECD	3.27	7.8
Total non-OECD	2.34	32.0
World	6.51	6.4

Notes:

1 Due to the collapse of the communist-based economies and the removal of hugely wasteful energy usage.

2 These proposed figures will have to be supported by finance and industry ministers in each country, as well as by environment ministers. This will not be easy, as many of the negative emissions reduction strategies face a further period of CO_2 growth before policies begin to bite. Environment ministers are particularly opposed to the idea of flexible instruments, such as carbon trading and joint implementation, being used to undermine national climate strategies.

against measurement. So the compromise at Kyoto is to include only those terrestrial sinks of soil uptake and biomass growth caused by land-use change since 1990. This rules out the shallow coastal seas, whose sequestration may be as much as 1.5 GtC per year. Intriguingly, China has one of the most absorptive shallow seas on the globe. It is only a matter of time before the seas become part of the equation. The politics of sinks will be part of the climate change scene for a decade. Already the USA believes it can discount up to 1% of its 7% reduction because of post 1990 land-use change.

Joint implementation

The UNFCCC introduced the notion of activities implemented jointly, or joint implementation. Danny Harvey and Elizabeth Bush (1997) provide a fine summary of the history, purpose and politics of JI. The idea is simple. Any molecule of CO_2 has the same global warming potential wherever it is emitted, and by whatever means. So if it is cheaper for one nation or firm to 'buy' CO_2 reduction from elsewhere, this meets the efficiency principles of Pareto optimum. Or so the theory goes. Various firms, backed by trade and aid money from well-meaning Annex I countries, have tried out various ingenious options. Table 7.7 outlines the state of the trades so far.

Table 7.7: Recent agreements under the joint implementation pilot phase, following the Berlin Mandate.

Investing country	Host country	Project type	Location/Project title
United States	Belize	Forest management	Rio Bravo
United States	Costa Rica	Wind energy	Plantas Eólicas S.A.
United States	Costa Rica	Forest protection	ECOLAND
United States	Costa Rica	Forest protection/expansion	CARFIX
United States	Czech Republic	Coal-to-natural gas efficiency upgrade	Decin
United States	Honduras	Rural solar electricity	ENERSOL
United States	Russia	Reforestation	Saratov
United States	Costa Rica	Reforestation	Klinki Forestry
United States	Nicaragua	Geothermal power	El Hoyo-Monte Galán
United States	Costa Rica	Wind energy	Aéroenergía S.A.
United States	Costa Rica	Reforestation	BIODIVERSIFIX
United States	Russia	Methane capture from pipelines	Rusagas
United States	Honduras	Biogas power	Bio-Gen
United States	Costa Rica	Hydroelectric power	Dona Julia
United States	Costa Rica	Wind energy	Tierras Morenas
Netherlands	Malaysia	Reforestation	
Netherlands	Ecuador	Reforestation	
Netherlands	Uganda	Reforestation	
Netherlands	Czech Republic	Reforestation	
Netherlands	Russia	Capture of landfill methane	Daskovka
Netherlands	Russia	Energy-efficient greenhouse	Tyumen, Siberia
Netherlands	Hungary	Diesel-to-gas bus conversion	
Netherlands	Hungary	Municipal energy efficiency	
Norway	Mexico	Energy-efficient lighting	ILUMEX
Japan	China	Ironworks efficiency upgrade	

Note: Many other projects are in various pre-approval stages.

Source: Harvey and Bush (1997, 18). Compiled by the authors from various issues of *Joint Implementation Quarterly*.

The problems with JI are as follows.

- Verification is very difficult to achieve.
- Guarantees of long-term continuation of the schemes cannot be politically delivered.
- The science of the sink is in dispute.
- There is as yet no proper legal framework for monitoring and compliance.
- The transaction costs of compliance could exceed 10% of the total costs, notably with developing countries, where surveillance is more likely to be necessary.
- Comparability of technical transfers, reforestation schemes, end-use efficiency measures and the like are almost impossible to justify.
- JI appears to let Annex I countries off the hook. This raises important issues of equity. And if JI takes the place of development aid, then there are also accusations of blackmail.
- Any JI with developing countries that involves technological transfer will almost certainly require associated deals on capacity building and on the reduction of perverse energy and forestry subsidies. These will be hotly contested, emphasising the Rayner–Malone point that climate change cannot be extricated from a host of tricky political and economic contexts.

As a pilot project the Norwegian government has signed a deal to convert boilers in a large Slovakian lumber yard and nursery from coke and natural gas to biofuels. This is designed to save 50 000 tonnes C by 2028, and will cost 145 000 ECU. The Norwegians see this as part of an aid/development package as much as a JI deal. More such arrangements will follow.

Carbon trading and hot air

Carbon trading is a cherished US dream. The idea is essentially that of JI but relates to specific bilateral deals, nation to nation, or firm to firm, or a mixture of the two, so that an offset of carbon dioxide increase in a legitimate trade can be credited against a carbon excess. But the rules of such carbon trading are not yet set, and it is imperative that a formal arrangement is in place to provide technological and financial guidelines that all parties follow.

Two major issues over carbon trading will plague future negotiations:

- The total amount of trading that can be offset against domestic action. The EU wants a cap of 50% at maximum. Other Annex I countries look for more. The USA will resist, at least until the developing countries show signs of creating their own 'domestic actions' as the Kyoto Agreement requires.
- Any trades with Russia and Ukraine could be made against carbon rights that existed in 1990 but do not exist now because the emissions have subsequently ceased. This is called the 'hot air' agreement, as the CO_2 offset is not 'saving' any actual CO_2 emissions in recipient countries.

These barriers to successful implementation are very severe. Right now it is difficult to imagine any carbon trading regime without international validation of the toughest kind, and 'shared accountability'. The latter means that recipients and donors will have to play by the trading rules, or find their trade invalidated by the monitoring authority. Such a deal would have to stick.

This whole arena of trading and JI is called *flexible instruments* in the Kyoto jargon. The flexibility comes not just from the options on offer (intrinsic) but also from the interpretation of the effectiveness of these options (politically discretionary). Any proposal in this arena will be largely self-serving and advantage-accumulating, so the struggle for reliable and robust mechanisms, and independent validation of these mechanisms, will be long and hard. It is no wonder that the USA is pushing for these mechanisms, as it is highly unlikely that it has the governmental structure to deliver the equivalent of 35% CO_2 reduction over a decade. As Konrad von Moltke (1996, 332) shrewdly observes:

> It is difficult to visualise any federal measures in the United States which would be reasonably certain not only to reduce the emission of greenhouse gases but to keep them to a pre-determined level. In the absence of overwhelming public concern, the two available solutions to this dilemma – novel forms of federal–state co-operation or a change in the balance of powers between federal and state levels – are well beyond the reach of American political consensus.

In May 1998, the International Petroleum Exchange proposed a 'futures market' in carbon permits coupled to a secondary market in derivatives. This would begin in the UK but would extend to the whole of the European Union and eventually beyond. What this kind of initiative signals is the willingness of business to push out the boundaries of trading in advance of political commitment domestically, and before the post-Kyoto regulatory arrangements are in place. This is a good example of pre-emptive agenda setting of the kind outlined in Chapter 3.

For adventurous souls, the carbon permit is exciting. It provides a 'market' and it changes the property right in CO_2 emissions from public to private, but with a sustainability duty attached. It also enables any unit of economic activity to 'shadow' the Kyoto process and get credited for so doing. Box 7.7 summarises just how a carbon trading regime might work for the UK. Note that the implications for personal behaviour are rather awesome. This will take a lot of political and education 'selling', as noted in the conclusions to Chapter 19.

Box 7.7 Carbon trading and consumer lifestyles

The Lean Economy Initiative (1998) has prepared a case for a domestic tradable quota/scheme. This is designed to give everyone equal *per capita* entitlements of carbonates to cover basic domestic and transport needs. Any additional carbon requirements would have to be purchased from a pool of sellers, who could be industry or government, or non-governmental groups.

The scheme would look as follows:

- Each kilogram of carbon would be classed as a unit, linked to all activities of production and consumption and a total quota set in terms of national obligations to the Kyoto Protocol.

- The holdings of every individual would be held on a computer database, known as Quota Co.

- An 'entitlement' of units would be issued, together with a 'tender' for higher values of units. Anyone not using their full entitlement would sell to those seeking a tender through an electronically operated trading market.

- The tender would also be the carbon units allocated to industry, government and the public sector, issued through direct credit systems. The revenue would go to carbon resource charities.

- All tenders would then be available on the secondary market. All fuel prices would include the surrender of units, or the purchase of units in addition to the fuel price if the buyer did not have the credits to hand.

Box 7.7 (continued)

Retailers buying fuel would purchase credits which they would sell to retailers.

- Primary energy providers would surrender units to Quota Co when buying or importing fuel, but would be given units when retailing to the generators.

Figures 7.7.1 and 7.7.2 show how the market in quotas might lower overall domestic emissions through a period of phased quota reductions. The actual quota market is portrayed in the second diagram. Table 7.7.1 translates the emission levels into fuels.

This kind of proposal seems most unconventional and unworkable to the inexperienced eye.

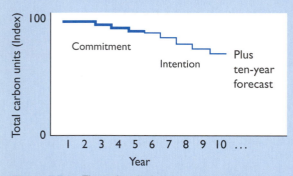

Figure 7.7.1: The carbon budget – 'commitment' and 'intention'.

Table 7.7.1: Translating emissions into fuels.

Fuel	Carbon units (kg CO_2 released)
Natural gas	0.2 per kWh
Petrol	2.3 per litre
Diesel	2.4 per litre
Coal	2.9 per kg
Grid electricity (night)	0.6 per kWh
Grid electricity (day)	0.7 per kWh

We have not tried it, so we do not know it. The lessons from the SO_2 tradable quota scheme under the US Clean Air Act of 1990 is that for large producers it does work, but that it has to be coupled to strong framework regulations and regional variations. There is no point, for example, in a New England utility buying SO_2 rights from a Midwestern counterpart when the acid deposition concentrations are already above threshold in New England.

But for CO_2, the regional perspective is not a matter of concern. What is such a matter is the political will to tackle it. The idea smacks of rationing and imposition over private enjoyments. In the USA, such a scheme would be tested in the courts for its constitutionality by civil rights groups. In Europe, where the tradition of social responsibility in both behaviour and private property rights is higher than in the USA, there is a

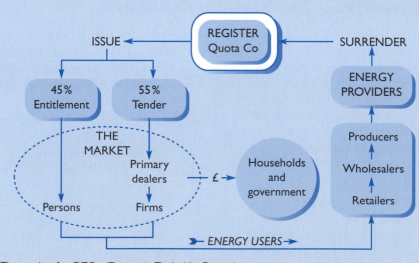

Figure 7.7.2: The market for DTQs (Domestic Tradeable Quotas).

Box 7.7 (continued)

chance to change the notion of the property right to link civil responsibility to consumer choice.

Carbon trading at the household level will be very difficult to put in place in the foreseeable future. The stumbling block is fainthearted politicians who do not wish to upset the middle-income voters, consumers who cannot decide how much to be citizens, ill-defined property rights which do not place a global obligation on the purchase of carbon, and technology which is not quite ready for a highly sophisticated market scheme of quota trading. Of these, the technol-ogy is probably the easiest to crack and the politics the most difficult. Until there is a serious national commitment to climate change, all the political decisions will be far removed from the hearts and minds of the 'consumer lifestyle'. And until that 'lifestyle' is altered by freshly minted hearts and minds, then the real carbon challenge will continue to be ducked. This is where citizenship training from sustainability comes in (Chapter 19) but where it is, sadly, being let down by the higher education sector.

Source: Fleming (1998).

Clean development mechanism

One of the big battles fought at Kyoto was the demand by the USA to require the South to be in on greenhouse gas reductions from the 2008–12 window. This is, in principle, what was expected in the UNFCCC. However, on equity grounds, the South refused point-blank to be drawn in, not until there was demonstrable sign of US commitment. Part of the reason for this intransigence was the failure to deliver to the South any significant additional money or capacity building to ease the cost of the transition to sustainability. The South feels badly let down, and used this grievance as a bargaining weapon.

As a means of softening the blow for eventual Southern involvement, the protocol calls for a clean development mechanism (CDM). The history of the CDM lies in the proposal by Brazil for a penalty fund to be levied on Annex I countries for failing to meet the UNFCCC commitments. The Kyoto deal converted this into a positive investment for improving the efficiency of technology, and for building capability for technological innovation generally. The interesting issue now is that the CDM may deal more with private investment than with with government-to-government transfers. This will raise a question mark on intellectual property rights, normally waived for such agreements, as well as within-company teaching where business straddles North and South. In principle the CDM could become an important private financing arrangement, though its potential throws into doubt the whole concept of joint implementation and the financing mechanisms under the Global Environment Facility. This provision is particularly unclear at the moment, as the package was signed only in the late stages of marathon negotiations. The idea is to create a pool of money from a levy on fossil fuels to finance energy saving and technology transfer developments in the South. These could include efficiency investments but more likely will cover combined heat and power schemes, energy from waste schemes, and renewable energy schemes, notably using biomass and the Sun.

It is far too early to assess the scope for CDM. So much depends on the terms:

- What will be the principles on which CDM is based?
- Who will pay, and through what mechanism?
- What can the North legitimately offset against CDM?
- Will the CDM be an excuse to offload any further public sector finance, of aid and of trade?
- Will CDM, like JI generally, remove legitimate aid and trade deals?

The evolution of the CDM poses critical questions for equity and for freedom of development action for the poor countries. In effect, the CDM denies the right of Third World nations to select their own CO_2 futures. The property right of their CO_2 'envelope' is being traded away for development opportunities in a narrow area of investment (i.e. renewable energy), which may be by no means commensurate. This is a kind of 'ecological colonialism' by another name, yet skilful negotiators could just tie CDM to improved trade and aid packages, where real reforms are urgently needed (see Chapters 2 and 19). There is much political sensitivity at present over CDM, so the issue will have to be handled very carefully. An accessible summary of the CDM can be found in Lanchberry (1998) and Forsyth (1998). As this book goes to press, many of these questions will be better understood. Readers looking for further detail should read *International Journal of Environment and Pollution* (1998).

Contraction and convergence

A single NGO, the Global Commons Institute (GCI), has initiated an ingenious approach to CoP 4 and beyond, namely the implementation of contraction and convergence as an ethically based deal for climate futures. The aim is to reduce global emissions by 60% by 2020 in order to meet the UNFCCC objective of 'stabilisation of greenhouse gases in the atmosphere at a level that would prevent dangerous anthropogenic interference with the climate system' (Article 2) (see Robotham (1996) for a fuller legal interpretation).

The GCI looks at the historic and likely future emissions of CO_2 up to 2030 and seeks a mechanism to contract these to 3 GtC, i.e. to less than half present levels – the 60% criterion – and to do so by contracting *per capita* emissions to an equal level – the equity criterion. The ambitious aim is to obtain global agreement on an objective and a mechanism before the 'sub-global' flexibility instrument deals take hold. The GCI, and its friends in the South, believe such deals will be destabilising and self-serving for the North. It looks for a stable concentration of CO_2 at 480 ppmv, well below the non-precautionary 550 ppmv indicated by the EU (GCI, 1998, 8).

Figure 7.7 plots the GCI proposals. It will be seen that the historic dominance of the USA, European and Russian emissions all but disappear by 2075, leaving the bulk of CO_2 to come from the redeveloping South. This is based on the *per capita* emissions convergence, though some population cap is also envisaged. What the GCI looks for is a trading regime with a purpose, not an *ad hoc* series of deals. Thus, under contraction and convergence it would be quite proper for 'hot air' deals to be promulgated, so long as the *per capita* convergence property right was agreed as an objective. Right now, the proposed post-Kyoto trades, according to the GCI (1998, 10), are 'arbitrary and asymmetrical'. Under the 'C–C' regime, developing economies would have 'hot air' and 'no air' tradable permits, which they can cash in for agreed sustainability investments. In essence this creates a global trade–aid deal with sustainability built in.

The GCI proposals are currently being actively promoted by India, China and Africa. They are the brainchild of a single-person NGO, Aubrey Meyer. The journalist Geoffrey Lean (1994, 13) places a human face on the man:

The story is a remarkable triumph of stubbornness, obstreperousness and sheer bloody mindedness – all orchestrated from a tiny back room from a ground floor flat in No. 42 Windsor Road, Willesden in North London. Meyer sleeps above a poky study filled with files, lap tops, two fax machines and a colour printer ... He admits to being bloody rude, disruptive and confrontational and maturely obsessive.

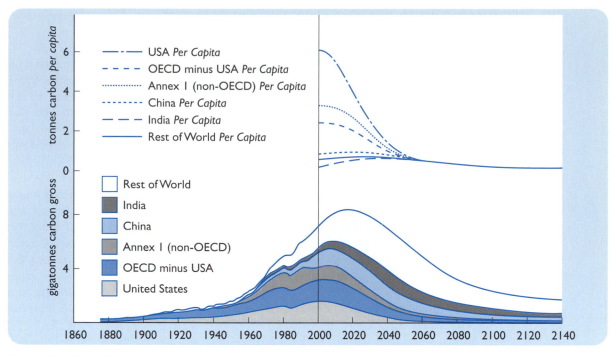

Figure 7.7: Global Commons Institute proposals for reducing carbon emissions. *Source:* GCI (1998, 10).

The point about the GCI proposal lies in its pursuit of equity. The IPCC Working Group III looked at this with some difficulty (Banuri *et al.*, 1996). It looked at two broad principles of equity:

- Procedural equity, in the form of inclusive participation and empowerment (see Chapters 1 and 4), and equal treatment. In the latter case comes Nozick's (1974) notion of a 'sphere of moral right' in which no-one, including the state, is allowed to interfere, irrespective of the consequences that might arise. This sphere pertains to liberty and also to freedom of speech and association, though in principle it could apply to minimum livelihood quality, including a right to a certain CO_2 equivalent.

- Consequentialist equity, or the distribution of benefits and losses in relation to any allocation of resources. The basis of this distribution could be *parity* (the GCI objective), or *proportionality* (the North objective), or *priority* (the South objective) in terms of need, or *utilitarianism* (the greatest good for the greatest number), or *Rawlsian* (equal distribution unless unequal distribution operates to the benefit of the least advantaged).

The GCI proposals verge on parity, procedural fairness and Rawlsian justice. The trouble is that power in the world does not operate so ethically. So the struggle for the GCI is the age-old one between inequitable power domination and equitable rights to survive into a warming world. Because eventually the UNFCCC is a global convention, and because eventually the South will become the biggest emitters, there is a reasonable chance that some variation of the GCI proposals will prevail subject to advantageous carbon-trade–aid deals between North and South. This in turn requires a co-operative morality in a competitive framework, and a recognition that vulnerability could be counterproductive even to the seemingly least affected interests.

Integrated assessments, policy futures and inclusive participation

The next phase of global modelling lies in the fusion of climate change scenarios with policy futures, with means for ensuring genuine consensus among key stakeholders. This is the essence of Holling's evolutionary science. Models facilitate, they do not dominate. Science unfolds the mind, it does not narrow perspectives. Outcomes are learned and are created by sequential decisions, where each decision responds to a pattern of inputs over which consensus shapes the next stage. All this was spelled out in Chapter 1. Here is a golden opportunity to put it to the test.

Integrated assessments are models that combine scenarios of future physical states of global change with policy options as to societies and economies that might evolve. The state of the art is explained by Jäeger *et al.* (1998) and Rotmans and Dowlatabadi (1998). Jäeger and his colleagues explore the transition from single-actor, single-objective decision models to the complex interactive learning modes of many actors, outlined in Chapter 1 and also introduced in Chapter 17. They promote the cause of 'Aristotelian understanding of democratic policy as the exercise of social rationality' (Jaeger *et al.*, 1998, 205), namely the establishment of trust-building and empowering procedures of conflict avoidance. Optimisation follows a non-optimising search through the maze of uncertainty.

Rotmans and Dowlatabadi look for multidisciplinary approaches to modelling that provide additional value as well as real input for policy analysts. They show how scenario building can couple to prediction to create interactive models through which players can guide their own futures. The keys to the true integrated model are:

- plausible scenarios based on realistic but distinctive assumptions as to economic and social behaviour;
- clear metaphors of nature, along the lines proposed by Holling;
- a visual display of sequential outcomes for a set of parameters selected by stakeholders as meaningful and interpretative;
- software that allows the novice to 'play' without fear of the technology;
- post-simulation discussion arrangements that permit genuine empowerment.

These conditions are not yet in place. The models are now arrested by scenarios of the kind developed by Gallopin and Raskin (1998). The nature metaphors are being loaded into cultural theory (O'Riordan and Jordan, 1999), though this is an early stage. The links between nature metaphors, cultural theory and the new institutionalism were made more explicit in Chapter 2. The participatory approaches are still at the research stage. These need to be related to institutional arrangements like LA21, or some form of community visioning (see Chapter 1). Nevertheless, this is one promising way forward if climate change is to be located in a society searching for sustainability on all fronts.

An example of this approach, and one that influenced the US negotiating position at Kyoto, is the case for delay, through which more cost-effective strategies could be devised in the near term. The idea is to use the delay to offset avoidable emissions elsewhere, through JI and CDM, and to create scenarios of temperature and CO_2 trajectories that allow a series of more cost-effective strategies to be developed. These scenarios are based on assumptions of 'climate sensitivity', namely the response in the climate regime to a doubling of CO_2 concentrations. Box 7.8 summarises the

Box 7.8 Climate sensitivity

Climate sensitivity is a measure of the response of global average temperature to a change in the carbon dioxide concentration in the atmosphere. For a doubling of CO_2 equivalent from its pre-industrial levels (which will be reached by about 2030), the range in the early models was from about 1.5 to 6.0°C, though the highest value has now been rejected following IPCC discussions. In the case of the former value, we could burn even more fossil carbon in future years than at present between now and 2030, yet atmospheric warming would not exceed 2.5°C. If the higher values apply, then even if no more carbon were to be added between now and 2030, the temperature rise could still reach 2.5°C as the world comes into equilibrium with current warming. This measure of climate sensitivity is thus the most critical variable in our estimates of future warming, certainly far more important in the short term than any range of estimates of future carbon dioxide output, and current research is attempting to acquire evidence for the best value to use in our calculations. Most projections show several outcomes utilising different sensitivities: currently, typically the following values are used: low (1.5°C), best guess (2.5°C) and high (4.5°C). Sensitivity is affected by such issues as the increased cloudiness of a warmer Earth and the time it will take to warm the deep ocean. It is also strongly affected by the cooling, resulting in a low figure for sensitivity. A limited cooling effect moves the value towards the high end of the range. Figure 7.8.1 shows the likely effect of the Kyoto Agreement on the middle-range sensitivity. The desired target range is 450–550 ppmv CO_2.

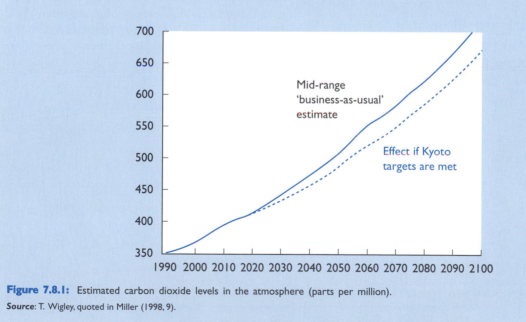

Figure 7.8.1: Estimated carbon dioxide levels in the atmosphere (parts per million).
Source: T. Wigley, quoted in Miller (1998, 9).

issues here. Box 7.9 reviews the somewhat complicated pattern of interactive economic and political analysis. The point here is that three different CO_2 trajectories were presented, two temperature trajectories calculated, and carbon-reducing technologies costed on the basis of payoff. The final costs of delay could then be set against the benefits, with the net gains available for JI and CDM measures. For better or for worse, this is an example of integrated modelling that may well shape the destiny of Kyoto.

Box 7.9 The politics of delay in climate change response

Two US economists, Bob Repetto and Duncan Austin (1997), have looked at the range of assumptions built into a variety of economic models to explore the link between future economic prospects as influenced in part by the nature and timing of climate change response measures. Their conclusions are portrayed in Figures 7.9.1 and 7.9.2, based on two principal models. These are the Wigley, Richels and Edmonds scenario (WRE) and the Working Group I (WG-I) scenarios. WRE paths allow for significantly higher emissions in the near term than their WG-I counterparts before requiring more stringent reductions later to meet the same final concerntration level. The issue at stake is the pattern of CO_2 concentrations and the relative costs of removal of excess CO_2. According to the Repetto and Austin analysis, 80% of the variation of future models can be based on eight policy topics, namely:

- the scope for alternative, non-carbon fuels;
- the efficiency of the macro-economic response;

- the degree of substitution for energy and economic goods;
- the extent of joint implementation;
- the use of carbon tax revenues (i.e. to offset other distortions, or to recycle to the economy);
- the degree of aversion of associated air pollution damage (caused by removing other noxious gases by switching vehicle fuel);
- the degree of climate change damage averted in the future;
- the size of the eventual CO_2 emissions reduction.

Of course, these are all massively indeterministic policy areas. Once again, Holling's evolutionary approach to interdisciplinary science may have to be invoked to cope with climate change on something other than a programmatic basis. This will require politics of great vision and democracy that is more participatory than at present is the case.

Source for figures: Austin (1997, 3, 4, 8, 27).

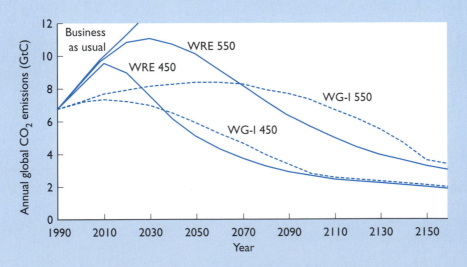

Figure 7.9.1: Alternative emission paths to reach final carbon dioxide concentrations of 450 ppmv and 550 ppmv (parts per million by volume).

Box 7.9 (continued)

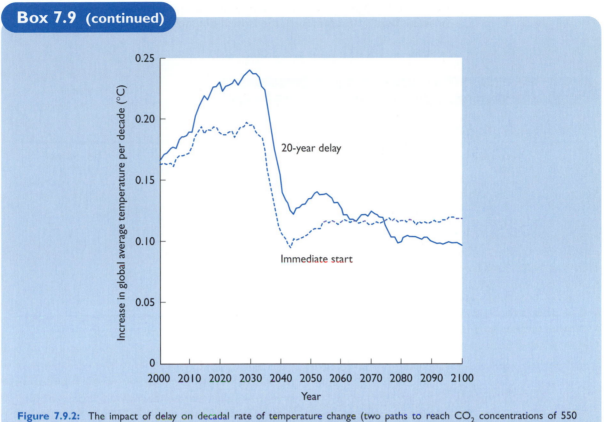

Figure 7.9.2: The impact of delay on decadal rate of temperature change (two paths to reach CO_2 concentrations of 550 ppmv calculated by the IMAGE model).

Conclusions

This has been a long and complicated chapter. Climate change is one of the ultimate integrating themes. By way of a summary, here are five main conclusions which relate science to perception to policy to action.

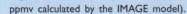 The role of climate change science is changing from networks of collaborating, but separately operating, natural and social scientists, to integrated modelling and the emergence of cross-scientific interdisciplinary assessments. This is increasingly the norm of inter-research council funding arrangements. There is nothing better than the institution of 'grantsmanship' to fuse science. For example, there is a proposal to create a 'virtual research centre' for climate change studies across the three main UK research councils (the Biological and Biotechnology Science Research Council, the Natural Environment Research Council, the Economic and Social Research Council). This would unite a wider group of research teams into a mix resembling the working groups of the IPCC, namely science, adaptation, mitigation and preparation for new climate futures.

● It is now widely recognised that this interdisciplinary science must become inextricably part of the policy machinery and not remain aloof from it. This means penetrating business, government, civic groups and schools, so that no nook and cranny of culturally acquired global outlooks is left untouched by the process. Box 7.10 summarises the British dilemma in this regard.

Box 7.10 The UK dilemma for CO_2 removal

Table 7.10.1: Emission trends and targets, in million tonnes of CO_2 equivalents.

	Baseline emissions, 1990	1996	2010 emissions on current trends	Kyoto/EC target, 2010	UK target, 2010
CO_2	615	593	622	–	493
CH_4	93	78	59	–	–
N_2O	67	59	44	–	–
HFCs	15*	16	6	–	–
PFCs	0.6*	0.5	0.8	–	–
SF_6	0.7*		1.1	–	–
Total	791	747	733	692	–

* Baseline for these gases is 1995, as allowed by the Kyoto Protocol. Emissions in 1995 totalled 16.3 mtC; emissions in 1990 were 15.6 mtC.

Table 7.10.1 indicates how far the promised Blair government target of 20% reduction of CO_2 emissions diverged from the likely projections of CO_2 based on current policies. To achieve the proposed cuts, analysts adopt the kinds of strategies picked up by the analyses outlined in Box 7.9, namely a combination of energy taxes, carbon trading, fuel switching, efficiency measures through the tax credits, and tougher regulation. There are also serious proposals to charge motorists to enter towns during rush hours and to allow local authorities to keep the revenue for alternative transport investments. These might include setting up 'mobility trusts', carbon-removal charities operating on the community level and designed to cut individual car use through a series of measures such as safe cycleways, community taxis and more consumer-specific public transport. It will take a minimum of a decade for any of these measures to work through both the political hurdles and individual behaviour and outlooks.

UK industry is deeply divided over the prospect of energy taxes to promote energy efficiency, and the prospect for trading between companies (as opposed to the greater scope within companies) appears, for the moment, to be receding (ENDS, 1998, 28–33). Yet the UK has successfully managed to raise the real price of petrol by levying a tax worth nearly $2 billion annually for investments in road repair, public transport and urban traffic calming. The ultimate prize is to allow local authorities to raise their own revenues from car parking and other charges. But that is a policy beyond political acceptability for the time being. The optimistic 20% target is likely to be just that: optimistic.

Figure 7.10.1 shows petrol prices, including the tax component, in a number of countries. Taxes account for a large proportion of petrol prices in most countries: in only two countries in the chart, the USA and New Zealand, do they make up less than half of the price of a litre of premium unleaded fuel. Taxes also explain most international variations in prices. The highest pre-tax price in the chart (Austria's 30.2 cents) is less than two-thirds greater than the lowest (France's 18.5 cents). But the top price after tax (Norway's $1.15) is more than three times the lowest (the USA's 33.3 cents). Thanks to cheaper oil the dollar price of petrol has in the past three years fallen in most countries. The exception is Britain, where tax increases have helped to push up prices in dollar terms from 84.9 cents a litre to $1.08.

Box 7.10 (continued)

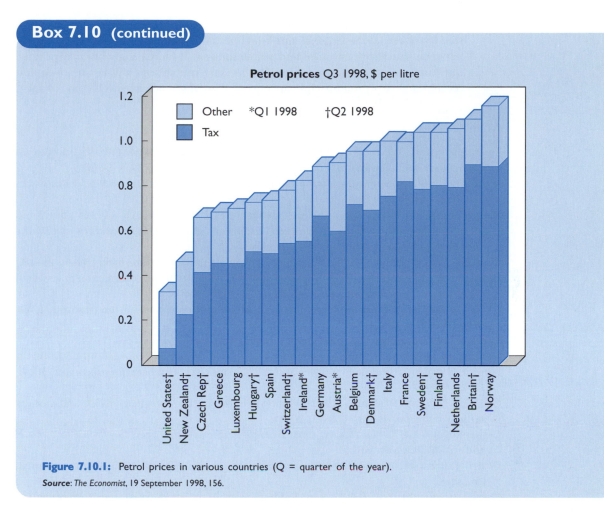

Petrol prices Q3 1998, $ per litre

Legend:
- Other   *Q1 1998   †Q2 1998
- Tax

Figure 7.10.1: Petrol prices in various countries (Q = quarter of the year).
Source: *The Economist*, 19 September 1998, 156.

● The complexities of the legal agreements linked to the global climate agreements mean that the politics and economics of policy options will have to become transmuted through hard and soft law (see Chapters 1 and 19). Over time, all science, no matter how evolutionary, will become enmeshed in this engulfing process of bargaining and compromise. In this process, ethical and equity considerations may come to dominate the science. This is the message of the contraction and conveyance scenarios, and the associated North–South political bargaining proceeding from CoP 4. The climate science of the future may well be more 'truly' interdisciplinary, as outlined in Chapter 1.

● As these negotiations over strategy and policy reveal themselves, so climate change science will become incorporated into a much wider array of policy change and technological advance – tax reform, business sustainability innovations, social welfare issues, North–South aid and capacity building transfers, and the winning of 'hearts and minds' of consumers. This means that pure climate change negotiations, complicated as they already are, will enter an even more demanding phase of multi-policy bargaining and horse-trading. Over time, any distinctive climate change politics that remain may continue to disappear in a stormy sea of international diplomacy and local action.

● The ethics, justice and distributional consequence of options and strategies will steadily overtake the original and fundamental climate change science. The link between causality, resistance and vulnerability will increasingly be incorporated into the scientific investigations. The Holling metaphors will be exposed and recalibrated. Truly interdisciplinary environmental science may well come of age.

References

Agarwal, A. and Narain, S. (1991) *Global Warming in an Unequal World: a Case of Environmental Colonialism.* Centre for Science and Technology, New Delhi.

Austin, D. (1997) *Climate Protection Policies: Can We Afford to Delay?* World Resources Institute, Washington D.C.

Banuri, T., Goran-Maler, K., Grubb, M., Jacobsen, H.K. and Yamin, E. (1996) Equity and social considerations. In J.P. Bruce (ed.), *Economic and Social Dimensions of Climate Change.* Cambridge University Press, Cambridge, 79–124.

Clark, W. C. and Jäger, J. (1997) The science of climate change. *Environment, 39*(9), 23–28.

Demerett, D. and Rothman, D. (1999) Figuring the costs of climate change: an assessment and critique. *Environment and Planning A,* **31**(3), 389–408.

ENDS (1998) Energy taxes and emission permits get mixed response from business. *ENDS Magazine,* **284**, 28–33.

Ezard, J. (1998) Antarctic ice sheet about to melt. *The Guardian,* 18 April, 7.

Fleming, D. (1998) Your climate needs you. *Town and Country Planning,* **67**(9), 302–304.

Forsyth, T. (1998) Technology transfer and the climate change debate. *Environment,* **40**(9), 16–20, 39–43.

Gallopin, G.C. and Raskin, P. (1998) Windows on the future: global scenarios and sustainability. *Environment,* **40**(3), 6–11, 26–31.

Gelbspan, R. (1997) *The Heat is On: the High Stakes Battle over the Earth's Climate.* Addison-Wesley, New York.

Global Commons Institute (1998) *Climate Change and the G8.* GLOBE International, Brussels.

Harvey, D. and Bush, E. (1997) Joint implementation: an effective strategy for combating global warming. *Environment,* **39**(8), 14–20, 36–43.

Holling, C.S. (1995) What barriers? what bridges? In L.H. Gunderson, C.S. Holling and S.S. Light (eds) *Barriers and Bridges to the Renewal of Ecosystems and Institutions.* Columbia University Press, New York, 14–34.

Houghton, J.T., Jenkins, G.J. and Ephrams, J.J. (eds) (1990) *Climate Change: IPCC Scientific Assessment.* Cambridge University Press, Cambridge.

Hulme, M. and Downing, T. (1996) *Climate Change in Southern Africa: an Exploration of Some Potential Impacts and Implications for the SADC Region.* Climatic Research Unit, University of East Anglia, Norwich.

Intergovernmental Panel on Climate Change (1997) *Climate Change 1995: Summary for Policymakers, and Technical Summary of the Working Groups Report.* Cambridge University Press, Cambridge.

International Journal of Environment and Pollution (1998) *EU Climate Policy: The European Commission Policy/Research Interface for Kyoto and beyond.* UNESCO, Paris, 345–522.

Jäeger, C.C., Renn, O., Rose, E.W. and Webler, T. (1998) Decision analysis and rational action. In S. Rayner and E. Malone (eds), *Human Choice and Climate Change,* Vol. 3. Battelle Press, Columbus, Ohio, 141–216.

Jäger, J. and Ferguson, H.L. (eds) (1991) *Climate Change: Science, Impacts and Policy.* Proceedings of the Second World Climate Conference. Cambridge University Press, Cambridge.

Jäger, J. and O'Riordan, T. (1996) The history of climate change science and politics. In T. O'Riordan and J. Jäger (eds), *Politics of Climate Change: a European Perspective.* Routledge, London, 1–31.

Jasanoff, S. and Wynne, B. (1998) Science and decisionmaking. In S. Rayner and E. Malone (eds), *Human Choice and Climate Change,* Vol. 1. Battelle Press, Columbus, Ohio, 1–77.

Kates, R.W. (1997) Impacts, adaptations and mitigation. *Environment*, **39**(8), 29–33.

Kates, R.W. and Clark, W.C. (1996) Environmental surprise: expecting the unexpected? *Environment*, **38**(2), 6–11, 28–34.

Kowolak, M.E. (1993) Common threads: research lessons from acid rain, ozone depletion and global warming. *Environment*, **35**(6) 12–20, 35–38.

Lanchberry, J. (1997) What to expect from Kyoto. *Environment,* **39**(9) 4–12.

Lanchberry, J. (1998) Expectations for the climate talks in Buenos Aires. *Environment*, **40**(8), 16–20, 42–45.

Lean, G. (1994) Maverick musician could put a stop to global warming. *Independent on Sunday,* 11 May, 13.

McGibbon, B. (1990) *The End of Nature.* Viking Press, New York.

Miller, A. (1998) The gas that no-one wanted. *Review (BP)*, April–June, 9–11.

Nozick, R. (1974) *Anarchy, State and Utopia.* Basic Books, New York.

O'Riordan, T. (1997) Economic and social dimensions. *Environment*, **39**(9), 34–39.

O'Riordan, T. and Jordan, A. (1999) Institutions climate change and cultural theory, towards a common framework. *Global Environmental Change*, **9**(2).

Ott, W. (1998) The Kyoto Protocol: finished and unfinished business. *Environment,* **40**(7).

Pearce, D.W., Cline, W.R., Achanta, A.N., Frankhauser, S., Pachauri, R.K., Tol, R.S.J. and Vellinga, P. (1996) The social costs of climate change: greenhouse damage and the benefits of control. In J.P. Bruce (ed.), *Economic and Social Dimensions of Climate Change.* Cambridge University Press, Cambridge, 183–224.

Rayner, S. and Malone, E. (eds) (1998a) *Human Choice and Climate Change*, 4 volumes. Battelle Press, Columbus, Ohio.

Rayner, S. and Malone, E. (eds) (1998b) *Human Choice and Climate Change: Ten Suggestions for Policymakers: Guidelines from an International Social Science Assessment.* Battelle Press, Columbus, Ohio.

Repetto, R. and Austin, D. (1997) *Climate Protection Policies: Can we Afford to Delay?* World Resources Institute, Washington DC.

Robotham, E.J. (1996) Legal obligations and uncertainties in the Climate Change Convention. In T. O'Riordan and J. Jäger (eds), *Politics of Climate Change: a European Perspective.* Routledge, London, 32–51.

Rotmans, J. and Dowlatabadi, H. (1998) Integrated assessment modelling. In S. Rayner and E. Malone (eds), *Human Choice and Climate Change,* Vol. 3. Battelle Press, Columbus, Ohio, 291–376.

Skea, J. (1996) Packing knowledge for policy consumption: does IPCC make a difference? In J. Smith (ed.), *Institutions for Global Decisionmaking: Europe and Climatic Change*, Department of Geography, University of Cambridge, Cambridge 7–11.

Victor, D.C. and Salt, J. (1994) From Rio to Berlin: managing climate change. *Environment*, **36**(10), 6–15, 25–31.

Von Moltke, K. (1996) External perspectives on climate change: a view from the United States. In T. O'Riordan and J. Jäger (eds), *Politics of Climate Change: a European Perspective.* Routledge, London, 330–337.

Von Weizsäcker, E., Lovins, A.B. and Lovins, L.H. (1998) Factor 4: *Doubling Wealth, Halving Resource Use.* Earthscan, London.

Chapter Eight

Managing the oceans

Michael P. Meredith, Ian P. Wade, Elaine L. McDonagh and Karen J. Heywood

Topics covered

- Oceans and climate
- Modelling ocean currents and gyres
- Monitoring ocean change
- Minerals on the sea bed
- Fisheries management
- *Brent Spar* politics

Editorial introduction

Humans have depended upon the seas for many thousands of years. Remains of marine organisms found in human settlements along the Baltic Sea shorelines date back as far as the Mesolithic Era some 10 000 years ago. These people's reliance on seafood is evident not only in the large piles of shells and fishbones they discarded but also in the tools they left behind, such as barbed fishhooks and serrated harpoons. The 'Maglemosians', as these societies were called, were probably the first true maritime people.

As our economies have developed, we have progressively found new ways to use the oceans' resources to our own advantage. Civilisations which have exploited both land and sea resources have given themselves added security over their exclusively terrestrial counterparts. In modern times, in addition to food, we have increasingly come to depend on the seas for energy, in the form of non-renewable sources such as fossil fuels, and renewable sources such as wave and tidal power. Ship-borne cargoes are vital to our commercial interests, as are the mineral resources we extract from the seabed. The sea also has value to us as an area for tourism and recreation, and, controversially, as a disposal site for industrial wastes. The oceans thus constitute a vast resource and, as such, require effective management to ensure their exploitation remains sustainable and non-damaging. This chapter describes some of the oceanic processes affecting our exploitation of the seas and the issues involved in managing that exploitation. Special emphasis is given to the issues of global fisheries and deep-sea waste disposal, and some of the legal aspects which are necessarily associated with them.

In addition to their commercial and industrial value, we have recently begun to understand how reliant we are on the oceans for their profound role in controlling our climate and its variability. Although still a relatively young science, the study of the oceans has already revealed much about the ways the sea distributes heat and greenhouse gases and forms coupled systems with the atmosphere and cryosphere (ice), which are manifested as natural variability in climate. However, it is clear that considerable further research is needed to understand these processes fully. Also in this chapter, we outline the present state of our understanding of how the oceans affect our climate, and some of the ways in which that understanding is being advanced.

This book covers aspects of ocean management in a variety of ways. It is relevant to Chapters 1, 2, 5, 7, 10, 14 and 19. The science of the ocean requires huge amounts of investment in monitoring its physical movement, its chemical composition, its temperature and salinity, and the micro-processes that appear to regulate clouds and land weathering. Ocean science is one of the most evolving parts of global environmental change research. While climate change has been the top topic for a decade, ocean health may well take its place in the next. Yet the key to this lies in that precious commodity, namely international co-operation and reliable, prolonged expenditure on recording and modelling. It remains to be seen if the international scientific organisations are prepared to commit themselves to such a task over the coming decades. This has not been the case in astronomy, so one wonders if the full scientific benefits of the present monitoring effort will be adequately recognised.

The oceans are a battleground of competing national ownership, commons property, and large commercial investment in fishing, mining and shipping. This is the most endangered regime where all the proto-commitments to managing the commons outlined in Chapter 19 will be put to the test. The Law of the Sea Conferences have worked wonders in regulating shipping, dealing with liability for pollution and waste, and beginning the arduous task of allocating property rights, as is discussed in this chapter. Sadly, the much promised Convention on the Oceans, designed to be one of the quintet of Rio agreements (climate, biodiversity, desertification and forests being the others) has still to materialise, along with its sister on forests. This delay is partly due to the youthfulness of the science of the oceans as a totality of biogeochemical systems. But it is very much a function of the development struggle between North and South, and the huge power of the commercial interests at stake. Eventually there will be a convention. One would like to think that the science discussed here will critically play its part. Any future for the world's oceans that does not fully appreciate the life-support functions, and the climate-linked responsiveness of ocean and ice systems, will be doomed to let down the well-being of our offspring.

The *Brent Spar* episode reveals how difficult it can be to try to separate 'science' from wider policy and ethical issues. This point is dealt with in detail in Chapter 1 and recently emerged when the Environment Agency requested ministers to integrate social and moral aspects of regulating the Sellafield complex, rather than leave the matter solely to the agency. In point of fact, the *Brent Spar* case will be relearned by business time and time again, as is happening recently to Monsanto (see *The Ecologist*, 1998) over the production of genetically modified crops. The lesson is: 'do not make a decision, then try to defend it.' Business is now listening to that message, but old attitudes die hard, in science as well as in the boardrooms.

The ocean is the last planetary frontier: it is as mysterious as outer space, and in Gaian terms it is indispensable. Not to support a holistic ocean science would do an enormous disservice to the effort now being put in by the scientists represented by the authors of this chapter.

The role of oceans in climate change

Solar radiation, the primary energy source for life on Earth, falls unequally on the Earth's surface. The relatively low level of incident solar radiation in polar regions compared with equatorial regions sets up meridional (north/south) temperature gradients in each hemisphere. The circulation of the atmosphere and ocean is driven by this meridional temperature gradient and the tendency for each of these fluids to transport heat down-gradient. The ocean has always been known to contribute to this flux, but whereas 20 years ago this contribution was thought to be at most a quarter of the total flux, recent modelling and observational efforts put it at more like one-half. The atmospheric and oceanic circulations interact at all scales. The largest of these interactions are the wind-driven gyres; these are large rotational circulations the size of an ocean basin, stretching over thousands of kilometres and dominating the subtropical oceans. At the other extreme, bubbles as small as a few millimetres in diameter burst at the ocean's surface and inject water and salt particles (which ultimately encourage cloud growth) into the atmosphere. The climate, often thought of as a long-term average of atmospheric patterns, is therefore intimately interdependent on the processes occurring in the ocean.

The oceans and atmosphere react in very different ways to changes in heating. This is not least because of the high specific heat capacity of water compared with dry air (four times larger), and a difference in density of three orders of magnitude. The oceans can therefore act as a much more effective store of heat, over perhaps decades or centuries, whereas the atmosphere reacts to heating or cooling on timescales of weeks. Heating the ocean not only allows for a transfer of energy from the atmosphere to the ocean but also causes evaporation and a freshwater flux from the ocean to the atmosphere. The increase of water vapour in the atmosphere has two immediate effects or feedbacks. First, the water vapour acts as a greenhouse gas, i.e. it is transparent to the incoming short-wave solar radiation but reflects and scatters the long-wave radiation which is re-emitted from the Earth, trapping energy in the troposphere and warming it (see also Chapter 7). The global effect of water vapour as a greenhouse gas is difficult to estimate because of its high temporal and spatial variability. However, it is certainly the largest contributor to the 'natural' greenhouse effect, with a global effect thought to be two to three times that of its nearest rival, carbon dioxide. If this positive feedback were left unchecked then the situation would run out of control, with a continually warming troposphere encouraging more evaporation from the ocean's surface, increased water vapour and an enhanced greenhouse effect. The increased water vapour in the atmosphere has another major effect – an increase in cloud formation. Cloud acts to increase the albedo of the Earth, i.e. the amount of incident solar radiation reflected back into space. This reduction of incident energy to the ocean/atmosphere system acts to curb the effect of the initial increase in energy and move it closer to its original state. This is an example of negative feedback. By heat and freshwater fluxes at the air–sea interface, the ocean can crucially act to moderate changes in climate.

Oceanographers tend to divide the ocean circulation into a wind-driven component (forced directly through friction from the atmosphere) and a buoyancy-driven component (forced by the difference in density of water at different locations). Although this division is somewhat artificial, it is useful in describing certain aspects of the circulation. The part of the ocean circulation driven by buoyancy (heat and freshwater) fluxes between the ocean and the atmosphere is generally known as the *thermohaline circulation* (see Box 8.1).

Box 8.1 The great ocean conveyor

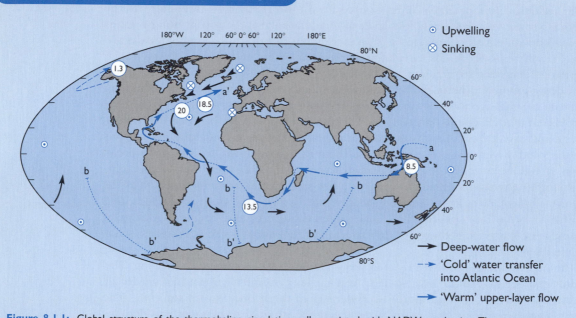

Figure 8.1.1: Global structure of the thermohaline circulation cell associated with NADW production. The warm water route shown by the solid arrows, marks the proposed path for return of upper-layer water to the northern North Atlantic as is required to maintain continuity with the formation and export of NADW. The circled values are volume flux in 10^6 m^3/s, which are expected for uniform upwelling of NADW with a production rate of 20×10^6 m^3/s. These values assume that the return within the cold water route, via the Drake Passage, is minor significance.
Source: Gordon (1986).

Description of the thermohaline circulation is often conceptualised as the *Great Ocean Conveyor Belt* (Figure 8.1.1), which overturns on a timescale of hundreds of years. The formation of North Atlantic Deep Water (NADW) is initiated when cold salty surface water undergoes deep convection, i.e. strong downwards mixing. This occurs in the boreal winter, when surface waters are made very dense by a large heat flux to the atmosphere. This deep water moves south and eventually meets northward-moving bottom water formed in the Antarctic (Antarctic Bottom Water, AABW). AABW, denser than NADW, is made very saline by ice formation in the Weddell Sea. (Salt tends to be expelled from sea water as it freezes, a process known as 'brine rejection'). The NADW exits the South Atlantic as part of the eastward-flowing Antarctic Circumpolar Current. Some of this NADW enters the Indian and Pacific Oceans (Figure 8.1.2). NADW is progressively altered, through mixing and upwelling, until it

becomes intermediate and surface waters. These surface waters then return to the South Atlantic and become incorporated into the subtropical gyre. The southward flow of deep water and the northward flow of upper-layer water put the South Atlantic in the unique position of having an equatorward heat flux. The warm return flow of the thermohaline circulation crosses the equator in the western Atlantic and is passed to the northern North Atlantic by the subtropical gyre.

The path by which the warm return flow enters the South Atlantic continues to be a topic for debate, as indeed are many other elements of the thermohaline circulation. Apart from a small flow from the Pacific to the Atlantic Ocean via the Arctic, inputs to the Atlantic Ocean must pass either south of South America or south of South Africa. The 'cold water path' passes south of South America to complete the thermohaline circulation. Water which passes south of South Africa to return to the North Atlantic takes the

Box 8.1 (continued)

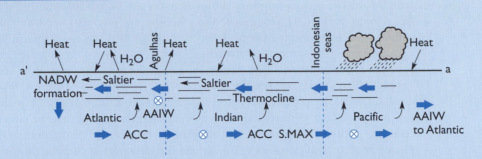

Figure 8.1.2: Schematic representation of the thermohaline circulation cell associated with NADW production and the warm water route along line *a–a'* shown in Figure 8.1.1. The upper-layer water within the main thermocline begins its passage to the North Atlantic in the Pacific as low-salinity water. It enters the Indian Ocean via the Indonesian seas, where its salinity and volume flux increase by excess evaporation and further upwelling of NADW, respectively. The thermocline water enters the Atlantic south of Africa and spreads to the northern Atlantic, continuing to increase in salinity and volume flux.
Source: Gordon (1986).

'warm water path'. The majority of the water travelling via the warm water path is contained in discrete closed loops of water around 200 kilometres in diameter, known as *Agulhas rings*. As well as having been observed through temperature and salinity measurements taken from ships, these can be seen in satellite maps of ocean temperature, from which they can be observed to form every few months and travel into the subtropical South Atlantic. This type of inherent variability in the thermohaline circulation causes problems in estimating exactly how the ocean transports heat and freshwater, and, as a consequence, affects climate variability.

Although this is a useful description, recent studies of the thermohaline circulation have derived circulation schemes significantly more complicated than the idea of an ocean conveyor belt. Hydrographic sections have been taken at various places in the oceans. These comprise measurements of the temperature and salinity of the water to the full depth of the ocean, often right across an ocean basin, and provide some of the best present estimates of heat and freshwater fluxes in the ocean. When data from a number of different sections are assembled, they can be used to solve the equations which govern the ocean circulation. Currently there are more variables in the equations than there are data, so various assumptions need to be made. Also, there are problems with the time-varying nature of the flux. For example, a hydrographic section which crosses an Agulhas ring would give different results from one which did not, even though they may have been performed at exactly the same places. The most recent and sophisticated computer models of the ocean are now powerful enough to be 'eddy-resolving', thus permitting features such as the Agulhas rings to be represented. Comparison of these models with their less advanced predecessors (which could not include such rings) shows very different heat flux characteristics in the South Atlantic. Decreasing the resolution of the model leads to the near extinction of the warm water path and a lower temperature for the South Atlantic. Therefore the inclusion of these finer-scale features in computer models of the oceans is essential to produce accurate estimates of the oceans' capability to transport heat.

The linking of the thermohaline circulation and climate is clearly observed by examining how the two have changed in the past. For example, around 11 000 years ago (the

'Younger Dryas'), part of the thermohaline circulation was suppressed due to a reduction in the amount of NADW being produced. During this period, the thermohaline circulation was consequently very different, with the North Atlantic becoming much colder than at present. This rapidly led to a build-up of glaciers over some higher areas of countries adjacent to the Atlantic. It has been suggested that such thermohaline 'catastrophes', where the circulation switches from a strong to a weak state, are likely to occur very rapidly but will take rather longer to recover. For example, the Younger Dryas event developed over just a few decades but lingered for centuries. It may be that the North Atlantic is, at present, close to one of these sudden switches of state. It is thus vitally important that we research as fully as possible the effects that the oceans have on the climate, particularly what changes are likely to be induced by our own actions.

Fluxes of gases and particles at the air–sea interface are of climatic importance, with the ocean acting as both a source and a sink. Perhaps one of the most important of these fluxes is the ocean take-up of atmospheric CO_2. Carbon dioxide contributes the most to the greenhouse effect after water vapour. Although carbon dioxide is well mixed in the troposphere, its chemical uptake by the ocean has great regional variability, dependent on the solubility of the gas in the underlying water. The solubility of a gas in water depends on the water's temperature and pressure, and slightly upon its salinity. Therefore the areas where deep waters form, and surface waters are very cold and dense, make ideal regions for taking up CO_2. Not only is the solubility of CO_2 high in these waters, but the CO_2-laden waters are continuous taken away from the surface and stored in the deep ocean. The dynamics associated with the uptake of CO_2 are not simple, and include the gas being involved in the reaction which creates soluble ions from the gas in the atmosphere. The ocean is currently acting as a sink to CO_2 by this purely chemical mechanism, although upwelling at the equator facilitates an ocean-to-atmosphere flux of CO_2 as the pressure, and therefore solubility, of carbon dioxide decreases as the water approaches the surface.

In addition to the purely chemical sink for carbon dioxide described above, the oceans also sequester CO_2 from the atmosphere via a key biological process, photosynthesis. This process is mediated by microscopic plants called phytoplankton, which float freely in surface ocean waters, and require nitrate, phosphorus, trace elements and light in addition to dissolved carbon dioxide. The uptake of CO_2 by phytoplankton causes an undersaturation of CO_2 in surface ocean waters, which drives further transfer of CO_2 from the atmosphere to the oceans. Much of this biogenic carbon is degraded in the surface oceans. However, some escapes to the ocean interior and represents a sink for carbon. In the interior most of the biogenic carbon is remineralised to an ionic form of carbon dioxide. This additional CO_2 is taken up throughout the water column. A small portion of this biogenic carbon falls to the sea floor and returns the fixed carbon to the solid Earth reservoir. The exact magnitude of the biological sink is a matter of some debate, but it is generally agreed that the chemical sink is substantially larger. The uptake of CO_2 due to the combination of the biological and chemical sinks is estimated as being approximately one-quarter of the additional CO_2 added to the atmosphere each year, namely a subtotal of some 2 billion tonnes of carbon, but the uncertainty in this number is 50%. Half of the CO_2 added to the atmosphere remains there, which leaves an unknown sink of CO_2 of comparable size to that described for the ocean. At least some of the mechanisms proposed for this additional sink are marine.

The ocean is not as significant a sink for other greenhouse gases, such as chlorofluorocarbons (CFCs). For methane (CH_4) and nitrous oxide (N_2O), the ocean actually acts as a source. Methane and nitrous oxide form when biogenic material decays in the absence of oxygen (anaerobically). This may occur in regions of large phytoplankton biomass, where the decay of biogenic material means that oxygen levels are sufficiently reduced. The oceanic contribution of methane input to the atmosphere is around 1–2%. The ocean input to the atmosphere of nitrous oxide, an increasingly important greenhouse gas, is around 25%.

Another climatically important gas, for which biological activity within the ocean is a source, is *dimethyl sulphide* (DMS). Appreciable quantities of DMS have been measured in association with plankton blooms, particularly after the peak of the bloom when zooplankton grazing begins to dominate. Although DMS is consumed within the ocean via several mechanisms, a significant amount does escape to the atmosphere. In the atmosphere, the DMS is oxidised to particles, or aerosols, which scatter and absorb radiation. They also encourage cloud growth by acting as cloud condensation nuclei. Sulphate aerosols have limited life spans and are therefore only detectable in the vicinity of their terrestrial or marine source, the most striking manifestation being in the acidifying of rainfall.

Other aerosols are injected into the air when waves whitecap and break. During this process, spray is liberally thrown into the air, and much of this is whisked away, adding to the store of water and condensation nuclei in the atmosphere. When waves break, bubbles of gas are injected into the ocean surface. Often before the gas can dissolve, the bubble rises to the surface and bursts, injecting a plume of droplets into the air which are swept away. The salt particles injected into the air by wave breaking and bubble bursting constitute the largest quantity, by mass, of cloud condensation nuclei in the maritime atmosphere, and a significant amount over land. The resulting cloud growth has a first-order effect on the climate and therefore affects all of the processes (physical, chemical and biological) within the ocean which moderate climate.

Other important particles are the terrestrial aerosols which moderate the ocean's role in climate. These are particularly important for biological processes. Phytoplankton photosynthesis in the open ocean is generally limited by the availability of nitrate. One source of nitrogen compounds for the upper ocean is from atmospheric deposition. The source of these aerosols is predominantly terrestrial. More often, nitrogen from deeper waters mixes up into the near-surface waters of the photic zone (the region of the ocean into which light can penetrate). However, there is now evidence that in large areas of the oceans, species of phytoplankton capable of synthesising their own nitrate may be much more significant than previously thought. If this is the case, the total export production (and by implication the biogenic CO_2 sink) may be much higher than previously calculated. There is also good evidence that iron limits phytoplankton production in some parts of the oceans (Box 8.2).

Box 8.2 Iron limitation of phytoplankton production

This phenomenon can be seen in nutrient-rich areas of the equatorial Pacific, where chlorophyll *a*, a pigment in phytoplankton, can be observed only in the wake of islands. Here, terrestrial dusts containing iron are present, stimulating the growth of the phytoplankton. The iron limitation hypothesis was tested by experiments in the equatorial Pacific. Half a tonne of iron was injected into the surface water of the ocean and initiated a phytoplankton bloom which drew down 100 tonnes of CO_2 and caused a 3.5 times increase of DMS in the atmosphere. Theoretically, this process could have a large impact on climate change. In addition to causing an overall cooling by reducing the amount of CO_2 (a greenhouse gas) in the atmosphere, further cooling would be initiated by the cloud formation associated with the effects of aerosols

Box 8.2 (continued)

which are a product of DMS. Around one-fifth of the world's oceans are considered to be rich in nutrients while lacking in plankton growth. These areas are predominantly in the tropical Pacific and Southern Oceans. The experiments carried out in the Pacific suggest that, although the seeding of the ocean with iron has a large immediate effect on the drawdown of CO_2, the long-term climatic effect of such seeding would be limited. This is due to the stratified nature of the water column, where nutrients would eventually be used up without being replenished and become limiting. This is thought not to be the case in the Southern Ocean. Here, deep convection causes the surface water to be regularly replenished, so the nutrient supply will not become limiting, and injecting iron into the system would lead to a significant and long-term drawdown of CO_2. Theoretically, a continuous supply of iron to the Southern Ocean could cause a 10% reduction of projected atmospheric CO_2 levels for the year 2100. It has also been suggested that it was such an input of iron which contributed to the low atmospheric levels of CO_2 which kept the planet cool during the last ice age. This theory is supported by the presence of iron-rich dusts in Antarctic ice cores from this time.

Monitoring changes in ocean climate

As has been seen, there are at present large uncertainties in our predictions of the magnitude and nature of climate change likely to be caused by human action. Effective management of the oceans with regard to predicting, making provision for and even limiting anthropogenic climate change is therefore being significantly hampered. One of the main problems is the lack of appropriate oceanic observations – without these, there is no basis for testing or improving existing climate models, and consequently no way of accurately predicting how our climate will change under the various different possible scenarios. To help address the problem, scientists involved in the World Ocean Circulation Experiment (WOCE) have been directly measuring various properties of sea water (for example, temperature, salinity, dissolved oxygen concentration, CFC concentrations) along dozens of transects across the major ocean basins. This will enable important properties such as heat transport to be calculated at various places in the oceans and give new insight into the nature of the present thermohaline circulation. The measurements were all made between 1990 and 1997, thus providing a 'snapshot' of the state of the world's oceans during this period against which future changes in ocean climate can be determined. However, since the WOCE will only provide the starting point from which climate changes can be observed, reliable monitoring of the evolving state of the oceans over long time periods clearly needs further efforts to be made.

How this long-term monitoring can best be performed is not straightforward. Intuitively, one might think that regularly repeating temperature measurements at the WOCE locations would clearly reveal any changes, and there is indeed much value in performing such repeats. However, the WOCE has been the largest oceanographic experiment ever undertaken, and its huge cost (approximately $1 billion) precludes the repeating of the vast majority of its measurements. An additional problem is that any anthropogenic changes in ocean temperature are likely to be small compared with the natural variability in temperature at any given location. For example, naturally occurring 'mesoscale' ocean fluctuations (which are often associated with eddies and meanders of ocean currents, and have typical scales of 100 km and

100 days) can have temperature variabilities of around 1°C, many times greater than the variabilities likely to be associated with greenhouse gas-induced temperature changes over the same period.

Measuring long-term changes in sea level at tide gauge sites can provide useful information about climatic signals in the oceans. A problem with tide gauges, however, is that they measure only local sea level. Even with a comprehensive network of tide gauges at continental and island sites such as is maintained by the Permanent Service for Mean Sea Level (PSMSL), the measurements can be taken only where there is a coastline, leaving the vast majority of the ocean as an unknown quantity. Satellites are proving useful in addressing this problem, with radar altimeters such as the joint US–French TOPEX/POSEIDON mission now providing measurements of sea level to an accuracy of a few centimetres over the whole world ocean. Also, ocean temperatures are now routinely measured from space, with instruments such as the UK's Along-Track Scanning Radiometer (ATSR) regularly producing global maps accurate to a few tenths of a degree Celsius. Although restricted to measuring just the very surface of the ocean, such satellite-borne instruments will be fundamental to long-term monitoring of the ocean climate (see also Box 7.3).

A promising project for studying changes in the large-scale temperature patterns of the oceans is the Acoustic Thermometry of Ocean Climate (ATOC) programme, which uses the nature of sound transmission through sea water to avoid the problem of natural variability swamping the anthropogenic signal (see Box 8.3). By transmitting an acoustic pulse from one point across an area of ocean, and measuring the time between transmission and arrival at a listening hydrophone, it is possible to derive the *mean* temperature of the water between the two points. The great potential of ATOC is due to the fact that the time taken for the acoustic pulse to make its journey will decrease if the intervening water is warmed, and so by repeating the acoustic measurements over long periods it should be possible to monitor the long-term trends in ocean temperatures. Fortunately, the oceans are good propagators of sound, thus the acoustic sources and receivers can be placed at separations enabling the averaging of temperature over very large distances.

Concerns voiced about the possible impact of the sound sources on marine life have highlighted conflicting interests in ocean management. It is known that many species such as whales and dolphins communicate, locate prey and even navigate using acoustics and the sound channel. Some marine biologists and environmental groups fear that the sound sources will add unreasonably to an ocean already over-polluted with noise from ships, oil and gas exploration, military testing, and so on, and that additional human-made noises could further disrupt and even permanently damage the hearing of these animals. This could then compromise their ability to survive and reproduce, endangering the future of their species. The true degree of potential danger is obviously difficult to assess, since the possible long-term effects of additional noise pollution will only reveal themselves after many years of acoustic transmissions. It has been argued that whales already withstand very loud natural noises, for example from earthquakes, and that by comparison the effects of the acoustic sound sources will be small. However, the reverse argument has also been made, that many whale species are already endangered, and adding to natural harmful effects with ones of our making is not acceptable. Clearly there is a need for care to be taken in such climate monitoring programmes.

It is important to note that ATOC will measure *all* climate variability, both natural and anthropogenic combined. Distinguishing one from the other in the measurements of acoustic travel time is not straightforward, since both can occur on similar timescales. However, it is known that the spatial relationships of these changes are likely to differ,

Box 8.3 Acoustic propagation in the oceans

The speed of sound in sea water depends primarily on the temperature and pressure of the water, with salinity playing a much smaller role. For most of the world's oceans (excluding the polar regions), sound speed initially decreases with depth due to the associated decrease in temperature away from the surface. However, it increases again in the deeper waters due to the increased pressure there. This leads to the refraction of the transmitted sound back towards the axis of minimum sound velocity, which effectively focuses the sound away from the ocean's surface and bottom, where significant scattering and attenuation would otherwise occur. This axis of minimum velocity – the so-called sound channel – is what enables sound to be transmitted over huge distances in the ocean. For example, in January 1991, coded acoustic signals transmitted from a source near Heard Island in the southern Indian Ocean were able to be detected up to 16 000 km away (Figure 8.3.1). Acoustic rays transmitted to monitor temperature changes follow the paths of 'refracted geodesics'. These are almost great circles but diverge slightly due to the horizontal refraction of sound caused by horizontal temperature gradients in the ocean, and also the flattening of the Earth at the poles.

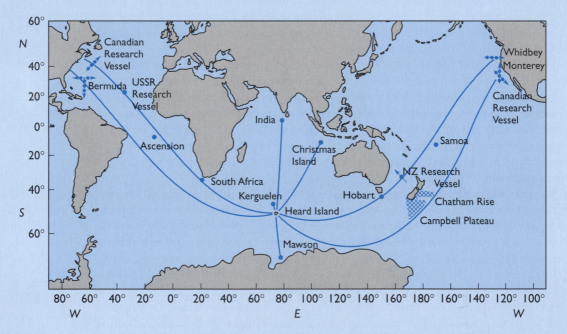

Figure 8.3.1: Ray paths from source to receiver sites are refracted geodesics, i.e., great circles corrected for Earth flattening and horizontal sound-speed gradients. The source array was suspended from R/V *Cory Chouest* 50 km south-east of Heard Island. Single dots indicate sites with single receivers. Dots connected by horizontal lines designate horizontal bottom-mounted arrays, vertical lines designate vertical arrays, and slanted lines designate arrays towed in the direction of the arrow. Signals were received at all sites except for the vertical array at Bermuda (which sank) and the Japanese station off Samoa. **Source:** Munk *et al.* (1994, 2331).

implying that combining ATOC with other techniques, such as different observations and computer modelling, is still needed. It is clear that only by studying both natural and anthropogenic climate variability jointly can the full picture of climate change be gained and a proper assessment of how to limit the effects of human influence be made.

The impact of natural climate variability

Probably the most well-known example of the oceans' effect on natural climate variability occurs in the tropical Pacific. This is the *El Niño* phenomenon ('the Christ Child' in Spanish), the name given to the roughly periodic eastward shifting of warm surface waters towards the western coast of South America. Despite being predominantly thought of as a process local to the Pacific Ocean, El Niño has huge physical and socio-economic impacts which extend around the globe. Obviously, it is vital to understand the causes and mechanisms of El Niño if we are to successfully accommodate its effects, and to understand its role in determining climate.

El Niño is, in fact, linked to one extreme of an atmospheric cycle known as the *Southern Oscillation*. This is commonly defined by the Southern Oscillation Index (SOI), the difference in barometric pressure between Tahiti and Darwin, Australia. The pressure difference is a measure of the basic driving pressure gradient force causing the trade winds across the Pacific. Approximately every 3–4 years, the pressure gradient (and hence the easterly trade winds) drops below its usual level. This raises sea level in the eastern Pacific, lowers it in the western Pacific and suppresses the usual upward movement of colder water from beneath the surface of the equatorial Pacific (Figure 8.1). Sea-surface temperatures can increase by up to 5°C. Eventually, the ocean adjusts over the whole basin until a balance is reached and the coupled system becomes stable. This stability is finally disrupted by the appearance of cold water in the central Pacific, which acts to reverse the El Niño by re-establishing the easterly trade winds.

In general, the upwelling of colder, nutrient-rich waters at the eastern boundary of the Pacific near Peru, Ecuador and Chile supports one of the most bountiful fisheries in the world. However, when these waters are replaced by warmer ones during El Niño, the anchoveta fishery collapses, sometimes with catastrophic effects for the local fishermen. El Niño also impacts on other local wildlife. For example, the marine iguanas of the Galapagos Islands starve during El Niño events, having been deprived of the algae on which they feed. Equatorial islands which are usually arid are transformed by rainfall, but at the cost of their seabird populations.

The El Niño events of 1982–83 and 1995–97 were the most severe so far recorded. The abnormal atmospheric circulations led to the west coast of the Americas being subjected to torrential rain and storms, while large parts of Australia were hit by drought. Seasonal rainfall over parts of Africa can be affected following El Niño through a lower than average number of convective storms. Reduced rainfall during the Indian monsoon can also be associated with El Niño, and in extreme circumstances can result in drought. However, not all consequences of El Niño are bad – in parts of southern Africa, rainfall and hence crop yields are highest during such events, though local flooding assisted by sand degradation (Chapter 13) can be very damaging.

Despite being the best-known and probably best-understood example of natural interannual climate variability, El Niño is not the only such phenomenon. For example, there is the North Atlantic Oscillation (NAO), a characteristic cycle in the strengths and positions of the high and low atmospheric pressure centres over the North Atlantic, which affects the climates of northern Europe and North America. There is also the Antarctic Circumpolar Wave (ACW), a naturally occurring coupled mode which affects ice distributions as well as the properties of the oceans and atmosphere (Box 8.4).

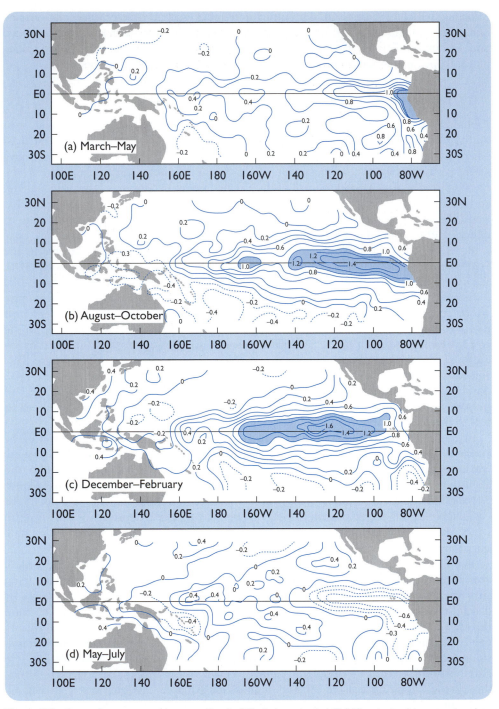

Figure 8.1: Sea-surface temperature anomalies (in °C) during a typical El Niño obtained by averaging data for the episodes between 1950 and 1973: (a) March to May after the onset, (b) the following August to October, (c) the following December to February, and (d) May to July more than a year after the onset. *Source*: Rasmusson and Carpenter (1982).

Box 8.4 The Antarctic circumpolar wave

The Antarctic Circumpolar Wave (ACW) is a disturbance in sea-surface temperature, barometric pressure, wind speed and sea-ice extent which propagates around Antarctica with a period of 4–5 years. There are two positive anomalies and two negative anomalies in one complete circuit, thus each takes roughly 8–10 years to circle Antarctica (Figure 8.4.1). The oceans around Antarctica are not zonally bounded by continents like all the other major oceans, enabling the world's largest ocean current (the Antarctic Circumpolar Current, or ACC) to flow continuously clockwise around the continent. The ACC can thus transport anomalies in sea-surface temperature clockwise around the continent into all three major oceans. Furthermore, this cycle in ice and ocean temperature could lead to changes in the rate of deep and bottom water formation, and thus impact on the atmospheric and oceanic heat budgets. An intriguing possibility which has been suggested is that the ACW is linked to El Niño through signals propagating to these high latitudes through the atmosphere. This was suggested by correlations between sea-ice anomalies and the SOI, though a *direct* link awaits proof – there is the possibility of intermediate processes being involved. The ACW is a relatively recent discovery, and it is clear that computer models which do not adequately represent it may perform poorly in predicting future climate.

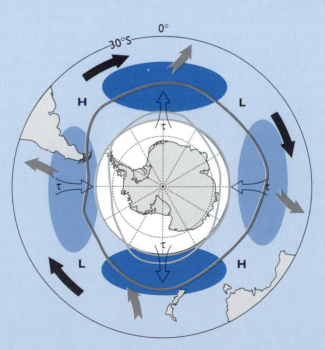

Figure 8.4.1: Simplified schematic summary of interannual variations in sea-surface temperature (light blue, warm; dark blue, cold), atmospheric sea-level pressure (bold H and L) meridional wind stress (denoted by τ), and sea-ice extent (light grey lines), together with the mean course of the Antarctic Circumpolar Current (dark grey). Heavy black arrows depict the general eastward motion of anomalies and other arrows indicate communications between the circumpolar current and the more northerly subtropical gyres.

Source: White and Peterson (1996, 701).

Although El Niño tends to recur roughly every 3–4 years, the cycle is not a simple regular one. For example, some decades feature no El Niño event at all. This makes it very difficult to predict when the next El Niño will occur, with associated uncertainties as to what fish stocks might be available, what crop yields could be hoped for, and in extreme circumstances what drought or flood provisions need to be made. There is thus a clear need for understanding and ultimately predicting El Niño if attempts at managing the ocean's climate and its effects are to be successful. However, the underlying 'trigger' for El Niño remains elusive. Possible influences that have been suggested include solar variability, volcanic eruptions, anomalously high snowfall over Tibet weakening the Indian summer monsoon, and hurricanes generated at the equator in the western Pacific.

That aside, predictions of El Niño are becoming more sophisticated as we learn more about it. For example, some climate models were able to predict the 1995–97 event a year in advance. Ultimately, the intention is to develop methods to routinely forecast the onset, development and magnitude of such events. Present problems include the limited amount of data with which to compare computer simulations, and the generally complex nature of the coupled system. However, once established, such predictions will allow improved ocean management in a number of ways. Foreknowledge of drought, crop failure and extreme weather conditions will allow at least provision to be made, if not actually prevention. Accurate predictions will enable better management of existing fish stocks by ensuring that they are not depleted to levels where they cannot recover from El Niño events. On this point, it is important to bear in mind that fish species present in the oceans today have survived countless El Niño events already; it is only with the combined effects of increased fishing pressure and additional anthropogenic changes in their environment that their existence may become threatened.

Fisheries management

Three thousand years ago, the Peruvian fishing industry was heavily dependent on the shellfish gathered close to shore and not on the fish that teemed in the offshore waters. It was around this time that our societies started to grow rapidly and to experience periodic depletions of these nearshore resources for the first time. A need for a 'harmony with nature' developed in most fisheries during these early times. Any society not developing such harmony would simply not have survived, and would have been forced to move or to concentrate on terrestrial food supplies, which were not always plentiful. Problems associated with resources close to shore encouraged early fishermen to develop methods enabling them to go further afield in search of fish.

Improvements in fishing methods and technology

Today, fishing techniques are highly varied and utilise a wide diversity of methods, though some of these have changed little since prehistory. Even in well-developed countries, some simple methods are still used in some nearshore areas (for example, trap fishing for species such as lobster and crab). Changes in fishing methods and equipment had been very gradual until the last few centuries, since when the development of major commercial fisheries has moved hand in hand with the development of methods that could easily be employed on the larger scale in the open ocean (Figure 8.2). For example, Dutch fishermen are known to have used long-line fishing on the open sea during the seventeenth century. This long-lining involved the baiting of thousands of hooks and was thus on a much larger scale than the previous inshore fishing.

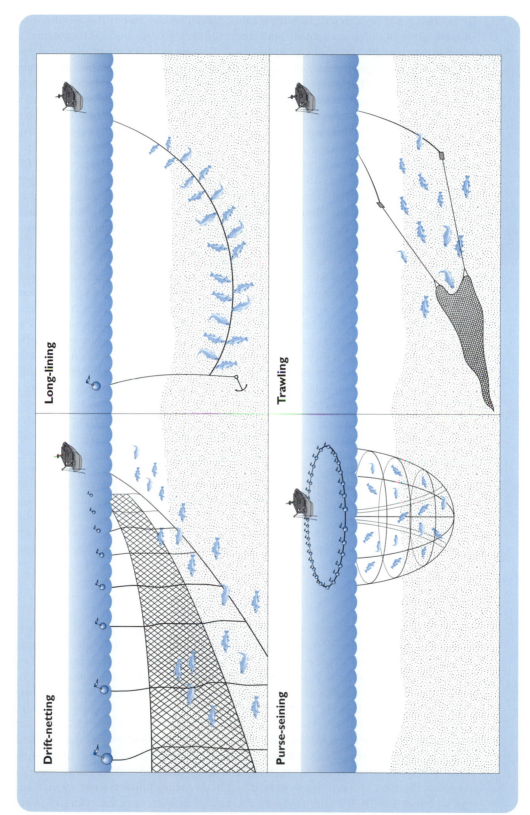

Figure 8.2: Main fishing methods.

Source: Coull (1993, 2).

By far the most productive fishing methods involve the use of nets, which can be shaped to suit different situations and, importantly, lowered on lines to greater depths and towed through the water column. Drift netting is still commonly used in large parts of the world, especially in the tuna fisheries of the tropical oceans, where the nets often exceed a kilometre in length. This method relies on the fish swimming into a passively placed net, unlike the more active methods of purse-seining and using trawl nets. Trawl nets have been in common use since the early nineteenth century and are now perhaps the most versatile of all fishing methods. Purse-seining involves encircling groups of densely packed fish, and is usually more energy-efficient than trawling. However, it was largely limited to nearshore waters until the Second World War and requires considerable labour in order to close a net of any significant size. The refinement of net fishing, especially with the use of powerful machinery for hauling in far heavier loads than could be brought aboard by manual labour, has revolutionised fishing this century. New materials for stronger, lighter nets also allow larger catches to be removed in a single haul.

Alongside the more efficient equipment, improved techniques for locating fish stocks have been developed. Specialised echo locating equipment can give fishermen valuable information concerning fish stocks, with shoals and even individual fish being detected. The 'most likely' areas and depths for catches are then trawled. Fishermen have long known that their catches are often better in the vicinity of oceanic and shelf-break fronts. In the same way that some coastal regions are highly productive, due to upwelling of nutrient-rich waters from below the thermocline (for example, the Peruvian coast in non-El Niño years), ocean fronts are also often highly productive areas. Here, vertical and horizontal water motions either 'draw up' nutrient-rich deeper waters or concentrate phytoplankton blooms in convergent zones, ultimately leading to increased fish numbers. Since 1978, satellite-borne instruments have made the locating of such areas routine by producing maps of sea-surface temperature and phytoplankton pigment concentration over very large areas of ocean. These data can then be processed and supplied as 'fishing probability' maps to the fishermen (Figure 8.3).

The efficiency that we have achieved in removing fish stocks from the oceans has increased massively since the Industrial Revolution. In the last few decades the availability of larger fishing vessels, improved techniques, increased knowledge of fish behaviour and life cycles as well as 'state of the art' technology have put unprecedented pressure on these stocks, and fisheries regulation has never been as important as it is today.

Fisheries policy and methods of regulation

By the early 1950s, fisheries management developed at the turn of the century was beginning to have some effect in reducing the strain on fish stocks. However, during the late 1950s and 1960s the production of far more effective nylon nets, outboard motors and heavy lifting gear, as well as the more advanced fish locating equipment, led to serious depletions in fish stocks from a number of locations worldwide. By the 1970s, total world catch had levelled off. The vast fleets of factory ships had massively overfished the North Atlantic between the 1950s and 1970s, with single ships capable of lifting in excess of 500 tonnes in a single haul. The massive capacity of these ships and the availability of refrigeration meant that they could stay at sea much longer than before. In 1974, record catches were recorded in the North Atlantic, but 'catches per unit of effort' were declining markedly. Also, the size of fish caught had been getting smaller since 1968. Catch per unit of effort, or CPUE, is the standard method of determining the effect of fishing on

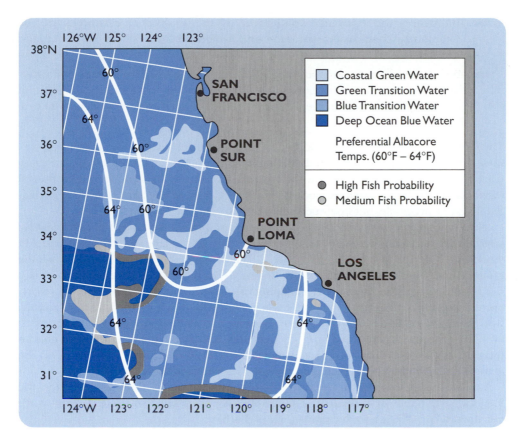

Figure 8.3: Sea-surface temperatures off California.

Source: Orbital Imaging Corporation, Dulles, VA.

stock levels. If doubling the fishing effort increases yields by only 10%, the CPUE is decreasing and the fish stock is being overexploited. But how do we measure CPUE and the total catches, and how do we control one nation's fishing effort when another nation could simply step in and continue to overexploit the fish reserves?

Marine resources management is a vastly complicated process. Fisheries managers and policy makers have to address a number of concerns from various sources:

- human needs, values and social equity;
- biological conservation and resource productivity;
- economic productivity and efficiency;
- feasibility;
- political acceptability.

To attempt to balance these concerns into a coherent international policy is by no means a trivial task. With the added problem of natural variability of the fish stocks, the task of pleasing all of the people all of the time becomes virtually impossible. International disputes occur constantly over fishing rights. Foreign incursions into domestic fisheries and the lack of faith in the enforcement of international fisheries policy prompted nearly all of the coastal nations of the world to enforce 200-mile exclusive economic zones (EEZs) in the 1970s. With the establishment of the EC fishing zone, the EC became actively involved in matters of conservation and resource management. This new responsibility led to the Common Fisheries Policy (CFP), which had first been proposed a few months earlier. The principal objectives of the CFP are:

- To ensure 'the optimal exploitation of the biological resources of the Community Zone' in the medium- and long-term interests of fishermen and consumers.

- To ensure 'the equitable distribution of these resources between member states' while maintaining 'as far as possible the level of employment and income in coastal regions which are economically disadvantaged or largely dependent on fishing activities'.

The CFP, in its current form, was adopted in 1983 and is designed to run until the end of 2002. The EC has signed 25 bilateral agreements with 15 African and Indian Ocean countries, nine with North Atlantic and one with a Latin American country, thus the EC participates in fisheries management well outside EC waters. The aims of policies such as the EC's CFP are noble, but how such aims are best achieved is open to debate.

Throughout the 1940s and 1950s, the theory and practice of *maximum sustainable yield* (MSY) became widespread. MSY had its origins in the simple biological model of stocks of single species of fish (see Box 8.5). When fishing begins on a virgin stock the equilibrium is upset and the stock size begins to decrease. This decrease results in a net increase in growth rate as the remaining fish have less competition and, therefore, more food (Figure 8.4(a)). If fishing pressure is stabilised at a low level a new population biomass equilibrium is reached. However, if fishing pressure continues to increase, the stock size continues to decrease and the growth rate continues to increase as the stock tries to replace the weight of individuals lost. At some point the growth rate will ultimately reach a maximum. This corresponds to the MSY.

Beyond the point of MSY, the rate of growth decreases (usually rapidly) and the level of sustainable yield decreases. Once MSY is approached or exceeded the catch per unit effort (CPUE) also decreases (Figure 8.4(b)).

Box 8.5 Fish life cycles

Many species of fish migrate over distances of hundreds or even thousands of miles. Adults migrate between spawning and feeding grounds and back in regular cycles. Larvae drift passively from the spawning grounds to settle in nursery areas (Figure 8.5.1). As the larvae mature they migrate to the adult feeding grounds to join the adult stock (recruitment).

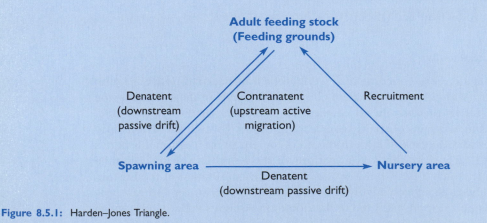

Figure 8.5.1: Harden–Jones Triangle.

Box 8.5 (continued)

An understanding of these migratory patterns has important implications for commercial fisheries. The adult population biomass can be increased by the growth of individuals within the population, or by the recruitment of young fish from the nursery grounds. Biomass is decreased by both natural mortality due to predation, starvation, disease or simply old age, and by fishing mortality. The population biomass will be in equilibrium if gains due to growth and recruitment to the adult stock balance losses caused through natural and fishing mortality (Figure 8.5.2).

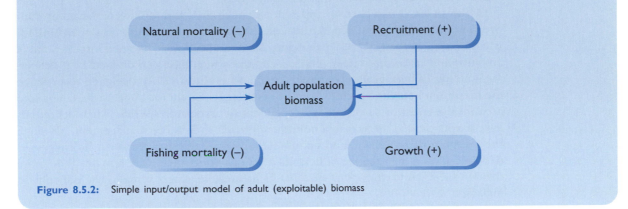

Figure 8.5.2: Simple input/output model of adult (exploitable) biomass

The CPUE shows the problems of allocating fishing quotas based on catches alone. Catches can continue to increase if fishing effort is increased (initially suggesting that fish stocks are coping, as mentioned earlier). The catch rate, or catch per unit effort, is often used as a measure of fish stock density. While the MSY model was worked during the 1950s and 1960s, other scholars began to study the economic implications of fishing policy. They suggested that overfishing was as much an economic problem as a biological one. The MSY concept was coming under attack at this time as being fundamentally flawed in its assumption that controlling the fishing effort on certain fish stocks would keep their numbers and age-class composition in some kind of biological equilibrium. It also gave no weight either to the value society places on that yield or to

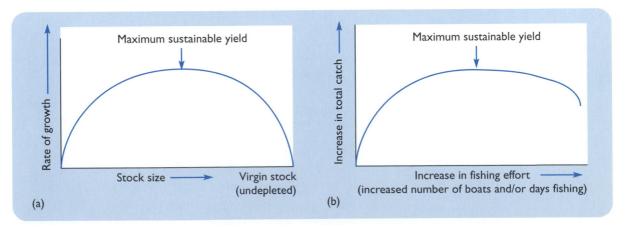

Figure 8.4: (a) Theoretical curve of rate of growth vs. stock size; (b) theorectical curve of catch vs. fishing effort.

the cost of obtaining that yield. In the 1950s and 1960s, economic considerations led to the idea of a *maximum economic yield* (MEY). The idea behind the MEY was to find the greatest net economic yield, i.e. the point where the difference between the value of the catch and the cost of obtaining that catch was at a maximum (Figure 8.5).

Although economically highly desirable, the MEY idea proved an elusive goal, with the holding of the level of fishing at the point of MEY rarely being achieved. The fishing industry has always been prone to problems caused by the 'open access' ideal. In the absence of a single governing body to hold the fishing effort at that point, the tendency is for the effort to increase as more boats join what is now an economically viable industry until the profit margins are eroded.

By the 1970s, it had become apparent that both MSY and MEY were flawed by their oversimplicity. Biologists, economists and fisheries managers were now searching for an alternative to the MSY/MEY paradigms. MSY and MEY were blended into the idea of an *optimum sustainable yield* (OSY), a concept which has been defined as 'a deliberate melding of biological, economic, social and political values designed to produce the maximum benefit to society from stocks that are sought for human use, taking into account the effect of harvesting on dependent or associated species'. Such definitions have hardly cleared the way to a uniform approach to fisheries management. Although the merging of biological and economic factors is the way forward, its success is dependent on a high degree of flexibility from all concerned. OSY remains little more than an ideal, and critics point out that its greatest weakness remains the difficulty in making it work.

At the most basic level, fisheries management aims to solve two fundamental problems. The first is conservation of the fish stock; deciding what exploitation can be sustained. This is decided using criteria of MSY *and* MEY in what is sometimes termed OSY. The second is allocation of those stocks. Several basic strategies are currently employed to address these problems, and these are discussed in the following subsections.

Figure 8.5:
Theoretical curve of expenditure/ profit vs. fishing effort.

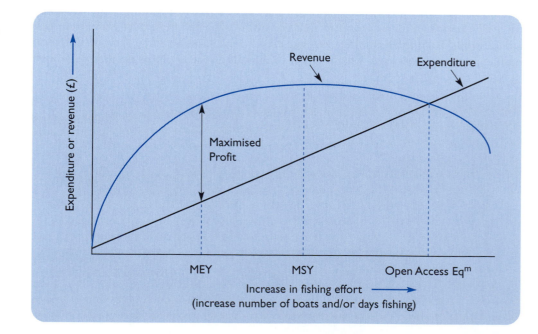

Closed areas and/or seasons

These are generally imposed when fish stocks are already under threat. The main objective is conservation through protection from overexploitation. Protected areas are usually where there are large aggregates of spawning stocks or juveniles. The major advantage of this method is its ease of implementation and its extreme flexibility, i.e. the policy can be implemented at very short notice. In practice, however, fishermen will often attempt to maximise capacity during the open seasons, which may result in overcapacity fishing. Closing areas has the disadvantage of shutting down an area to all stock removal, thus preventing our exploiting other stocks that are not in danger. This policy needs to be implemented alongside continual monitoring of catches, and this is often expensive. Closed areas/seasons are a vital tool in overall fishing policy. However, on their own, they are limited and possibly even detrimental to the overall goals of fishery management. They are most useful when used in conjunction with other strategies designed to reduce the development of overcapacity.

Equipment/technology restrictions

Restrictions usually take the form of minimum mesh sizes on trawl and other gill nets, and escape routes in fish traps or lobster/crab pots. This ensures a certain minimum age of first capture and works well in regions where single species of fish are being targeted (for example, the North Sea or North Atlantic). However, such restrictions are of little use in regions such as tropical seas, where the catches are largely multi-species. Where mesh restrictions are valid, their implementation has time-dependent effects on yield. Initially yields decrease as smaller fish are released. In theory, however, these smaller fish should then grow and increase future yields. The fall in short-term yield is problematic, and persuading people whose living depends on today's catches to adopt short-term sacrifices for long-term gains is never easy.

The attempts to control the age of first capture can be supplemented by size limit regulations. This technique has been used extensively in the lobster and scallop fisheries, where undersized captures can mostly be returned to the sea alive. Such procedures are less successful for other fisheries and offer little protection to most fish species, since mortality of undersized fishes returned to the water is generally high.

Establishment of aggregate quotas

In the 1970s and 1980s the setting of limits on total catches became the most common method of controlling fishing. These limits are now known as *total allowable catches* (TACs, e.g. Box 8.6) and are accompanied by national allocation of shares of the total quota. Management of TACs requires considerable scientific and technical manpower and is therefore not suited to smaller, poorer nations. Determining a TAC is analogous to determining an MSY. The calculations themselves are simple and based on many years of observational data. However, in order to remain a viable method of fisheries regulation, considerable monitoring is required. The problems arise in obtaining detail about the size and biological structure of a targeted stock. Although more costly than indirect methods, direct data gathering, from research vessels analysing random catches or from scientists working directly with the fishermen aboard fishing vessels, is far more accurate in practice. Acoustics and remote sensing from aircraft and satellites are used in conjunction with these research catches. Without proper thought to social and economic aspects, the TAC system has some of the same pitfalls as the MSY concept. The establishment of accurate TAC values from the scientific background is then in the hands of the international policy makers and governments, who have to share and enforce the TACs equitably at the fishing fleet level.

Box 8.6 1980s North Sea herring collapse

Annual catches of North Sea herring were stable between 1947 and 1963, ranging between 500 000 and 700 000 tonnes. In 1964 and 1965, catches increased suddenly to in excess of 1 million tonnes. However, by 1972 the catch was just 500 000 tonnes, despite increasing fishing effort, and it continued to fall after 1973. In 1974, TACs were agreed upon to reverse this trend. By 1976, it was obvious that TACs were not being enforced and that the fishery was heading towards a crisis. Demands for a closure of the North Sea herring fishery were ignored and stocks collapsed. With the extension of the EEZs to 200 miles in 1977, the responsibility for managing the fisheries passed to the EC. From 1977 to 1979 the fishery was closed in order to allow it to recover. A total herring fishing ban from 1977 to 1979 reduced the exploitation rate and thus preserved sufficient spawning potential. The policy in 1980 was to allow the spawning stock to rebuild to at least 800 000 tonnes. This figure was the estimated tonnage required to avoid a risk of a future recruitment failure. In 1983, herring fishing was allowed in parts of the North Sea for the first time since 1977. In the first few years after opening the fishery, the total catches still exceeded TAC limits. In 1982 and 1983, approximately half of the catches were still not officially reported. However, despite illegal fishing, stocks recovered in the next few years. From 1983, to 1986, TACs were almost doubled and by 1987 the spawning stock had reached the set 800 000 tonnes. While the recovery of the North Sea herring in the 1980s is a success story, the managers were very fortunate that nature provided good recruitment in those years. Illegal fishing and misreporting could have caused major problems. Tighter controls on misreporting of catches are constantly being incorporated.

Stimulation of fisheries growth or control of fishing effort through monetary measures

To improve the health of their fisheries, many governments offer financial incentives in the form of subsidies or tax incentives. Subsidies take many forms, such as fuel cost subsidies, loans for replacement of old equipment, support for fish prices and occasionally even funds to pay fines of fishermen violating the fisheries of other nations (primarily to prevent fishermen being used as pawns in international disputes). Uncontrolled subsidies, however, can easily have the wrong effect and rapidly lead to overcapacity fishing. Subsidies, used sensibly, can correct imbalances in distribution and give small-scale fishermen more chance to compete. Fishermen are taxed in a number of direct and indirect tax schemes (i.e. payments on landings, licences, berthing fees) and reduction of some of this financial burden can encourage new growth in a developing fishery. Conversely, increasing taxes can be used to inhibit overfishing. However, taxation is rarely popular and fishermen are reluctant to see taxes from fishing used elsewhere and not in, say, the improvement of the aquatic habitat. Taxation can be effective in the very long term, but with problems requiring immediate action tax schemes alone are rarely of use. In many cases it is often considered more desirable to reduce subsidies than to increase taxes.

Limiting entry and/or property rights

Limiting the access to a fishery appears, perhaps, the most effective overall method of preventing overcapacity. Such schemes are highly controversial in that some governing body has to decide who will qualify for entry to a particular area. There are generally three categories for limiting entry to a fishery. First, there are licence requirements that limit the total number and size of vessels in a region. Second, there is quota allocation, and third, there is limiting entry through monetary means such as taxes. Licensing is by far the most commonly used and is most effective when a fishery is developing. However, licences that simply limit the number and/or size of vessels cannot by themselves prevent the problem of potential overcapacity. The distribution of licences is based on considerably more than economics and involves politics, industrial relations and international law. Quotas are usually determined (as TACs) and 'sold off' to various parties, a method highly biased in favour of the larger bidders. Taxation, as mentioned previously, is often unable to respond quickly enough to changing conditions.

Property rights are based on the assumption that certain groups of fishermen will consider a particular fish stock or grounds their property. The advantage is that they may then be more inclined to act in the long term to maintain the stock and their industry. Often such fishermen have imposed stricter restrictions on their own fishing effort than external bodies have. Such voluntary restraint reduces the cost of management. Deciding who gets rights to a particular fishery is difficult, and giving rights to some may be denying rights to others who also have interests in that fishery.

The distribution of licences and the introduction of the EEZs involves the whole question of jurisdiction over marine resources. This is a contentious problem which has led to various legal aspects now playing important roles in determining how we manage the oceans, with ownership of fish stocks and mineral resources frequently the subject of controversy.

Law of the sea

In Roman times, the oceans were considered as belonging to the international community and therefore were not controlled by anyone. This communal approach persisted for many centuries. 'The high seas should be free for the innocent use and mutual benefit of all', claimed the Dutch jurist Hugo Grotius in 1609, in his treatise called *Mare Liberum* ('Freedom of the Seas'). However, as nations began to develop empires based on international maritime trade, they became more possessive. In 1702, a document named *De Dominio Maris* ('Control of the Seas') stated that countries had complete sovereignty and control in their local territorial waters. Such local waters were defined as being within one league (one cannon shot) or three nautical miles offshore, because this was the distance that could be defended by cannons.

United Nations Law of the Sea conferences

During the 1940s, North and South American countries declared jurisdiction over their continental shelves, including the associated fisheries and petroleum resources, up to 200 miles. The main reason for this was that they had realised that such resources were not inexhaustible. There was international uproar from other maritime nations, and this led eventually to a series of conferences which addressed jurisdiction over marine resources. The first was in 1958, the second in 1960, and the third started in 1973 and met every year until 1982. The last of these involved more than 150 countries and led to the United Nations Convention on the Law of the Sea (UNCLOS), opened for signature on 10 December 1982. This treaty stated that:

- territorial waters were defined up to 22 km (12 nautical miles) from shore. Countries have complete control over their territorial waters just as they do over land, but the treaty decreed that ships would have right of free passage through straits which are narrower than 22 km;

- each country would establish an EEZ 200 nautical miles (370 km) from the coast. Within the EEZ the nation has sovereignty over natural resources, economic ventures and environmental protection. This includes mineral extraction, drilling for oil and gas, fishing and control of pollution;
- the treaty confirms the freedom of the high seas;
- the treaty established an 'International Seabed Authority' to oversee the extraction of mineral resources from the deep sea (e.g. manganese nodules, found outside any country's EEZ). Private exploitation of seabed resources may proceed under the regulation of this authority. Mining sites would be developed in pairs, one run by a private company and one by a United Nations company which gives its profits to developing nations.

There was initially some opposition to the Law of the Sea Treaty. This was led by the USA and came largely from those nations which planned extensive seabed mining operations. Despite this, the UNCLOS came into force on 16 November 1994, 12 months after the date on which the sixtieth country ratified it.

A further document, the United Nations Agreement on Fish Stocks, is still under debate (see also Chapter 2). This 'elaborates on the fundamental principle that states should cooperate to ensure conservation, and promote the objective of the optimum utilisation of fisheries resources both within and beyond the EEZ'. Specifically, it attempts to promote effective management of fish stocks by establishing minimum international standards for their exploitation. It sets out to ensure that measures taken in areas under national jurisdiction are compatible with those in the adjacent high seas, and that there are effective mechanisms for their enforcement. The agreement gives particular attention to the conservation and management requirements of developing states, in addition to their requirements for development and participation in fisheries. The agreement was opened for signature on 4 December 1995 and will come into force 30 days after the thirtieth country has ratified it.

About one-third of all the world's oceans lies within the EEZ of a coastal state. The North Sea is totally divided among the nations which surround it, and the USA controls an ocean area similar to its land area. Developing countries in particular are often highly suspicious of foreign oceanographic research ships requesting permission to take measurements in their EEZ. They see such scientific studies as threatening the future exploitation of their mineral or fishery resources. The problem has been exacerbated by some nations which have pretended to be undertaking oceanographic research when actually they were spying or on military operations.

Exploitation of ocean resources

We saw in the previous section the complexity of managing the world's fisheries. Such problems are nowadays as much the realm of the lawyer as the marine scientist. Fish do not know about economic zones and may spawn in the EEZ of one country and spend most of their life in another. The exploitation of mineral resources can be equally contentious, particularly if the industry of one country may endanger the natural resources of another.

Minerals have been mined from beneath the seabed for centuries. Coal mines often extend outward from land, penetrating kilometres from the coast. The dredging of coastal waters for sand and gravel can be cost-effective if the alternative is transporting such building supplies long distances by land. In the last few decades, mining of manganese nodules from the seabed in the deep abyssal plains has been the subject of much research. These nodules are extremely numerous, found scattered throughout the deep ocean, and rich in a variety of minerals as well as manganese. Techniques to mine them have been developed, but the industry has not become as widely established as was originally hoped. Indeed, it was primarily this industry which caused such debate during the UNCLOS meetings, but in the 1990s relatively few countries are continuing the research and development for manganese nodule mining. Because they are found in international waters, developing countries felt they should benefit as much from such potential exploitation as those countries which have the technology to mine the manganese nodules.

Oil and gas contribute about 90% of the value of minerals mined from the ocean floor. Recently, the search for oil and gas has extended beyond the already well-developed North Sea continental shelf region. Since 1990, the technology has advanced to enable extraction from depths greater than 500 m. The Atlantic Frontier, the deep waters to the north-west of the UK, is currently being explored. At the Foinaven and Schiehalion fields, in UK waters between Shetland and the Faeroe Islands, oil and natural gas have been discovered and rigs are being installed. During August 1997, the Greenpeace environmental group drew attention to this new exploitation of the Atlantic Frontier by occupying a rig on the Foinaven field for a week. It believes that such oilfields might endanger the wildlife of the region, particularly whales and seabirds, through oil spills and operational discharges (see Chapter 14). Greenpeace also regards this explanation as a test case to leave oil under the Earth's surface in the battle to combat climate change. Such disputes show that the protection of the ocean in its natural state is a matter of concern to the general public. Politicians and government advisory bodies need to balance the need for fuels and employment against their responsibilities for conservation and environmental protection. The next step for the multinational oil and gas companies is probably the exploitation of the reserves in the Arctic and in the Southern Ocean around the Falkland Islands. The exploration of the Atlantic Frontier will develop and prove the technology which may be applied in the future.

Antarctica is a special case because it is internationally recognised as a wilderness area. Although many countries lay claim to various sectors, they abide by strict rules to limit pollution (for example, the husky dogs which accompanied scientists on expeditions across Antarctica are now forbidden because they may transmit diseases to indigenous species). In 1988, there was an agreement to regulate mining exploration around Antarctica. This will extend the original 1959 Antarctica Treaty signed by 20 nations, which banned all military activity and permitted scientific research. The 1988 treaty stated that no commercial activities would be allowed if they caused 'significant changes' in marine, terrestrial or atmospheric environments. Since Antarctica holds large reserves of fossil fuels and minerals, the exploitation of the Southern Ocean and continent of Antarctica is likely to be a contentious political issue in the next few decades.

Deep-sea dumping

The idea that the deep oceans might be a safer repository than land-based sites for industrial waste materials started with the disposal of munitions and chemical weapons at sea during and after the Second World War. At that time, the long-term effect of unexploded shells or poisonous gases on the marine environment was not a high priority as countries struggled to cope with the aftermath of wartime devastation of the countryside and cities. Over the past few decades, it has become increasingly apparent that the world is generating more waste than it knows how to cope with. Although there is now a long overdue move towards recycling and the reduction in the production of waste, the safe disposal of waste is always going to be an important issue. Disposal of waste into coastal seas, such as the dumping of sewage sludge, is discussed in Chapters 14 and 18. Here we shall consider whether substances such as fly ash from incinerators, nuclear waste and other by-products of heavy industry could be deposited in the deep ocean rather than being stored on land, where there is always a danger of accidental discharge into the atmosphere.

International law (for example, the London Dumping Convention) is, at present, moving towards prohibiting the use of the deep ocean for waste disposal. This view has primarily arisen because of the pollution problems caused in coastal seas by the discharge of sewage and industrial contaminants. The need to regulate and clean up coastal seas such as the North Sea is not disputed. However, the claim has been made that the deep ocean is a very different environment and should not be covered by the same prohibitions.

The ocean's abyssal plains are about 4 to 5 km deep, and away from narrow straits and the western boundaries of basins the currents are relatively slow, usually about a centimetre per second. The thermohaline circulation discussed earlier transports waters on a timescale of hundreds of years, so waters from the deep ocean will not approach surface waters for centuries. It is the near-surface waters that are generally most important for both man and marine creatures. During their passage towards the surface, there will be mixing (and therefore dilution) and breakdown by geochemical and sedimentary processes. On the abyssal plains, there is very sparse living biomass, so toxins are unlikely to bioaccumulate. We do not fish any of the species which inhabit the abyssal plains.

Heavy metals are one of the main causes of concern in coastal waters (see Chapter 14). However, in the open ocean it is unlikely that waste contaminated with such substances would be a problem. We have recently discovered the existence of hydrothermal vents in the mid-ocean ridges which have communities of benthic fauna living around them. The vents emit large amounts of heavy metals into the sea water, yet the animals have evolved to thrive in such an environment. The fact that the plumes of heavy metals were not detected by scientists until recently is testimony to the rapid mixing of the metals into the surrounding waters.

Radioactive waste is an emotive subject, and it does indeed need to be minimised. Even if no further waste were to be generated, however, there would still exist a substantial stockpile of contaminated waste and redundant weapons. Mankind needs to consider carefully the safest place to store such waste. Radioactive half-lives of some of the contaminants are so long that the waste will essentially be hazardous for ever. If it is stored on land, it needs to be outside any tectonically active region to avoid damage of the store by earthquake or volcanic eruption. It must be inaccessible to terrorists, and should remain intact in the event of bombing. It may well be that, if sufficient care is taken in choice of site and mechanism for storage, the deep ocean could be the safest place to dump the waste.

In 1983, all dumping of radioactive waste in the open ocean was banned internationally. Before that, there were a small number of licensed sites where drums of low-level radioactive waste were routinely dumped. One such site was in the Bay of Biscay. Recently, some drums were recovered and found to be leaking. However, no contamination of the surrounding sediments could be detected, with only minimal effects on the ecology of the area. Further studies are needed to consider more fully the actual and potential environmental impacts of these sites.

The Brent Spar

In June 1995, the issue of dumping industrial wastes into the oceans received almost unprecedented media attention. This was caused by the decision of the Shell Oil Company to dispose of the 14 500-tonne *Brent Spar* oil platform (Figure 8.6) by sinking it on the North Feni Ridge, about 150 miles west of the Outer Hebrides. The *Brent Spar* was originally deployed in 1976 about 120 miles north-east of Shetland to serve as a tanker offloading and storage system for the North Sea Brent Field platforms. In 1991, it was taken out of service to be decommissioned. Its operators, Shell (UK) Exploration and Production, examined a number of options for its disposal, with both offshore abandonment and onshore disassembly being considered. It eventually settled upon its favoured option of sinking the structure in about 2300 m of water on the margin of the deep North Atlantic.

Greenpeace strongly opposed this action on the grounds that the full environmental effects of such deep-sea disposal were not sufficiently well known, and that it would set a dangerous precedent. It argued that with over 200 rigs in the British sector of the North Sea currently in operation, the cumulative effect of allowing direct deep-sea disposal would inevitably be damaging to the ocean environment. Furthermore, it believed that it would undermine the existing bans on dumping radioactive and industrial wastes at sea. Instead, it favoured the dismantling and recycling of such installations onshore on the grounds that it would safeguard the marine environment. Greenpeace drew attention to the fact that the *Brent Spar* is made of 6700 tonnes of high-grade steel, which is increasingly in demand by scrap steel mills, plus smaller amounts of other metals and haematite ballast. It argued that since it takes four times the energy to manufacture steel as to recycle it, the environmental benefits of dismantling onshore stretch beyond simply preventing the deep-sea sinking.

Shell argued that the potential environmental risk caused by towing the *Spar* through the North Sea for disposal on land was, in fact, far greater than simply sinking it at sea. The structure had been weakened due to an accident during its initial deployment, and

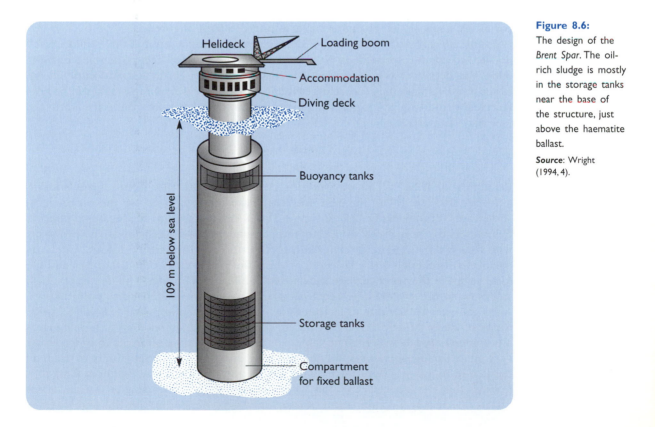

Figure 8.6:
The design of the *Brent Spar*. The oil-rich sludge is mostly in the storage tanks near the base of the structure, just above the haematite ballast.
Source: Wright (1994, 4).

although no oil spill occurred at the time, Shell believed that it was too dangerous to bring back to land. Media attention became strongly focused on the debate as Greenpeace campaigners occupied the *Brent Spar* to prevent its disposal in the deep ocean. In June 1995, following huge public pressure, and despite government support, Shell decided to postpone disposal of the rig pending further consultation. Accordingly, the platform was towed to Erfjord, Norway, while other options were considered.

Support for Shell's case came from an unexpected quarter – a number of marine biologists, who many people intuitively expected to support the arguments of Greenpeace, actually agreed that deep-sea disposal was likely to be the least harmful option in the case of the *Brent Spar*. While stressing that they were not agreeing with a general policy of dumping at sea, they pointed out that the levels of radioactivity (from isotopes such as polonium-210, lead-210, radium-226 and thorium-232) were sufficiently low for it to be classed as low-level waste. Similarly, the concentrations of heavy metals in the *Brent Spar* scale/sludge (mercury, cadmium, lead, copper, nickel and zinc) were not very large. They argued that with the proposed deep-sea disposal site being near the outflow for cold deep waters from the Norwegian Sea, it would take hundreds of years for the toxic leakage to return to the surface, by which time it would have been hugely diluted.

In April 1996, the first report of the Natural Environment Research Council (NERC) scientific group set up to study the decommissioning of offshore structures was published. It noted that both Shell's and Greenpeace's views of the nature of deep-ocean disposal were overly simplistic, in that they contained some elements of the truth but exaggerated them and omitted others. The report also noted that the area chosen for disposal was unusual in that the bottom currents are quite fast and variable, and that biodiversity in the deep ocean can be unexpectedly high. However, it was concluded that the effect on the global environment of dumping the *Brent Spar* would be comparatively small, similar to the probable impact of a large shipwreck.

Eventually, following two years of deliberation, costing a reputed $25 million, Shell decided to create a quayside in Norway using the *Brent Spar* steel (Greenpeace, 1998, 206). This decision has been extended to analyses of many North Sea platforms and rigs to see if recommissioning (i.e. refurbishment for reuse) or decommissioning can be both materials-conserving and job-creating for unemployment hotspots around the North Sea (see also Chapter 17).

Conclusions

Examples such as collapsing fisheries clearly illustrate that ocean management, although instituted with the best intentions, can often fail to solve the problem. In some cases it can actually make it worse. This is due as much to the legal and political troubles as to the inability to predict the consequences of overfishing. The problems with fisheries management seem likely to recur for the foreseeable future. A delicate balance is continually being striven for, where the livelihoods of the fishermen and the food supplies they produce are maximised, while simultaneously protecting the fish stocks for future exploitation. Legal safeguards are clearly needed, since political pressures can vary greatly due to factors aside from implementing what is believed to be the best course of action. The very nature of the oceans requires that these safeguards are implemented as widely as possible. For example, there is no point in one country ceasing its fishing operations if a neighbouring one seizes the opportunity to catch all the fish in its place. Efforts such as the Law of the Sea Treaty are essential first steps towards this, but it is clear that this will need extending and revising as circumstances dictate. Generally a combination of many methods, depending on the objectives, is necessary. In the future, fisheries management will continue to evolve and will need to be continually fine-tuned.

In fisheries, and indeed in any form of ocean management, it is essential to learn from the mistakes made. However, in itself, this is not sufficient. In addition, it is vital that as much information about the biological, physical and chemical processes of the ocean is gained so that policy decisions are made and implemented with the fullest possible understanding of their consequences. This is by no means a trivial task – the ocean–atmosphere system is hugely complex, with subtle feedbacks and teleconnections that we are only now beginning to understand. With regard to the climate question, the measurements needed to do this are only just being completed as the observational phase of the WOCE project nears its conclusion. What is due to follow is an intensive period of examining these measurements and using them to learn as much as possible about the way that the ocean, atmosphere and cryosphere interact. These measurements, and the processes they reveal, will then be used to design and implement better climate prediction models. These will allow us to examine various future possible climates under different scenarios of our own actions, and, importantly, to know how much faith we can put in such predictions. Key to the running of these models and the interpretation of their results is a deep understanding of the natural variability in climate that occurs, and how it is likely to interact with any anthropogenically induced changes. For example, the comparatively recent discovery of the ACW has suggested that the rate of deep and bottom water formation near Antarctica may be modulated by naturally occurring interannual climate variability. Among other things this would, for example, affect the rate at which greenhouse gases are removed from the atmosphere into the deep ocean, with consequent effects for global warming. In the general case, we now know that the changes in climate we observe are due to natural variability added to long-term trends caused by both mankind's actions and other forcings. It is necessary to understand the nature of both the natural varia-bility and the anthropogenic variability if we are going to be able to predict future climate successfully and implement policies to minimise successfully our impact upon it. Only then can the kind of integrated assessments of response to climate change, outlined in Chapter 7, fully begin to reach their interdisciplinary potential.

References

Coull, J.R. (1993) *World Fisheries Resources*. Routledge, London.

The Ecologist (1998) The Monsanto files: can we survive genetic engineering? *The Ecologist,* **28**(5), 251–316.

Gordon, A.L. (1986) Interocean exchange of thermocline water. *Journal of Geophysical Research*, **91**(C4), 5037–5046.

Greenpeace (1998) *The Turning of the Spar*. Greenpeace, London.

Munk, W.H., Spindel, R.C., Baggeroer, A. and Birdsall, T.G. (1994) The Heard Island feasibility test. *Journal of the Acoustical Society of America*, **96**(4), 2330–2342.

Rasmusson, E.M. and Carpenter, T.H. (1982) Variations in tropical sea surface temperature and surface wind fields associated with the Southern Oscillation / El Niño. *Monthly Weather Review*, **110**, 354–384.

Schmitz, W.J. (1996) On the interbasin-scale thermohaline circulation. *Review of Geophysics*, **33**(2), 151–173.

White, W.B. and Peterson, R.G. (1996) An Antarctic circumpolar wave in surface pressure, wind, temperature and sea-ice extent. *Nature*, **380**, 699–702.

Wright, J. (1994) All change at Brent Spar. *Ocean Challenge*, **5**(3), 4–5.

Further reading

An excellent introduction to the science of the oceans is given by C. P. Summerhayes and S. A. Thorpe (1996), *Oceanography: an Illustrated Guide* (Manson Publishing). This introduces various aspects of oceanography, including ocean resources, satellite oceanography, ocean diversity, and the physical and chemical processes of the ocean and their effects on climate. Also worth reading is *Towards Sustainable Fisheries*, by the Organisation for Economic Co-operation and Development (OECD, 1997). Some details of the *Brent Spar* affair can be found in the first report (April 1996) of the Scientific Group on Decommissioning Offshore Structures (Natural Environment Research Council); this is available by contacting the Southampton Oceanography Centre. More detail on the oceans' effects on our climate is provided by G. R. Bigg (1996), *The Oceans and Climate* (Cambridge University Press), which includes discussions of their role in both natural and anthropogenic climate change. Various good textbooks exist which discuss the El Niño phenomenon, such as *El Niño, La Niña, and the Southern Oscillation* by S. George Philander (1990, Academic Press). However, there is currently considerable research being undertaken on many aspects of climate variability, so such books can quickly become out of date.

For those with access to the Internet, a good starting point for finding information about the oceans is the World Wide Web Virtual Ocean Library at the University of East Anglia (*http://www.mth.uea.ac.uk/ocean/oceanography.html*). This gives links to most of the research institutes and universities around the world currently engaged in oceanographic research. Information on ATOC is available at *http://atocdb.ucsd.edu/summarypg.html*; information on El Niño is available at *http://www.pmel.noaa.gov/toga-tao/el-nino/home.html*; information on the United Nations' Law of the Sea is available at *http://www.un.org/Depts/los*. One side of the *Brent Spar* debate is outlined on Greenpeace's website at *http://www.greenpeace.org/~comms/brent/brent.html*. These web addresses may change with time; however, a general web search will reveal numerous pages containing large amounts of information on all these subjects.

Chapter Nine

Coastal processes and management

Timothy O'Riordan, Julian Andrews, Greg Samways and Keith Clayton

Topics covered

- Coastal processes
- Vulnerability and resilience in coastal processes
- Sea-level rise and adaptation
- Integrated coastal zone management
- Redesigning management institutions
- Public–private partnerships in coastal defence
- Planning responsibilities for the coastal zone

Editorial introduction

The coast is an aggravatingly awkward zone to manage. Yet it is of crucial significance for the future of humanity and the ecosystem processes upon which we all depend (see Figure 9.1).

Over half of the present global population lives in this zone, an increasing number of whom are vulnerable to flooding, sea-level rise, destruction of wetlands and freshwater systems, and collapse of nearshore fisheries. Over two-thirds of marine biological activity takes place near the coasts, notably in estuaries and around coral reefs and mangroves. In crucially significant nutrient-rich zones, desperately vulnerable to toxicity from incoming river flows and atmospheric aerosol fallout (see Chapter 14), such as the Waddensee in the southern North Sea, well over half of the commercial fish population spawn and develop their productive life cycle. To lose that area would be a major blow to North Sea marine ecology. This is why 'hotspots' in biodiversity are so important to identify and to nurture.

Mangrove swamps in the coastal tropics play indispensable roles in regulating tides and floods, trapping nutrient-rich sediment and providing refuges for fish and invertebrates. These ecosystems, along with the equally diverse coral reefs, are under threat from coastal mismanagement. This includes everything from poaching coral for commercial sale, through polluting discharges from hotels and coastal residences, to altering marine currents and sediment flows as a result of ill-thought-through port schemes, marine developments, or well-meaning but ecologically inappropriate coastal defence structures.

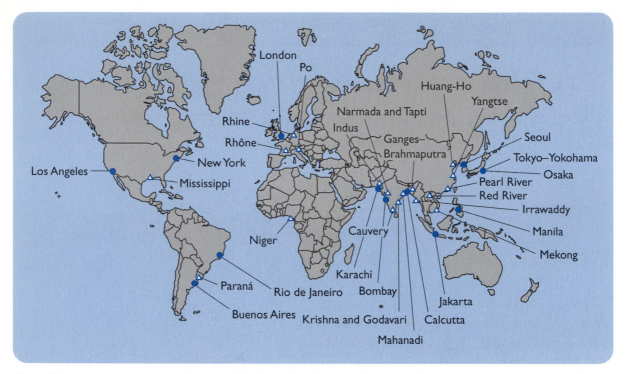

Figure 9.1: Coastal populations at risk from sea-level rise. Thirteen of the 20 most populated cities are located near the coast. Well over half of the inhabitants of 30 small island nation states could be inundated by a predicted sea-level rise of up to 20 cm by 2050 and 100 cm by 2100. Over two-thirds of the Bangladesh population is also at risk, at least 71 million people. All in all, 180 million people are uniquely vulnerable, one billion are at risk, and $1.2 trillion of coastal assets are at stake. *Source*: Rahman and Huq (1998, 147 and 167).

The coastal zone is awkward to manage because it covers three 'territories', three administrative regions and three 'styles' of management. The territories are as follows:

- *the offshore waters*, beyond low tide and within the normal 12-mile national territorial jurisdiction. This is the zone for shipping, fishing and onshore–offshore sediment flux.

- *the coastal margin* between high and low water marks, including beaches, offshore sediment nourishment and estuaries. This is the zone for marine ecological sensitivity, tourism, and natural and artificial coastal defence.

- *the littoral landward zone*, including headlands, cliffs, dunes and the land adjacent to the actual tidal reaches. This is the zone of development for settlement, facilities found and surface freshwater flows, wetlands generally, and tourist-related economic activity.

Normally, each of these three 'territories' is handled by a different collection of management and regulatory agencies in every country. Partly this is organic: agencies emerge and remits change. Part of the problem, too, is the sheer linearity of the zone, making co-ordinated political and administrative responsibility more cumbersome. Much, however, is due to a general lack of appreciation of the precise interconnections between physical and ecological systems, as outlined for saltmarsh in this chapter.

The other administrative problem is the mix of international, national and local governmental regions, each with different rules, budgets, accountability patterns and allegiances to coastal processes and their inhabitants. Offshore, the long reach of planning regulations, including environmental assessments, is just beyond grasp. Specific applications for sand and gravel extraction are given full planning treatment. But it is difficult for planners and their specialist advisers on wildlife and sediment transport to be able to provide precautionary and strategic insight into the cumulative and long-term significance of a host of such extractions. This is a common dilemma for planning, also on the landward zone. Hence the interest in this chapter in strategic, ecocentred approaches to coastal management, and the increasing use of deliberative consultation procedures. The precise role of planning in integrated coastal governance remains to be resolved.

On the actual coast, there is a renewed 'tension' between the 'techno-oriented' style of coastal defence, forced upon the engineering fraternity by expenditure limitations and narrowly framed cost–benefit justifications. The consequence is an institutionally framed, crisis-driven approach based on hardware and independent, competitive tendering. This kind of stop–start investment is inflamed by the lobbying of coastal landowning interests, including conservation agencies and voluntary organisations which own and manage coastal ecosystems, as well as the ever powerful tourist industry. Their aim is to safeguard against a stormy and rising sea, come what may. On the landward side the management style is primarily facilitative, allowing development to proceed with incomplete regard to flooding and possible sea-level rise, and by no means fully taking into account the special demands of the 'eco-protective' approach, designed to safeguard vulnerable peoples as well as ecosystems (see Chapter 6). The tools are then in the planning kit boxes. The reason why these have not been deployed as sensitively as they should be is that the interdisciplinary skills are not available. Nor is the management mix suitable for promoting these approaches, nor is there adequate institutional delivery.

All this raises the perennial issue in environmental management, namely that of *policy integration*. Despite years of interdepartmental, interstate and intersectoral co-ordination, the coastal zone is still a policy battleground. In the United States, subsidies encourage the overuse of chemicals in the catchments, flooding the coastal areas with nutrients and toxins, destroying food chains and creating tourist-repellent algae (see Chapter 14). Flood hazard safeguards can be made to stick, so that ill-sited or ill-protected properties do not get bailed out by insurance companies and coastal defence authorities. But politically it is very difficult to walk away from a coastal disaster in this day and age of instant television coverage. In South Carolina, for example, Hurricane Hugo struck Folly Island, off the coast of Charleston, destroying or damaging 89 of 290 properties (Platt *et al.*, 1991, 7). Despite carefully crafted legislation to limit rebuilding, an existing coastal protection act was changed to enable reconstruction in zones known to be vulnerable to inundation and erosion. Federal tax provisions do not take into account the sort of ecological economics covered in Chapters 4 and 6. So it pays to develop wetlands, and to receive tax reduction in so doing, while the local value of the coastal wetland is lost because there is no tax subsidy for its survival. These paradoxes are slowly changing, as this chapter reveals, simply because the development of GIS (see next chapter) and ecosystem economics is coming of age. Equally important is the role of the insurance industry: its unwillingness to cover avoidably vulnerable property is becoming a hugely influential matter, if politically very charged, in this age of sea-level rise.

Policy conflicts on the coastal margin are very troublesome. They are a product of political bias (Chapter 3), of inadequate interdisciplinarity, of narrowly framed evaluation and of incomplete deliberation. The beneficiaries of coastal protection win because the politicians and the courts cannot face them down, technical solutions are always easier than moral ones, and in any case, coastal vulnerability is still genuinely unpredictable. This chapter shows how all this is changing for the better. Humans do learn and do accept, but unlike their non-human brethren in the course of evolution, they also like to make a fuss about it.

Coastal processes

Coasts are formed by waves and currents. Waves in turn are generated by wind, which transfers its kinetic energy to the sea surface. The longer the fetch or distance of wind-disturbed sea, the more powerful the waves. Surfing is always most popular on open ocean coasts. As waves approach shallow water they dissipate their energy in sediment disturbance, erosive power and broken water. The shape of the coast influences the pattern of energy. Converging orthogonals, the force of energy at right angles to the wave crest, focus wave power, while diverging orthogonals disperse it. With satellite technology it is possible to map potential wave energy for different wind directions, and thus to explain the pattern of erosion and accretion zones for subsequent planning purposes. Adding an understanding of the processes at work to historical records of coastal change allows the recognition of persistently hazardous coastal sectors. How far that will in time be converted into planning restrictions will depend on the credibility of the science and the political influence of planning and management agencies.

As waves usually approach any beach at an angle, part of their energy is converted into lateral movement of sediment. Even where wind patterns are diverse, the importance of fetch will ensure that this drift will be predictable and often significant; for example, beach sediment moves southwards along the California coast at a rate of about 300 000 m³ per km each year. Severe storms, maybe occurring with a probability of once in 10 or 20 years, will account for an extraordinary amount of sediment disturbance. Whole beaches can form or disappear overnight. This episodic convulsion not only gives rise to much property loss and human suffering: it is the very essence of coastal formation. In years to come, climate change may well mean an increasing frequency of such events. This is a more important prospect for coastal management than sea-level rise, though obviously the two outcomes can combine in severe tidal surges. Figure 9.2 shows how the coast of England and Wales has prograded and eroded over the past 100 years.

In addition to wave-generated sediment movement, tidal currents sort sediment and help to shape the geomorphology of the coastline. This is dominant in deeper areas offshore and especially important in estuarine areas protected by offshore bars or spits. From a management point of view, estuarine tidal-induced sediment movement will offset dredging and can move toxic material buried below the surface. In densely populated areas where river waters may carry sediment from polluting industrial zones, such as the Rotterdam harbour at the mouth of the Rhine, this can be a severe problem – so severe that the Dutch government is contemplating an ingenious variant of a tradable permit. The proposal is to establish a property right in effluent discharge by issuing a covenant to the polluting firm. This is a form of voluntary payment by the identified polluter to pay for sediment removal, or stabilisation, in lieu of court proceedings and the possibility of a heavy fine. The scheme relies on proof of effluent discharge and sediment transport. 'Fingerprinting' of the geochemistry of the sediment, and sophisticated current modelling, can provide this service. The courts are now prepared to back such evidence, which allows more strategic and proactive management to take place.

Beaches are forever readjusting. Their prime function is to buffer the coast from wave and tidal energy. Without this protection, soft coastlines would disappear very rapidly. Beaches form from a 'sediment store' collected from the sea floor, from eroded headlands or from riverborne materials. Shingle tends to be found where wave energy is concentrated, and sand or mud where it is more dispersed. Figure 9.3 portrays this

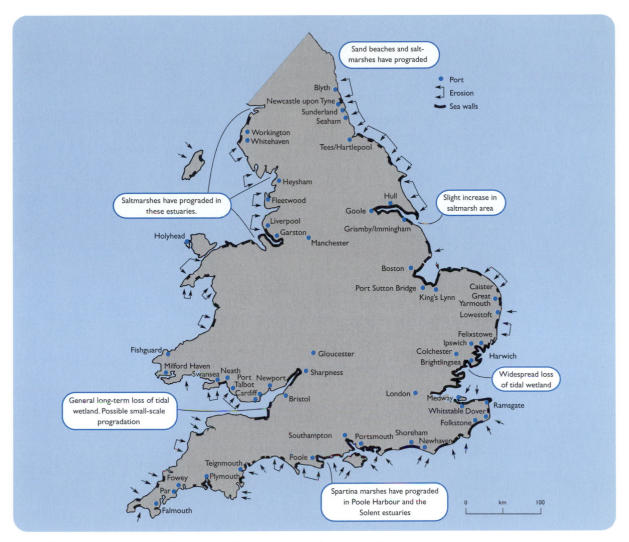

Figure 9.2: Coastline changes in England and Wales. The arrows show zones of erosion, and the heavy lines indicate zones of accretion. Development on the former leads to 'hard' coastal defence walls and other protective arrangements, thus cutting off the natural sediment flow to the latter. Offshore dredging can also limit sediment availability. The rise of strategic shoreline management, plus GIS technology, is beginning to recreate more natural processes. But the political response is not always favourable.

relationship. Dunes are especially important in this arrangement for they feed beaches when they erode but re-establish naturally during periods of relative calm. Dunes may be stabilised by long-rooted grasses such as marram. This not only binds the sand, forming a thin organic soil on the surface; it also lowers wind speed, thereby encouraging greater sand settlement.

Dunes can be damaged more by the trampling of visitors than by natural factors: hence the need for tough dune management schemes to keep people off vulnerable or damaged areas. More often than not, wardens and public information programmes are necessary to ensure that trampling is actually kept to a minimum. Better public understanding of the ecological and geomorphological role of dunes in beach formation and coastal protection has to be actively promoted. This will become even more of an issue as dune reconstruction and new dune construction is undertaken as part of the strate-

Figure 9.3:
Coastal terminology:
(a) the beach in
profile; (b)
composite sand
beach profile; (c)
composite shingle
beach.

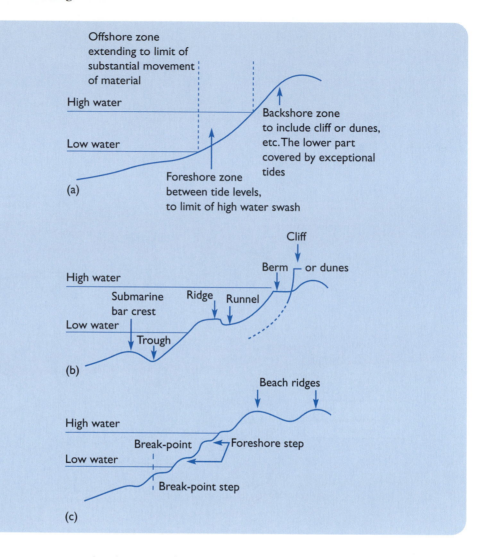

gic response to sea-level rise. In the Netherlands, for example, dune protection is a major component in coastal management. All existing dunes are safeguarded by law and by sympathetic management practice. Dune restoration zones have also been identified. The cost–benefit justification of this programme includes estimates of the recreational value of the dunes using the travel cost method and economic–ecological approaches as presented in Chapter 4. Without these techniques, it would have been politically more difficult to justify major investments in the restoration programme. The dune complexes are now used as sources of fresh groundwater supply. This also has a significant ecological–economic benefit (see Chapter 4).

Figure 9.4 illustrates a typical set of beach profiles. The profile has an enormous influence on the erosion potential of a beach, as well as on the likely shape of the beach during a period of persistent sea-level rise. Under what is known as the *Bruun Rule*, as devised by a Norwegian engineer,

$$\text{Shoreline erosion } (R) = \frac{\text{profile width} \times \text{sea-level rise } (S)}{\text{profile depth } (Z)}$$

as illustrated in Figure 9.5.

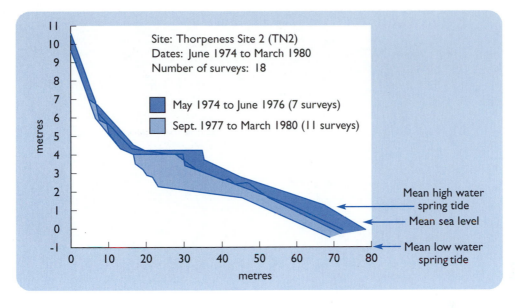

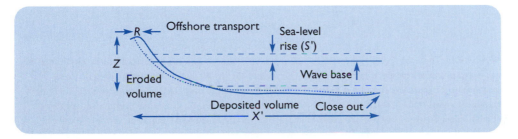

Figure 9.5: The Bruun Rule, describing the development of an equilibrium coastal profile during landward movement of the shoreline as a result of sea-level rise. Bruun assumed no long-term changes in energy input or sediment volume.

The rate of erosion can be modelled as follows:

● record wind direction and speeds over as long a time as possible;

● model the relationship between wind speed and direction and wave height from both observation and wave tank experimentation;

● in turn model wave diffraction on the basis of the beach profile, using Snell's Law of wave refraction.

Application of this approach, plus sophisticated rise of GIS techniques and modelling, allows reasonable calculations of the speed and direction of beach and littoral sediment. Such calculations are vital for the prediction of what might happen to beach erosion or aggradation if particular coastal defence structures are proposed (see Figure 9.6).

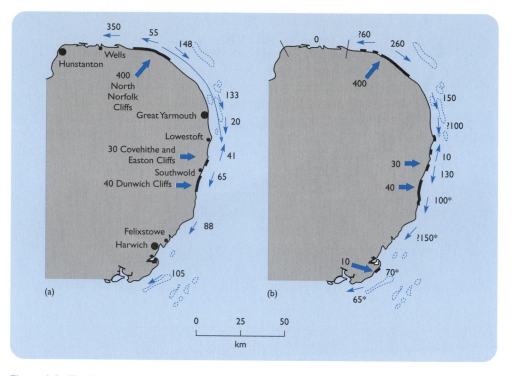

Figure 9.6: The East Anglian sand budget. Cliff inputs and littoral drift values in 1000 m³/Yr. (a) Computed net longshore sand transport values over 13 years (1964–76). (b) Most probable values derived from examination of gross values and net values for longshore transport over varying periods of time, using both computed values and those calculated from wave observations, 1974–79. These are regarded as relatively reliable values (over a period of 20 years or more), but for those marked? the value remains uncertain, although the direction is certain. The asterisks note theoretical values not reached due to lack of sand and/or lack of exposed beach at all states of tide. The main offshore banks are indicated on both maps.

Significance of saltmarshes

Coastal saltmarshes in temperate climatic zones occur high in the intertidal zone, grading seawards into intertidal mud- or sandflats (Figure 9.7). In subtropical and tropical areas saltmarshes are replaced by mangrove swamps. Saltmarshes (and mangroves) are sites of very high biological activity, converting and recycling nutrients – derived from both land and sea – into halophytic plants. These plants may help to trap or bind fine-grained marine sediments by slowing water flow over the marsh surface, often resulting in the upward building (accretion) and sometimes seaward building (progradation) of marshes. It has recently been recognised that mature saltmarshes (those that have built upwards to the point that they are flooded only by high spring tides) may act as natural sea defences, in the way that their rough surfaces and complicated creek systems slow down flooding tides and dissipate the energy of breaking waves. For example, a mature saltmarsh in front of a clay seawall might dissipate the energy of the incoming sea to such an extent that both the height and routine maintenance of the wall can be reduced.

While saltmarshes typically support a low diversity of plant life (usually dominated by just a few plant genera), they can supply nutrients to the coastal zone, thereby supporting offshore productivity. They are also very important feeding grounds for birds. For example, many east of England saltmarshes are important overwintering or stop-off sites for migrant birds. Saltmarshes also offer protection for juvenile fish stocks.

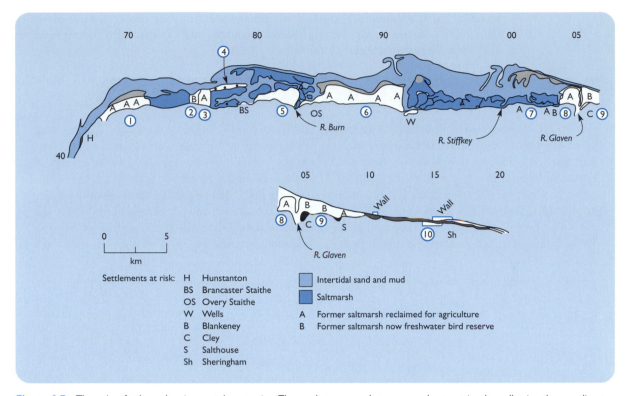

Figure 9.7: The role of saltmarshes in coastal protection. The modern approach to managed retreat involves allowing the coastline to return from freshwater marsh to aggrading saltmarsh. Where the formerly protected freshwater marsh is an internationally recognised wetland under the EC Habitats Directive or Ramsar Convention, then national governments are required to retain 'overall conservation status'. This means recreating wetland reserves elsewhere on the coast, as well as managing the newly forming saltmarsh for bird conservation. Such decisions involve complex processes of ecological assessment and facilitated negotiations of the kind introduced in Chapter 1, as well as the ecological economics of Chapter 4. Such decisions should involve genuine interdisciplinarity.

The ability of saltmarshes to function healthily is thus important, both from a human perspective and in terms of balanced coastal-zone ecosystems.

As with many other natural resources, human attitude towards saltmarshes and their significance has been ambivalent. On the one hand, conservationists, ornithologists and wildfowlers have viewed these places as areas of natural wilderness, important in their own right, or in a recreational sense: a resource to be protected. On the other hand, farmers have viewed saltmarshes as areas of potentially fertile farmland ripe for reclamation; industrialists have been keen to exploit their coastal position for development; while local authorities have often viewed these areas as little better than wasteland of low economic value, suitable only as dumping grounds. On balance, human impacts have been considerable and about 50% of saltmarshes have been lost from the UK coast by reclamation since Roman times, but mostly in the past 300 years. For example, in the Humber estuary about 250 km^2 of saltmarsh has been reclaimed during this time period.

These traditional attitudes towards saltmarshes have started to change, mainly driven by the increased public awareness of sea-level rise and the threat of flooding, coupled with the very real problem of maintaining sea defences. Planners are now considering, and in some places implementing, the 'managed retreat' of sea defences. This involves letting the existing engineered defences fail and encouraging the sea to rebuild natural marshes, often on areas of formerly reclaimed saltmarsh. The new marsh is encouraged to act as a baffle, dissipating the energy of the flooding sea. These so called 'soft engineering' options sound simple, but they can create problems of their own, sometimes threatening valued resources such as managed nature reserves and Sites of Special Scientific Interest (see Box 9.1)

Box 19.1 Managed retreat and saltmarsh recreation in Essex

Since 1970, annual losses of saltmarsh from the Essex coast have been up to 12% and are increasing. With the prospect of sea-level rise, future losses could be as much as 40% by 2050. An innovative programme of conversion of arable land to saltmarsh restoration has begun to take place, as part of an overall shoreline management strategy. The key is to simulate tide and sediment flow by means of bunds, lagoons and sluices. The recharge of sediment is, however, proving to be a problem, as the bunds limit natural flows, while sources of biogeochemically imprinted sediment are very difficult to find. The alternative is to set the coastal defence back, beyond previously grazed land, and allow natural processes to restore the saltmarsh as a genuine coastal defence. This means devising a

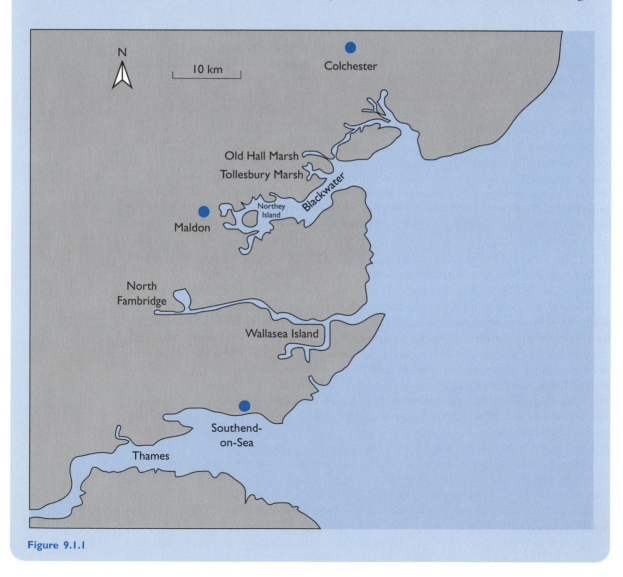

Figure 9.1.1

Box 9.1 (continued)

scheme for compensation for the landowner. Three possibilities now exist, each of which involves innovative legal arrangements and 'cross-over' budgets:

● *habitat management scheme* of 20-year set-aside of previously agricultural land, for an annual payment, likely to continue beyond the 20 years with forthcoming changes to the Common Agricultural Policy;

● *conservation reserve* by lease or by purchase to a wildlife management body;

● *transferable wildlife rights scheme* whereby a recreated saltmarsh is paid for in lieu of the loss of another wildlife habitat else-

where, due to sea-level rise, under the EC Habitats Directive. This requires designated sites to be replaced by alternative sites of equivalent conservation status, through an imaginative use of flood defence funds (for the new marshes *are* a flood defence), or through agri-environmental management scheme payments.

Figure 9.1.1 shows the stretch of the Essex coast in which saltmarshes form part of the managed-retreat coastal defence. It is this kind of imaginative coupling of budgets and administrative 'styles' that is characterising integrated coastal management in the modern age.

Source: Boorman and Hazelton (1995, 179).

Saltmarshes also have a biogeochemical role to play in cycling and potentially storing both natural and anthropogenic compounds and elements. At the local level, the mineral grains, and organic matter from dead vegetation, might act as a 'sponge' soaking up and fixing pollutant elements like lead and cadmium. On a global scale, saltmarshes constitute a large store of carbon (as buried organic matter): the estimated global area of saltmarsh is around 3.8×10^7 hectares (roughly the combined area of Germany and Belgium), while the area of global mangrove forest is about 2.4×10^7 hectares (roughly the area of the UK).

Biogeochemistry of saltmarshes

On timescales of tens to hundreds of years, saltmarshes probably act as long-term carbon (C), nitrogen (N) and phosphorus (P) stores, as these nutrient elements become trapped and buried. However, subsequent reworking of the sediment by erosion or biogeochemical processes can cause element remobilisation. For example, the bacterial decomposition of plant-derived particulate organic matter may release dissolved inorganic or organic C, N and P to pore waters, from which the nutrients may (1) escape to coastal waters, (2) re-precipitate/adsorb onto particle surfaces, or, (3) be reused by living plants and be temporarily stored in plant biomass. These recycling processes affect different nutrients in different ways: phosphorus is bound to particles under oxygenated conditions but may be released under reducing conditions, while nitrogen concentrations can be modified by nitrogen fixation, nitrification and denitrification. In general, saltmarshes are probably net nutrient element stores, although biogeochemical processes and different or changing environmental conditions may result in marshes acting as sources or stores at given times.

Pollutant metals such as lead, cadmium and arsenic are often released into coastal or estuarine waters as a result of industrial activities. Some of these elements naturally bond to electrically negatively charged surfaces of clay minerals and fine-grained particulate organic matter (both abundant components of saltmarsh sediment). In this way, the saltmarsh sediments can act as a natural 'sponge' for some pollutant metals, removing them from the water and locking them into the marsh. Their lifetime or residence time in the marsh 'store' depends in part on the geochemical conditions in the sediment (e.g. whether oxidising or reducing) and whether the marsh is accreting or eroding over decadal–centennial timescales.

Measuring nutrient and metal biogeochemical fluxes and storage

As in many other areas of environmental science, measuring natural biogeochemical fluxes in and out of saltmarshes is not easy. The most informative data should come from studies that integrate all of the processes affecting a marsh during all stages of the tidal cycle. However, problems in measuring discharges, incomplete understanding of transport mechanisms and the unknown effects of large storms on the 'steady state' values are all inherent errors. These problems are magnified when estimates based on small areas are scaled up to represent whole marshes or regional coastlines. Despite the problems, informative data come from tidal creek measurements, which integrate processes over a known marsh area. Measurement of the concentration of a chemical species and the discharge during a complete tidal flood and ebb cycle allow the calculation of import of that species during the flood and export during the ebb. Repeating this process at monthly intervals allows a seasonal picture to be built up, and over a 12-month period a net budget can be calculated. The seasonal perspective is important because some saltmarshes, for example in the Netherlands, have been shown to be net sinks for phosphate during the winter months but net sources of phosphate during summer months (and released at a time when phosphate concentrations in coastal waters are at a minimum, i.e. when nutrients are most needed in the biological system).

Flux calculations are more or less instantaneous snapshots of processes going on today. Storage calculations are effectively time-averaged data and on the face of it easier to make. Measuring the concentration of the chemical species is relatively easy, and scaling these values upwards to be representative of a mass of saltmarsh sediment is simple. The problems come in understanding the age of the sediment, or the length of time it has taken for that mass of sediment to accumulate. To do this we need an accurate measure of time, which is typically expensive, technologically complicated or labour-intensive. Knowing when a saltmarsh formed and how quickly it grew in area over the last 50 years or so is relatively easy based on aerial photographs and high-quality maps, and the presence or age of individual marshes can sometimes be decided on the basis of old maps and charts, and in some cases, archaeological artefacts buried in or under the marsh. However, knowing the average rate of accumulation on a yearly basis is only possible using more sophisticated techniques. Direct measures of sediment accumulation on known surfaces (e.g. patches of dye or metal plates) give snapshots of modern accumulation rates. Geochemical methods using radioisotopes such as ^{210}Pb and ^{137}Cs can also give estimates of sediment mass accumulation over 100-year and 40-year timescales, respectively. Box 9.2 illustrates a case example from Welwick Marsh in the Humber estuary on the eastern seaboard of the UK.

Box 9.2 Welwick Marsh, Humber estuary, UK

Welwick Marsh, located in the north-west corner of Spurn Bight (north shore of the outer Humber estuary), is one of the few saltmarshes in the region that has apparently been accreting continuously during this century and is still accreting today (Figure 9.2.1). Elsewhere in the estuary, saltmarshes have been extensively eroded because of continuing reclamation of the estuary margins over the last few centuries. This has squeezed the marshes into narrow strips between the sea defences and the main channel. The squeezing means that slight changes in the course of the channel result either in erosion, forming a saltmarsh cliff, or in deposition of pio-

neer marsh at the base of older saltmarsh cliffs. The resulting deposits are not continuous and so do not record the history of storage (e.g. carbon storage or metal contamination).

High-resolution records of saltmarsh accumulation rates on Welwick Marsh have been acquired using natural (^{210}Pb) and anthropogenic (^{137}Cs) radionuclides, which accumulate in the sediments during deposition. The natural radionuclides are produced at known rates, and the anthropogenic radionuclides were deposited at known times, e.g. from atmospheric atomic bomb testing in the 1960s, the Chernobyl accident in 1986, and discharges from nuclear reprocessing sites like

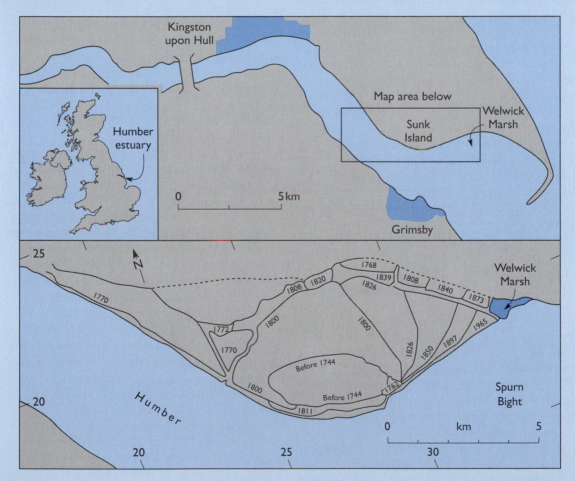

Figure 9.2.1: Map of the Humber estuary, UK, showing location of Welwick Marsh and reclamation history of the Sunk Island area.

Box 9.2 (continued)

Sellafield. Using this 'event' knowledge it is possible to date sediments deposited after the 1950s. By collecting cores from a range of mudflat, pioneer lower saltmarsh and mature upper saltmarsh sub-environments on Welwick Marsh, average accumulation rates of sediment have been calculated over longer timescales than conventional surveying techniques allow.

Welwick Marsh formed in an embayment between three seawalls constructed between 1873 and 1965. A sequence of aerial photographs has been used to reconstruct the position of the vegetated marsh front from 1946 to the present day, in order to calculate the average progradation rates between each successive photograph. The change in calculated progradation rates clearly shows a rapid decline, from about 15 m/year during 1946–51 to less than 3 m/year during 1984–94, as the marsh front approached equilibrium with the adjacent marsh front to the west and east of it.

Knowing the changing position of the marsh front through time, it is possible to reconstruct a sequence of marsh-front profiles back to 1946 by simply projecting a surveyed profile of the modern marsh surface back through time. This backward projection assumes that the equilibrium marsh-front profile has not changed during the period 1946 to the present, which is not unreasonable given that the primary depositional processes and conditions, and overall sediment supply regime, have probably not changed significantly over that period. This projection allows subdivision of the marsh sediments into timeslices, which can then be characterised in terms of their organic carbon and metal contaminant content, in order to calculate the postwar history of organic carbon and metal storage in the marsh.

Given the slowdown in progradation rate since 1946, the volume of sediment and hence the volume of organic carbon (C_{org}) storage have progressively declined as shown in Table 9.2.1.

This means that the creation of the embayment in the seawalls provided the accommodation space for sediments to accumulate quickly, and bury, somewhere in the region of 30 000 tonnes of

Table 9.2.1: Decline in volume of organic carbon (C_{org}) storage at Welwick Marsh.

Time interval	C_{org} / yr (tonnes)
1946–51	1205
1951–69	622
1969–76	426
1976–84	466
1984–94	298

organic carbon since 1946. This demonstrates the importance of coastal marshes as sinks for organic carbon; had the carbon not been fixed in the marsh by plants, the CO_2 used during photosynthesis would otherwise still be in the atmosphere and contributing to global warming. These data give an interesting perspective on the impact of reclamation in coastal environments too. For example, the reclamation of the Sunk Island area of the Humber (west of Welwick Marsh) has been going on since 1744, preventing natural sediment deposition over an area of 35 km². Because we know that on average Humber sediments accrete at about 1 mm/yr as sea level rises, it is simple to estimate that the deposition of 18 million tonnes of sediment has been prevented since 1744, which equates to more than 300 000 tonnes of organic carbon. This is another important mechanism by which anthropogenic activity, in this case reclamation, has disrupted the local carbon cycle.

In addition to evaluating the organic carbon storage history of Welwick Marsh, it is also possible to estimate the storage of contaminant metals that have been discharged into the estuary from various industrial installations over the years. In the case of zinc, attributed to discharges from the Tioxide plant on the south shore of the estuary, the sedimentary record on Welwick Marsh suggests a steady decline in storage in recent years. In 1970, the zinc concentration in the sediments was 100 µg/g above the pre-industrial background zinc content in the estuary sediments (about 100 µg/g). When multiplied by the sediment deposition in 1970 this equates to roughly 1.5 tonnes of zinc deposition in that year, whereas by 1995 the

Box 9.2 (continued)

zinc concentrations were down to 25 µg/g above background, equating to about 0.15 tonnes of zinc deposition during 1995. Overall, assuming a roughly linear decline in zinc concentrations between 1970 and 1995, it appears that about 20 tonnes of zinc was deposited on Welwick Marsh between 1970 and 1995. The decline in zinc deposition recorded in the saltmarsh probably reflects a decrease in discharge practice in a nearby industrial plant which occurred in the mid-1980s. This study, using the saltmarsh sediments as a kind of 'tape recorder of events', clearly demonstrates the impact of improved pollution management in the estuary.

Sea-level rise and coastal vulnerability

According to the Intergovernmental Panel on Climate Change (1997, 35), over the last 100 years global sea levels have risen by 10–25 cm. The more recent models have been able to filter out the natural causes of land-to-sea isostatic changes, so the volume of the oceans has now been shown to have increased due to temperature rise. The panel estimates that 2–7 cm of the recent rise is due to thermal expansion, and 2–5 cm to the melting of glaciers and ice caps. Chapter 7 carries more of these data.

Under the IS92 temperature scenarios, the models predict a sea-level rise of 13–94 cm by 2100 (see Figure 9.8). The major unknown here is the rate and timing of thermal expansion after an early period of inertia. This is the period we are in now. Once the thermal inertia effect has been overcome, sea-level rise would be continuous and very prolonged, even if greenhouse gas emissions are actually reduced. The actual rise in sea levels will not be uniform and will depend on ocean currents (see previous chapter) and local tidal conditions.

Sea-level rise offers particularly dangerous prospects for coastal communities with varying degrees of resilience and vulnerability. Neil Adger addresses this explicitly in Chapter 6. Some 24 million inhabitants in Bangladesh alone are already vulnerable to coastal flooding. Given a 100 cm sea-level rise and a 15% increase in rainfall, 71 million people out of a total population of 200 million (by 2100) will be in danger. To cope means a combination of managing the river feeding catchments and redesigning coastal settlement and flood warning schemes. This means that future freshwater management, including interbasin transfers, will be a crucial part of coastal management.

Rahman and Huq (1998, 195–198) also point out that non-literate rural peoples will require very particular recognition and coping strategies if they are to be brought into proactive management in a constructive manner:

In a world where communities will be differentially affected ... new approaches to conflict resolution will be needed. In most situations of stress, such as natural disasters, the informal and local level and (often) traditional community-based institutions may offer better opportunities for survival, coping, redressing social imbalances, problem sharing and conflict resolution. Social scientists and politicians must better understand these decision making processes.

(Rahman and Huq, 1998, 197)

Sea-level rise, increased storminess, vulnerability and episodic catastrophes on the coast point up all too clearly the inadequacy of proactive, participatory management approaches that incorporate ecological economics, deliberative science and integrated tools of evaluation, zoning, compensation and flood accommodation. This is the task

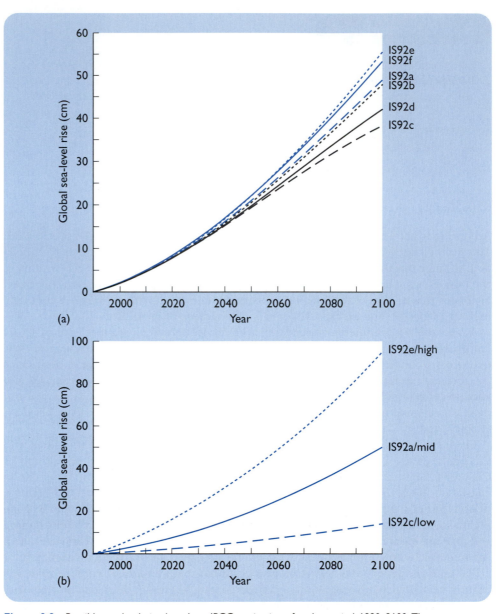

Figure 9.8: Possible sea-level rise based on IPCC projections for the period 1990–2100. The range depends on climate sensitivity, i.e. the increase in temperature associated with a doubling of CO_2 pre-industrial base levels, ice-melt estimates and the nature of thermal inertia. (a) Projected rise for the full set of IS92 emission scenarios. A climate sensitivity of 2.5°C and mid-value ice-melt parameters are assumed. (b) Projected rise extremes: the highest sea-level rise curve assumes a climate sensitivity of 4.5°C, high ice-melt parameters and the IS92e emission scenario; the lowest a climate sensitivity of 1.5°C, low ice-melt parameters and the IS92c emission scenario; and the middle curves a climate sensitivity of 2.5°C, mid-value ice-melt parameters and the IS92a scenario. Clearly for planning and coastal management purposes, notably in low-lying, populated shorelines, this range, though helpful, is insufficient as a guide. Precautionary measures may have to be taken, though democratic consent for a retreat-and-adapt strategy may prove very tricky to achieve.

Source: Intergovernmental Panel on Climate Change (1997, 35).

of integrated coastal zone management, the promise of the age, but a chalice still beyond our grasp.

Integrated coastal zone management

Integrated coastal zone management is now all the rage. This is the coupling of adaptive management techniques, as outlined in Chapter 1 and as illustrated further in Chapter 7, to the interconnected social and biogeochemical processes that characterise the coast. Figure 9.9 outlines the broad principles involved, and Figure 9.10 illustrates some of the applied management relationships. Note the pattern of integration:

- that natural processes of defence and protection should be encouraged, costed properly and fully incorporated into any plan or management scheme;
- that natural zones essential to this purpose, such as headlands, dunes, saltmarshes and wetlands, should be adequately protected by law, cleared of existing settlement, with compensation if necessary, and carefully monitored for their continuing role;
- that coastal defence works should always be designed sympathetically to encourage the retention of a natural beach, and that cost–benefit analyses should recognise the essential interlinkage between the two;
- that land-use planning formally take into account the vulnerable areas of coast subject to sea-level rise and increased storminess, so that no new settlement or economic activity is permitted in such areas, and, where possible, existing buildings are left unprotected, again with compensation where necessary.

Integrated coastal management is not a single-minded outcome. It is a process with many possible pathways, depending on the physical, institutional and political circumstances of coastal protection. It would be most unwise to lay down a blueprint planning guide. Three criteria can be used to assist the evaluation of good performance in integrated coastal management.

- *Optimisation of multiple objectives*. Multiple objectives cannot be met without some losing while others gain. The aim of integrated coastal management should be to devise structures of administration and public involvement geared to achieving agreement. This may mean suboptimal realisation of objectives for each special interest group, but at least respect for and support towards a common goal. It also means flexible and adaptive management geared to shifts in information, scientific experimentation and shifting public interest demands as all this innovation proceeds. Clearly this also indicates the need for imaginative and innovative cost–benefit analysis backed by informed public support.
- *Maintenance of life-support processes*. A well-managed coastal project should demonstrably enhance the life-support systems of the coastal zone, by empathetic design, by creating and protecting habitats, and by ensuring that there is room for ignorance because of the necessarily limited understanding of ecological resilience. This requires a sound scientific basis of the workings of the coast, good ecological and hydrological modelling as far as knowledge allows, and honest acceptance of the limits of knowledge. Uncertainty is not an ogre if it is properly taken into account.
- *Responsive management*. The 'cost' of any coastal zone scheme must be seen in terms of its success in balancing inherently contradictory objectives and in guar-

Figure 9.9: The integrated coastal zone management regime. The key to this is improved understanding of coastal physical and biogeochemical processes, coupled with regular GIS updating, applying the use of ecological economic techniques. Cost–benefit analysis requires the application of values for maintaining natural processes, as well as the social gains of well-managed ecosystems, to be permanently in place. For this to be most effective, deliberative and inclusive processes of consultation have to be invoked.

Source: Bower and Turner (1997, 2).

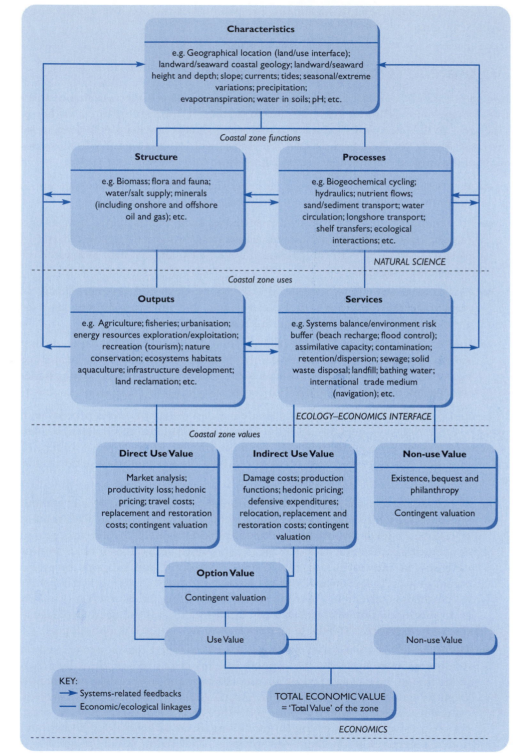

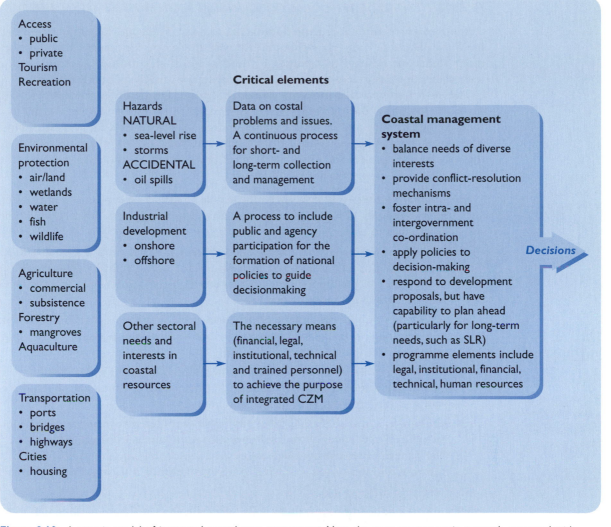

Figure 9.10: A generic model of integrated coastal zone management. Note the more programmatic approach compared with Figure 9.9, but the application of techniques of criticality, vulnerability and resilience as laid down in Chapter 6. The design of problem options becomes the basis for adaptive management. Continuity of funding, coupling of agency effort and guarantees of stakeholder support help to sustain reliable management regimes.

Source: Rahman and Huq (1998, 188).

anteeing ecological viability. Where these objectives are evident, the 'benefits' are partly measured by the procedures that achieve such success. Responsive management translates into the satisfaction of trust in a scheme that is well grounded, well monitored and well communicated to those who have a stake in its outcome. Participation is not a democratic symbol: it should be a process of guidance and continuous readjustment to a changing set of optimised objectives that become compatible only by the revelation of shared interest.

Administering integrated coastal zone management

This approach requires very fresh approaches to coastal defence, beach management, land-use planning, property rights tenure (with the scope for transferable development rights introduced in Chapter 5), and political-stakeholder communication techniques. For this to happen, a special administrative arrangement is required.

Public–private partnerships are one such option. Originally termed the private finance initiative by the UK Conservative government in the early 1990s, the PPP is a method for financing and managing complex large schemes over long periods of time. The idea is that government puts up a set sum of money, and a consortium of private interests run the project and absorb the overall responsibilities and risks involved. The advantages of this approach are as follows:

- *reliable funding*, usually sufficiently flexible to attract co-funding from co-operative agencies. This is because reliable funding gives other bodies the confidence to co-invest;

- *reliable tendering* in that every subcontractor is hired on a reasonably long-term basis, subject to explicit procedures for accountability. The value for money should, in theory, be easier to justify if the tenders are from a small number of well-managed companies;

- *Investment in experimental schemes*, ecological management techniques, soft engineering and proactive maintenance are all possible, because of the long-term nature of the financing mechanism;

- *full and regular monitoring* of all elements of investment and experiment to ensure that value for money is gained, that objections are met, and that all relevant interests are informed of all developments;

- *deliberative and inclusive processes* of stakeholder involvement of the kind introduced in Chapter 1, touched on in Chapter 7 and reinforced in Chapter 17. The pattern of reliable funding ensures the reasonable likelihood of community support, though funds will have to be available for a formal (and informal) facilitative role.

Public–private partnerships are still in their infancy in the UK. There is still much to be learned. By no means are the early versions in the coastal defence arena suitable for the sophistication of the ICZ treatment outlined in Figures 9.9 and 9.10. The main drawbacks are as follows:

- *lack of appropriately integrated approaches* to the management regime and the primary tendering process. This is still an engineering contract with little scope for integrating to planning and more expansive forms of valuation;

- *inability of governance* in the broadest sense of the term to deliver a full range of linked services and management options. The coastal zone administrative regime is still far too fragmented;

- *inadequate knowledge* of the coastal zone processes themselves. This is the arena of uncertain science. Hence the need for a duty of care through the precautionary principle, and full participatory consultative approaches.

In England, the responsibility for managing the coast falls to two ministries and two political jurisdictions. The coast itself is the responsibility of the Ministry of Agriculture and its agency, the Environment Agency. The Environment Agency is divided into ten regions and directed in its work by local flood defence committees. These are composed, for the most part, of landowners in flood-prone areas and local authority councillors, more often than not representing coastal constituencies, as well as those reflecting conservation interests (see Figure 9.11). In Scotland, the pattern is different with unitary local authorities holding the local government slot, plus the single Scottish Office agency of agriculture and fisheries. The coastal zone landward of the sea in England is the responsibility of the Department of the Environment, Transport and the Regions, and through it, the district councils bordering the coast. They aim to control erosion and to stabilise the coastal zone. They receive grants for approved coastal protection schemes from the MAFF but generally have to meet the cost of repairs themselves.

Cash bottlenecks are also common for flood protection schemes, known as sea defence. In the UK, the county councils must contribute about a quarter of the Environment Agency sea defence capital works programme. That money eventually comes back from the Department of the Environment, Transport and the Regions a year later. In the meantime, the county councils must levy their ratepayers for the capital when they also have to raise revenue for a host of other requirements, such as school buses, social services and roads. The formula on which county council spending is assessed does not take into account specific sea defence needs. In practice, the counties are regularly faced with a bill that exceeds their actual spending allocation. Even though the money is eventually returned, it is always less than the actual annual spend. In any case, the county is usually restricted in its total expenditure by central government allocations based on theoretical but standardised spending assessments.

Consequently, there is a tug-of-war between the coastal counties and the local land drainage committees, on which they are well represented, and the specific needs of the Environment Agency. Engineering works rarely receive funding to the degree to which finances are required, and each year of shortfall builds up a capital spend backlog that can prove politically embarrassing. The MAFF, in turn, is controlled by the Treasury, so both the grant aid that is made available and the acceptability of the cost–benefit justifications are enormously influenced by macro-economic policy determined from the centre. The actual circumstances of coastal protective need are miles away.

This in turn leads to an economically inefficient and biogeochemically insensitive regime of spending. In 1996, the MAFF published a 'points scoring scheme' for justifying flood defence expenditure based on a formula of crisis (i.e. need), value of property to be saved (urban and only extremely valuable habitat), degree of protection required, and interconnectedness to other schemes. What this actually means is that flood defence expenditures are only allocated on a critical basis to the high-profile areas, while needy investments, usually of a valuable proactive nature, are placed into the spending plans for subsequent years. This is why the public–private partnership approach is beginning to receive so much attention.

The time has come to re-examine the whole mix of governmental responsibilities depicted in Figure 9.11. This was done by the House of Commons Agriculture Select Committee (1998) in a powerful and comprehensive report. The backdrop to the investigation was a falling expenditure pattern of coastal defence and flood protection due to government macro-economic policy, and a major problem of poor co-ordination around the shoreline management plans. The main conclusions reached by the committee were as follows:

● *Coastal protection must accommodate to sea-level rise, and not try to hold the line*. Managed retreat, coupled with strategic erosion and accretion, plus ecological restoration and renewal, must be regarded as an important element of shoreline management plans. This in turn means that key nature conserva-

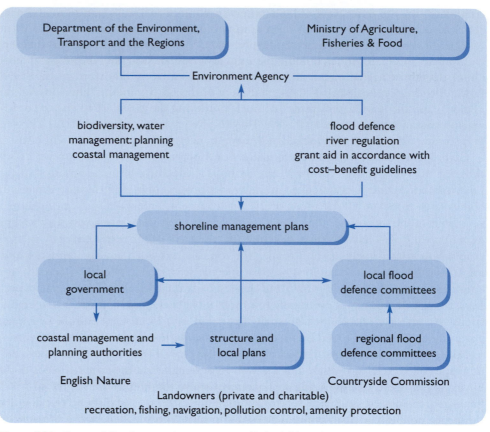

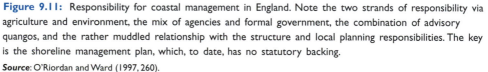

Figure 9.11: Responsibility for coastal management in England. Note the two strands of responsibility via agriculture and environment, the mix of agencies and formal government, the combination of advisory quangos, and the rather muddled relationship with the structure and local planning responsibilities. The key is the shoreline management plan, which, to date, has no statutory backing.
Source: O'Riordan and Ward (1997, 260).

tion sites must be replaced by equivalent conservation investment in part of integrated coastal management. Compensation by management agreement and special funding will also be necessary. The matter of compensating landowners for deliberately allowing land to erode for coastal defence purposes was confronted by the committee. Sediment inputs for coastal safeguard via natural fluxes have been reduced by 50% over the past 100 years. So increased erosion is necessary. The committee (para. 104) was of the view that a 'robust financial mechanism for the reimbursement of property holders whose assets are sacrificed for the wider interests' must be put in place. Eventually, some of the revenue for such a mechanism could come from a carbon tax, though the committee did not address this.

● *Reliable funding for local authorities*. The committee was well aware that the cash available for proactive maintenance was simply not to hand. 'To safeguard life and property in the long term, the piecemeal, reactive approach to maintaining defence works, endemic in the current system, must be replaced by a considered national strategy' (para. 46). The committee argued passionately for this reform on sustainability grounds.

- *Project justification*. The committee also pressed hard for more deliberative approaches to allow the incorporation of 'soft' social and ecological values into sustainable coastal management (para. 60). This accords with the recommendations of the Royal Commission on Environmental Pollution (1998, para. 7.28) that the two environment agencies should incorporate such opinion on an inclusive basis. This is a major strike against the programmatic and economistic approach to project appraisal currently followed by the Ministry of Agriculture, under pressure from the Treasury.

- *Governance of the coastline*. The committee favoured the strengthening of the regional flood defence committees (see Figure 9.11), with an element of independent funding and executive powers, rather than any new institutional structure. The package proposed includes:

  - a strengthened Environment Agency, with regional disbursement funds, and with new powers to stop inappropriate property development, in line with democratically agreed shoreline management plans;

  - regional devolution of executive authority to the regional flood defence committee, along with the regional development agencies, coupled with more flexible and reliable funding for the local authorities;

  - a more effective co-ordinative role for shoreline management plans, linked to a more durable and transparent consultation process. In essence, the shoreline management plans would become coastal management plans, with formal statutory status akin to local and structure plans. The consultation arrangements linking the local interests to local coastal authorities would also be given formal statutory powers. Landowners occupying vulnerable land should set aside funds for their protection, and all areas subject to possible flooding should contain such information in title deeds and in insurance policies.

These are wide-ranging recommendations, introduced here to show how the committee responded to the principles of integrated coastal-zone management outlined earlier. Of course a select committee is only advisory, and MAFF may well ignore it. But the substance of the proposals will eventually have to be adopted if the coastline is to survive as a functioning physical system that still provides wealth and beauty for residents and visitors for generations to come.

Prospect

The five chapters preceding this one create the justification for the science, the valuation and the strategy for coping with one of the most troubled zones of human existence on the planet. Integration of science, valuation, policy, administration and consultation/deliberation will place a huge demand on a widening and deepening environmental science for the next millennium. There are far too few practitioners with the skills to do the job as adequately as it must be accomplished. This is why new statutory, regulatory and participatory structures, of the kind suggested by the Agriculture Select Committee, need to be put in place the world over, according to the cultural, political and social norms of each area and the particularities of the coastal processes themselves. This will become the true environmental science for integrated coastal management.

References

Agriculture Select Committee (1998) *Flood and Coastal Defence. Sixth Report*. The Stationery Office, London.

Boorman, L. and Hazelton, S. (1995) Saltmarsh creation and management for coastal defence. In M.G. Healey and J.P. Dooley (eds), *Directions in European Coastal Management*. Samara Publishing, Cardigan, 175–184.

Bower, B.T. and Turner, R.K. (1997) *Characterising and analysing benefits from integrated coastal zone management*. Working Paper GEC 97–12, CSERGE, University of East Anglia, Norwich.

Intergovernmental Panel on Climate Change (1997) *Climate Change 1995: the Science of Climate Change*. Technical summary for policy makers. Cambridge University Press, Cambridge.

O'Riordan, T. and Ward, R. (1997) Building trust in shoreline management: creating participatory consultation in shoreline management plans. *Land Use Policy*, **14**(4), 257–276.

Platt, R., Beatley, T. and Miller, H.C. (1991) The folly at Folly Beach and other failings of US coastal erosion policy. *Environment*, **33**(9), 6–9, 25–32.

Rahman, A. and Huq, S. (1998) Coastal zones and oceans. In S. Rayner and E. Malone (eds), *Human Choice and Climate Change*, Vol. 2. Battelle Press, Columbus, Ohio, 145–201.

Royal Commission on Environmental Pollution (1998) *Setting Environmental Standards. Twenty-first Report*. The Stationery Office, London.

Further reading

A good start can be made in Carter, R.W.G. (1988), *Coastal Environments: an Introduction to the Physical, Ecological and Cultural Systems of Coastlines*, Academic Press, London. For the geomorphological story, see Clayton, K.M. (1979), *Coastal Geomorphology*, Macmillan Educational Books, London. For a useful perspective on how various countries approach this topic, see Organisation for Economic Co-operation and Development (1993), *Coastal Zone Management: Integrated Policies*, OECD, Paris. For a perspective on ecological planning, see Salm, R.V. and Clark, J.R. (1984), *Marine and Coastal Protected Areas: a Guide for Planners and Managers*, International Union for the Conservation of Nature, Gland, Switzerland. *The World Coasts Conference* (1995), RIM, The Hague, provides a comprehensive and up-to-date review, as does the valuable collection of case studies in Healey, R.G. and Dooley, J.P. (eds) (1995), *Directions in European Coastal Management*, Shambala Press, Cardigan.

Chapter Ten

GIS and environmental management

Andrew Lovett

Editorial introduction

The first edition of this book concentrated on remote sensing from satellite imagery. This is an important tool, but it is only a small part of the rapidly expanding application of geographical information systems. This chapter describes the growth of GIS and explains the technology of gathering data and of analysing its distributional meanings. This is a hugely innovative technique, driven on by a combination of technological uplift and greater demands for more sophisticated displays of vital data. Arguably, no practising environmental scientist should be without this tool, so the websites and valuable introduction to the methodology that follows will act as a practicable enticement.

GIS is part of the modern revolution in knowledge management. There is a view, as yet unproven, that a sustainable society is more likely to emerge with the explosive opportunities of information technology than by any other means. This may be so. A more likely prognosis is that GIS will co-evolve with fresh approaches to deliberative science and democracy as communities shrink in physical space but expand their social and psychic space.

The advent of the Internet has certainly revolutionised environmental management. Now it is possible to display all removals of the wild forest biome on the web, and in real time. Soon it will be possible to do the same with the other critical resources such as coral, mangroves, wetlands generally, and estuaries. Patterns of stress from water shortage, contamination by pollutants and shifting climate patterns can also be recorded. So, too, can toxic waste dumps, both legal and illegal, and damage to prime conservation sites. But one should be wary. GIS is only as good as the data available, and can be a victim of both secrecy and manipulation. In the USA, the Freedom of Information Act of 1957 helps to ensure a legally enshrined data availability. This is not the case in the UK, where agencies either routinely claim confidentiality or charge a very high price for data access. In principle, nothing is hidden from the prying eye of GIS data presentation on the web. It is no wonder that non-governmental groups are seeking to use this ubiquitous and extraordinarily versatile technology for campaigns and for testing the claims of government, business generally, and other community organisations. It is equally no wonder that they are often thwarted in doing so!

All this is opening up the possibility for a new approach to *civil accountability* in environmental management. This is the promotional, or oppositional, collaboration with regulatory agencies and voluntary compact holders over the accuracy of their claims to be in compliance and truthful. The protection of a habitat by a progressive ecological management scheme depends on full co-operation between project director, representing the landscape 'whole', and the landowner, representing the private interests. The modern data management regime can now use the web to record compliance and non-compliance to all who need to know. This should put everyone on their mettle. Other examples abound: for example, the distribution of nutrient additions to a lake can be mapped by satellite imagery, verified on the ground and then converted to a web-based GIS for all to see.

Civil accountability also extends to informal governance and deliberative processes. Thus the future of an urban complex in terms of waste generation or energy or transport demand can be examined by a series of interactive GIS images. Players then use the technology to forecast the reactions of their own representative groups, as well as those of others, to debate just what should be their behaviour (political and household) to various policy options of regulation and economic incentive. Already the technology for this is well advanced, as is the database for assessing plausible scenarios of various future sustainability states. So this particular approach, now in use in climate modelling, will surely take off as more deliberative processes come into vogue.

A word of caution is necessary. The computer may be becoming ubiquitous, but not to everyone. The educationally disadvantaged may be left behind, as may be those vulnerable to disempowerment and injustice generally. Building their capability to prepare them to enter the new democracy, as outlined in Box 1.4, will become a companion process to the expansion of civil accountability. To fail on that coupled responsibility would deal a double blow to social justice.

Introduction

Twenty-five years ago the term 'geographical information system' (GIS) would hardly have been used outside a few research institutes. Today, there is a global market for products and services related to GIS which involves an enormous range of applications. Table 10.1 provides some examples; clearly there are few areas of the environmental sciences where GIS is not being utilised. There is also little doubt that the variety of applications will increase still further in the future. This chapter therefore aims to explain what a GIS is, discuss the types of questions it can be used to answer and consider some possible future developments. Several case studies are included to illustrate how the use of GIS can assist in environmental management.

What is GIS?

There are several different ways of defining GIS. One common approach is to consider it in terms of the functions performed, e.g.

a system for capturing, storing, checking, integrating, manipulating, analysing and displaying data which are spatially referenced to the earth.

(Department of the Environment, 1987, 132)

Table 10.1:
Examples of GIS
applications.

Address matching	Groundwater protection
Archaeology	Highway maintenance
Biodiversity evaluation	Hydrological modelling
Catchment area definition	In-car navigation
Census analysis	Infrastructure planning
Conveyancing	Land contamination assessment
Crime pattern analysis	Landscape change analysis
Crowd control	Local government
Customer targeting	Marketing
Development control	Mining
Ecological research	Oil exploration
Ecosystem management	Pollution monitoring
Emergency planning	Precision agriculture
Environmental impact assessment	Property records management
Environmental monitoring	Retail outlet planning
Epidemiological studies	Risk assessment
Estate management	Sales planning
Facility location	Service delivery planning
Fisheries control	Soil erosion studies
Flood modelling	Tourism management
Forestry	Transport planning
Geological modelling	Water management

Source: Association for Geographic Information (1996).

The phrase 'spatially referenced' in the quotation is significant because one character-istic which distinguishes a GIS from a traditional computerised database is the manner in which both the positions and attributes of real-world features are stored. This link-age between the two types of data makes it straightforward to generate maps of attributes (e.g. pollutant concentrations recorded at sampling points), but a GIS is more than a means of producing cartography by computer. Technologies such as com-puter-aided design (CAD) and automated mapping have certainly contributed to the development of GIS, but the latter is a distinct advance in terms of the ability to inte-grate data from different sources (e.g. compare sightings of a rare bird species with the boundaries of protected areas) and undertake a wide range of analytical operations. In this sense, a GIS can be regarded as a *toolbox* for converting raw data into usable information. Some examples of the questions that can be investigated using a GIS are given in Table 10.2.

Table 10.2:
Typical questions
that a GIS can be
used to answer.

Type of question	*Example*
Identification	What is at a particular location?
Location	Where does a certain type of feature occur?
Trend	Which features have changed over time?
Routing	What is the best way to travel between two points?
Pattern	Is there a spatial association between two types of feature?
What if?	What will happen if a particular change takes place?

Sources: based on Rhind (1990); Kraak and Ormeling (1996).

A second perspective on a GIS is to regard it as a model (i.e. simplified representation) of the real world. For any selected geographical region only a subset of real-world features can be stored in the database, and even these will undergo some generalisation or distortion in terms of how positions are defined and attributes measured. To illustrate, there are often gradual transitions from one type of soil or geology to another, but for many practical purposes it is usually necessary to define such boundaries in a GIS as a definite line. Whenever possible, therefore, it is important to represent phenomena in a GIS so that key characteristics are preserved and loss of information is minimised.

Another common approach to defining GIS is as an organisational entity. This perspective focuses on the components involved, typically distinguishing between computer hardware, software, data and operating personnel. Over time, the importance of these elements as constraints on the use of GIS has changed considerably. Access to suitable hardware was a major problem until the development of personal computers during the 1980s, and it is only in the 1990s that there has been a proliferation of GIS software. A recent international survey listed over 230 organisations with GIS software products, and more than 100 of these reported an environmental specialism (Adams Business Media, 1998). At the present time, there are still shortages of suitably trained personnel (particular those with both technical GIS skills and knowledge of application areas), but it is probably issues regarding the availability of data that are of greatest current concern. These are discussed in more detail in the following section.

Data sources for GIS

It is common for database construction to be the most time-consuming and expensive part of any GIS project (often over 70% of total costs). In many initial applications of GIS most digital data were obtained by digitising or scanning details from paper maps. These techniques require specialist equipment (e.g. digitising tables or large document scanners) and need considerable care from the operator to produce reliable results (Kraak and Ormeling, 1996). Both digitising and scanning are still used extensively today, but during the 1990s reliance on these methods has been reduced by three developments. These are the use of global positioning systems (GPS), improvements in satellite remote sensing and the increasing availability of digital map databases.

GPS provides a means of recording geographical locations, elevations and any associated attributes using a portable receiving unit. The receiver can detect signals from a constellation of satellites following precise orbits around the Earth, and by calculating the distances involved (based on elapsed time and the speed of light) is able to triangulate its position. The accuracy of recorded locations depends on factors such as atmospheric conditions, topographic or ground features that restrict the view of the sky, and deliberate degradation of signals for civilian receivers (the most widely used satellite system is operated by the US Department of Defense). But with a single unit it is usually possible to compute a horizontal position to within 100 metres (Kennedy, 1996). In practice, accuracy is often rather better than this and, by using differential corrections from a second receiver at a known location, it can be further improved to around 5 metres (or even to the sub-metre level with more expensive equipment). Plans have been announced to phase out the degradation of the civilian signal from the US satellites by 2006. When this happens, single receivers costing several hundred pounds should be able to calculate positions to within 10–15 metres.

The importance of GPS is that data can be recorded in the field and then directly transferred into a GIS. This flexibility has proved beneficial in many environmental mapping exercises, and GPS is now widely used in contexts such as habitat delineation, pollution monitoring and the tracking of animal movements (Carver *et al.,* 1995; Johnston, 1998). Another significant development is precision farming, where a GPS receiver on a tractor controls applications of fertilisers or pesticides so that they are sensitive to any known variations in soil type or other characteristics across a field (Lachapelle *et al.,* 1996).

Imagery from satellites such as Landsat and SPOT has been an important source of land cover data in GIS applications for over 20 years, but the 1990s have seen a substantial increase in the range of Earth-surface parameters that can be measured from space. In part, this is due to the launch of systems with microwave (active) sensors (e.g. Radarsat, and ERS 1 and 2). These sensors differ from the passive devices for measuring reflected electromagnetic radiation carried by Landsat and SPOT in that they are not impeded by cloud cover, and their imagery has proved particularly useful for mapping topographic or geological features (Lillesand and Kiefer, 1994). Satellites have also improved in their spectral sensitivity (the ability to measure reflected radiation at different points along the electromagnetic spectrum) and spatial resolution (the minimum size of area from which reflectance is sampled). Both of these characteristics are important for feature identification. Current Landsat imagery has a ground pixel size of 30 by 30 metres, while on a SPOT panchromatic scene (with poorer spectral sensitivity) the resolution is 10 metres. The end of the Cold War, however, has seen the marketing of 2-metre imagery from former Soviet military satellites, and in the USA there are active plans to launch commercial satellites with 1-metre ground resolutions during 1999 (Stoney and Hughes, 1998). Once such imagery becomes widely available, it will certainly represent an alterative source of basic geographical data to all but the most detailed traditional maps and could have particular advantages for many planning or management purposes because of the ease with which coverage can be updated.

The third key development in data availability during recent years has been the growth in digital map databases. Global coverage of elevation and land cover information now exists at a 1-kilometre grid resolution, and many other variables (e.g. climatic and soil characteristics) have been compiled at coarser scales. Much of this information is freely available via the Internet (see Box 10.1). An important complementary product is the Digital Chart of the World, which is usually supplied on four CDs and provides details of features such as coastlines, political boundaries, main roads, rivers, urban areas and selected contours at a 1:1 000 000 scale. It should be emphasised that these databases are too generalised to be appropriate for analyses at the local level, but they constitute a major resource for GIS applications at the regional and international scales and consequently are particularly valuable in contexts such as evaluating the impacts of global environmental change scenarios.

Progress in generating large-scale digital map databases varies greatly both between and within countries. Coverage in the UK is relatively good, with the Ordnance Survey having completed conversion of its paper map series to produce a national digital database at a scale of 1:10 000 or better (1:1250 in cities). In the USA, the most detailed digital maps routinely available from the US Geological Survey are at a 1:24 000 scale, but because such information is regarded as being in the public domain there is a substantial commercial sector which enhances and resells such data. This contrasts strongly with the situation in the UK (and many other European countries), where copyright protection and pricing of digital data from publicly funded organisations have tended to reflect high cost-recovery targets. Suitable digital data for a project such as an environmental impact assessment might therefore exist, but access to the information could be restricted or the charge for acquiring it be more than was affordable.

Box 10.1 Data sources for GIS on the Internet

One of the most prominent technological developments since the first edition of this book has been the growth of the Internet. Many organisations supplying data for use in GIS now have their own Internet sites, often providing examples of applications and sometimes with data that can be downloaded. The following list contains URLs (universal resource locators) for a number of sites concerned with digital map databases or satellite imagery. You will need to type the addresses into browser software (such as Internet Explorer or Netscape) in order to access the sites. Please also note that the Internet is a very dynamic entity and, although the URLs are correct at the time of writing, they could possibly change in the future.

Many national mapping agencies have Internet sites. Good examples include:

Ordnance Survey   *http://www.ordsvy.gov.uk/*
US Geological Survey   *http://www.usgs.gov/*

An excellent site for global environmental data is the EROS Data Center in the USA. This is part of the United Nations Environment Programme Global Resource Information Database (UNEP/GRID), and the address is:

http://grid2.cr.usgs.gov/

From within this site it is also possible to access global coverage of elevation and land cover data. A more direct URL for these resources is:

http://edcwww.cr.usgs.gov/landdaac/

Further details of the Digital Chart of the World are provided at the following location:

http://ilm425.nlh.no/gis/dcw/dcw.html

Examples of existing satellite imagery and planned developments can be found at:

Landsat   *http://geo.arc.nasa.gov/esd/esdstaff/landsat/landsat.html*
SPOT   *http://www.spotimage.fr*
ERS   *http://gds.esrin.esa.it/*
Radarsat   *http://radarsat.space.gc.ca/*
EOS   *http://eospso.gsfc.nasa.gov/*
Eosat   *http://www.spaceimaging.com*

Another interesting site is the Microsoft TerraServer project, which combines 2-metre resolution imagery from Russian satellites and USGS data in a format which allows locations around the world to be selected and viewed. The URL is:

http://terraserver.microsoft.com/

Such problems do not encourage good environmental management, and there seems little doubt that improvements in public access to information would do much to facilitate wider and more effective use of GIS in countries such as the UK (Pipes and Macguire, 1997).

Storing and analysing data

Data are most commonly stored in a GIS by separating different types of real-world phenomena into layers. Details of the individual layers are usually represented in one of two main types of data structure, raster and vector. Raster structures involve assigning values to cells in a grid, while in vector approaches the positions of features such as points, lines or areas are defined by sets of co-ordinates (see Figure 10.1). Many refinements to methods of data storage are possible (see Laurini and Thompson, 1992, or Burrough and McDonnell, 1998, for further discussion), but a key point is that different structures tend to be more appropriate in some circumstances than others. For example, if precise boundaries, shapes or networks are important then a vector structure will usually be the best option, while a raster technique often has advantages if continuous phenomena (e.g. temperature and elevation) are involved or analysis of multiple layers (as in statistical modelling operations) is planned. Ideally, data structures should be selected to minimise distortion when creating a digital representation of reality and maximise analytical or presentational options given the intended uses of the data. Sometimes choices may be restricted by factors such as the format of an existing digital database or limitations in the GIS software available (e.g. only certain data structures are supported), but it needs to be recognised that the manner in which digital representations of environmental phenomena are created can have a substantial influence on the conclusions reached from any subsequent analyses or any policy recommendations formulated.

A useful typology of basic data analysis methods within a GIS is presented by Berry (1987, 1993). This approach identifies four main classes of operation, namely:

- reclassifying attributes on an existing data layer to create new information;
- overlaying two (or more) map layers to create new sets of boundaries;
- measuring distances or connectivity between features on maps to define outputs such as buffer zones, suitable routes or fields of view;

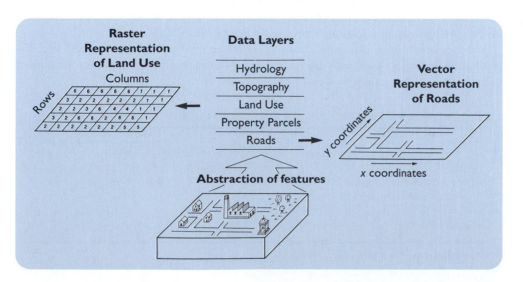

Figure 10.1: Representing real-world phenomena as raster or vector data layers.

● characterising neighbourhoods so that the value assigned to a location depends on attributes of the surrounding area (e.g. calculating slope angles from elevation data).

Each of these techniques will now be discussed in greater detail, with examples chosen to illustrate how they can be applied in environmental management.

Reclassification essentially involves the redefinition of data on an existing map layer. It may take the form of converting numerical values to categories (e.g. grouping individual elevations into height ranges) or the amalgamation of existing classes (e.g. the conversion of a complex land cover classification into a simpler scheme). A particularly common operation is the conversion of existing data to a binary form so that the output map indicates where a specific condition is met (e.g. land below a set elevation threshold and so defined as more vulnerable to flooding).

The value assigned to each location after an overlay operation reflects the attributes of that point on the input data layers. New values are usually defined either by comparing existing features (e.g. to check for the geographical coincidence of certain characteristics) or through arithmetical processing of variables. The combination of reclassification (especially to a binary form) and overlay is widely used in suitability assessments or siting studies, where the objective is to identify locations that meet a number of criteria. An example of such an analysis is provided by the work of Openshaw *et al.* (1989), who used a GIS to identify potential sites for a radioactive waste repository in Great Britain. In one scenario for a deep repository the constraints examined were that sites should be:

● within an area characterised by a suitable hydrogeological environment;
● inside an area with a population density below 490 persons per km^2;
● within 3 km of the rail network;
● outside a designated nature conservation area;
● at least 4 km^2 in size.

Table 10.3 indicates how the cumulative intersection of map layers representing these criteria served to reduce the potentially suitable locations to less than 10% of the British land area (mostly locations around the coast or in a zone stretching across the Midlands and up the east coast of England). Carver (1991) presents an extension of this investigation by evaluating the possible sites from the perspective of different interest groups such as the nuclear industry or environmentalists. Taken together, these studies provide an excellent illustration of the way in which GIS can be used to investigate a range of 'what if' questions in a flexible and efficient manner.

Another example of a GIS application is environmental management, where reclassification and overlay operations are widely used is biodiversity gap analysis. Overlays of map layers are undertaken to identify areas where:

Table 10.3:
Areas satisfying cumulative deep radioactive waste repository siting criteria.

Criteria	Area (km²)	Percentage of Britain
Geological	56 951	25.03
+ low population density	55 612	25.44
+ near to rail network	26 385	11.60
+ outside conservation areas	22 582	9.93
+ sufficient size	22 455	9.87

Source: Openshaw *et al.* (1989).

- species diversity is high;
- habitats are vulnerable (due to development pressures);
- areas lack protected status.

Regions where such conditions coincide are regarded as priorities for conservation initiatives. Gap analysis is therefore a screening tool for identifying sites for new reserves or where changes in land-use practices might be merited. Such techniques are now being widely used in the USA (see Scott *et al.*, 1996, or the Internet site of the National Gap Analysis Program at *http://www.gap.uidaho.edu/gap/*) and have been applied on a global scale by organisations such as the International Council for Bird Preservation (Bibby *et al.*, 1992). A more localised example is provided by Smith *et al.* (1996), who mapped lemur species richness in part of western Madagascar (Figure 10.2(a)), compared it with the results of a threat analysis (Figure 10.2(b)) and concluded that the principal reserve (Andranomena) was not optimally located and that future conservation effort would be better focused on the north-east of the region (Figure 10.2(c)).

The third category of analysis operations is the assessment of distances or connectivity. Distances may be measured on a straight-line basis or in metrics such as travel time or cost. It is common to categorise distances to define buffer zones (or corridors) around map features. These buffers, in turn, may be utilised in overlay analyses to determine the characteristics of particular phenomena close to (or further away from) specified locations. Applications of such a methodology would include the identification of land uses around a river or the population living within a particular distance of

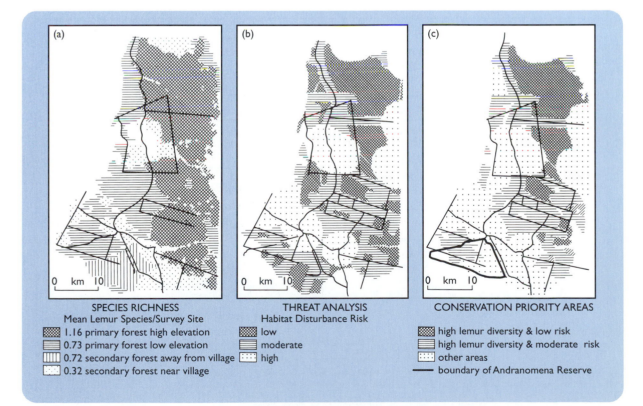

Figure 10.2: Analysing lemur species richness and habitat threats in western Madagascar.
Source: after Smith *et al.* (1996, 507).

a road. Deriving these types of measure can also be valuable in the investigation of routing issues, and an example is provided by Lovett *et al.* (1997) in the context of hazardous waste transport in southern England. Details of the origins and destinations of liquid waste consignments were available, but the routes travelled were not recorded. A GIS was therefore used to model the likely pattern of movements under different scenarios such as the minimisation of travel time, avoidance of densely inhabited areas and the reduction of accident risks. Once routes had been predicted, buffer zones were defined around them and characteristics such as resident populations and

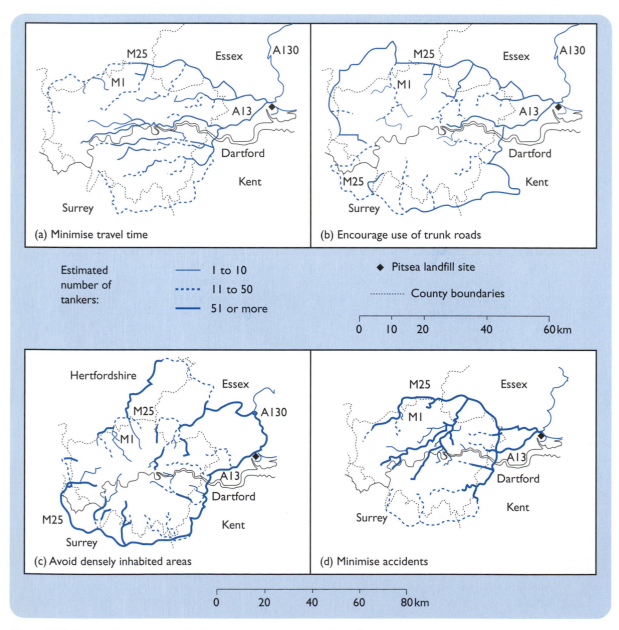

Figure 10.3: Results of routing scenarios for waste consignments to Pitsea, Essex.

Source: after Lovett *et al.* (1997, 631).

Table 10.4:
Results of routing
scenarios for liquid
waste transport to
Pitsea.

Routing scenario	Average length of journey (kilometres)	Population within 500 metres of chosen routes	Tanker traffic over zones of extreme groundwater vulnerability (tanker – km)	Time interval (in years) between cargo threatening accidents
Minimise travel time	58	1 629 442	1460	60
Encourage use of trunk roads	78	1 175 625	4040	62
Avoid densely inhabited areas	91	1 061 145	4740	42
Minimise accidents	64	1 376 026	1290	68

Source: Lovett et al. (1997).

groundwater vulnerability were assessed. Figure 10.3 shows four sets of predicted routes for consignments sent from Greater London to the Pitsea landfill site (one of the largest in the region), and Table 10.4 presents summary statistics for each scenario. A feature of these results is risk trade-offs, since avoiding residential areas resulted in the diversion of shipments out of London onto less suitable roads, with an outcome of more traffic across areas of high groundwater vulnerability and a higher accident risk. The findings also suggested that the environmental risks associated with hazardous waste transport could be mitigated by some simple controls on routing, and the research thus illustrates how GIS can be applied to aspects of risk management.

A different type of connectivity between locations is intervisibility. If data on elevation and other landscape features such as buildings or trees are available, then a GIS can be used to calculate lines of sight and determine the viewshed of a point (i.e. what can be seen from there). This type of analysis has proved useful in several planning contexts, particularly the assessment of visual impacts from proposed developments such as wind farms (see Selman et al., 1991, or Sparkes and Kidner, 1996, for examples). It has also been employed in environmental economics as part of a study estimating monetary values for the negative impacts of road development (e.g. noise and visual intrusion). This research was undertaken in the city of Glasgow by Lake et al. (1998) and involved the use of GIS to compile a large number of structural, accessibility, socio-economic and environmental variables for a sample of some 4000 properties whose sale price was known. By controlling for other relevant factors it was hoped to isolate the influence of characteristics such as the composition of the view on the sale price (a technique known as hedonic pricing, introduced in Chapter 4). Viewscores were calculated for each property reflecting the prominence with which different types of land use were visible (see Figure 10.4) and then incorporated into a statistical analysis. Environmental variables were found to be a significant influence on property prices even after the other types of factor had been taken into account, with visibility of features such as parks having a positive effect and roads, railways and industrial land uses all having negative impacts.

Neighbourhood (or context) operations create output maps where the value assigned to a location depends on its surroundings on another data layer. Examples of

Figure 10.4:
Deriving measures of the visual prominance of a land use from a property.
Source: after Lake *et al.* (1998, 127).

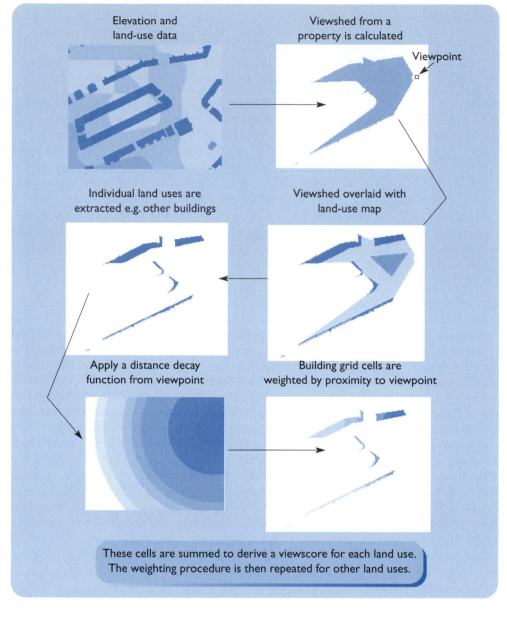

Elevation and land-use data

Viewshed from a property is calculated

Viewpoint

Individual land uses are extracted e.g. other buildings

Viewshed overlaid with land-use map

Apply a distance decay function from viewpoint

Building grid cells are weighted by proximity to viewpoint

These cells are summed to derive a viewscore for each land use. The weighting procedure is then repeated for other land uses.

such methods include the calculation of habitat diversity indices from vegetation distributions, or of slope steepness and aspect (the direction in which the slope faces) from elevation data. Measures such as slope and aspect are, in turn, widely used in the modelling of soil erosion or hazards such as landslides (see Wang and Unwin, 1992, or Burrough and McDonnell, 1998) and are central to the definition of watersheds (i.e. catchment areas) around river systems. Overlays can be undertaken to integrate watershed boundaries with land cover data and so provide a basis for investigating how changes in agricultural practices or new urban developments may influence parameters such as river discharge or water quality (Joao and Walsh, 1992; McDonnell, 1996).

One final point which should be made regarding analysis operations is that the manner in which techniques are implemented does vary between GIS software products and is not always particularly well documented. Several articles have been published demonstrating contrasting results from the same data (e.g. Fisher, 1993; Moore, 1996). This emphasises the need for care when interpreting GIS output and the importance of being aware *how* particular results have ben derived.

Extending the capabilities of GIS

Traditional GIS databases have limited ability to cope with three- or four-dimensional data. For some environmental applications this is a serious restriction. Changes in attributes with elevation or depth are important in a range of contexts such as meteorology, geophysics and oceanography (Mason *et al.*, 1994). Similarly, representing the dynamism of phenomena like coastal spits, river channels or pollution plumes is essential for enhancing both the understanding of processes and the formulation of strategies for effective environmental management (Raper and Livingstone, 1996). Improving the scope for spatio-temporal modelling in GIS is not a straightforward matter (Peuquet, 1994), but significant progress has been made in this direction during the past few years.

A second aspect of enhancing realism concerns issues of error and uncertainty in geographical data. Conventional GIS analyses are good at producing results to a high level of precision (i.e. many decimal places), but it has been increasingly recognised that the underlying accuracy (i.e. closeness to 'true' values) may be rather different, and often less satisfactory. The degree of accuracy in a data layer can depend on many factors, e.g. the scale of source maps or methods of storage and processing, but techniques have been developed to take such imperfections into account by expressing the results of analyses in probabilistic form. For instance, viewsheds have been defined in terms of the chance that a point is in view rather than as a simple yes or no outcome (Fisher, 1994), results of a radioactive waste repository siting exercise have been presented as likelihoods that locations meet specified criteria (Openshaw *et al.*, 1991), and the impacts of sea-level rise have been estimated using probabilities that different stretches of coastline are above a particular elevation threshold (Hunter and Goodchild, 1995). From the perspective of adopting a precautionary approach to environmental management, this type of development is clearly very valuable. There is also research interest in modelling boundaries between features as fuzzy transitions rather than abrupt changes, which will do much to improve the representation of phenomena such as soil types or vegetation communities (Burrough and Frank, 1996; Burrough and McDonnell, 1998). As yet, few commercially available GIS include facilities for error or uncertainty handling, but examples such as Idrisi (Eastman, 1997) do exist and more can be expected in the future.

The capacity of GIS to integrate information from a variety of sources makes them well suited to supporting decision-making processes that must take account of multiple factors, e.g. facility siting or land-use zoning. Given potential difficulties in establishing objectives or measuring relevant factors, it is important not to overstate the benefits of using GIS. Jones (1997) notes:

> *Decision support tools in GIS should therefore be regarded as there to assist in the decision-making process, not to make the decisions. The value of the tools is in helping the decision-makers to evaluate alternatives and to explore certain possibilities, hopefully in more depth and with greater precision than might otherwise have been possible.*

Several techniques have attracted particular interest as means of enhancing the decision support capabilities of GIS. Examples include expert systems (e.g. Lam and Pupp,

1996) and multi-criteria evaluation. The latter is of particular interest in relation to environmental management, since it provides a means of using information gained from interviews or questionnaire surveys to derive sets of weights for the factors under consideration. It is then possible to generate 'suitability maps' for different stakeholders within a GIS and compare them to see if some form of compromise solution is possible. An illustration of such an approach is provided by Emani *et al.* (1997), who modelled views on a hypothetical landfill development for four interested parties (site operator, town council, citizens' action group and the state department of environmental protection). Carver *et al.* (1997) have further extended the possibilities by developing a multi-criteria application for siting a radioactive waste disposal facility in Great Britain that can be accessed via the Internet. This site is available at *http://karl.leeds.ac.uk/mce/mce-home.htm*, so you can try using it for yourself. It certainly demonstrates very effectively how a GIS might be employed to encourage more participatory decision making.

Improvements in spatial multimedia constitute a related way in which the use of GIS for environmental management is being enhanced. By incorporating sources of information such as photographs, sound recordings and animations, it is possible to convey a much more realistic and 'immersive' sense of a location (Raper, 1997). A particularly exciting innovation is the ability to produce virtual environments from GIS databases that can be viewed over the Internet and explored in an interactive manner. Research at the University of East Anglia is currently using such techniques to visualise present and potential future landscapes in part of Oxfordshire, and then investigating the opinions of farmers and other stakeholder groups with respect to the different outcomes and the agricultural policies required to achieve them. Figure 10.5 shows a view (through an Internet browser) of part of the study area near the River Thames, while Figure 10.6 compares a photograph looking north across the floodplain with a view of the virtual landscape from the same point. It is evident that the latter is somewhat simplified, but hopefully the correspondence with the photograph is apparent and it should be noted that further improvements in computing power and visualisation techniques are likely to greatly enhance the realism of such landscapes in future years. Other examples of such simulations include the Virtual Field Course project (see *http://www.geog.le.ac.uk/vfc*) and a

Figure 10.5: A virtual landscape model viewed through an Internet browser.

(a)

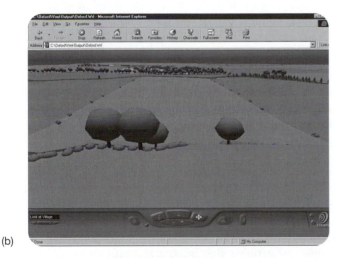

(b)

Figure 10.6: (a) view across the River Thames floodplain; (b) view of the virtual landscape from a similar vantage point.

series of applications with an urban planning focus at University College London (see *http://www.casa.ucl.ac.uk* and Doyle *et al.*, 1998).

Conclusions

This chapter has described the main types of function a GIS can perform and illustrated a variety of environmental applications. There is little doubt that GIS is an important tool (i.e. a means of problem solving), but there has been much debate as to whether it also constitutes a science (i.e. involves research on more fundamental and general questions) (Wright *et al.*, 1997). Personally, I would tend towards the latter view, since in my opinion making effective use of a GIS requires more than awareness of the appropriate commands in a piece of computer software. As I hope would be apparent from comments earlier in this chapter, thinking about and investigating issues such as how to model spatial phenomena and characterise data uncertainty are very important in a context such as the environmental sciences.

Another issue related to the tool/science debate which has been extensively discussed in recent years is the social implications of GIS (Pickles, 1995). Certainly it is fair to say that the use of GIS can encourage a rather technocratic approach to environmental problems. There may also be concerns about who provides the data (most often the state), the types of variables considered (usually those that can be readily quantified), and whether access to the technology gives an advantage to some sections of society (e.g. developers or other large organisations) at the expense of others (e.g. community action groups) in situations such as arguments over planning decisions. Consideration and awareness of these matters is obviously important as GIS becomes ever more widely used in environmental management, but it is also essential not to overreact and conclude that the use of GIS is completely inappropriate or dangerous. Examples earlier in this chapter indicated the scope for using GIS to increase public participation and involvement in decision making, and as Openshaw (1997, 8) notes:

> *GIS is not a totally objective, neutral and value-free technology but this does not mean that it is either random or corrupt or evil. Seldom, if ever, is there likely to be a single 'correct' universally true answer, but some 'solutions' will always be judged to be better than others, no matter how 'better' is measured or who does the judging.*

It is hopefully apparent from all of the above that GIS is no universal panacea for environmental management. A careful and responsible attitude is needed when applying GIS to practical problems, but the technology has already greatly extended our ability to analyse geographical data and can be confidently expected to provide the basis for further contributions to improved environmental management in coming years.

References

Adams Business Media (1998) *1998–1999 GeoDirectory: Products and Services Purchasing Guide*. Adams Business Media, Arlington Heights, IL.

Berry, J.K. (1987) Fundamental operations in computer-assisted map analysis. *International Journal of Geographical Information Systems*, **1**, 119–136.

Berry, J.K. (1993) Cartographic modelling: the analytical capabilities of GIS. In M.F. Goodchild, B.O. Parks and L.T. Steyaert (eds), *Environmental Modelling with GIS*. Oxford University Press, Oxford, 58–74.

Bibby, C.J., Collar, N.J., Crosby, M.J., Heath, M.F., Imboden, C., Johnson, T.H., Long, A.J., Stattersfield, A.J. and Thirgood, S.J. (1992) *Putting Biodiversity on the Map: Priority Areas for Global Conservation*. International Council for Bird Preservation, Cambridge.

Burrough, P.A. and Frank, A.U. (eds) (1996) *Geographical Objects with Indeterminate Boundaries*. Taylor & Francis, London.

Burrough, P.A. and McDonnell, R.A. (1998) *Principles of Geographical Information Systems*. Oxford University Press, Oxford.

Carver, S. (1991) Integrating multi-criteria evaluation with geographical information systems. *International Journal of Geographical Information Systems*, **5**, 321–339.

Carver, S., Heywood, I., Cornelius, S. and Sear, D. (1995) Evaluating field-based GIS for environmental characterization, modelling and decision support. *International Journal of Geographical Information Systems*, **9**, 475–486.

Carver, S., Blake, M., Turton, I. and Duke-Williams, I. (1997) Open spatial decision making: evaluating the potential of the world wide web. In Z. Kemp (ed.), *Innovations in GIS 4*. Taylor & Francis, London, 267–278.

Department of the Environment (1987) *Handling Geographic Information: Report of the Committee of Inquiry Chaired by Lord Chorley*. HMSO, London.

Doyle, S., Dodge, M. and Smith, A. (1998) The potential of web-based mapping and virtual reality technologies for modelling urban environments, *Computers, Environment and Urban Systems*, **22**, 137–155.

Eastman, J.R. (1997) *Idrisi for Windows Version 2.0*. Clark Laboratories for Cartographic Technology and Geographic Analysis, Clark University, Worcester, MA.

Emani, S., Eastman, J.R., Jiang, H. and Johnson, A. (1997) Environmental conflict resolution. In *Applications of Geographic Information Systems (GIS) Technology in Environmental Risk Assessment and Management*. Clark Laboratories for Cartographic Technology and Geographic Analysis, Clark University, Worcester, MA, Chapter 5. Report available at *http://www.clarklabs.org/10applic/risk/start.htm*.

Fisher, P.F. (1993) Algorithm and implementation uncertainty in viewshed analysis. *International Journal of Geographical Information Systems*, **7**, 331–347.

Fisher, P.F. (1994) Probable and fuzzy models of the viewshed operation. In M.F. Worboys (ed.), *Innovations in GIS 1*. Taylor & Francis, London, 161–175.

Hunter, G.J. and Goodchild, M.F. (1995) Dealing with error in spatial databases: a simple case study. *Photogrammetric Engineering and Remote Sensing*, **61**, 529–537.

Joao, E. and Walsh, S. (1992) GIS implications for hydrologic modelling: simulation of non-point source pollution generated as a consequence of watershed development scenarios. *Computers, Environment and Urban Systems*, **16**, 43–63.

Johnston, C.A. (1998) *Geographic Information Systems in Ecology*. Blackwell Science, Oxford.

Jones, C.B. (1997) *Geographical Information Systems and Computer Cartography*. Longman, Harlow.

Kennedy, M. (1996) *The Global Positioning System and GIS*. Ann Arbor Press, Chelsea, MI.

Kraak, M.J. and Ormeling, F.J. (1996) *Cartography: Visualization of Spatial Data*. Longman, Harlow.

Lachapelle, G., Cannon, M.E., Penney, D.C. and Goddard, T. (1996) GIS/GPS facilitates precision farming. *GIS World*, **19**(7), 54–56.

Lake, I.R., Lovett, A.A., Bateman, I.J. and Langford, I.H. (1998) Modelling environmental influences on property prices in an urban environment. *Computers, Environment and Urban Systems*, **22**, 121–136.

Lam, D. and Pupp, C. (1996) Integration of GIS, expert systems, and modeling for state-of-environment reporting. In M.F. Goodchild, L.T. Steyaert, B.O. Parks, C. Johnston, D. Maidment, M. Crane and S. Glendinning (eds), *GIS and Environmental Modelling: Progress and Research Issues*. GIS World Books, Fort Collins, CO, 419–422.

Laurini, R. and Thompson, D. (1992) *Fundamentals of Spatial Information Systems*. Academic Press, London.

Lillesand, T.M. and Kiefer, R.W. (1994) *Remote Sensing and Image Interpretation*. John Wiley, Chichester.

Lovett, A.A., Parfitt, J.P. and Brainard, J.S. (1997) Using GIS in risk analysis: a case study of hazardous waste transport. *Risk Analysis*, **17**, 625–633.

Mason, D.C., O'Conaill, M.A. and Bell, S.B.M. (1994) Handling four-dimensional georeferenced data in environmental GIS. *International Journal of Geographical Information Systems*, **8**, 191–215.

McDonnell, R.A. (1996) Including the spatial dimension: using geographical information systems in hydrology. *Progress in Human Geography*, **20**, 159–177.

Moore, I.D. (1996) Hydrologic modeling and GIS. In M.F. Goodchild, L.T. Steyaert, B.O. Parks, C. Johnston, D. Maidment, M. Crane and S. Glendinning (eds), *GIS and Environmental Modelling: Progress and Research Issues*. GIS World Books, Fort Collins, CO, 143-148.

Openshaw, S. (1997) The truth about ground truth. *Transactions in GIS*, **2**, 7–24.

Openshaw, S., Carver, S. and Fernie, J. (1989) *Britain's Nuclear Waste: Safety and Siting*. Belhaven Press, London.

Openshaw, S., Charlton, M. and Carver, S. (1991) Error propagation: a Monte Carlo simulation. In I. Masser and M. Blakemore (eds), *Handling geographical information*. Longman, Harlow, 78–101.

Peuquet, D.J. (1994) It's about time: a conceptual framework for the representation of spatiotemporal dynamics in geographic information systems. *Annals of the Association of American Geographers*, **84**, 441–461.

Pickles, J. (ed.) (1995) *Ground Truth: the Social Implications of Geographic Information Systems*. Guilford Press, New York.

Pipes, S. and Macguire, F. (1997) Green door: has public access to environmental information improved? *Mapping Awareness*, **11**(10), 28–29.

Raper, J. (1997) Progress towards spatial multimedia. In M. Craglia and H. Couclelis (eds), *Geographic Information Research: Bridging the Atlantic*. Taylor & Francis, London, 525–543.

Raper, J. and Livingstone, D. (1996) High-level coupling of GIS and environmental process modelling., In M.F. Goodchild, L.T. Steyaert, B.O. Parks, C. Johnston, D. Maidment, M. Crane and S. Glendinning (eds), *GIS and Environmental Modelling: Progress and Research Issues*. GIS World Books, Fort Collins, CO, 387–390.

Rhind, D.W. (1990) Global databases and GIS. In M.J. Foster and P.J. Shand (eds), *The Association for Geographic Information Yearbook 1990*. Taylor & Francis, London. 218–223.

Scott, J.M., Tear, T.H. and Davis, F.W. (eds) (1996) *Gap Analysis: a Landscape Approach to Biodiversity Planning*. American Society for Photogrammetry and Remote Sensing, Bethesda, MD.

Selman, P., Davidson, D., Watson, A. and Winterbottom, S. (1991) GIS in rural environmental planning. *Town Planning Review*, **62**, 215–223.

Smith, A.P., Horning, N. and Moore, D. (1996) Regional biodiversity planning and lemur conservation with GIS in western Madagascar. *Conservation Biology*, **11**, 498–512.

Sparkes, A. and Kidner, D. (1996) A GIS for the environmental impact assessment of wind farms. *Proceedings of the 1996 ESRI European Users, Conference (CD)*. ESRI (UK), Watford.

Stoney, W.E. and Hughes, J.R. (1998) A new space race is on. *GIS World*, **11**(3), 44–46.

Wang, S.Q. and Unwin, D.J. (1992) Modelling landslide distribution on loess soils in China: an investigation. *International Journal of Geographical Information Systems*, **6**, 391–405.

Wright, D.J., Goodchild, M.F. and Proctor, J.D. (1997) GIS: tool or science? *Annals of the Association of American Geographers*, **87**, 346–362.

Further reading

The following books provide good introductions to GIS and a range of applications.

Burrough, P.A. and McDonnell, R.A. (1998) *Principles of Geographical Information Systems*. Oxford University Press, Oxford.

Johnston, C.A. (1998) *Geographic Information Systems in Ecology*. Blackwell Science, Oxford.

Jones, C.B. (1997) *Geographical Information Systems and Computer Cartography*. Longman, Harlow.

Martin, D. (1996) *Geographic Information Systems: Socioeconomic Applications*. Routledge, London.

A very comprehensive (though expensive) reference source on GIS is:

Longley, P.A., Goodchild, M.F., Maguire, D.J. and Rhind, D.W. (eds) (1999) *Geographical Information Systems*. Wiley, New York.

Chapter Eleven

Soil erosion and land degradation

Michael Stocking

Topics covered

- Soil as a resource
- Consequences of soil degradation
- Processes of soil degradation
- Assessment of soil erosion
- Prediction and explanation of soil erosion
- Soil and water conservation
- Community perceptions of land degradation

Editorial introduction

The Rio Conference was criticised by many developing countries on the grounds that it was hijacked into concentrating on climate change and biodiversity losses. The poor nations believe that the wealthy commandeer these issues because they have a vested interest in long-term climate stability, and have a huge guilt trip over their prolonged period of greenhouse gas emissions. As for biodiversity, well, that begins abroad: it is regarded as a moral issue linked to the civil and property rights of indigenous peoples. Biodiversity enhancement in nations with a long history of land transformation is yet to be recognised as a high priority.

This is obviously a simplistic analysis of a sophisticated argument. For the world's poor, health, soil erosion, loss of fuelwood supplies, inadequate sanitation and reduction of cultivable land area are the major environmental concerns. Arguably land degradation is the single most pressing current global problem. As a result of remote-sensing evidence we know that, since 1945, 1.2 billion ha, an area roughly the size of China and India combined, has been eroded at least to the point where the original biotic functions are impaired. At the very least, it will be costly and time-consuming to rehabilitate them. Of this area, about 9 million ha is very severely damaged, to the point of unreclaimability, and 300 million ha is so damaged that cultivation is all but impracticable. Much of this is in regions that are already overpopulated in relation to food production capacity.

Agenda 21 placed a lot of emphasis on re-establishing sustainable livelihoods for the 1–2 billion who are impoverished by the lack of water, land cover, soil fertility and fuelwood. This is a critical issue of both human care and environmental well-being.

The long-term consequences of removing land cover from huge areas of the tropics and semi-arid lands are frankly not known. But one can be sure that there will be cumulative catastrophic failure of both the land and its peoples if these processes are allowed to continue.

Agenda 21 argued for both capacity building and the promotion of ecologically and culturally sensitive schemes of land restoration. Capacity building means expanding the vernacular science of local knowledge and tradition with the Western science of soil and water management, monitoring of rainfall and groundwater movement, and analysing changes in soil fertility following different treatments. This is slow work, long in preparation, involving scientists who are as much anthropologists and communicators as they are soil specialists, hydrologists or ecologists. There is no blueprint for land restoration. The process will always be dictated by the rhythms of cultural experience, political expectation and investment capital from aid agencies and philanthropic foundations.

There is also a body of thought that regards soil erosion and land degradation as a process of continual negotiation between culture and the land. Careful monitoring of the tree or savanna margins in semi-arid areas reveals a fluctuating interface, punctuated by changes in rainfall, movement of people and shifts in agricultural techniques and cropping patterns. Soil erosion is by no means a purely physical phenomenon. It cannot solely be measured by sediment movement, nutrient status and surface seepage. True, these are measures of physical status: but they may have limited cultural meaning for societies that prefer to co-operate only when the need arises, that share property as well as food in times of need or plenty as part of established peer-group ritual, and that appreciate systems of water harvesting that rely on low technology coupled with communal action. Such responses are not unusual. It is always dangerous to generalise. This is why the land degradation theme is so dependent on both vernacular science and capacity building for local land managers and community leaders.

For example, Harold Brookfield and Christine Padoch (1994, 9) look at the changing fortunes of *agrodiversity*, namely the management of cropping to suit both biodiversity and local conditions of soil, land tenure and accessibility. They believe there are many hundreds of thousands of variations of shifting agriculture. While it is true that migration, commercial acquisition and population pressures undermine such practices, nevertheless the scope for maintaining a locally centred community agriculture, based on the principles of diversity and adaptability, remains less examined than should be the case. Yet the use of indigenous knowledge, promoted in this chapter, is now back in vogue. The main donor organisations are much more willing to go to the smaller scale, and to assist farmers to retain a huge variety of tiny plots on scattered holdings, and to ensure some financial support. This will require carefully selected and calibrated research programmes. One such project is the People, Land Management and Environmental Change Project promoted by the Global Environmental Facility, executed by the United Nations University and implemented by the UNEP (Brookfield and Padoch, 1994, 42). This has established a network of field-based projects across the tropics, moving away from simple explanations to the rich multi-theoretical and multi-practice examples that are waiting to be examined with an open mind (see Brookfield and Stocking, 1999).

Paalberg (1994, 40) sums up an optimistic recommendation:

Fundamental political and social change in rural areas should be the goal. New and less abusive power relations must evolve between the landed and the landless, between indigenous and non indigenous peoples, and between citizens and governments. Ordinary peasant farmers need to gain control over the resource base on which they and their children depend.

We shall see in Chapter 19 that aid and debt relief may begin to be tied to reforms in such aspects of local politics and anti-corruption practices. There is no ideal or quick solution, as rapid movement across a sweep of terrain may well result in violence and even more inequitable outcomes. This is why participatory and holistic environmental science must co-evolve with policy shifts and institutional redesign.

The trick is to devise schemes that promote integrated community-led development yet do not alienate the machinery of local government or national administrative agencies in agriculture, forestry conservation, tourism and regional planning. This will not be easy. Experimentation on a modest scale, but tried out in a host of circumstances, may be the best bet.

Soil – a resource under threat?

Soil is essential to support life on Earth: 90% of all human food is produced on land, from soils varying widely in quality, nature and extent.[1] The livelihoods of most rural people are directly supported by soil; and all of us, whether in recreation or occupation or subsistence, depend upon soil as the hub of all life-support processes. As an essential component of terrestrial ecosystems, soil sustains the primary producers (living vegetation) and the decomposers (micro-organisms, herbivores, carnivores). It also provides the major sinks for heat energy, nutrients, water and gases. (See the useful introductory text by Wild, 1993.)

Soil is a harmonious microcosm of many realms, having hydrological, biological, physical and chemical environments, each of which is bound intricately with the others, and all combining to produce a multi-purpose medium for plants, buildings, water and carbon storage, buffering of the atmospheric climate and many other vital processes. Yet soil is an innocuous substance – trampled, abused and wasted; it is something to be dug, not nurtured, to be exploited rather than sustained. This chapter is about the dichotomy of a scarce and valuable but unloved and ignored natural resource, and how the soil's beguiling semblance of complex harmony is a challenge to the environmental manager.

The soil resources of landscapes vary widely in their suitability for use. Each soil type has its limitations, and each agro-ecological zone its climatic factors restricting crop-growing seasons. For example, in the humid tropics plant stresses are mostly the result of nutrient deficiencies exacerbated by leaching and surface removal rendering the soil acid and nutrient-poor. In contrast, in the vast areas of seasonal tropics of South Asia, Africa and South America, where rainfall is concentrated into part of the year only, the lack of plant-available water capacity is the major handicap. What masquerades as drought is often nothing more than a reduction in the ability of soils to hold sufficient water for plant growth between quite normal gaps in rainstorms.

Are the world's soils under threat? If official and influential statements are to be believed, most certainly both the scale of degradation and its impact suggest that we are heading for catastrophe (see Box 11.1). Soil quality is apparently diminishing at an

---

[1] About 10% of the Earth's land surface is covered by ice. Of the 13 000 million hectares of ice-free land, 15% is too cold, 17% too dry, 18% too steep, 9% too shallow, 4% too wet and 5% too poor for cultivation. It is on these lands, often used for extensive grazing, that most of the world's soil erosion and land degradation occurs. Only 3% of the Earth's surface has highly productive soils, 6% moderately productive and 13% slightly productive. Erosion also occurs, in absolute quantitative terms to a lesser extent than on marginal lands, but the economic impact of loss in soil productivity is much greater (see Hurni et al., 1996).

Box 11.1 Crisis narratives on soil erosion and land degradation

The literature, especially from United Nations sources and environmental agencies, is replete with 'crisis narratives' – that is, sensational and sweeping statements of the extent of environmental problems. Here is a small sample. Note the difference in estimates and their provenance between the first three, all for Africa:

> *About 321 million hectares [of sub-Saharan Africa] are moderately, severely, or extremely degraded and a further 174 million hectares are lightly degraded.*
> (World Bank – Cleaver and Shreiber, 1994, 21)

> *About a half-billion hectares in Africa are moderately to severely degraded, correspond-*

> *ing to one third of all cropland and permanent pasture on the continent.*
> (United Nations – UNEP, 1997, 26)

> *More than 80% of sub-Saharan Africa's productive drylands, some 660 million hectares, are affected by desertification.*
> (environmental NGOs – WRI/IIED, 1987, 59)

> *Each year, 75 billion metric tons of soil are removed from land by wind and water erosion.*
> (Pimentel *et al.*, 1995, 1117)

See Reij *et al.* (1996, 2–3) for further quotes under the heading 'The power of numbers'.

alarming rate; crop production is becoming more expensive as soil fertility, water-holding capacity and depth decline; and badlands are increasingly evident in vulnerable places (Figures 11.1 and 11.2).

Figure 11.1:
Severe erosion: sheet erosion in Sukumaland, Tanzania (**Photo**: M.A. Stocking)

The most important study on land degradation designed to generate analysis at a continental scale is GLASOD (Global Assessment of Land Degradation: Oldemann *et al.*, 1990). It depends on the 'expert judgement' of soil professionals in 21 regions to evaluate changes they have seen since the Second World War using systematic criteria – so possibilities of exaggeration as demonstrated in Box 11.1 still remain. Nevertheless, GLASOD estimates that nearly 2 billion hectares or 22.5% of all productive land has been degraded since 1945 (Box 11.2). Globally, GLASOD calculates that about half the productive area is under forest, of which 18% is degraded; 3.2 billion hectares is under pasture, of which 21% is degraded; and nearly 1.5 billion hectares is in cropland, of which 38% is degraded. Cropland is most threatened in Africa with 65% degraded. However, to appreciate the threat, it is necessary to know what is expected of this land and how many people the soil can support.

Estimates of the productive potential of land are difficult to make because of the enormous variety of land uses, levels of technology, land management standards and population pressures. The largest study of its kind, entitled *Potential Population Supporting Capacities of the Lands in the Developing World*, was conducted by the UN Food and Agriculture Organisation (FAO, 1982). Although dated, its predictions are widely used and give a measure of the global scale of the impact of degradation. At the base year 1975, 54 developing countries with a total population of 460 million did not have land of sufficient quality at low levels of inputs to feed their own populations. Overloading of some lands was estimated to be threatening 2450 million hectares or 38% of the total land area. According to Norse (1992), 64 of 117 countries surveyed would be unable to support more than half their year 2000 populations on current

Box 11.2 Soil resource sustainability – a global assessment

According to assessments undertaken for the World Map of Soil Degradation and made possible by satellite telemetry and co-ordinated teams of environmental scientists worldwide, nearly 2 billion hectares, or 22% of the Earth's vegetated and potentially cultivable surface, has been degraded to a measurable extent. Some 300 million hectares is so badly damaged that it has essentially lost its biological function. Rehabilitation is possible but would be far more expensive than farmers can afford. While world agricultural production is at an all-time high, for 64 countries future food production increases will enable less than half the projected year 2000 population to be fed (Oldeman *et al.*, 1990; Norse, 1992).

Table 11.2.1: Extent of major productive land uses affected by soil degradation (millions ha).

	Agricultural land			Permanent pasture			Forest and woodland		
	Total	Degraded	%	Total	Degraded	%	Total	Degraded	%
Africa	187	121	65	793	243	31	683	130	19
Asia	536	206	38	978	197	20	1273	344	27
S. America	142	64	45	478	68	14	896	112	13
C. America	38	28	74	94	10	11	66	25	38
N. America	236	63	26	274	29	11	621	4	1
Europe	287	72	25	156	54	35	353	92	26
Oceania	49	8	16	439	84	19	156	12	8
World	1475	562	38	3212	685	21	4048	719	18

Source: FAO (1990); Oldeman (1994).

technologies, while 18 countries remain critical even at the highest projected levels of agricultural inputs.

Much of the evidence is, however, confused. IFAD (1992, 17) notes: 'It is difficult to give an indication of the extent of land degradation in sub-Saharan Africa, as there is no simple or commonly agreed measure of degradation. Nor are consistent data available about the state of land in each country.' Ambiguities abound (see Dahlberg, 1994) and biophysical indicators such as loss of vegetative cover are contradictory. For example, Fairhead and Leach (1996) document a compelling case for viewing the West African forest–savanna zone as a landscape 'half-filled and filling' with trees, rather than as forest 'islands' as the last vestiges of once pristine forest cover. Similarly, social and economic indicators such as poverty, vulnerability to environmental change, inequity, marginalisation, conflict and migrations show remarkably different situations in different places. Major agencies such as the International Food Policy Research Institute talk instead of 'hotspots' of degradation (Scherr and Yadav, 1996) to signify the patchiness of the problem, especially in the developing world, and 'bright spots' to counter undue pessimism (see Box 11.3). Certainly, soil erosion and land degradation are serious in some places; equally, their degree and extent have been overstated, often for political ends (see case of Africa: Stocking, 1996). Rhetoric on soil erosion can be a useful device to displace people, point fingers of blame, promote the cause of environmental agencies and the professionals who work in them, and mobilise international aid.

Box 11.3 'Hot spots' and 'bright spots': East and Southeast Asia

East and Southeast Asia are expected to experience continuing economic growth to 2020, which, together with rapid urbanisation and industrialisation, will change the focus of degradation concerns. Environmental issues will become much more important . . . land scarcity will continue to create pressure for expansion of cultivation into marginal lands,

particularly the hillside areas. The potential for sustainable, increased agricultural production in the irrigated areas will depend on successfully combating salinisation, improved input use, and scientific advances to increase biological yield potentials.

(Scherr and Yadav, 1996, p.17)

Table 11.3.1

'Hotspots'	'Bright spots'
Agricultural lands:	
• salinisation: irrigated lands of moderately populated NE Thailand and China	• high-yielding varieties and judicious use of chemicals: irrigated rice production throughout South-east Asia
• waterlogging and nutrient imbalances: irrigated rice in densely populated Java, Philippines and Vietnam	• sustainable intensification: densely populated parts of fertile Indonesia and south China
• erosion and decline in fertility: sloping lands throughout SE Asia	• high-value production, such as horticulture and aquaculture: Thailand, Philippines, Malaysia, China
• infestation by *Imperata cylindrica*: low-density areas of Philippines and Vietnam	• perennial plantation crops: in low-density, erosion-prone parts of Philippines
• wind erosion: northern China	• conservation farming systems: sloping area technologies in Philippines, Thailand
• nutrient depletion and water erosion: NE Thailand and remote uplands	
Related environmental impacts:	
• biodiversity loss and forest depletion: through high population growth in Indonesia, Malaysia, Cambodia, Laos	• sedimentation: all coastal areas and deltas
• devegetation of grazing lands: through shifting cultivation in Burma	• community forests: low population areas of Thailand and south China
• pollution by agrochemicals: around Bangkok, Jakarta, Manila	• changing property rights: encouraging greater investment in land improvement in Vietnam, Laos, Cambodia
• acid sulphate soil problems: coastal Indonesia, Malaysia, Thailand, Vietnam	• recreational area development: habitat conservation in China, Vietnam, Indonesia and Papua New Guinea

A question of definition and perspective

This chapter examines the processes of soil and land degradation, looks at the extent of soil erosion, ways it may be assessed and its impact on production and productivity, and concludes with an overview of methods of soil conservation. Many books use the terms 'erosion' and 'degradation', and 'soil' and 'land', interchangeably. Yet this hides some important distinctions. *Soil degradation* is defined as a decrease in soil quality as measured by changes in soil properties and processes, and the consequent decline in productivity in terms of immediate and future production. Some authors specify that soil degradation encompasses a range of processes, including water (or soil) erosion and nutrient depletion, resulting in a deterioration in physical, chemical and biological attributes of the soil (e.g. Lal, 1990). *Land degradation* is a composite term signifying the temporary or permanent decline in the productive capacity of the land. It is the aggregate diminution of the productive potential of the land, including its major uses (rain-fed arable, irrigated, rangeland, forestry), its farming systems (e.g. smallholder subsistence), and its value as an economic resource. Some types of land degradation are considered as effectively irreversible (large gullies and extreme salinisation), while most soil degradation is reversible by adding nutrients or soil amendments such as lime, re-establishing vegetation and other management practices. Enters (1997) defines land degradation as reduction in the capacity of the land to produce goods and services. He describes it as 'more than just a physical or environmental process. Ultimately, it is a social problem with economic costs attached as it consumes the product of labour and capital inputs into production' (Enters, 1997, 4). A topsoil can be created in a decade, moderate salinity can be corrected in two or three years, and soil water problems reversed in an even shorter timescale with the right management. *Soil erosion* is one of the main processes of degradation and consists of physical detachment of soil particles by wind and water and their transport to other parts of the landscape, to rivers and the sea.

The impact of soil erosion and land degradation is usually considered in terms of *off-site* and *on-site* effects. Off-site includes all the damage that is created by sediment, such as sedimentation of the lower slopes of catchments, reduction in capacity of reservoirs, pollution of water supplies and disruption to waterways. While off-site effects may often damage major investments such as hydroelectric power generators, the on-site impacts are often greater in total value. However, they are insidious: a relentless and slow decline in the life-support functions of soil. In challenging future production, degradation processes reduce soil quality, nutrient status, water-holding capacity, fertility and productivity. *Soil fertility*, while often mistakenly related solely to nutrient status, is the soil's ability to produce and reproduce. Indicators of fertility include organic matter content, micro-organisms, soil air and water, and the adequate balance and availability of nutrients. *Soil productivity* is a composite term to describe the overall productive potential of a soil arising from all aspects of its quality and status, such as its physical and structural condition as well as its chemical content. Conceptually, productivity is close to *sustainability* in representing the intrinsic quality of the soil resource. It should not be confused with *production*, the actual output (e.g. biomass or crop yield) from the soil–plant system, which is affected not only by the resource quality but also by applications of technology such as fertiliser, improved seeds, irrigation and the like.

The productivity and sustainability of the world's soils is, by most estimates, declining, even though aggregate production (in all continents except Africa) is being maintained or even increased. Because of population growth, the amount of cultivable and usable land per person inevitably declines (Box 11.4). Examples of intensive and productive land use are, however, abundant and being increasingly identified in indigenous and locally devised technologies, such as the *ngoro* pits in the Matengo Highlands of

Box 11 4 Projected decline in potentially cultivable land, 1990–2025

Apart from a few vast countries with somewhat small but nevertheless expanding populations, such as Brazil and Zaire, most developing countries have essentially reached the limit of new cultivation – at least with the technologies, resources and costs they can afford. As populations grow, therefore, so the amount of cultivable land *per capita* will fall.

Table 11.4.1

| | Land per person (ha) | |
	1990 actual	2025 projected
Central and South America	2.00	1.17
Sub-Saharan Africa	1.60	0.63
West Asia and North Africa	0.22	0.16
South and South-east Asia	0.20	0.12

Source: Norse *et al.* (1992).

Tanzania. New technologies, changes and adaptations by society, as well as alternative economic opportunities, have enabled the decline in land per person to be accommodated (Figure 11.4). They also prove environmental crisis-mongers wrong – the question now is whether the doomsters merely got their timing awry or whether society (and its

Figure 11.3: Productive land – pearl millet grown on terraces, Loess Plateau, China (**Photo**: M.A. Stocking).

environmental managers) really does have coping mechanisms to continue adjusting to resource availability and abundance. The record so far is quite good. (See the fascinating study *More People, Less Erosion* by Tiffen *et al.*, 1994.)

As optimists are now tending to gain advantage from recent evidence of the greater resilience of soil resources – even since the first edition of this book in 1995 – it is instructive to note that there has always been a wide variety of perspectives as to the significance of soil erosion and land degradation. Surveys, in both developed and developing countries, have shown that farmers often do not see erosion as a problem. Some may even see it as an opportunity – see Box 11.12 later in this chapter. Perhaps the most perverse yet engaging perspective on soil degradation is that of the Burungee in Tanzania – what we see as degradation they see as soil formation (Box 11.5). To understand and appreciate these various viewpoints is a necessary part of the work of today's environmental managers.

Consequences of soil degradation

The consequence of erosion and degradation can best be demonstrated in three ways. First, soil has a *resource value*. Nutrients are depleted, taken in the sediment and runoff. Generally, those nutrients associated with organic matter (N and P) and the cation exchange of soil colloids (K and Ca) are most at risk. Nutrients in sediment are approximately ten times the quantity of those dissolved in runoff. Using what is known as the 'replacement cost approach', values may be attached to these lost nutrients

Figure 11.4:
Productive land – conservation tillage with chisel plough, southern Brazil (**Photo**: M.A. Stocking).

Box 11.5 Burungee perspectives on soil formation

The Burungee people of Kondoa district, Dodoma Region, Tanzania, have an interesting way of seeing soil formation and degradation. Even though the signs of erosion are everywhere, the evidence before *their* eyes leads them to very different conclusions as to what is happening. In common with much of semi-arid Africa, gullies punctuate extensive sheet-eroded pediments. Arable lands give meagre crops and the range supports emaciated cattle. How do the Burungee perceive their environment?

The Swedish anthropologist Wilhelm Östberg (1991) collected these observations:

How soil is created:

● 'If it rains heavily, it results in thick layers of soil; if the rains are poor the soil layers become thin.'

● 'Have you not held a hailstone in your hand and allowed it to melt? Do you not get small particles of soil in your hand? There you can see that the rain contains soil.'

● 'When rain falls, water sinks to the ground. It reaches a point where it cannot go further down. Water accumulates there forming a soil layer as it dries up. this layer will be stained by the colour of the water ... The soil layer moves up, and next time it rains a new soil layer will form below the one that was last created. In this way, new layers of soil are continuously being formed.'

Why the surface is littered with stones:

● 'Look, the land is coming up.'

Land has life:

● 'If you cut the branch from a tree, it is dead. But put it back into the soil and it will grow again. It draws its life from the land.'

through the equivalent cost of fertiliser containing the same amounts of elemental N, P or K (Box 11.6).[2]

Another aspect of resource value is the water lost to the soil and to plant growth in runoff. As soil degrades, its infiltration capacity worsens. Typically, with surface sealing and crusting on a tropical soil, water losses in runoff may increase from 20% to 50% of total rainfall. For a farm in, say, the rain-fed maize zone of Zambia (800–900 mm mean annual rainfall), this is equivalent to moving the farm to a semi-arid rainfall regime (400–500 mm), suitable only for low-value crops such as millet. Relative land prices could be used as one method for attaching a cost to this degradation.

A third way of viewing resource value is to consider the selectivity of the process of erosion. Because water erosion takes the finer, more fertile fraction of the soil, the eroded sediments are always richer in nutrients and organic matter than the soil from which they were taken. Known as the *enrichment ratio* (ER), this measure compares the concentration of a key nutrient in the sediment with its concentration in the topsoil. In some storm events, especially after fresh soil has been exposed by ploughing, ERs can be as high as 10:1. Long-term average values of 2.5:1 have been measured for several soil types in Zimbabwe. Such measures indicate that erosion involves losses of a soil's productive capacity that are very much more than the removal of a topsoil slice.

[2] This approach assumes that (1) the nutrients would be replaced by fertiliser – an unlikely situation on most resource-poor farms, and (2) the nutrients lost by erosion and those replaced by fertiliser are in the same form and have the same availability to plants – again, not true, because most N lost is bound in organic matter, while N replaced is in immediately available form such as ammonium nitrate. Netherthless, the replacement cost approach is simple to apply when nutrient loss data are available. Used by many authors (e.g. Bojö, 1996, for sub-Saharan Africa; Predo *et al.*, 1997, for the Philippines), fertiliser prices make a useful 'financial handle' to value nutrients and put a cost on erosion.

Box 11.6 The cost of soil erosion in Zimbabwe: a resource value approach

In Zimbabwe, 2000 storm soil loss events over five years on four soils were monitored, and losses of nitrogen, phosphorus and organic carbon measured. Extrapolating these losses to all soil types and the major land-use systems, Zimbabwe loses annually 1.6 million tonnes of N, 0.24 million tonnes of P and 15.6 million tonnes of organic C.

If the value of N and P were to be translated into money by the cost of the equivalent amount of fertiliser (at 1984 prices) containing those nutrients, Zimbabwe's annual financial drain because of erosion amounts to US$1.5 billion.

On a per hectare basis, the financial cost of eroded nutrients varies according to actual erosion rates:

- $20–50 per ha per year on arable lands;
- $10–80 on grazing lands.

Arable lands alone lose by erosion about three times the N and P which is applied as fertiliser each year.

There are biophysical and economic assumptions behind these calculations. But nevertheless the potential loss is staggering compared with Zimbabwe's GNP. It amounts to a cost per person of about US$200 – unsustainable for any country, let alone a poor one.

Source: Stocking (1988a).

The second demonstration of the consequences of degradation is to consider the loss in *production value*. Most data derive from the United States, where yield reductions and associated costs have been closely monitored (Box 11.7). Erosion affects crop growth and yields directly through losses of nutrients and water, and indirectly through reductions in soil fauna populations, poor seed germination, delayed and more difficult farming operations, and increased impacts of risks such as drought, pests and diseases. Pimentel *et al*. (1995) calculate that 10% of the energy used in US

Box 11.7 The cost of soil erosion in the United States: a production value approach

Assume at least 7 cm of topsoil has been lost throughout the Mid-west USA, equivalent to a cumulative erosion of 900 tonnes/ha, a total easily attained without soil conservation on arable lands in two to three decades. Or take a net average erosion rate by water and wind of 17 tonnes/ha/year. How is production affected? These are alternative ways of looking at the production value of erosion:

- *without replacement inputs*: 6% yield reduction per centimetre of soil loss; 42% total yield loss, equivalent to average US maize yield of 6.5 tonnes/ha/year falling to 2.7 tonnes, or a financial loss of about $480/ha/year at current grain prices;

- *with inputs to maintain production*: 75 mm of irrigation water or $30/ha/year; 462 kg of nutrients or $100/ha of fertiliser;

- to restore yields to pre-erosion levels: based on the 920 kcal of inputs expended in the USA per kg maize produced, fossil energy inputs of 2.5 million kcal/ha (or 10 500 MJ). At current fuel energy prices, $25/ha/year, or the food energy required to keep one person alive for 1000 days (over $500 without Big Macs).

Summing on-site and off-site costs, David Pimentel and associates at Cornell University calculate that erosion costs the United States $196 per hectare. As the above figures show, costs vary according to the perspective adopted, even when starting with the same basic assumptions about erosion and the same production value approach.

Source: estimates adapted from data in Pimentel *et al*. (1993; 1995).

agriculture goes to offsetting the effects of erosion. The most direct measure, however, is the loss in yields over time or over cumulative amounts of erosion. Recent data from two soils in southern Brazil (Figure 11.5) show the typical marked decline in yields as erosion progresses. The implications of this curve are that much of the production value of the soil is lost in the early stages, when erosion might not be seen as a problem. When erosion is obvious, the damage is done and little more production will be lost. A further implication, pursued later in this chapter, is that conservation efforts are needed on our good soils. To concentrate only on badlands and seriously degraded landscapes is not only technically difficult but also brings few economic returns.

A third way of highlighting the consequences of degradation is to consider how long a soil may continue to be used with current practices. Since the best-fit relationship in Figure 11.5 is logarithmic, using the 'half-life' of a soil in a similar way to measuring radioactive decay gives a figure for the length of time taken for yields to fall to half their initial value. Table 11.1 indicates that a good cover of vegetation and careful management can increase the half-life of a Brazilian cambisol (a typical subtropical brown earth) 10–16 times, whereas for the ferralsol (an iron-rich, high-clay but acid soil) good management prolongs the life by only three to four times. The reason for this difference lies in the specific sensitivity of the ferralsol to yield decline with erosion, whereas the cambisol is moderately resilient – see later in this chapter a discussion on 'resilience' and 'sensitivity' that complements Chapter 6. The implications for soil management are that for the cambisols, maintaining a good cover of vegetation in all seasons is sufficient to reduce the impact of erosion to acceptable levels. Farmers at Itapiranga (Table 11.1), for example, tend not to use any physical measures of conservation, relying instead on good land husbandry and protecting the fertility of the soil by organic means. In contrast, for the ferralsols, additional measures such as physical structures (earth banks and terraces) are essential to address the inherent lack of soil resilience. Environmental managers need to appreciate such differences between soils in order to plan protective measures, but sadly the data are not often available except in a few specific localities such as southern Brazil.

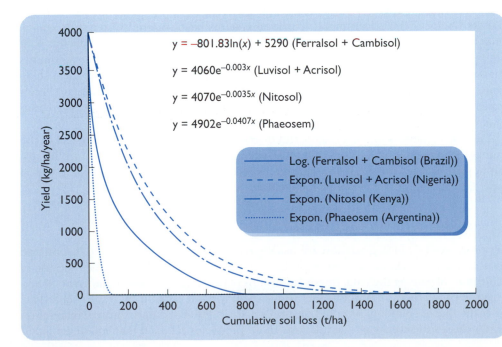

Figure 11.5: Relationships between soil loss and maize yield for six major soil types. Initial grain yield on virgin land set to 4000kg/ha.

Source: Tengberg and Stocking (1997).

Equations shown in figure:

$y = -801.83\ln(x) + 5290$ (Ferralsol + Cambisol)

$y = 4060e^{-0.003x}$ (Luvisol + Acrisol)

$y = 4070e^{-0.0035x}$ (Nitosol)

$y = 4902e^{-0.0407x}$ (Phaeosem)

Legend:
— Log. (Ferralsol + Cambisol (Brazil))
- - - Expon. (Luvisol + Acrisol (Nigeria))
—·— Expon. (Nitosol (Kenya))
········· Expon. (Phaeosem (Argentina))

Table 11.1: Soil 'half-life': southern Brazil.

Site	Soil, slope & mean annual rainfall	Initial production: maize yield (kg/ha)	Half-life: poor management (years)	Half-life: good management (years)
Itapiranga, Santa Catarina State	Ferralsol; 16% slope; 1850 mm	5500	2	36
Ponta Grossa, Paraná State	Cambisol; 22% slope; 1750 mm	3000	4	39
Chapecó, Santa Catarina State	Ferralsol; 10% slope; 1300 mm	2000	1	4
Campinas, São Paulo State	Cambisol, 12% slope; 1450 mm	2500	6	16

Source: Tengberg et al. (1998).

Although on-site damage to soil resources is quantitatively most significant to land users, off-site impacts are also substantial. Sediments derived mostly from agricultural lands cause damage to canals, water storages, irrigation schemes, ports and hydroelectric power plants. In the USA, 10–25% of new reservoir storage capacity is built to accommodate sediment rather than water (Clark, 1985). Damage in developing countries is equally serious. Five major dams in the hill country of Sri Lanka, built with aid donor finance, supply electricity to Colombo and water to the dry-zone irrigation schemes of Mahaweli. Not only has the erosion of fertilisers applied to tobacco in the upper catchment caused eutrophication in the reservoirs, but also damage to the turbines has caused power cuts, incurred economic costs in lost production and reduced the credibility of the government. Irrigation systems have suffered water shortages at critical growing periods as the Polgolla Diversion Barrage at Kandy repeatedly fills with sediment. Ironically, such impacts are worsened by the water storage investments themselves. People displaced by the reservoirs now have little choice but to farm steep lands in the catchment, while others, seeing the opportunity of wage labour and better access to water, have also flocked to the area, thereby increasing land pressure and encouraging further erosion.

The consequences of soil erosion and degradation are, therefore, biophysical in the damage done to the soil, economic in on-site and off-site costs, and social in the disruption to society. The degree of damage, cost and disruption varies according to soil type, land use and the process of degradation.

Processes of soil degradation

Six processes of soil degradation are usually recognised (Box 11.8). Identification and measurement of each process poses problems for the environmental manager because of difficulties of gaining the data, interrelationships between the processes and errors inherent in measurement methods – see next section for soil erosion by water. Table 11.2 gives the most commonly accepted measures and UN Food and Agriculture Organisation (FAO) classes of severity of degradation. These processes act such that soil degradation

Box 11.8 Processes of soil degradation

- **Water erosion.** Splash, sheet and gully erosion, as well as mass movements such as landslides.

- **Wind erosion.** The removal and deposition of soil by wind.

- **Excess of salts**. Processes of the accumulation of salt in the soil solution (salinisation) and of the increase of exchangeable sodium on the cation exchange of soil colloids (sodication or alkalinisation).

- **Chemical degradation.** A variety of processes related to leaching of bases and essential nutrients and the build-up of toxic elements; pH-related problems such as aluminium toxicity and P-fixation are also included.

- **Physical degradation.** An adverse change in properties such as porosity, permeability, bulk density and structural stability; often related to a decrease in infiltration capacity and plant-water deficiency.

- **Biological degradation.** Increase in rate of mineralisation of humus without replenishment of organic matter.

will challenge plant productivity in a number of ways simultaneously. For example, sodication is the greatest single factor in the tropics rendering soils more erodible: when dry, the soil becomes massive and hard; when wet, it loses cohesion and erodes alarmingly. Similarly, water erosion results in loss of structure, surface sealing and breakdown of water-stable aggregates; on high-clay soils in Zimbabwe, once organic carbon content goes below a threshold of 2%, erodibility suddenly increases. Such interrelationships underline the vulnerability of many soils and farming systems, especially in the tropics, to soil degradation. On a few deep soils with large reserves of weatherable minerals (e.g. vertisols and nitosols), the impact of degradation may be slight. However, on soils that lose their fertility quickly through intensive cropping without artificial replenishment of organic matter and nutrients, the impact can be disastrous with knock-on effects to adjacent lands which have to take up the pressure of feeding the population. Figure 11.6 compares two soils (A and B) and how degradation over 40 years interacts with management of the soil to give yields above and below that which is economical.

Take the sandy, easily worked alfisols (FAO classification: luvisols) of the savanna zones of Africa, South Asia and South America. They provide for much of the subsistence smallholder farming, where hand tools are used for cultivation and chemical inputs are scarce. Nutrients are concentrated in the top few centimetres. Consequently, erosion has a devastating impact on crop production. In the example from Brazil in Figure 11.5, not only are nutrients concentrated in the topsoil but erosion also acidifies the soil. The drop in pH is only small, but this has the knock-on effect of fixing phosphorus, rendering it unavailable and freeing aluminium, which is extremely toxic to plants. This double effect supports the compelling evidence that degradation processes have a far greater severity of impact in terms of yields on tropical and subtropical soils than on temperate soils.

Assessment of soil erosion

Erosion assessments are usually made from standard soil loss and runoff plots as illustrated in Figure 11.7. Nearly every country has some of these plots on agricultural research stations in order to test the erosion hazard of different permutations of cropping system, soil, slope and management (see Figure 11.8). Dimensions vary from

Table 11.2: Soil degradation processes, units and classes according to the FAO Methodology of Soil Degradation Assessment.

Degradation process	Code	Definition	Units of measurement	CLASSES			
				None to slight	Moderate	High	Very high
Water erosion	E	Soil loss	t/ha/yr	<10	10–50	50–200	>200
			or mm/yr	<0.6	0.6–3.3	3.3–13.3	>13.3
Wind erosion	W	Soil loss	t/ha/yr	<10	10–50	50–200	>200
			or mm/yr	<0.6	0.6–3.3	3.3–13.3	>13.3
Excess of salts:							
salinisation	Sz	Increase of electrical conductivity of saturated paste at 25°C in 0–60 cm layer	mmho/cm/yr	<2	2–3	3–5	>5
sodication	Sa	Increase in exchangeable Na% in 0–60 cm layer	% per year	<1	1–2	2–3	>3
Chemical degradation:							
acidification	Cn	Decrease of base saturation in 0–30 cm layer, if:					
		(a) BS <50%	% per year	<1.25	1.25–2.5	2.5–5	>5
		(b) BS >50%		<2.5	2.5–5	5–10	>10
toxicity	Ct	Increase in toxic elements in 0–30 cm layer	ppm per year	not yet used – not mappable at 1:5 m			
Physical degradation	P	(a) Increase in bulk density in 0–60 cm layer Initial level (g/cm^3):	% change per year				
		<1		<5	5–10	10–15	>15
		1–1.25		<2.5	2.5–5	5–7.5	>7.5
		1.25–1.4		<1.5	1.5–2.5	2.5–5	>5
		1.4–1.6		<1	1–2	2–3	>3
		(b) Decrease in permeability Initial level:	% change per year				
		rapid (20 cm/h)		<2.5	2.5–10	10–50	>50
		moderate (5–10)		<1.25	1.25–5	5–20	>20
		slow (5)		<1	1–2	2–10	>10
Biological degradation	B	Decrease in humus in 0–30 cm layer	% change per year	<1	1–2.5	2.5–5	> 5

Source: FAO (1979).

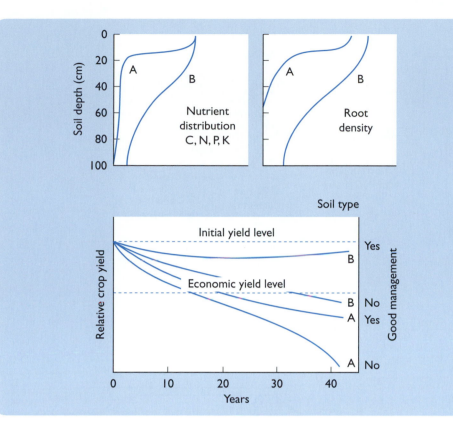

Figure 11.6:
Soil productivity
(crop yield) in
relation to different
soils (A and B), level
of management and
soil degradation
over 40 years.

country to country, but the length of bounded plots ranges from 6 to 10 m and the width from 1.5 to 3 m. Sediment and runoff are caught in a downslope trough and led to a set of storage tanks. To cope with the exceptionally heavy storm, which causes the greatest erosion, there are usually two or more tanks, separated by a divisor arrangement by which only a proportion of the runoff is taken to the lower tanks. Runoff is measured by the level of water in the tanks, and soil loss by taking a thoroughly stirred sample of water and sludge from each tank followed by drying and weighing. Errors are frequent, and the sampling method for sediment is known to underestimate actual soil loss from the plot, because of inadequate stirring. To obviate this, the scientists at the Institute of Agricultural Engineering in Zimbabwe use a total weighing technique, but this does mean that after a heavy thunderstorm they have huge quantities of water and sediment to handle.

Geomorphologists commonly employ other assessment methods such as erosion pins to measure the lowering of the ground surface. Gerlach troughs, of the approximate dimensions of a hand dustpan, catch soil and runoff as they move downslope. At a broader scale, small catchments of 0.5–2 ha can be monitored using a flume and sediment sampler. Hydrologists often measure sediment loads from large catchments by sampling, followed by extrapolating total sediment loss from the stream hydrograph and then quoting the result as sediment loss per unit area of catchment. More sophisticated methods use fluorescent or radioactive tracers, or monitor the concentration of isotopes such as caesium-137. What is crucial to note is that the measurements from different techniques give different results because they assess different things.

The commonest error is to forget the scale effect. Small, bounded plots give the highest measured soil losses per unit area. This is because each soil particle that is

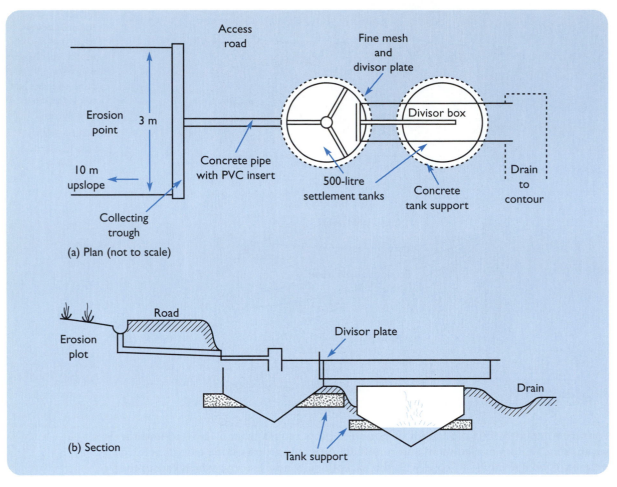

Figure 11.7: Soil loss plots as used at the Institute of Agricultural Engineering, Harare, Zimbabwe, to measure eroded sediments and runoff.

detached by erosion and starts to move is caught and weighed. As the area for assessment increases, there is a greater likelihood of storage of sediment within the bounded area (and hence there being no recording of its movement at the downslope end of the plot). In real field conditions, as much as 90–95% of eroded soil is redeposited elsewhere within the landscape. Consequently, catchment sediment yields are typically only a small fraction of measured losses from field plots. Hydrologists express this as the sediment delivery ratio. For the environmental manager, it means that all measurements of soil loss must be treated with extreme caution, and both the technique and the scale of measurement must be quoted. In the following discussion, all rates of soil loss will be from the standard field plot as shown in Figure 11.7.

Plot studies show the immense variability in measured rates of erosion. Just removing a complete cover of vegetation increases erosion by about three orders of magnitude (1000 times or more). The USA and Australia have the largest record of plot results displaying the effect of land use and site condition on soil loss (see Table 11.3). Specific losses in exceptional events can be much larger, such as the single storm in Western Australia falling on freshly seeded ground where 350 tonnes/ha soil loss was recorded. The type and degree of canopy cover is the single most influential factor in

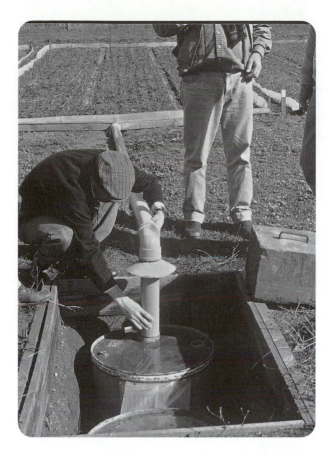

Figure 11.8: Soil loss and run-off plots at Chillan, Chile; part of the FAO Erosion–Productivity Network to measure yield declines with erosion. (*Photo*: M.A. Stocking).

Table 11.3: Selected soil loss rates from plot experiments in USA and Australia.

Land use	Location	Soil loss rates (t/ha/yr)
USA		
Cropland	Average of all	18.1
	Mid-west, deep loess	35.6
	Southern high plains	51.5
Australia		
Bare	New South Wales	31.3–87.0
Cropland	New South Wales	0–16.0
	Queensland	7.0–36.4
Pasture/range	New South Wales	0–1.9
	Queensland	0–21.1
Mine sites	Hunter Valley, NSW	0.4–11.8
	Jabiru, NT	20–102

Sources: Pimentel *et al.* (1993); Edwards (1993).

determining erosion rates. Inevitably in cropping systems, the land is bare or inadequately covered for the early part of the season. If storms occur, not only will the rate of erosion be high but also the potential impact on productivity.

In tropical environments, there is good evidence that soil loss rates in agricultural systems are even higher (see Figure 11.9). From an experiment in Java to measure the impact of erosion on hill rice, 900 tonnes/ha of soil were eroded in just over a year. Similar alarming figures are available from throughout the humid tropics where the natural vegetation on steep slopes has been replaced by annual cultivation. In the seasonally wet-and-dry tropics, rates of 100–200 tonnes/ha are common. Again, there is significant variability, and with a good cover of foliage erosion rates decline to less than 1% of the rate without vegetation – an aspect crucial when considering soil conservation.

Prediction and explanation of soil erosion

An analysis of erosion rates and immediate site causes demonstrates that a number of factors can explain the variations. *Soil type* is a determinant of soil erodibility. In the tropics especially, erodibility is greatly enhanced not only by the physical and chemical properties of the soil but also by management factors which allow the soil to crust, increase in bulk density or decrease in organic matter. *Topographic factors* such as slope steepness, length and shape are also influential. Physical measures such as contour bunding (which cuts the slope into shorter lengths) or terracing (which alters the effective slope steepness) are ways by which the influence of topography may be reduced in land use. *Rainfall* is an important factor giving rise to erosivities

Figure 11.9: Soil loss rates on the Loess Plateau, China, are reputed to be the highest in the world. Extensive bench terraces barely control the erosion. (**Photo**: M.A. Stocking).

varying according to the amount, intensity and seasonality of storms. The only way to protect the soil from the erosivity of rainfall is to ensure a *cover of vegetation* to intercept the kinetic energy of raindrops and absorb it harmlessly into leaves and surface organic matter.

These factors may be combined into empirical models to predict rates of erosion under any combination of soil, rainfall, slope and vegetation. One such model, used by the conservation services in southern Africa for nearly 20 years, is shown in Figure 11.10. Predictions are only as good as the database used to give physical values to the control variables. In developing countries, this places severe limitations on the use of more sophisticated models that need large research resources to support the required experiments.

It is tempting to stop at the site factors to explain soil erosion. Prediction models calculate how changes in land use or management may alter the erosion hazard, and most technical studies and consultant reports take only these obvious, tangible factors into account. The socio-economic, cultural and political contexts are, if considered at all, merely externalities that may marginally modify the core technical analysis. This view of the importance of site-specific technical analysis has unfortunate side-effects. Because several of the variables can be controlled by land users and because soil conditions and vegetation cover are much affected by mismanagement of the land, prediction models are not only useful in designing conservation strategies but are also, by implication, able to point the finger of blame for erosion at the land user. If that person is a resource-poor farmer, gaining a subsistence on a steep slope, poor soils and deficient vegetation, so much the better – the culprit for erosion is all the more obvious (the Mafuta family in Box 11.9 is so blamed). But is this a fair or even adequate explanation for erosion?

No. Explanations of soil erosion and land degradation are possible at a number of levels. For example, a valid question is why should farmers erode their soil when many know full well that their practices jeopardise their future well-being. Part of the answer lies in the decision making and prioritisation of income opportunities of different households. For developed country farmers, it may not be worthwhile economically to

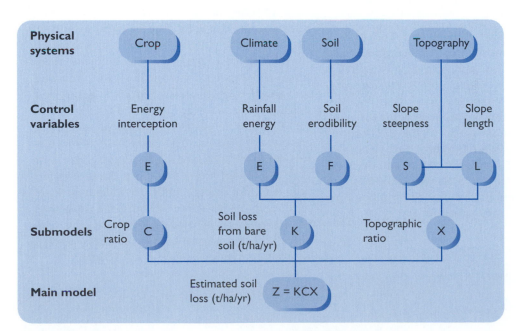

Figure 11.10: The soil loss estimation model for southern Africa.

Source: Elwell and Stocking (1982).

Box 11.9 Why does soil conservation so often go wrong? A true story of how to blame a Zimbabwean farmer for soil degradation

In his youth Watch Mafuta worked at the mines. Now, at age 40, he is tired, and his yearning for his communal ancestral home has brought him back to the family *shamba* ... A thriving home garden supplies vegetables ... Watch cultivates cotton and maize on an extensive plot ... Declining yields force him to plough more land each year. Weeding is a real problem on the extra land ... Concerned because of reports of land degradation, the government agricultural officer, a local extension assistant and a foreign aid worker call on the Mafutas.

The outcome

● *Identification of the problem*. These technicians see the degrading arable land, the overgrazed range, the poor crop stands ... and not a single measure they could call 'conservation' ... . Erosion is confirmed as serious; soil conservation and land reclamation are urgently needed!

● *Planning of control measures*. Soil loss rates are calculated and packages of remedial measures designed to reduce degradation ... 'We only need cooperation from the community now', they say.

● *Implementation of the plan*. The plan is explained to the Mafutas. Encouragement and persuasion, even subtle threats are employed. Typically, the household is shown photographs of eroded land, statistics on soil loss, and embellished descriptions of the dire consequences of allowing erosion to continue. Appeals based on patriotism, their custody of the land for future generations, and the security of the state are launched. Demonstration plots and field days show what can be done. Hoe in hand, the Mafutas set to.

Headlines in the press

● *Soil conservation project hits teething problems*

● *Targets not met in conservation plan*
● *Government minister urges unity to fight menace of erosion*
● *Lazy farmers blamed for erosion*
● *Heavy rains destroy contour terraces*
● *Aid agency pulls out*

The innocent and guilty are thus exposed. Foremost among the innocent is the aid agency. Didn't it do its best? The government minister and the experts too are blameless. Didn't they warn of erosion's dangers? Watch Mafuta, his family, and the millions of households like theirs that failed to heed the warnings and do the necessary work: they are the guilty; they now suffer the consequences. If nothing else, justice at least is done.

Lessons

The Mafutas demonstrate how small is the room for manoeuvre of farmers and how easily blame can be pinned to the least powerful in society. At a global level, economic and political forces encourage developed country aid agencies to channel some money – many say far too little – to developing countries. But the agency wants its say in how the money is spent. Capital goods (bulldozers, tractors), technical inputs (professional experts, chemicals) and training (advanced techniques, often in a developed country institution) are the easiest aid items to mobilise and the preferred means of addressing the erosion problem. The immediate problem is not, however, a lack of machinery or experts or even knowledge – it is that the Mafutas have too little land, not enough labour, a difficult environment and precious few resources to enhance their livelihood. Working harder, digging contours and drains, and being told they are lazy, ignorant or uneducated are not solutions.

Source: abridged from Stocking (1988b).

undertake soil conservation. The immediate investments may be too costly in contrast to benefits in future production, which may accrue only in succeeding generations. A change to organic farming can be extraordinarily demanding in labour and may depress initial yields, even though the long-term benefit is reduced production costs, a premium on market prices for the produce and a lessened dependence on inputs. If society wants the comfort of knowing its land resources are in caring hands, then the land user cannot be expected solely to take the burden. For resource-poor farmers in developing countries, the issue may be starker. It may rest between a choice of starvation now or starvation in the future. Usually, however, it is less dramatic. The 'reproduction squeeze' described by Blaikie (1989) is a good example of how rural peasantry may be caught between maintaining current livelihoods and looking after conservation works – livelihood will usually win. Typically a small farmer must produce primary products for sale to purchase commodities for production (e.g. a hand hoe) or for consumption (batteries for the radio). As the relative value of local production falls in comparison with the cost of purchases – a worldwide phenomenon – the small farmer must reduce the cost of production and/or increase primary product output. The result is an impoverished peasantry working ever harder on poorer, degraded and more distant soils. Labour essential for investment in soil improvement or maintenance of conservation structures must be diverted to the immediate goal of primary production.

A further valid question is how have this world's economic and political systems developed to the state whereby it is the rational choice of land users to degrade their land. In such a milieu, market prices, the terms of trade, the structure of agrarian society, international economic relations, competing political interests and global politics become parts of the explanation of why the Mafutas (see Box 11.9) struggle, and, despite their valiant efforts and the arguments of well-meaning professionals, they fail to conserve their soils.

Explanation of soil erosion thus involves a wider political economy and a complexity of non-technical issues. Blaikie (1989) provides a model, a 'chain of explanation', to combine these issues into an analytical framework (see Figure 11.11) which can relate why a gully in Africa could be linked to a desk-bound bureaucrat in Brussels. The finger of blame turns to a broad sweep of the hand.

For the environmental manager, inevitably some explanatory issues can be addressed and some cannot. Where natural systems of soil, water and vegetation are resilient, only major disturbances may cause degradation that has a substantial impact on production. This natural resilience can be assessed. Couple this with relatively stable economies in societies that are not only willing and able to subsidise land users to ensure conservation but also have a high degree of empowerment and local decision making, then land management is relatively easy. But in a poor developing country with inequitable access to resources, corruption and many other pressing problems, it would be a brave land manager who might contemplate tackling international economic relations. Nevertheless, an understanding of the varying levels of explanation for soil degradation will enable purely technical solutions to be set in a real-world context and will allow assessment of the likelihood of innovations being acceptable to the land user. An explanatory framework also provides alternative avenues to address the soil erosion problem – a manipulation of producer prices, perhaps, to favour good cover crops; or subsidies and incentives to implement conservation practices; or a change in rights of access and ownership of land. All these may, in the right circumstances, have the potential to change a situation of increasing degradation to one of rehabilitation. A matrix of guidance to the land manager is offered in Box 11.10 using two useful terms – resilience and sensitivity – each of which can be considered in both physical and human terms. Individual soils can be assessed

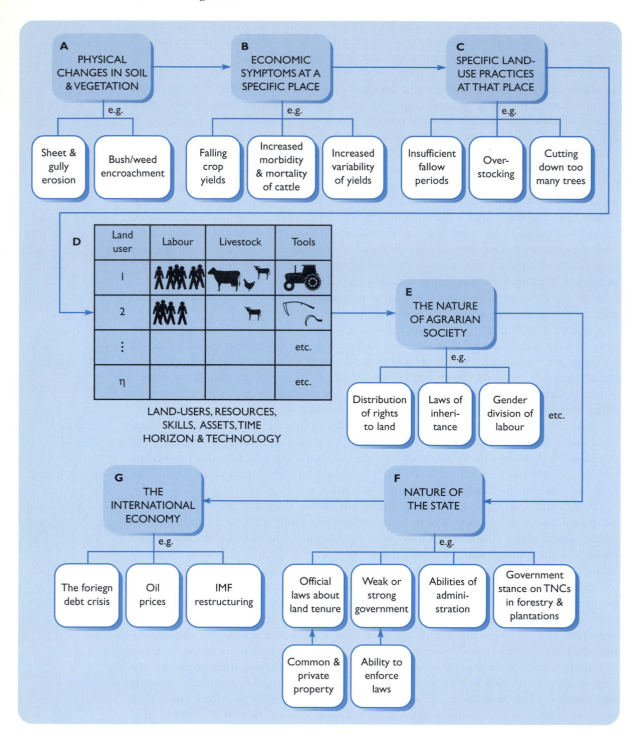

Figure 11.11: Chain of explanation for soil erosion.
Source: after Blaikie (1989).

Box 11.10 Resilience and sensitivity: a matrix for environmental managers

Resilience is a property that allows a land system to absorb and utilise change; or resistance to a shock. For soils, resilience describes the ability to resist being eroded. *Sensitivity* is the degree to which a land system undergoes change due to natural forces, following human interference; or how readily change occurs. For soils, sensitivity describes the impact erosion has on soil properties or crop yield.

Examples

● **Ferralsols and acrisols: *low to very low resilience; moderate sensitivity.*** Typical humid tropical rainforest soil that, once vegetation has been removed, degrades very quickly and irreversibly. Strongly acid, free aluminium and P-fixation. If used, degrada-

tion should be addressed by combinations of structures and biological measures.

● **Nitosols: *moderate resilience; low sensitivity.*** One of the safest and most fertile soils of the tropics and subtropics, with only small problems of increased erodibility if organic matter declines. Biological conservation methods effective to address both erosion rate and fertility decline.

● **Phaeozems: *high resilience; high sensitivity.*** A soil with a good structure, but when degraded it suffers large yield losses per unit of soil loss. Under good management can give consistently high yields. The key to the sustainable use of phaeozems is good vegetation cover.

Source: Tengberg and Stocking (1997).

Table 11.10.1: The sensitivity – resilience matrix.

| | | SENSITIVITY | |
		High	Low
RESILIENCE	**High**	Resists erosion well but, if allowed to degrade through poor management, yields and soil quality rapidly decline. Nevertheless, capability can be restored by appropriate land management	Only suffers degradation under very poor management and persistent mismanagement. Biological conservation methods adequately maintain production and protect soil
	Low	Easy to degrade, with disastrous effects on production – loss of nutrients and plant-available water capacity. Difficult to restore, and should be kept under natural vegetation or forestry	Soil can suffer considerable erosion and degradation without apparent influence on production or overall soil quality. Physical structures indicated to control off-site impacts

for their resistance to degradation and the on-site impact that occurs with degradation, the particular combination of resilience and sensitivity indicating appropriate management and remedial measures.

Soil and water conservation

The response of society to land degradation is *soil conservation*, defined as any set of measures which control or prevent soil erosion, or maintain soil fertility. *Water conservation*, especially in drier zones, is closely related to soil conservation, so that measures to conserve water act through increasing plant-available water and maintain-

Box 11.11 Sustainability quotients: a way of analysing land use

Sustainability quotient (SQ) is the fraction of present net farm income per hectare of all land including fallow that is *not* obtained at the expense of extracting soil nutrients. It expresses that part of production obtained from renewable sources and that part of income not reliant on using the soil 'capital'.

An example from Mali

Agricultural practices in southern Mali are contributing to a severe decline in soil fertility. Different cropping systems and application rates of fertiliser are being tried not only to maintain yields and farmer income but also to restore soil fertility.

How far can fertilisers contribute to sustainability? The two main depleted nutrients, N and K, could be restored by a maximum of 27% and 12% respectively if fertiliser application were doubled. If erosion were to be reduced by half, the N and K deficits would be reduced by 20% and 33%. In other words, in this dry environment soil conservation is at least as effective as increasing fertiliser use. Furthermore, increased production through water conservation, which comes along with soil conservation practices, will also occur.

What are the differences between land uses?

● Overall: SQ = 0.57, i.e. 43% of farmer income is based on monetary value of nutrients (N, P, K, Ca, Mg) leaving the soil through erosion, leaching, cropping and not being replaced by artificial or natural inputs (e.g. fertilisers, N-fixation).

● Recommended cotton–maize–sorghum rotational package: SQ = 0.95.

● Actual cotton–maize–sorghum package as practised by farmers: SQ = 0.73.

● Traditional groundnut–millet–millet rotation: SQ = 0.39.

● Traditional millet–fallow rotation, the cycle for which is under threat because of increasing population: SQ = 0.01, or 99% of farmer income through 'mining' the soil.

Source: van der Pol (1992).

ing soil fertility.[3] The objective of soil and water conservation is to achieve consistent and lasting production from land while keeping soil loss at or below the soil's rate of renewal. Since subsoil formation rates are in the region of 0.5–1 tonne/ha/year – which is equal to natural rates of soil erosion – soil conservation strategies should not allow greater soil losses than those occurring under a complete natural vegetation. Because any disturbance of the soil causes an increase in soil loss, we have a dilemma: all agricultural and land-use systems are theoretically unsustainable. In practice, therefore, 'tolerable' soil losses are taken to be either some arbitrary figure which is known to be achievable in practice (the US approach, where soil loss tolerance values are quoted in the range 5–10 tons/acre/year) or the level at which soil fertility can be maintained with inputs of technology in the medium term, say up to 30 years. The latter approach to setting target levels of allowable soil loss in a conservation programme is gaining ground. It recognises that topsoil formation rates can easily reach ten times subsoil formation rates, it acknowledges that farmers utilise their soil 'capital' of nutrients at certain times, restoring them later by fallowing (see Box 11.11), and it allows for the role of the land manager in compensating for nutrient and organic matter losses by specific interventions such as green manuring, cover crops and fertiliser.

[3] While soil conservation and water conservation may often be treated as one for practical purposes, a useful distinction exists for structures that are *primarily* for water. Known as 'water harvesting', these techniques collect and concentrate runoff from small catchment areas. The *teras* system of Sudan, for example, has bunds on three sides, with the upslope side kept open to accept runoff from the gently sloping central plains of Sudan. See Critchley *et al.* (1992) for many other examples.

Soil and water conservation is nothing new. Civilisations have been based on water harvesting and soil management technologies (Mesopotamia and the Andean Incas, for example) and Homer, Virgil and Plato make reference to environmental problems and the need for conservation. Nevertheless, it is only during this century that soil conservation has been recognised as a necessary strategy for restoring degraded lands and protecting production. The American Dust Bowl conditions of the early 1930s were the trigger which set in train the largest single institution of its kind, the Soil Conservation Service (SCS) (in 1996 renamed the Natural Resources Conservation Service) of the US Department of Agriculture and its associated Agricultural Research Service (ARS). The influence of SCS–ARS has been large, leading to similar organisations in other countries – notably Australia – and to a high-technology/high-input/research-based approach to erosion control. Partly in response to the difficulty of applying high-cost solutions in resource-poor environments, an understanding of indigenous responses to land degradation has developed along with a recognition of the conservation value of many traditional systems in developing countries (Critchley *et al.*, 1994). Silt harvesting is one such technique which has an obvious livelihood rationality for smallholder farmers in India (Box 11.12). Tables 11.4 and 11.5 contrast the two principal approaches to soil and water conservation, the differences between which can be summarised as:

- developed country vs. developing country
- subsidies/incentives from society vs. farmer pays
- introduced technology vs. indigenously developed technology
- erosion control vs. erosion prevention
- structures and mechanical measures vs. biological means
- bulldozers and tractors vs. hand hoes and seeds
- high cost vs. low/no cost.

These differences between countries should not be exaggerated. Small-farm situations arise in developed countries where low-cost biological methods of soil improvement are appropriate; and large-scale estate and commercial production occurs in developing countries where machinery and conservation structures are the obvious first choice of land managers. Clashes between the approaches do, however, happen where a conservation strategy developed in one set of resource, human and environmental circumstances is forced on quite another situation – the Mafutas' circumstance in Box 11.9 is a classic example. Identification and analysis of these cultural and technological conflicts is essential to support the land manager and planner; see, for example, Atampugre's (1993) analysis of villagers' views of soil and water conservation techniques promoted by Oxfam in Burkina Faso. Communication and community-based planning is now seen by many to be the primary challenge (see Box 11.13 for Australia's way of dealing with conservation).

The term *land husbandry* has recently been coined to try to overcome the different perspectives of those attempting conservation and to get over the realisation that soil and water conservation is really only a part of agricultural production and environmental management. Rather than seeing conservation as a separate strategy with its own approaches and institutions, land husbandry fosters the idea that integrated, grassroots approaches to the land involve the total production cycle, land users' constraints and opportunities, access to land, labour and capital, and the technical appropriateness of solutions. Although lauded as a new philosophy by Hudson (1992) – and perhaps it is new for agricultural engineers – land husbandry does not directly address soil erosion. Others see today's approaches to soil and water conservation as *Continued on p.317.*

Box 11.12 Silt harvesting in India

In semi-arid India, land users deliberately leave their upper lands bare and overgrazed in order to initiate sediment transport and removal of the soil to valley bottoms, where it is trapped by *nala gode* or 'wall across the nala'.

A *nala gode* is a loose rock check dam constructed in a *nala* (gully or ephemeral natural channel) to harvest soil eroded in the upper catchment and bring the *nala* bed into cultivation.

Figure 11.12.1 shows a series of silt harvesting structures built by Vitu and Monu Somla, farmers of Limbu mini watershed, Karnataka, India.

Why did they construct them? Three reasons: (1) prevention of widening and deepening of the gully and promotion of deposition of nutrient-rich sediments; (2) reduction in volume and velocity of runoff to protect lower lands; and (3) to gain a field where none existed before. This last is the most important reason, because using these small fields so created in a land-scarce community, high-value rice production is enhanced, often with a second, dry-season crop of pulses or vegetables.

Source: Premkumar (1994).

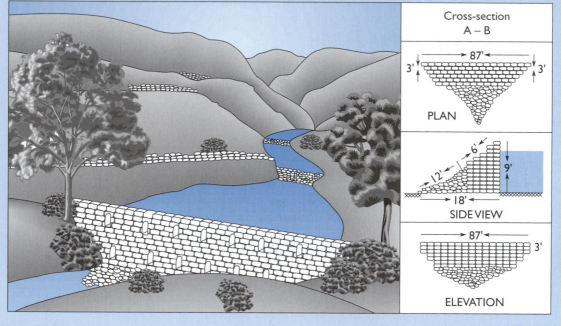

Figure 11.12.1

Table 11.4: Modern techniques of soil and water conservation.

Technique	Description	Resource implications
Tillage practices:		
Strip tillage	Conditioning soil along narrow strips in or adjacent to seed rows, leaving rest of soil undisturbed	Special equipment; possible crusting/infiltration problems in untilled parts
Basin listing, or tied ridging	Formation of contour bunds (earth banks) with constructed 'ties' or banks between bunds. Listing done by machinery and tied ridging usually by hand	Need for dedicated machinery and/or considerable extra labour. Danger of waterlogging
Conservation tillage	Technique of light harrowing and retention of crop residues at surface	Special machinery. Crop residues not available for other uses, e.g. forage, fuel
Minimum/zero tillage	Use of herbicides, then direct drilling into residues. Very little disturbance to the soil	Expensive chemicals and machinery. Danger of pollution and soil compaction. Saving on conventional tillage costs
Land formation techniques:		
Contour bunds	Earth banks up to 2 m wide across slope to form barrier to runoff and break slope into shorter segments Varieties: narrow or wide-based; contour or graded	Additional labour and/or equipment Varieties have different costs, e.g. narrow-based loses about 14% of plantable area
Terraces	Earth embankments and major reformation of surface. Three main types: diversion, retention and terrace	Large labour and equipment requirement. Continual maintenance
Terracettes	Small constructions usually to harvest water. Known by many local names, e.g. fish-scale terraces in China for rows of perennial crops; eyebrow terraces for tree planting in semi-arid zone	Labour and continuing maintenance requirement
Stabilisation structures:		
Gabions	Stone and rock-filled bolsters to protect vulnerable surfaces, e.g. bridges, culverts	High cost, transport of stone, continual maintenance
Gully control dams	Usually constructed of brushwood across a gully	Materials, labour and maintenance. Often fail in heavy storms

Table 11.5: Indigenous soil and water conservation (ISWC) techniques.

Country (region)	ISWC technique	Comment
Cameroon (Mandara Mts)	Bench terraces (0.5–3 m high); stone bunds	Variety of practices to retain soil on-site; depends on soil depth and stone availability
Mali (Dogon Plateau)	Cone-shaped mounds; planting holes; terraces; square basins; stone lines; millet stalk trash lines	A long tradition exists here of indigenous techniques which have enabled local people to survive drought
Burkina Faso (Yatenga Region)	*zay*, or small planting pits	Traditionally used to rehabilitate crusted land
Tanzania (Matengo Highlands)	*ngoro*, or cross-ridge pit system with weeds buried under ridge	System used to cultivate steep hill slopes for at least the last century
Nepal	Natural grass strips of *Imperata cylindrica*	Used across hill slopes to create small terracettes over time
India (Karnataka)	Farm boundary bunds; often of *Vetiver* grass or small trees	With small plots, boundary bunds retain large amounts of soil
Indonesia (Central Java)	Log lines across the slope	Felled trees retained and aligned along the contour
Pakistan (Baluchistan)	Inundation tanks (*sailaba*) – floodwater harvesting of monsoon behind large earthen bank	Traditional system to produce additional winter crop from stored water
USA (Arizona)	Diversion walls for floodwater; a type of wall technology	Used traditionally by Papago, Hopi and Najavo tribes; now defunct

Source: Critchley *et al.* (1994).

Box 11.13 Landcare in Australia

The Landcare movement has been embraced throughout Australia as the means to foster change towards a more sustainable use of soil resources. There are now over 2200 Landcare groups, involving nearly 30% of the farming community.

The movement is based upon five principles:

● *principle of local sustainability*: all communications about natural resource management further the core theme of Landcare. Local action is combined with stewardship, community responsibility and accountability;

● *principle of mutual advantage*: recognition of different specialist interests, different value positions; community groups and agency services;

● *principle of informed collaboration*: incorporate Landcare message into existing community, government and specialist communications through education, research and development, and information exchange;

● *principle of mutual respect and co-operation*: respect integrity of all contributors, especially independent community networks, formal reporting of government agencies, and the intellectual frameworks of specialist advisers;

Box 11.13 (continued)

● *principle of perseverance towards long-term goals*: long-term sustainability needs local and national scales of problem-solving, social, economic and environmental resources, and realistic projections.

Soil and water conservation as embodied in Landcare is essentially a citizens' movement. Its articulation and delivery have not been easy.

Source: Brown (1997).

'conservation by stealth' because techniques are promoted on the grounds of the individual economic interest and private rationality of land users. Their primary concern is production, risk minimisation and livelihood security: if these are achieved by whatever means (improved seed, new soil management techniques, intercropping, agroforestry, contour bunds), then soil conservation automatically follows. It is an appealing prospect – simple in theory, but extraordinarily difficult to practise.

In the final analysis, soil and water conservation is one of the major challenges of our time to counter what many see as the single most immediate threat to the world's food security, land degradation. Other threats – global warming or environmental pollution, for example – catch the catastrophists' imaginations. But for soil erosion and land degradation, not only have they been known for centuries but also people have adjusted their lives and practices to accommodate a declining resource base. Erosion is difficult to see; it is insidious and incremental. Technology has provided a buffer, buying time for society and making it easier for rich, developed nations to delude themselves into believing they hold the key to future production. Sooner or later the soil 'capital' must run out. It *has* run out in some places: parts of the Rif Mountains of Morocco are almost entirely depopulated because of land degradation; the problems of the Sahel regularly hit the headlines; less well known are the abandoned commercial farms in hilly areas of the United States, the so-called marginal lands where erosion renders agricultural land use uneconomic, and the Australian rangelands where wind erosion has stripped topsoils in some parts and covered formerly productive areas in sterile sand. The likely scenario is that such 'hotspots' of degradation will increase in number. This will place added burdens on adjacent lands, whose resilience will be sorely tested. To compensate for declining production, inputs will have to increase on the remaining lands.

What is the way out from the vicious circle? Frustratingly, the techniques are there for the asking. There is no shortage of measures for all permutations of environments, land uses and societies. Part of the answer may well lie in better matching of solutions to situations, but probably the only long-term answer is to allow soil resources to get worse and worse so that people are forced to change. After all, indigenous responses to land degradation have usually developed in the face of starvation and conflict. Developed and developing countries alike will have to develop their responses, as indeed they already are; land use will inevitably change; technologies will come and go; people will migrate, and some may die. Meanwhile, scientists may reinforce their armouries of techniques, develop new methods of analysis, find persons with the resources to heed their warnings and castigate those who ignore them. They may buy time before change is inevitably induced. There are, however, only two certain things: soil erosion and land degradation will continue; and societies will change and adapt to the new challenges. In that adaptation process, environmental managers play a vital role.

References

Atampugre, N. (1993) *Behind the Lines of Stone: the Social Impact of a Soil and Water Conservation Project in the Sahel*. Oxfam, Oxford.

Blaikie, P.M. (1989) Explanation and policy in land degradation and rehabilitation for developing countries. *Land Degradation and Rehabilitation*, **1**, 23–37.

Blaikie, P.M. and Brookfield, H.C. (1987) *Land Degradation and Society*. Methuen, London.

Bojö, J. (1996) The costs of land degradation in sub-Saharan Africa. *Ecological Economics*, **16**, 161–173.

Brookfield, H. and Padoch, C. (1994) Appreciating agrodiversity: a look at the dynamism and diversity of indigenous farming practices. *Environment*, **36**(5), 6–11, 37–45.

Brookfield, H. and Stocking, M. (1999) Agridiversity: definition, description and design. *Global Environmental Change*, **9**(2), 77–80.

Brown, V. (1997) *Landcare in Australia: talking local sustainability in policy, practice and place*. ODI Rural Forestry Network Paper 20e. Overseas Development Institute, London, 31–40.

Clark, E.H. (1985) The off-site costs of soil erosion. *Journal of Soil and Water Conservation*, **40**, 19–22.

Clark, R. (1996) *Methodologies for the economic analysis of soil erosion and conservation*. Working Paper GEC 96-13, CSERGE, University of East Anglia, Norwich.

Cleaver, K.M. and Schreiber, G.A. (1994) *Reversing the Spiral: the Population, Agriculture and Environment Nexus in Sub-Saharan Africa*. World Bank, Washington DC.

Critchley, W.R.S., Reij, C. and Seznec, A. (1992) *Water harvesting for plant production, Vol. 2: Case studies and conclusions for sub-Saharan Africa*. Technical Paper No. 157. World Bank, Washington DC.

Critchley, W.R.S., Reij, C. and Wilcocks, T. (1994) Indigenous soil and water conservation: a review of the state of knowledge and prospects for building on traditions. *Land Degradation and Rehabilitation*, **5**, 293–314.

Dahlberg, A. (1994). *Contesting views and changing paradigms: the land degradation debate in southern Africa*. Discussion Paper 6. Nordiska Afrikainstitutet, Uppsala, Sweden.

De Graf, J. (1993) *Soil Conservation and Sustainable Land Use: an Economic Approach*. Royal Tropical Institute, Amsterdam.

Edwards, K. (1993) Soil erosion and conservation in Australia. In D. Pimentel (ed.), *World Soil Erosion and Conservation*. Cambridge University Press, Cambridge, 147–169.

Elivell, H.A. and Stocking, M.A. (1982) Developing a simple yet practical method of soil loss estimation. *Tropical Agriculture*, **59**, 43–48.

Enters, T. (1997) *Methods for the economic assessment of the on- and off-site impacts of soil erosion*. Issues in Sustainable Land Management No. 2. International Board for Soil Research and Management, Bangkok.

Fairhead, J. and Leach, M. (1996). *Misreading the African Landscape: Society and Ecology in a Forest–Savanna Mosaic*. Cambridge University Press, Cambridge.

FAO (1979) *A provisional methodology for soil degradation assessment* (with mapping of North Africa at a scale of 1:5 million). Food and Agriculture Organisation of the United Nations, Rome.

FAO (1982) *Potential population supporting capacities of lands in the developing world*. Technical Report FPA/INT/513 and mapping at 1:5 million. Jointly with UNFPA and IIASA. Food and Agriculture Organisation of the United Nations, Rome.

FAO (1990) *Production Statistics*. Food and Agriculture Organisation of the United Nations, Rome.

Hudson, N.W. (1992) *Land Husbandry*. Batsford, London.

Hudson, N.W. (1997). *Soil Conservation* (3rd edn). Batsford, London.

Hudson, N.W. and Cheatle, R. (eds) (1993) *Working with Farmers for Better Land Husbandry*. IT Publications, London.

Hurni, H., Eger, H., Fleischhauer, E., El-Swaify, S.A., Östberg, W., Roose, E., Scharpenseel, H.W., Shaxson, T.F., Sombatpanit, S., Stocking, M.A., Trux, A. and Zweifel, H. (1996) *Precious Earth – from Soil and Water Conservation to Sustainable Land Management*. International Soil Conservation Organisation, and Centre for Development and Environment, Berne, Switzerland.

IFAD (1992) *Soil and water conservation in sub-Saharan Africa. Towards sustainable production by the rural poor*. Report prepared by Centre for Development Cooperation Services, Free University, Amsterdam. International Fund for Agricultural Development, Rome.

Lal, R. (1990) *Soil Erosion in the Tropics: Principles and Management*. McGraw-Hill, New York.

Lutz, E., Pagiola, S. and Reiche, C. (eds) (1994) *Economic and institutional analyses of soil conservation projects in Central America and the Caribbean*. World Bank Environment Paper No. 8. World Bank, Washington DC.

Norse, D. (1992) A new strategy for feeding a crowded planet. *Environment*, **34**(5), 6–12, 32–39.

Norse, D., James, C., Skinner, B.J. and Zhao, Q. (1992) Agricultural land use and degradation. In J. Dooge (ed.), *Agenda for Science for Environment and Development into the 21st Century*. Cambridge University Press, Cambridge.

Oldeman, L.R. (1994) The global extent of soil degradation. In D.J. Greenland and I. Szabolcs (eds), *Soil Resilience and Sustainable Land Use*. CAB International, Wallingford, 99–118.

Oldeman, L.R., Hakkeling, R.T.A. and Sombroek, W.G. (1990) *World Map of the Status of Human-Induced Soil Degradation: an Explanatory Note and 3 Maps*. International Soil Reference and Information Centre, Wageningen.

Östberg, W. (1991) *Land is coming up. Burungee thoughts on soil erosion and soil formation*. EDSU Working Paper 11, School of Geography, Stockholm University.

Paalberg, R. (1994) The politics of agricultural resource abuse. *Environment*, **36**(8), 6–9, 33–41.

Pimentel, D., Allen, J., Beers, A. *et al*. (1993) Soil erosion and agricultural productivity. In D. Pimentel (ed.), *World Soil Erosion and Conservation*. Cambridge University Press, Cambridge, 277–292.

Pimentel, D., Harvey, C., Resosudarmo, P. *et al*. (1995) Environmental and economic costs of soil erosion and conservation benefits. *Science*, **267**, 1117–1123.

Predo, C., Grist, P., Menz, K. and Rañola, R.F. (1997) Two approaches for estimating on-site costs of soil erosion in the Philippines: (2) The replacement cost approach. *Imperata* Project Paper 25-36. CRES, Australian National University, Canberra.

Premkumar, P.D. (1994) *Farmers are Engineers*. Pidow-Myrada and Swiss Development Cooperation, Bangalore, India.

Reij, C., Scoones, I. and Toulmin, C. (1996) *Sustaining the Soil: Indigenous Soil and Water Conservation in Africa*. Earthscan, London.

Scherr, S. and Yadav, S. (1996) *Land degradation in the developing world: implications for food, agriculture, and the environment to 2020*. Food, Agriculture and the Environment Discussion Paper 14. International Food Policy Research Institute, Washington DC.

Stocking, M.A. (1988a) Quantifying the on-site impact of soil erosion. In S. Rimwanich (ed.), *Land Conservation for Future Generations*. Department of Land Development, Bangkok, 137–161.

Stocking, M.A. (1988b) Socio-economics of soil conservation in developing countries. *Journal of Soil and Water Conservation*, **43**, 381–385.

Stocking, M. (1996) Soil erosion. In W.M. Adams, A.S. Goudie and A.R. Orme (eds), *The Physical Geography of Africa*. Oxford University Press, Oxford, 326–341.

Tengberg, A. and Stocking, M.A. (1997) *Erosion-induced loss in soil-productivity and its impacts on aricultural production and food security*. FAO (Land and Water Development Division)/Agritex Expert Consultation, Harare. Food and Agriculture Organisation of the United Nations, Rome.

Tengberg, A., Da Veiga, M., Dechen, S.C.F. and Stocking, M.A. (1998) Modelling the impact of erosion on soil productivity: a comparative evaluation of approaches on data from southern Brazil. *Experimental Agriculture*, **34**, 55–71.

Tiffen, M., Mortimore, M. and Gichuki, F. (1994) *More People, Less Erosion: Environmental Recovery in Kenya*. John Wiley, Chichester.

UNEP (1997) *Global Environment Outlook*. United Nations Environment Programme, Nairobi; Oxford University Press, New York.

Van der Pol, F. (1992) *Soil Mining: an Unseen Contributor to Farm Income in Southern Mali*. Bulletin 325, Royal Tropical Institute, Amsterdam.

Wild, A. (1993) *Soils and the Environment*. Cambridge University Press, Cambridge.

WRI/IIED (1987) *World Resources 1987: an Assessment of the Resource Base that Supports the Global Economy*. World Resources Institute and International Institute for Environment and Development. Basic Books, New York; Oxford University Press, Oxford.

Websites and further reading

The following websites present useful information, mainly on soil and water conservation, and sustainable agricultural development:

- *http://www.fao.org/waicent/FAOINFO/Agricult/AGL/projects.htm*
- *http://www.fao.org/waicent/FAOINFO/SUSTDEV/Welcome_htm*
  (a useful listing of current FAO Land and Water Development Division projects worldwide with short descriptions. FAO's Sustainable Development pages had a virtual library of over 1000 publications in February 1998, including the AFRICOVER land cover classification scheme)

- *http://www.nrcs.usda.gov/GenInfo.html*
  (Natural Resources Conservation Service of the United States Department of Agriculture – information on meetings, publications and activities)

- *http://soils.ecn.purdue.edu/WEPP*
  (USDA–Agricultural Research Service's National Soil Erosion Research Laboratory. This site contains the latest downloadable Water Erosion Prediction Project software)

- *http://www.landcare.cri.nz/aboutlcr/*
  (New Zealand's Landcare Research for the management of land resources for conservation).

- *http://www.environment.gov.au/land/landcare/landcare.html*
  (Australia's federal National Landcare Program)

Although dated, the standard text looking at degradation from a non-technical perspective is Blaikie and Brookfield (1987). With chapters on measurement, colonialism, common property resources, and economic costs and benefits, it was the first and is still the best at presenting the various causes of erosion and emphasising the scale-dependency of the process.

For those interested in pursuing the economic context of erosion, De Graf's (1993) book sets conservation technologies into a clear economic context with guidance on ways of including social costs and benefits, and many examples from developing countries. Clark (1996) has the most thorough review to date of methodologies for economic analysis of both erosion and conservation, while the World Bank paper edited by Lutz *et al.* (1994) and written by leading environmental economists has a wealth of good examples of the economic analysis of conservation projects.

For the more technically minded, the late Norman Hudson's (1992) book *Land Husbandry* presents the *mea culpa* of the engineer who, having published the standard text on soil conservation (3rd edition: Hudson, 1997), realised that maybe social and economic factors had been forgotten! *Land Husbandry*, the overall care and management of land resources, is still a technical approach but encompasses a wider range of factors, especially biological control processes.

Recent literature on conservation, especially indigenous techniques, has shown the diversity of management practices employed (and employable) by land users. Look especially at the excellent collection of African case studies edited by Reij *et al.* (1996), and Critchley *et al.*'s (1994) review of indigenous technical knowledge. Hudson and Cheatle (1993) is another set of case studies emphasising participatory resource management. For those who think that solving land degradation is merely a matter of population control, Tiffen *et al.* (1994) is a discomforting read.

Chapter Twelve

River processes and management

Richard Hey

Topics covered

- Natural channel processes
- Flood alleviation schemes
- River stabilisation
- River restoration

Editorial introduction

One of the less publicised aspects of environmental modernisation has been the slow but steady 'greening' of the professions. Engineering was one of the earliest to recognise that natural processes actually do engineering work, so it is wiser to design alongside such processes than against them. In the course of time, engineers have been followed by accountants, investment analysts, chartered surveyors, land agents and most recently the medical profession, with its grudging acceptance of complementary medicine.

This chapter surveys the progress made in ecological engineering for river management. It is of interest to examine how this wider perspective entered into the mindset of river engineers. There have always been ecologically minded visionaries in the trade, but their voices have not been heard until recently. One factor promoting this change of heart is the costly mistake of ignoring natural processes when designing dams, levees, and channel stabilisation and flood protection schemes. Invariably, lack of attention to ecological systems meant rapid siltation, which shortened the life of reservoirs or hydropower turbines; or channelisation created floods elsewhere; or defensive structures were undermined by sedimentation or erosional forces triggered by the engineered structures themselves. So one trigger was a costly learning curve. As the science of river flows and river bed movement became better modelled and understood, so the positive or complementary functional role of designing with nature was regarded as cost-effective.

Another factor was the rise of the environmental impact assessment. This was introduced in the US National Environmental Policy Act in 1969. Through this remarkable piece of legislation, which incidentally argued for a productive harmony with nature across all US federal policy, the formal environmental impact statement, or account, was born. The outcome was a generation of engineer-designers who simply had to know about ecology, about soil sciences, about environmental economics and ethics, and about planning and land-use change. This was a slow process, and even today too many EIAs are 'cook book studies' done in-house by shadowy teams to fulfil a legal obligation. Only recently has there been a move to improve the quality control of the various environmental science consultancies that have sprung up since EIAs became commonplace the world over. A recent study of EIAs in the UK showed that less than half were adequate and only a quarter could be regarded as comprehensive. There is a good job market in high-quality environmental assessment, within which the well-trained environmental scientist is an attractive employee.

A third propelling force was the rise of statutory duties of environmental care imposed on water management organisation by 'green' legislation. In the UK, this clause is sometimes referred to as the *amenity clause* because it sets the basis for a duty of care on environmental protection. Such a duty is non-statutory, supported by codes of practice and subject to scrutiny by advisory committees and environmental interest groups.

Water resources will become a critical aspect of sustainability politics in the years to come. Worldwide, water use is expanding to the point where extraction will be 40% higher in 2025 (UN Development Programme, 1998, 80). This could involve three-quarters of all freshwater supplies. By 2050, the number of people short of water could reach 2 billion, up from 135 million today.

One approach to dealing with the impending water crisis is to change the nature of the property right. South Africa has done just that in its National Water Act of 1998. Having been virtually the exclusive resource of white South African farmers and wealthy urban residents, South African water resources are now in the hands of all citizens, held in custodianship by the state (see Box 1.7 for a schematic interpretation of this move). The legal provision for this is section 25 of the South African constitution, dealing with property rights. This allows any property to be exploited for a 'public purpose', or in the 'public interest'. In the National Water Act, compensation to an aggrieved farmer, or other water user, is not payable unless the existing use is shown to be sustainable, and the denial of the right is also shown to have 'severe prejudice' on the economic circumstances of the user. This is, in effect, the precautionary principle in practice (see Chapter 1).

This fascinating change in the property right of water in South Africa is connected to a wholesale series of reforms in rights and redistribution in the post-apartheid era. The National Water Act creates three new institutional arrangements:

- a basic entitlement to all South Africans of 25 litres per day, free of charge;

- an environmental entitlement for natural systems to be allowed to survive. This is basically a biodiversity clause (see Chapter 5) and seeks to ensure that the ecological integrity of catchments, as well as water resources, is maintained;

- the creation of catchment management agencies comprising a representative and inclusive sample of stakeholders, whose job it will be to create deliberative processes of the kind introduced in Chapter 1 to ensure both the entitlements and subsequent tariff rates for future abstractions.

The National Water Act is supplemented by the South African National Environmental Management Act 1998. This also establishes fundamental principles of 'environmentally sustainable development' including 'creative empowerment' of the kind outlined in Box 1.8. The Act also creates a citizen's right to appear before the courts on the basis of upholding social and environmental well-being where this is threatened by a proposed administrative action.

These institutional innovations in water property rights and decision taking will take much time and thought actually to implement. But they do show that dramatic changes in legal structures can be enacted out of a visionary constitution through a fully participatory consultation process, including the sensitive application of the precautionary principle, that is now taking place in South Africa (see O'Riordan, 1998).

The New Zealand Resource Conservation Act of 1991 provides the image of times to come. This states as its purpose the promotion of sustainable management of natural and physical resources. Section 6 requires that all persons managing any resource must recognise the need to safeguard key sites, including sites important through Maori cultural traditions and property rights. This is a formal duty akin to the public trust doctrine outlined in Chapter 1. Furthermore, resource managers are expected to pay particular regard to the Maori notion of *Kaitiakitanga* or the exercise of guardianship and stewardship, efficiency of use, the maintenance and enhancement of amenity and of environmental quality generally, and the intrinsic values of ecosystems. These are slightly more discretionary obligations, but they do demand a formal statement of intent and of response for any resource management proposal. The extension of such legislation to the northern hemisphere would place a clearer burden on the developer to be environmentally accountable.

Finally, the emergence of ecological economics has begun to show that designing with nature is not only cost-effective but can also be justified on the basis of the social variations of wildlife and habitats or accessible amenities saved or reconstructed. So contingent valuation studies of the kind outlined in Chapter 4 are beginning to appear in the cost–benefit analysis of river management schemes and flood protection strategies.

However, as this book shows in Chapters 1, 4, 6 and 17, cost–benefit analysis, even involving imaginative use of environmental economics, is no longer sufficient for integrated river management. The South African example is worthy of emulation, albeit in appropriate national policy styles. The combination of shifting the property right to 'a joint account' with human needs and natural entitlements unleashes so much that is enriching over contemporary environmental management.

Introduction

For centuries we have harnessed rivers for our own purposes. They have been dammed for water resource development and hydroelectric power production, dredged, widened and straightened to aid navigation, improve land drainage and alleviate flooding, and stabilised to prevent the loss of buildings and bridges or to protect farmland. These works change the natural character of the river, which can lead to major instability problems unless the river is heavily engineered (Raynov *et al.*, 1986). All this can seriously impair the conservation and amenity value of the riverine environment.

Environmental impact legislation in the European Community, the USA and many other countries now makes it mandatory to carry out formal impact assessments of any major scheme likely to alter river systems. This includes dams and channel realignments. While this enables least damaging options to be identified and in some cases to be implemented, conservation bodies are advocating the adoption of more environmentally sensitive solutions that aim to preserve natural-type channels using bioengineering methods. Such approaches are certainly laudable, but arbitrary modifications to a river are likely to be no more than a temporary palliative. Poorly thought out, such alterations may actually jeopardise design objectives.

Major problems arise when unnatural conditions are imposed on a river, and it follows that more sympathetic approaches should be based on an understanding of the processes that control river morphology and its dynamic adjustment to changed conditions. By designing with nature, stable and environmentally sensitive solutions can be achieved. This maintains a natural range of in-stream and bankside habitats, which are required for fisheries, flora and fauna, preserves visual amenity and ensures that subsequent maintenance requirements are minimised. Environmental approaches are, therefore, likely to be more cost-effective in the longer term than traditional methods.

In order to understand why many traditional engineering works have caused instability and environmental degradation, and to identify more appropriate design procedures, consideration needs to be given to the factors controlling the shape and size of alluvial channels.

Natural channel processes

Basic research in river mechanics has identified the factors controlling the morphology of alluvial channels. Essentially, they adjust their overall, bankfull, width, depth, slope, plan shape and velocity in response to discharge, sediment load, calibre of the bed and bank sediment (gravel, sand, silt or clay), the bank's vegetation (grass, shrubs or trees) and the slope of the valley. Change in any one of these variables through flow regulation or land-use change, or by directly modifying the river, can destabilise the channel and promote erosion and deposition.

Empirical equations, based on field measurements and statistical analysis, which have been developed to predict the dimensions of stable alluvial rivers, can be used to identify the direction of change (Box 12.1).

If changes increase the sediment transport capacity of the river, then instability, manifested as either bed scour or fill and associated bank failure, is propagated upstream. Alternatively, if there is a rapid local change in sediment supply, instability will migrate downstream. Once initiated, a sequence of scour and fill phases will occur, each being of progressively smaller magnitude. Eventually a new equilibrium condition will be achieved, provided the controlling factors stabilise.

The principal design requirement for river engineering works is, therefore, to maintain the natural sediment transport regime of the river. This will prevent instability occurring and will obviate the need for expensive remedial works and long-term maintenance commitments.

In order to preserve or enhance fisheries and the conservation value of the river, it is essential that engineering works should, wherever possible, maintain or recreate natural channel features including pools, riffles, meander bends, point bars and natural steep banks. Where heavier engineering works are required, habitat improvements should be implemented to increase diversity. These need to be sensibly located as arbitrary modifications to a river are unlikely to be enduring.

Sufficient is now known about the hydraulic and sedimentary processes operating in meandering alluvial channels for ecologically appropriate measures to be incorporated into engineering design. Essentially, this identifies the large-scale erosion and deposition processes which are responsible for the formation of meanders, pools, riffles and bar deposits and for bank failure (Richards, 1982; Hey, 1986).

Flood alleviation schemes

Floods can be alleviated either by storing flood waters in specially constructed storage reservoirs or by modifying the river to accommodate the flood flows within the bank.

Construction of flood storage or detention reservoirs can provide adequate safeguards in the immediate vicinity of the reservoir, but their effect is reduced as the

Box 12.1 Stable channel dimensions and prediction of channel change

Empirical equations that define the dimensions of stable alluvial channels can be used to predict channel change. These equations are specific to a particular river environment, for example gravel-, sand- or silt-/clay-bed rivers, and should not be used out of context. For gravel-bed rivers, the following equations have been developed based on UK field data (Hey and Thorne, 1986).

Cross-section

Bankfull top width	$W = 4.33\,Q^{0.5}$ (m)	Vegetation type I	(1)
	$W = 3.33\,Q^{0.5}$ (m)	Vegetation type II	(2)
	$W = 2.73\,Q^{0.5}$ (M)	Vegetation type III	(3)
	$W = 2.34\,Q^{0.5}$ (m)	Vegetation type IV	(4)
Bankfull mean depth	$d = 0.22\,Q^{0.37}\,D_{50}^{-0.11}$ (m)	Vegetation types I–IV	(5)

Longitudinal section

Bankfull slope $S = 0.087\,Q^{-0.43}\,D_{50}^{-0.09}\,D_{84}^{0.84}\,Q_{s}^{0.10}$ Vegetation types I–IV   (6)

Plan section

| Sinuosity | $P = \dfrac{S_v}{S}$ | Vegetation types I–IV | (7) |
| Meander arc length | $Z = 6.31\,W$ (m) | Vegetation types I–IV | (8) |

(riffle spacing)

where Q = bankfull discharge (M³/S); Q_s = bankfull bedload discharge (kg/s); D_{50} = median bed material size (m); D_{84} = bed material size, 84% finer (m); Vegetation type I = bankside trees and shrubs absent; Vegetation type II = 1–5% trees and shrubs; Vegetation type III = 5–50% trees and shrubs; Vegetation type IV = greater than 50% trees and shrubs; S_v = valley slope.

River response to changed conditions can be identified by considering the direction of the imposed change on the river (increase +, decrease –) in the above equations. To achieve a new stable condition, the river must adjust the other variables in each equation (+ or –) to accommodate the imposed change.

For example, dam construction cuts off sediment supply to downstream reaches and reduces flood discharges (i.e. Q and Q_s decrease). Disruption to the sediment transport regime of the river causes erosion downstream from the dam with the following consequences:

From equation (6)

(imposed change)

$$\underline{S}\,\bar{Q} \propto \bar{Q}_s\,\underline{D}_{84}$$

(responses)

- channel slope reduced to point where bed material can no longer be transported (i.e. S decreases);

- fine bed material will be preferentially eroded leaving a coarser residual (i.e. D_{84} increases).

The remaining equations indicate that

- width is reduced;

- depth is reduced, in spite of incision, since higher banks are unstable;

- sinuosity increases provided valley slope does not change.

percentage of uncontrolled catchment increases. They are often employed to prevent flood runoff from urbanised areas, motorways and car parks. For large river basins, numerous reservoirs would be required in the headwater valleys or a number of major reservoirs on the main river if a significant reduction in flood flows were to be achieved in the lower reaches of the basin.

Reservoirs, by modifying the flow and sediment transport regime downstream from the dam, can cause significant bed degradation as releases from the dam are relatively sediment-free. Regime equations (Box 12.1) indicate that the decreased sediment load causes erosion and an associated reduction in channel gradient and an increase in the size of the bed material (Figure 12.1) until a new equilibrium is achieved. For example, on the Colorado River, USA, the bed has been eroded by 4.6 m immediately below Parker Dam and 50 km further downstream by 2.6 m. Degradation has been observed for 125 km below the dam (Figure 12.2).

Recent plans to construct major flood storage reservoirs at Serre de la Fare on the River Loire, on the Allier at Le Veurdre and on the Cher at Chambonchard to alleviate flooding in the upper and middle Loire have been turned down or deferred as a result of sustained environmental lobbying because cost-effective alternatives were shown to be possible without any associated environmental degradation (Purseglove, 1991).

Increasing in-bank flows was commonly achieved by widening and dredging the river (Figure 12.3). The aim was to increase the cross-sectional area of the channel in order to increase 'in-bank' discharge capacities. Much of lowland England has been affected by this course of action. Agricultural improvement schemes to drain low-lying floodplain land required the bed of the channel to be dredged by up to 1 m to lower the water table and widened to increase flood capacity. Much of this work was carried out in the early to mid-19th century, although this has been maintained and developed subsequently, particularly during the 'Plough for Victory' campaign in the Second World War. On these lowland rivers, the riparian and in-stream habitats were decimated, with adverse effects on invertebrates, fisheries, flora, fauna and bird life. Subsequent maintenance dredging in a 5–10-year cycle has maintained these artificial channels from siltation, which, inevitably, prevents ecological recolonisation.

On rivers transporting considerable bed-material load, often upland rivers, such a course of action can seriously destabilise the river as well as adversely affecting the riverine environment. Application of the regime equations (Box 12.1) indicates that erosion will occur at the head of the dredged section, because of increased gradients, and progress upstream (Figure 12.1), while much of this eroded material will be deposited in the lower part of the undredged reach, where the gradient has been artificially reduced (Figure 12.1). Should sediment loads downstream from the dredged reach be reduced, then further degradation could result (Figure 12.3). These sequences of changes have been observed on a dredged and widened section of the River Usk at Brecon (UK). Although headward erosion is prevented by a weir, 5000 tonnes of material has to be removed annually from the dredged section to maintain its design capacity (Figure 12.4). Even then, erosion is occurring further downstream.

Channel straightening has also been carried out to enhance flood capacities, since flow velocities are increased in the steepened reach. What has happened to the Mississippi is a classic example of how channel stability can be initiated by channel straightening. After serious flooding in 1927, sections of the lower river were altered to reduce navigation distances and enlarge flood capacities (Figure 12.5). The increased gradient resulting from channel straightening promoted bed erosion in the engineered section, and this progressed upstream and also affected tributary rivers (Figure 12.1). Enhanced sediment loads resulting from headward erosion subsequently caused serious

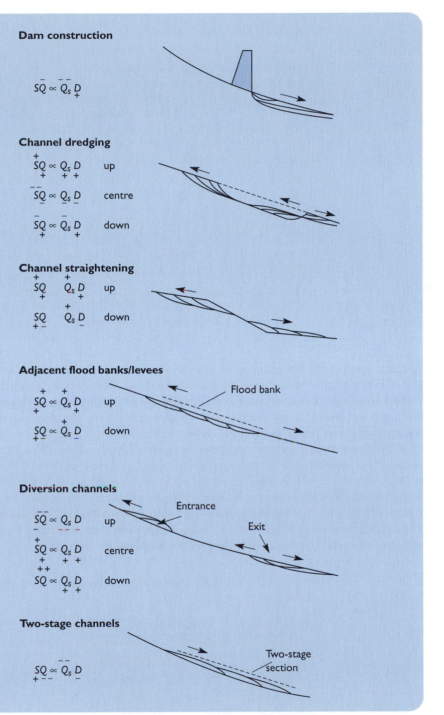

Figure 12.1:
Channel response to construction of flood alleviation schemes
(+, increase; −, decrease; above line, control; below line, response).

Dam construction

$$\bar{S}\bar{Q} \propto \bar{Q}_s \underset{+}{\bar{D}}$$

Channel dredging

$$\overset{+}{S}\underset{+}{Q} \propto \underset{+}{Q_s} \underset{+}{D} \quad \text{up}$$

$$\bar{S}\underset{\_}{Q} \propto \bar{Q}_s \underset{\_}{D} \quad \text{centre}$$

$$\bar{S}\underset{+}{Q} \propto \bar{Q}_s \underset{+}{D} \quad \text{down}$$

Channel straightening

$$\overset{+}{S}\underset{+}{Q} \quad \overset{+}{Q_s} \underset{+}{D} \quad \text{up}$$

$$\underset{+}{S}\underset{-}{Q} \quad Q_s \underset{-}{\overset{+}{D}} \quad \text{down}$$

Adjacent flood banks/levees

$$\overset{+}{S}\underset{+}{Q} \propto \overset{+}{Q_s} \underset{+}{D} \quad \text{up}$$

$$\underset{+}{S}\underset{-}{Q} \propto \overset{+}{Q_s} \underset{}{D} \quad \text{down}$$

Diversion channels

$$\overset{\_\_}{S}\underset{\_}{Q} \propto \underset{\_}{Q_s} D \quad \text{up}$$

$$\overset{+}{S}Q \propto \underset{+}{Q_s} \underset{+}{D} \quad \text{centre}$$

$$\overset{++}{S}Q \propto \underset{+}{Q_s} \underset{+}{D} \quad \text{down}$$

Two-stage channels

$$\underset{+}{S}\underset{-}{Q} \propto \overset{\_\_}{Q_s} \underset{\_}{D}$$

Figure 12.2:
Longitudinal profile of Colorado River downstream from Parker Dam at dam closure and 4, 13 and 37 years later.

Source: Williams and Wolman (1984).

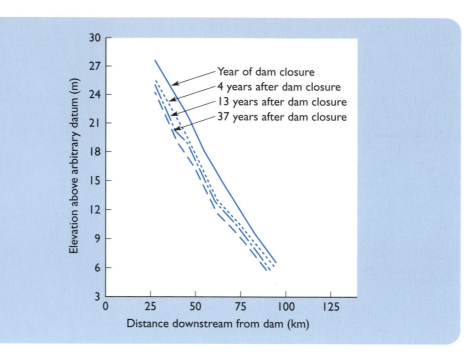

sedimentation in the straightened sections as the river attempted to recreate a meandering channel (Figure 12.1). Billions of dollars has since been spent dredging the river to maintain navigation depths and flood capacities, and building spur dykes to prevent bank erosion and bed sedimentation (Winkley, 1982). Significantly, the natural river required relatively little maintenance (Table 12.1).

Similar problems have been experienced in the UK. The Ystwyth in west Wales was regularly straightened to increase its gradient and enhance its flood capacity, but the meandering channel quickly re-established itself. In order to maintain the artificially straightened channel, some form of heavy revetment would be needed to stabilise the banks. This could be achieved by masonry walls, sheet piling, cellular concrete blocks, stone-filled wire baskets (gabions) or large blockstone slabs. Inevitably, this is visually very intrusive and can completely destroy the bankside environment.

Table 12.1: Maintenance dredging record: Greenville Reach, River Mississippi.

Period	Volume dredged (m³/km/yr)
Pre cut-off	137
Cut-off to 1950	10 360
1951–64	29 835
1965–73	62 833
1974–77	39 695

Source: Winkley (1982).

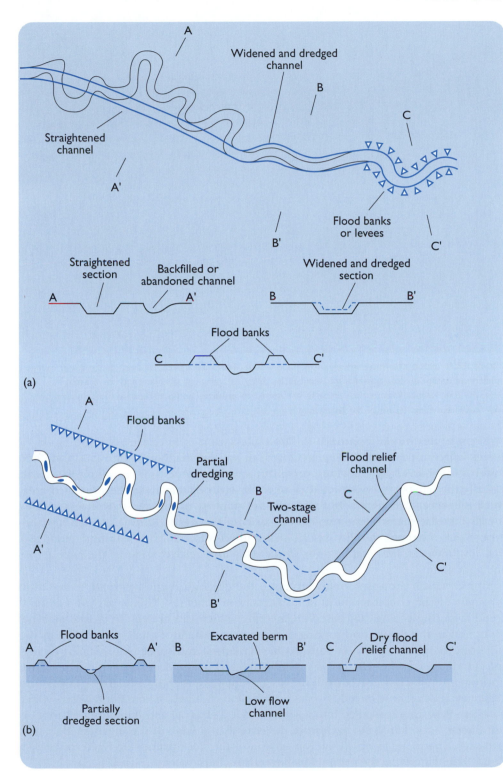

Figure 12.3:
Flood alleviation
schemes: (a)
traditional options;
(b) environmental
options.

(a)

A
Widened and dredged channel
B
Straightened channel
A'
C
Flood banks or levees
B'
C'

Straightened section
Backfilled or abandoned channel
A ——— A'
Widened and dredged section
B ——— B'

Flood banks
C ——— C'

(b)

A
Flood banks
Partial dredging
B
Flood relief channel
C
Two-stage channel
A'
B'
C'

Flood banks
A ——— A'
Excavated berm
B ——— B'
Dry flood relief channel
C ——— C'

Partially dredged section
Low flow channel

Figure 12.4: Flood alleviation scheme, River Usk, Brecon, Wales. This river has been dredged and widened to increase its flood capacity, which has reduced its ability to transport sediment and has created a sediment trap. Regular maintenance dredging is required to preserve its flood capacity. Note the masonry wall to prevent mass failure of the heightened bank.

Another traditional approach to flood alleviation is to construct flood banks or levees adjacent to the river (Figure 12.3). The natural channel is left intact, and flood control is achieved by allowing overspill from the channel but curtailing its extent. Usually the banks are constructed close to the river to maximise the protected area. Such measures can adversely affect channel stability on upland-type rivers as they locally increase the bed-load transport capacity of the river (Figure 12.1). On lowland-type rivers, channel stability can be maintained and bankside habitats enhanced. Visually, however, they can be quite intrusive as they restrict views of the river across the floodplain.

Conservation groups strongly criticised the ecological and aesthetic devastation wrought by the construction of flood alleviation schemes. This is part of a general trend favouring ecologically based engineering, which can be seen in coastal management and in the design and routing of motorways. Much of this is influenced by the kind of ecological economics outlined in Chapters 2 and subsequently incorporated into modified cost–benefit analysis. All this has led to the development of alternative approaches which were considered to be environmentally sensitive (Lewis and Williams, 1984). These include bypass or diversion channels, where flood waters above a certain flow are overspilled into a separate flood channel and the natural channel is left intact; flood banks set back from the river at the edge of the meander belt; and two-stage channels, in which the top section of the floodplain is dredged to create a flood channel, leaving the natural low-flow channel untouched (Figure 12.3).

Diversion channels have been successfully implemented on lowland rivers, as, for example, at Exeter on the River Exe (UK), without adversely affecting the conservation and fisheries value of the original river or its channel stability. However, on upland-

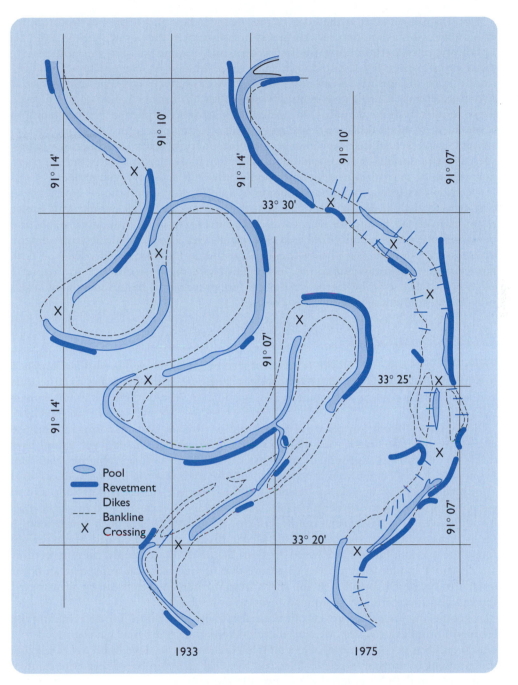

Figure 12.5:
Comparisons of Greenville Reach of lower Mississippi in 1933 and 1975.
Source: Winkley (1982).

type rivers the regime equations indicate that they would promote deposition at the entrance to the diversion and erosion at the exit (Figure 12.1). Similarly, two-stage channels are appropriate on lowland sites, especially as they can increase plant diversity. The flood berm would need to be managed to prevent it becoming overgrown, with consequent loss of flood capacity, and could form a linear park in built-up areas. On upland sites, two-stage channels would cause severe aggradation, as the transport capacity is significantly reduced (Figure 12.1).

Flood banks set back at the edge of the meander belt afford the best solution for flood alleviation as the natural channel is not disturbed (Hey *et al.*, 1990). Even on upland-type rivers the natural sediment transport regime will not be modified. In many cases, particularly for urban schemes, space limitations may preclude the adoption of such designs and alternative solutions will be necessary. For lowland sites, two-stage channels are a possibility provided they are properly managed, but even these require additional space. On upland sites, or where space is at a premium on lowland rivers, flood walls may need to be built to contain flooding. In these circumstances, it is essential that they have architectural merit and that in-stream structures, such as weirs, should be incorporated in the design to ensure that the sediment transport regime of the river is not disrupted and that habitat diversity is maintained.

Decisions regarding the implementation of a flood alleviation scheme are based on cost–benefit analysis. Environmentally sensitive approaches, by designing with nature rather than trying to impose man's will on the river, are inherently more stable and hence relatively maintenance-free, and can actually enhance the riverine environment. Consequently, they are likely to be less expensive than traditional approaches in both monetary and environmental terms. Techniques based on those outlined in Chapter 3 are now widely used. The recreational public are prepared to pay an identifiable extra price for an aesthetically pleasing river design.

River stabilisation

Rivers become unstable if there is an imbalance between the sediment load supplied to a reach of river and its ability to transport it. Regional instability refers to large-scale and systematic long-term degradation and aggradation resulting from natural or man-induced changes.

Reduced sediment supply due, for example, to dam construction or land-use change, or increased transport capacity resulting from a fall in sea level, raised land levels or channel straightening, amongst other things, causes channel incision. Eventually, a new equilibrium is achieved with the river flowing in a new, lower-level floodplain. The original valley is normally left as a terrace. To control erosion and to prevent it spreading to adjacent reaches, it is necessary to reduce the transport capacity of the river. This is best achieved by reducing the channel gradient by constructing a series of weirs or drop structures in the affected reach and at the head of the incised section to prevent further headward erosion (Figure 12.6). Typically, these structures are designed and located in order to ensure that no bed material transport can occur or that the material supplied to the reach from further upstream can be transmitted through the affected reach without any erosion or deposition.

A geomorphological approach offers an alternative procedure for determining the requisite number of weirs. Measurements taken at sections that have evolved to a quasi-equilibrium condition enable a simple equation to be developed for predicting the gradient of the stable reach. The size, location and number of gradient control structures can then be determined to provide the requisite gradient (Schumm *et al.*, 1984). Many streams and rivers in northern Mississippi have become dramatically unstable as a result of channel straightening for flood control purposes. Incision into the fine floodplain alluvium has increased channel widths from 2–3 m to over 40 m and depths from 0.5 m to upwards of 10–15 m. Gradient control structures have been constructed to stabilise these rivers, and calculations for Oaklimiter Creek indicate that a smaller number of weirs would be required using the geomorphologically based calculation procedure (six weirs) than the more traditional methods (15–20 weirs) and, hence, would be more cost-effective.

Figure 12.6: Grade control structure, Goodwin Creek, Mississippi, USA. This has been constructed to prevent headward erosion. The stilling basin lined with rockstone and baffle board dissipates energy below the weir.

Increased sediment loads can choke the river with material and promote braiding, which is manifest as a series of sediment bars and islands with multiple interconnected channels. The elevation of the river bed and its valley is raised in the process and a new floodplain is created. Such channels can be stabilised by controlling the sediment supply to the reach from upstream, through the construction of gradient control structures or afforestation programmes. Alternatively, the river can be engineered to transmit the sediment supplied from upstream. This is generally achieved by increasing stream power to enable the river to transmit the sediment load, by maximising the channel gradient and flow depth by straightening and narrowing the river into a single channel. In order to maintain the river in this unnatural state, the banks of the river have to be constructed of blockstone, or similar, to prevent bank failure (Figure 12.7). The formerly braided Alpine Rhine in Switzerland was extensively trained in the 18th century in this manner. Inevitably, the natural river with its diverse habitats is replaced by a uniform canalised channel of limited conservation and environmental value.

A more environmentally acceptable alternative would be to create an irregular sinuous channel. This would be designed to transmit the sediment supplied from upstream on a slightly reduced gradient to accommodate a meandering pattern. By creating a meandering channel with a variable bed geometry, a range of in-stream habitats could be created which are beneficial to flora, invertebrates and fisheries. Although harder bank protection measures would be required to stabilise the outer bank in meander bends, softer procedures, particularly those involving vegetation, could provide a more sympathetic approach elsewhere which is aesthetically more pleasing and provides cover for fish.

On a more local scale, instability problems are generally concerned with the prevention of bank failures and the maintenance of channel plan form. Traditionally, eroding

Figure 12.7: River stabilization, Alpine Rhine, Switzerland. The original braided river has been stabilized by narrrowing, deepening and straightening the river. The banks have been protected by blockstone to prevent erosion and maintain a straight channel.

banks have been protected by lining the bank with some form of protective cover to prevent both surface scour and mass failure. In extreme cases, this can take the form of a concrete or masonry wall, for banks exposed to extreme hydraulic loading, through cellular-type blocks, stone-filled wire baskets and stone riprap to geotextile mats and willow, reed and grass cover as loadings decrease. The softest possible options need to be chosen for each site according to local circumstances, and guidelines and flow-charts have been developed, based on field experience, which enable this to be achieved (Hey *et al.*, 1991).

All the aforementioned methods provide protection by increasing the erosional resistance of the bank. An alternative solution is to modify the flow adjacent to the bank such that the hydraulic loadings on the bank are reduced and bank failure prevented. This effectively would treat the cause of the problem, rather than the symptoms, and would enable the bank to remain in a natural state.

For meander bends, submerged vanes or hydrofoils can be installed on the bed of the river to generate secondary currents, or cross-flows, which oppose the main heli-coidal cell which results from flow curvature in the bend (Figure 12.8). As a result, downwelling of faster surface water, which normally occurs adjacent to the outer bank causing bed scouring, bank steepening and failure, now occurs inside the line of the hydrofoils (Figure 12.8). Bed scouring is now relocated in the cross-section and scoured material refills the pool adjacent to the bank. As a consequence, the toe of the bank is stabilised and further bank failure prevented. Should deposition cover the hydrofoils, the original secondary flow pattern would become re-established. As associated scouring would re-expose the hydrofoils, the system is self-regulatory. The systems have been successfully deployed for bank erosion control on the East

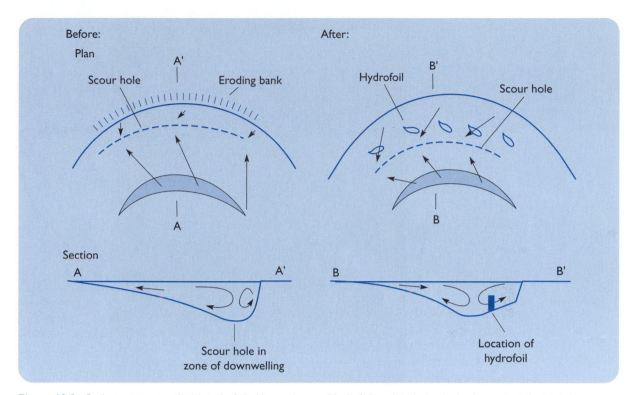

Figure 12.8: Bank erosion control with hydrofoils. Here submerged hydrofoils reduce hydraulic loading against the bank by suppressing development of secondary flows, which direct faster-flowing water against the bank, and prevent bank erosion.

Nishnabotna River, USA (Odgaard and Mosconi, 1987), and on the River Roding, Essex, UK (Paice and Hey, 1989, and Figure 12.9).

Knowledge of river mechanics can also be used to prevent local bed scour problems at bridge piers. Recent bridge failures on the River Towy at Glanrhyd (Wales), on the River Ness at Inverness (Scotland), on the Route 90 bridge over Schoharie Creek (USA) and on the River Inn at Innsbruck (Austria) have highlighted the need for improved methods of scour control. Traditionally, scour holes have been refilled with large stone riprap, but this is rarely successful as it effectively increases the size of the bridge footing, making the bed more prone to scour. In time, the stone is either washed away or buried. Observation of scour processes in scale models indicated that scouring occurred because of faster-flowing surface water descending the upstream face of the pier, which significantly increases near-bed velocities. It is the latter that is responsible for eroding the bed material adjacent to the pier. By installing a group of four cylindrical piles in a diamond-shaped formation into the river bed immediately upstream from the pier, the velocities at the pier are significantly reduced. Consequently, downwelling at the pier face is less pronounced and bed scour is reduced. Scale model tests indicate that local scour can be reduced by up to 70% (Figure 12.10), and field trials to evaluate the method further are currently being carried out on the Tavy viaduct across the Tavy estuary at Plymouth, the Over viaduct across the River Severn at Gloucester, (Figure 12.11) and the Coton viaduct crossing the River Tame at Tamworth (Paice *et al.*, 1993).

Figure 12.9: Hydrofoil installation to prevent bank erosion in meander bends.

River restoration

River engineering works have, over the centuries, adversely affected hundreds of kilometres of rivers (Brookes, 1988; Purseglove, 1988). A post-project appraisal of flood alleviation schemes indicates that unsympathetic engineering treatments consistently reduce plant species richness, when compared with natural channels, while environmentally sensitive approaches actually increased richness by creating a greater range of habitats (Hey *et al.*, 1990). The requirement, therefore, is to reinstate natural channels, provided that engineering objectives can be maintained, or to rehabilitate reaches that require a heavier engineering treatment.

Non-structural approaches refer to the reinstatement of natural features within a reach that had previously been dredged and/or straightened. It could involve, *inter alia*, the restoration of meanders, pools, riffles, vertical banks and dead zones. On upland-type rivers, there is generally only one restoration solution, which is prescribed by the controlling variables (Box 12.1). With lowland-type rivers, because flows are generally below bed material transport thresholds, there is more flexibility in the choice of restoration procedure.

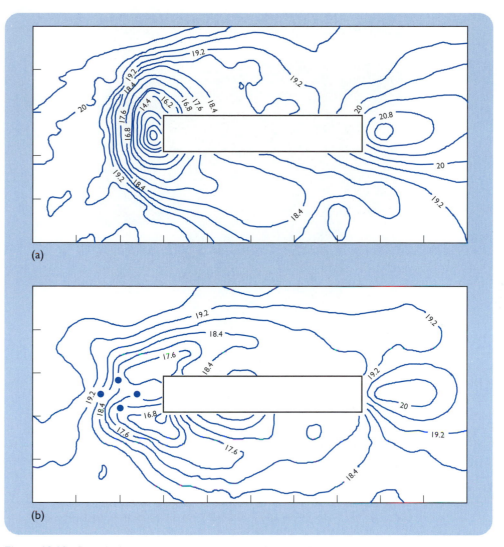

Figure 12.10: River bed scour at model bridge piers (blunt-nosed): (a) unprotected, local scour 80 mm, Q = 0.03 cumecs; (b) protected by pile group, local scour 33mm, Q = 0.03 cumecs. Contours in cm, initial bed level 20 cm.

Source: Paice and Hey (1993) The control and monitoring of local scour at bridge piers. In Hsieh Wen Shen, S.T. Shu and Feng Wen (eds), *Hydraulic Engineering '93*, 1. Reproduced by permission of ASCE.

A geomorphological survey of the reach is required in order to ascertain the nature of the existing channel morphology and substrate characteristics. This, together with old maps and aerial photographs of the site, enables the natural, pre-engineered and existing conditions to be compared.

For lowland rivers that have simply been straightened, it should be possible to reinstate the original meander pattern, with associated pools at meander bends and riffles at inflexion points between bends. Channel widths and depths can be determined from appropriate design equations (Box 12.1), or from the survey of adjacent natural reaches. Several lowland rivers in Denmark have been successfully reconstructed to recreate natural-type meandering channels using these techniques (Figure 12.12).

Figure 12.11: Bridge pier scour control, Over Viaduct, River Severn, Gloucester, England. A pile group has been installed upstream from the pier to reduce flow velocities and prevent scouring of the river bed. At this bridge, the width of the pier necessitated the use of six piles set in triangular shaped formation.

On rivers that retain their natural meander pattern but have been widened and the riffles removed, it is necessary to reinstate riffles and pools and to narrow the channel at strategic locations to accelerate natural accretion processes. Provided the riffles are drowned out during bankfull flows, the flood capacity of the river is unimpaired. This type of approach has been successfully carried out at Lyng on the River Wensum (UK) as part of a fish habitat improvement programme (Figure 12.13).

For channels that have been straightened and the bed elevation significantly lowered (by up to 1 m) by dredging, more drastic action is required. Simply recreating the meandering channel would produce a river that was too deep. Backfilling the channel to re-establish the pre-engineered bed elevations is not a viable option as it would raise the water table by a corresponding amount and adversely affect local land drainage. The solution is to excavate a corridor in the existing floodplain, by an amount the river bed has been lowered by dredging, to create a new low-level floodplain. The width of the corridor is determined by the amplitude of the meander pattern that is to be reinstated. Although the river will have a lower gradient than the

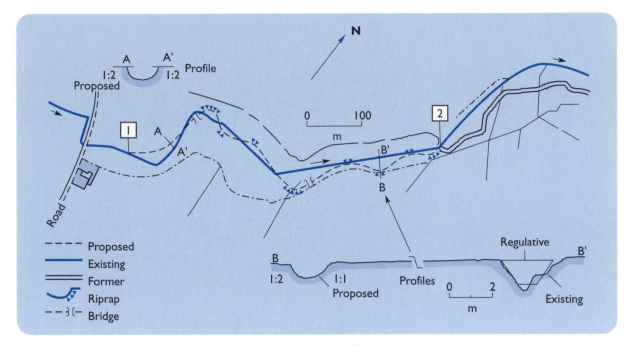

Figure 12.12: Restored course of Stensbaek stream in southern Jutland, Denmark.
Source: Brookes (1987).

straightened engineered channel, since it is sinuous, and a smaller cross-section, as the depth is reduced, its overall flood capacity is maintained within the new floodplain.

The River Blackwater in Norfolk (UK) was extensively dredged and canalised in the early to mid-19th century. In order to extend floodplain gravel workings, the opportunity was taken to restore the river to its precanalised condition. Maps of 1790 showed that the river was then much smaller and quite sinuous, and a geomorphological survey indicated that the river bed had been lowered by 0.8 m. A new low-level floodplain was excavated, 15–20 m wide and 0.8 m deep, containing a small sinuous channel, maximum width 5 m, with pools and riffles (Figures 12.14 and 12.15). Calculations showed that low water levels would not be raised, flood capacity below the existing floodplain level would be preserved, and the channel would remain stable. Effectively, it recreates the precanalised river, and its natural diversity will significantly improve the fisheries potential of this chalk-fed stream.

With river restoration on upland-type rivers, it is necessary to ensure that the sediment transport regime of the river is not impaired by the planned restoration works. The design of the diversion is, therefore, critical and it should be based on current design constraints, since the pre-engineered condition refers to an earlier situation which may not currently prevail. A geomorphological survey of the reach to be restored and adjacent natural sections will define existing conditions, and this, in conjunction with appropriate design equations (Box 12.1), will enable the new channel to be sensitively designed.

River diversions are similar to restoration schemes in terms of design procedures. Often engineers would advocate the construction of straight diversion channels where a river had to be moved to accommodate a new road. As this would increase the sediment transport capacity of the river locally, instability would result. It is

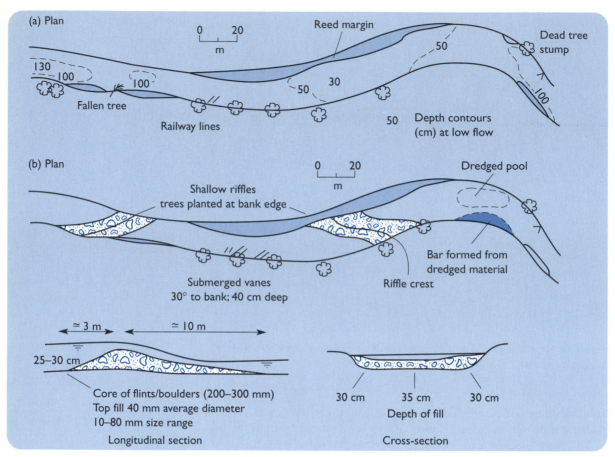

Figure 12.13: Recreation of pools and riffles at Lyng, River Wensum, England: (a) geomorphological map of original channel; (b) proposed riffle–pool sequence.

Source: Hey (1992).

Figure 12.14:
Design of diversion at Reymerston, River Blackwater, England.

Source: Hey (1992).

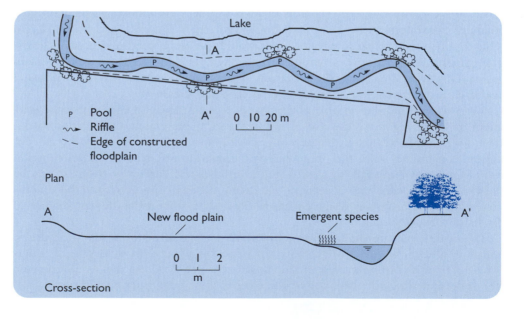

12.15: River diversion, River Blackwater, Reymerston, England. This shows the recreation of a small sinuous channel within a new low level flood plain. A variety of habitats had been restored which have since been colonized by instream and bankside vegetation.

necessary, therefore, to reinstate a natural-type diversion and, in particular, to maintain the channel length, and hence gradient, to avoid instability problems. Where the original channel is heavily engineered, opportunities should be taken to create a more natural channel.

River diversions on the River Neath, south Wales (UK), have been designed using these principles (Hey, 1992). A typical geomorphological map and design for part of one diversion is illustrated in Figure 12.16.

Structural methods, involving the installation of artificial in-stream structures, can be deployed to promote local aggradation or degradation, create ponded reaches and diversify substratum types. They are particularly useful for increasing habitat diversity in more heavily engineered sections of river. Before installing any structures, it is important to ensure that they are correctly located to enhance local in-stream processes and that they do not impair the sediment transport or flood capacity of the river. By harnessing the energy within the flow locally, either by overspill on weirs, by concentrating flows with deflectors, or by generating secondary flows with submerged vanes, erosion and deposition can be encouraged. These structures are particularly useful on lowland-type rivers, where stream power is generally insufficient to cause bed scour without some assistance. On upland rivers, they are best deployed for stabilising and creating habitat diversity in heavily engineered reaches. Non-structural methods are preferable on more natural channels.

The range and type of structures are illustrated in Figure 12.17. Submerged vanes are less intrusive than the other structures and also have the benefit of not trapping floating debris or adversely affecting the flood capacity of the river. They were originally developed for pool creation and maintenance (Hey, 1992), but they can also be deployed to cause bank erosion and meander initiation. A range of vanes have been successfully installed on the River Wensum (UK) at Fakenham and Lyng.

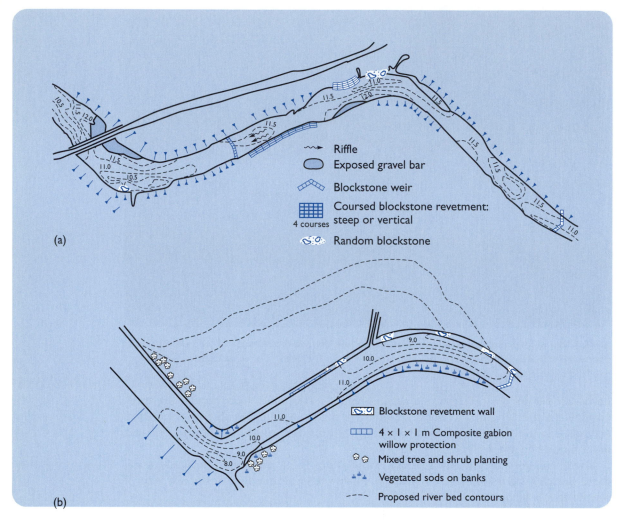

Figure 12.16: Design of diversion for section of River Neath, south Wales: (a) geomorphological map of original channel; (b) proposed diversion.

Source: Hey (1992).

Conclusions

Attempts to impose an unnatural condition on a river will, on upland-type channels that transport considerable amounts of bed material load, create major instability problems and long-term maintenance commitments. On lowland-type rivers, which transport relatively little bed material, instability may not be a problem, but the creation of uniform, sterile, canalised rivers totally destroys the riverine environment.

This brief review illustrates the key role of fluvial geomorphology in river management. It indicates that basic understanding of natural channel processes is a prerequisite for successful environmentally sensitive engineering design.

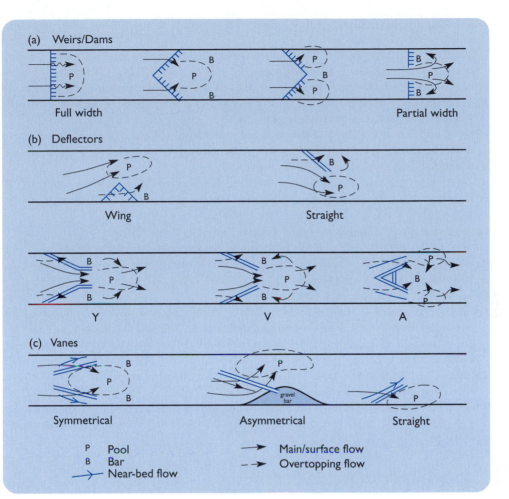

Figure 12.17:
River habitat improvement using structural measures. In part (b) Y, V and A represent the shapes of the deflectors.

Source: Hey (1992).

References

Brookes, A. (1987) Restoring the sinuosity of artificially straightened stream channels. *Environmental Geology and Water Science*, **10**, 33–41.

Brookes, A. (1988) *Channelized Rivers Perspectives, for Environmental Management*. Chichester, Wiley.

Hey, R.D. (1986) River mechanics. *Journal of the Institution of Water Engineers and Scientists*, **40**(2), 139–158.

Hey, R.D. (1992) River mechanics and habitat creation. In K.T. O'Grady, A.J.B. Butterworth, P.B. Spillett and J.C.J. Domaniewski (eds), *Fisheries in the Year 2000*. Institute of Fisheries Management, Nottingham; 271–285.

Hey, R.D. and Thorne, C.R. (1986) Stable channels with mobile gravel beds. *Journal of Hydraulics Division, American Society of Civil Engineers*, **112**(8), 671–689.

Hey, R.D., Heritage, G.L. and Patteson, M. (1990) *Design of Food Alleviation Schemes: Engineering and the Environment*. Ministry of Agriculture, Fisheries and Food, London.

Hey, R.D., Heritage, G.L., Tovey, N.K., Boar, R.R., Grant, A. and Turner, R.K. (1991) *Streambank protection in England and Wales*, R&D Note 22. National Rivers Authority, Bristol.

Lewis, G. and Williams, G. (1984) *Rivers and Wildlife Handbook*. Royal Society for the Protection of Birds, Sandy.

O'Riordan, T. (1998) Sustainability for survival in South Africa. *Global Environmental Change*, **9**(2), 9–20.

Odgaard, A.J. and Mosconi, C.E. (1987) Streambank protection by submerged vanes. *Journal of Hydraulic Engineering, American Society of Civil Engineers*, **113**(4), 520–536.

Paice, C. and Hey, R.D. (1989) Hydraulic control of secondary circulation in meander bend to reduce outer bank erosion. In M.L. Albertson and R.H. Kia (eds), *Design of Hydraulic Structures 89*. Balkema, Rotterdam, 249–254.

Paice, C. and Hey, R.D. (1993) The control and monitoring of local scour at bridge piers. In Hsieh Wen Shen, S.T. Shu and Feng Wen (eds), *Hydraulic Engineering '93*, **1**. American Society of Civil Engineers, New York; 1061–1066.

Paice, C., Hey, R.D. and Whitbread, J. (1993) Protection of bridge piers from scour. In J.E. Harding, G.A.R. Parke and M.J. Ryall (eds), *Bridge Management 2*. Telford, London, 543–552.

Purseglove, J. (1988) *Taming the Flood*. Oxford University Press, Oxford.

Purseglove, J. (1991) Liberty, ecology, modernity. *New Scientist*, 28 September, 45–48.

Raynov, S., Pechinov, D., Kopaliany, Z. and Hey, R.D. (1986) *River Response to Hydraulic Structures*. UNESCO, Paris.

Richards, K. (1982) *Rivers*. Methuen, London.

Schumm, S.A., Harvey, M.D. and Watson, C.C. (1984) *Incised Channels: Morphology, Dynamics and Control*. Water Resources Publications, Littleton, Colorado.

UN Development Programme (1998) *Human Development Report*. Oxford University Press, Oxford.

Williams, G.P. and Wolman, M.G. (1984) *Downstream Effects of Dams on Alluvial Rivers*. US Government Printing Office, Washington DC.

Winkley, B.R. (1982) Response of the Lower Mississippi to river training and realignment. In R.D. Hey, J.C. Bathurst and C.R. Thorne (eds), *Gravel Bed Rivers*. Wiley, Chichester, 659–681.

Further reading

This chapter briefly illustrates how knowledge of natural channel processes can be used to design environmentally sensitive river engineering works. Space precludes a full coverage of the river mechanics principles which underpin the design of these schemes. Consequently, it was not possible to provide detailed design guidelines.

For further information, the reader is referred to the following books and papers.

Rivers: processes and form

Hey (1986).
Knighton, D. (1984) *Fluvial Forms and Processes*. Edward Arnold, London.
Morisawa, M. (1985) *Rivers*. Longman, London.
Richards (1982).

Appraisal of engineering works and environmentally sensitive design procedures

Brookes (1988).
Hey, R.D. (1990) Environmental river engineering. *Journal of Water and Environmental Management*, **4**(4), 335–340.

Hey, R.D. (1993) Environmentally sensitive river engineering. In P. Calow and G.E. Petts (eds), *Rivers Handbook II*. Blackwell, Oxford, 337–362.

Hey, R.D. and Heritage, G.L. (1993) *Draft guidelines for design and restoration of flood alleviation schemes*, R & D Note 154. National Rivers Authority., Bristol.

Chapter Thirteen

Groundwater pollution and protection

Kevin Hiscock

Topics covered

- Sources of groundwater pollution
- Groundwater pollution and protection in developing countries
- Groundwater pollution and protection in industrialised countries
- Groundwater vulnerability mapping

Editorial introduction

Groundwater pollution could become one of the scourges of the age. Much depends on the spread of toxic and persistent chemicals from agricultural intensification the world over, from salt extrusion in disturbed dryland soils, and from the gentle rain from heaven contaminated with the fugitive emissions of millions of tiny sources, each one of which may be almost impossible to detect and to monitor. The natural repository of all these chemicals is the soil, groundwater and estuaries filling with toxic sediments, of which the most insidious is the contamination of groundwater by pathogens from sewage works discharges, nitrate pollution from excessive or inappropriate fertiliser use, heavy metals from rainfall, oil discharges from illegal dumping, solvent discharges from poorly managed waste disposal sites and rainfall, and sedimentation of rivers and estuaries from which seep all manner of stored pollutants.

Hydrogeology is coming into its own as a discipline and as a consultancy. Part of the reason is the steady growth in groundwater sources as new surface sources become opposed by amenity groups or for reasons of nature conservation, as introduced in Chapter 5. An equally important factor is the change in the law of contaminated land to a more formal status of strict liability. In principle, an owner is responsible for anything that occurs from his or her land that damages the interest of a neighbour. In the USA, this problem has to some extent been overcome by the introduction of the Superfund legislation, more formally known as the Comprehensive Environmental Response, Compensation and Liability Act 1981. This imposes a levy on all disposers of toxic waste into sites that could result in long-term groundwater pollution of mismanaged sites. Because many sites are abandoned from a previous industrial age, the cost of risk assessment and clean-up of these deserted sites is borne by the Superfund. The hazard assessment process generates a hazard ranking score for each site, and the most dangerous are placed on a national priorities list. About 35 000 preliminary assessments and 20 000 site investigations have been completed. Possibly as many as 3500 sites may be placed on the national priorities list out of a staggering 300 000 sites that need attention. Though Superfund has frozen some $15 billion, it is very likely that all of this and much more will be needed to do the job. The final cost of clean-up depends in part on the skills of the hydrogeologist, who has the tools to estimate where toxic pollution might enter the groundwater source if the site is not cleaned up. Similarly, any proposed landfill site will need the services of a trained hydrogeologist if the environmental impact assessment is to stand up.

An alarming development for both commercial companies and their financial backers in Europe is the probability of a European Community directive on the civil liability of wastes. This directive will impose a statutory responsibility on the landowner to clean up any site that is contaminated, or pay for any consequences arising from subsequent contamination arising from that site at any time in the future. Because the liability is both strict and severe, it will apply even when a landowner has been observing the pollution control regulations and will extend to anyone who has a financial interest in the site. Understandably, this has caused a flurry of activity in corporate boardrooms and among the insurance industry. Needless to say, any contemplated purchase of an old gasworks site or former electrical engineering plant would require the expensive advice of an experienced hydrogeologist before the site was acquired. It is not surprising that hydrogeology has become a lucrative source of environmental consultancy.

A notable case in the UK highlights the significance of applied groundwater studies in the tortuous process of civil liability for pollution. Cambridge Water Company is a private body supplying water to the city of Cambridge and its environs. In 1976, it purchased a borehole from which it supplied about one-eighth of all the water it provided for a quarter of a million people. To distribute this water, the company had to meet the European Commission drinking water directive. Among other requirements, this directive expected wholesome water to contain not more than one microgram per litre (1 µg/l) of organochlorine compounds. For the compound tetrachloroethane or perchloroethane (PCE) the maximum admissible concentration was to be 10 µg/l. When the company first assessed the borehole quality, such regulations were not in place. But by the mid-1980s, PCE concentrations were found to be 70–170 µg/l. The water company was stopped from pumping, so it looked for the source of the contamination to seek compensation.

That source turned out to be a tanning company known as Eastern Counties Leather plc. This firm degreased pelts with PCE. The PCE had been stored in drums on the site, and presumably some of this had either leaked or accidentally been spilled into the groundwater regime. The cost of the alternative borehole supply was close to £1 million, and this formed the basis of the compensation claim by Cambridge Water Company against Eastern Counties Leather.

The Appeal Court found in favour of Cambridge Water Company, not on the grounds of strict liability but on the grounds of nuisance. Strict liability is a common law tort which places the onus of responsibility on a landowner to be careful about the management of the site, irrespective of meeting regulatory requirements. Any discharge from a non-natural source, that is a source that would not be expected to arise naturally from the site, would form the basis of liability. The Appeal Court took another route and argued that Eastern Counties Leather was subject in nuisance in the sense that interference with a natural right of property is also a matter of strict liability. It is no defence to claim that pollution control standards came into force only after the spillage had occurred.

This case was finally settled in the House of Lords in December 1993. This is the highest court for matters of legal principle in the UK. Their lordships concluded that it was unreasonable to hold Eastern Counties Leather liable for activities undertaken so long ago, and where the company had taken reasonable safeguards against spillage. The judgement turned on the foreseeability of damage: because the storage of chemicals on an industrial site is an acceptable, or natural, use of land, spillage of these toxic substances was to be expected, so long as the spillages were kept small and appropriate safeguards were in place. In effect, the Lords' judgement returned the matter of liability to a narrower frame of reference, namely demonstrable and avoidable damage which any reasonable person would foresee. This finding was of much relief to the insurance world. But their lordships also pointed out strenuously that surveillance and enforcement measures under the modern regulations for groundwater protection should be considerably tightened up. This ruling will place even more emphasis on good-quality hydrogeological science in environmental regulation.

This case does not resolve the contentious legal disputes between strict liability *per se* and a strict liability interpretation of nuisance. But it does show that knowledge of groundwater hydrology is now essential if possible compensation claims of enormous magnitude are to be avoided. It is little wonder that a number of thoughtful insurance companies are avoiding cover for premises where there is any possibility of environmental risk. Indeed, cover generally for firms faced with environmental liability is increasingly difficult to obtain.

For the prospective consultant in hydrogeology, a number of themes covered in this book are increasingly pertinent. The principle of precaution is more systematically being applied in cases such as the one cited above. 'Due care' means that full knowledge of all possible groundwater conditions will be necessary before discharge or abstraction licences are granted. As water resources in eastern and southern England become decreasingly available, due to climate change and over-abstraction, so every new borehole application will require better understanding of the likely effects or quantities and qualities of the groundwater resources that remain.

Unlike the circumstances in South Africa (see Chapter 2), a natural entitlement to nature, and a citizen's right to intervene to defend that entitlement, do not formally exist in European Community law. But the EC Habitats Directive of 1992 does lay down a duty on conservation and planning authorities to ensure that key sites are not lost due to excessive abstraction. This is a weak version of a coupled property right involving private interest and 'Gaian responsibility'. Nevertheless, the groundwater hydrologist will be called upon to provide an assessment of what should be kept *in situ*. In this way, the hydrogeologist will steadily become involved with deliberative and interdisciplinary environmental science over the coming decade.

Introduction

Groundwater forms that part of the natural water cycle present within underground strata or aquifers. Of the global quantity of available freshwater, more than 98% is groundwater stored in the pores and fractures of rock strata. A natural asset of groundwater is the effect of rock strata to purify contamination entering from the ground surface, so offering a stable and wholesome source of drinking water. Today, in

England and Wales, groundwater abstraction accounts for 35% of the total public fresh-water supply, constituting approximately 75% of all abstracted groundwater. In the USA, groundwater provides more than half of the drinking water supply; in rural areas the figure is 96%. Groundwater is also an important source for industry and agriculture. Approximately 65% of groundwater withdrawn in the USA is used for irrigation.

Groundwater is not only abstracted for supply or river regulation purposes; it also naturally feeds surface waters through springs and seepages to rivers and is often important in supporting wetlands and their ecosystems. Removal or diversion of groundwater can affect total river flow. A reduction in either quantity or quality of the discharging groundwater can significantly influence surface water quality and the attainment of water quality standards. Surface water and groundwater are therefore intimately linked in the water cycle, with many common issues.

All too commonly, groundwater is considered out of sight and out of mind. If groundwater becomes polluted, it is difficult, if not impossible, to rehabilitate. The slow rates of groundwater flow and low microbiological activity limit any self-purification. Processes which take place in days or weeks in surface water systems are likely to take decades in groundwater.

The risk of groundwater pollution is the legacy of past and present land use and poor waste disposal practices and increasingly widespread use by industry and agriculture of potentially polluting chemicals in the environment. Pollution can occur either as discrete, point sources, such as from the landfilling of wastes, or from the wider, more diffuse use of chemicals, such as the application to land of fertilisers and pesticides and the deposition of airborne pollutants in heavily industrialised regions.

The volume and quality of groundwater must therefore be protected by proper management. It is better to prevent or reduce the risk of groundwater contamination than to deal with its consequences.

This chapter is concerned with the vulnerability of groundwater to contamination (see Box 13.1). This topic is dealt with from a consideration of hydrogeological principles, from a discussion of case studies illustrating various groundwater polluting activities, and from a review of approaches to protecting borehole supplies. By the end of the chapter, the reader should have gained understanding of the science and issues involved in understanding groundwater pollution and in formulating protection measures in industrialised and developing countries.

Box 13.1 Definitions of contaminated groundwater and groundwater pollution risk

Contaminated groundwater is groundwater that has been affected by human activities to the extent that it has higher concentrations of dissolved or suspended constituents than the maximum admissible concentrations formulated by national or international standards for drinking, industrial or agricultural purposes. The vulnerability of groundwater to pollution can be defined as a function of first, the accessibility of the saturated zone below the water table to the penetration of pollutants, and second, the physical and biochemical attenuation capacity of contaminants within the geological strata overlying the water table. The risk of groundwater pollution is determined when these two components of aquifer vulnerability interact with the mode of contaminant disposition in the subsurface, in particular the magnitude of any associated loading, and with the contaminant class in terms of its mobility and persistence.

Sources of groundwater pollution

Changes in groundwater quality may result from the direct or indirect influence of anthropogenic activities. Direct influences occur as a result of the introduction of natural or artificial substances into groundwater derived from human activities. Indirect influences are those changes of quality brought about by human interference with hydrological, physical and biochemical processes, but without the addition of substances.

The main contaminants of groundwater include toxic chemicals such as heavy metals, organic solvents, mineral oils, pesticides and fertilisers, and microbiological contaminants such as faecal bacteria and viruses. Figures 13.1 and 13.2 illustrate incidents giving rise to groundwater pollution. The enormous range of contaminants encountered in groundwater reflects the wide range of human economic activity in the world, as well as the incomplete measures for controlling pollution of groundwater the world over. The major activities generating contaminants are associated with the agricultural, mining, industrial and domestic sectors. Table 13.1 is a compilation of the sources and potential characteristics of groundwater contaminants. Box 13.2 discusses how contaminants are transported in porous media. The behaviour of toxic chemicals in groundwater is complicated by the physical and chemical characteristics of the contaminants involved. Boxes 13.3 and 13.4 describe the behaviour of non-aqueous phase liquids and heavy metals respectively in groundwater.

Figure 13.1: Point source pollution arising from the inadequate storage and subsequent rupture of barrels of lubricating oil additives. Such spillage of industrial chemicals contaminates land and is a serious threat to groundwater, particularly given that very low concentrations of synthetic toxic chemicals can render groundwater unfit for human consumption. In this photograph there is an apparently impervious hard surface, but in reality badly maintained concrete surfaces and surface drainage networks can permit direct access of aqueous pollutants to the groundwater environment.

Source: courtesy of R.P. Ashley.

Figure 13.2: An example of a pollution incident caused by the derailment of rail tankers containing oil. Noticeable warping of the tankers resulted from the ensuing fire. As with road transport accidents, runoffs either by gravity drainage to local surface water courses or by interception to soakaways are possible pathways for groundwater pollution.

Source: courtesy of R.P. Ashley.

Generally, many chemicals have had a long history of usage before becoming recognised as hazardous. During this time, handling and waste disposal practices have been frequently inadequate. Hence, it must be considered that any industrial site where hazardous materials have been used is now a potential source of contaminated land.

Groundwater contamination in Nassau County, New York

Infiltration of metal plating waste through disposal basins in Nassau County on Long Island, New York, has formed a plume of contaminated groundwater since the early 1940s (Figure 13.3). The plume contains elevated chromium and cadmium concentrations. The area is within an undulating glacial outwash plain, and there are two major hydrogeological units: the upper glacial aquifer of Late Pleistocene age; and the Magothy aquifer of Late Cretaceous age, which supplies all local municipal water supplies.

The upper glacial aquifer is between 24 and 43 m thick, with a water table from 0 to 8 m below ground level (Ku, 1980). The aquifer comprises medium to coarse sand and lenses of fine sand and gravel. Stratification of the deposits means that the vertical permeability is up to 5–10 times less than the horizontal permeability.

As part of the historical investigations in the South Farmingdale–Massapequa area, a number of test wells were installed in 1962 driven to depths ranging from 2.4 to 23 m below ground level. Water samples were collected at depth intervals of 1.5 m by hand pump during the driving operations. The results of this investigation are shown in

Continued on p.360.

Table 13.1: Potential sources of groundwater pollution arising from domestic, industrial and agricultural activities.

Contaminant source	Contaminant characteristics[1]
Septic tanks	Suspended solids 100–300 mg/l
	BOD 50–400 mg/l
	Ammonia 20–40 mg/l
	Chloride 100–200 mg/l
	High-faecal coliforms and streptococci
	Trace organisms, greases
Storm water drains	Suspended solids ~1000 mg/l
	Hydrocarbons from roads and service areas
	Chlorides or urea from de-icing
	Compounds from accidental spillages
	Bacterial contamination
Industry	
Food and drink manufacturing	High BOD, high suspended solids, colloidal and dissolved organic substances, odours
Textile and clothing	High suspended solids and BOD, alkaline effluent
Tanneries	High BOD, total solids, hardness, chlorides, sulphides, chromium
Chemicals	
acids	Low pH
detergents	High BOD
pesticides	High TOC, toxic benzene derivatives, low pH
synthetic resins and fibres	High BOD
Petroleum and petrochemical	
refining	High BOD, chloride, phenols, sulphur compounds
process	High BOD, suspended solids, chloride, variable pH
Plating and metal finishing	Low pH, high content of toxic metals
Engineering works	High suspended solids, hydrocarbons, trace heavy metals; variable BOD and pH
	Fly ash and flue gas scrubber sludges: low pH, disseminated heavy metals
Deep-well injection	Concentrated liquid wastes, often toxic brines; acid and alkaline wastes; organic wastes
Leakage from storage tanks and pipelines	Aqueous solutions, hydrocarbons, petrochemicals, sewage
Agriculture	
Arable crops	Nitrate, ammonia, sulphate, chloride and phosphates from fertilisers
	Bacterial contamination from organic fertilisers
	Organochlorine compounds from pesticides
Livestock	Suspended solids, BOD, nitrogen
	High-faecal coliforms and streptococci
Silage	High suspended solids, BOD 1–6×10^4 mg/l
	Carbohydrates, phenols
Mining	
Coal mine drainage	High TDS (total dissolved solids), suspended solids; iron; low pH; possibly high chloride
Metals	High suspended solids, possibly low pH, high sulphates, dissolved and particulate metals
Household wastes	High sulphate, chloride, ammonia, BOD, TOC and suspended solids from fresh wastes
	Bacterial contamination; on decomposition: initially TOC of mainly volatile fatty acids (acetic, butyric, propionic acids), subsequently changing to high molecular weight organics (humic substances, carbohydrates)

Note:

[1] BOD is biological oxygen demand, TOC is total organic carbon, pH is $- \log_{10}(H^+)$.

Source: adapted from Jackson (1980).

Box 13.2 Transport of contaminants in groundwater

The subsurface movement of a contaminant is influenced by the moisture content of the unsaturated zone and the volume of groundwater flow in the saturated zone below the water table, both of which are determined by climatic and topographical parameters.

The fundamental physical processes controlling the transport of non-reactive contaminants are advection and hydrodynamic dispersion. Advection is the component of solute movement attributed to transport by the flowing groundwater. Hydrodynamic dispersion occurs as a result of mechanical mixing and molecular diffusion, as illustrated in Figure 13.2.1.

The significance of the dispersive processes is to decrease the contaminant concentration with distance from the source. Figure 13.2.2 illustrates how a continuous pollution source will produce a

plume, whereas a single point source will produce a slug that grows with time as the plume moves in the direction of groundwater flow.

The idea of a homogeneous aquifer, in which the hydrogeological properties do not vary in space, is a simplification of the real situation in nature. Heterogeneities within the aquifer lithology will create a pattern of solute movement considerably different from that predicted by the theory for homogeneous material.

Reactive substances behave similarly to conservative solute species but can also undergo a change in concentration resulting from chemical reactions that take place either in the aqueous phase or as a result of adsorption of the solute to the solid matrix of the rock. The chemical and biochemical reactions that can alter contaminant concentrations in groundwater are acid–base

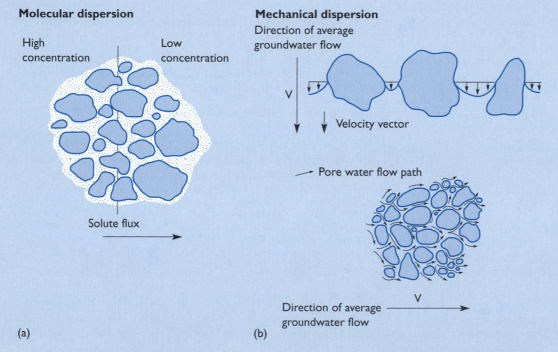

Figure 13.2.1: Diagrammatic representation of (a) molecular diffusion and (b) mechanical dispersion, which combine to transport solute within a porous medium by the process of hydrodynamic dispersion. Notice that mechanical dispersion results from the variation of velocity within and between saturated pore space and from the tortuosity of the flow paths through the assemblage of solid particles. Molecular diffusion can occur in the absence of groundwater flow, since solute transport is driven by the influence of a concentration gradient, while mechanical dispersion occurs while the contaminant is being advected by the groundwater.

Box 13.2 (continued)

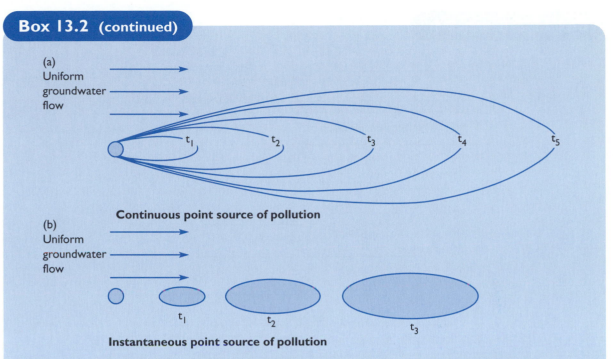

(a) Uniform groundwater flow

Continuous point source of pollution

(b) Uniform groundwater flow

Instantaneous point source of pollution

Figure 13.2.2: Dispersion within an isotropic porous medium of (a) a continuous point source of pollution at various times, *t*, and (b) an instantaneous point source of pollution. Spreading of the pollution plumes results from hydrodynamic dispersion, while advection transports the plumes in the field of uniform groundwater flow.

reactions, solution–precipitation reactions, oxidation–reduction reactions, ion pairing or complexation, microbiological processes and radioactive decay. Adsorption attenuates or retards a dissolved contaminant in groundwater.

The processes of advection, dispersion and retardation all influence the pattern of contaminant distribution away from the pollution source. If a pollution source contains multiple solutes and occurs within a heterogeneous aquifer, then there will be a number of contaminant fronts and the morphology of the resulting plume will be very complex indeed. Furthermore, a density contrast between the contaminant and the groundwater will affect the migration of the pollution plume, with a contaminant denser than water tending to sink steeply downwards into the groundwater flow field. Thus, the overall monitoring and prediction of the position of a groundwater pollution plume can be very difficult.

In fractured media, aquifer properties are spatially variable and are often controlled by the orientation and frequency of fractures. Information relating to contaminant migration in fractured rocks is limited. A common approach in field investigations and model simulations is to treat the problem as if it were an equivalent granular medium.

As Figure 13.2.3 shows, when contamination occurs in fractures, there is a gradient of contaminant concentration between the mobile groundwater in the fracture and the static water in the adjacent rock matrix. Under this condition, part of the contaminant mass will migrate by molecular diffusion from the fracture into the pore water contained in the rock matrix, so effectively removing it from the flowing groundwater. Such dual-porosity aquifers are notoriously difficult to remediate, since the contaminant stored in the matrix can gradually diffuse back into the moving groundwater in the fracture, long after the source of contamination has been removed.

Box 13.2 (continued)

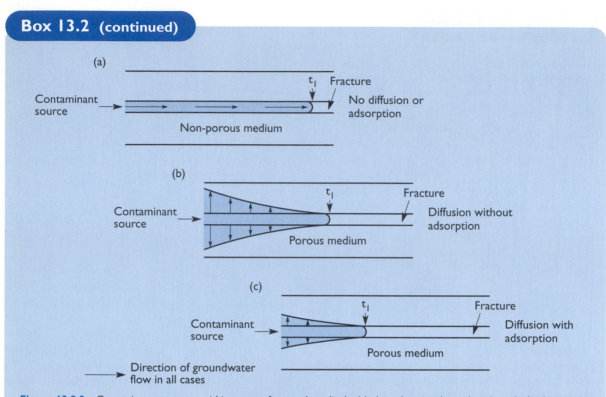

Figure 13.2.3: Contaminant transport within porous fractured media. In (a) the solute is advected, without hydrodynamic dispersion, with the groundwater flowing through a fracture where the matrix porosity is insignificant. In (b) the solute transport is retarded by the instantaneous molecular diffusion of the solute into the uncontaminated porous matrix. Further attenuation occurs in (c), where adsorption of a reactive solute occurs, accentuated by the greater surface area of contact resulting from migration of the solute into the porous matrix. The position of the leading edge of the contaminant front within the fracture is shown for time t_1 in each case.

Box 13.3 Non-aqueous phase liquid pollution of groundwater

Chlorinated solvents were developed in the early years of this century as a safe, non-flammable alternative to petroleum-based degreasing solvents in the metal processing industry. Until about 1970, trichloroethene (TCE) and tetrachloroethene or perchloroethylene (PCE) were predominantly used, the latter also in dry-cleaning application. Both solvents were recognised early on as potentially harmful to humans in both liquid and vapour forms, and in the 1960s both began to be replaced by the less toxic 1,1,1-trichloroethane (TCA) and 1,1,2-trichlorotrifluoroethane (Freon 113). From the mid-1970s, concern was expressed about the potentially carcinogenic effects of TCE, PCE and

carbon tetrachloride (CTC) at trace-level concentrations in drinking water, and the WHO set guide values of 30 µg/l for TCE, 10 µg/l for PCE and 3 µg/l for CTC.

Sources of solvent contamination arise during the delivery of solvents to the works, through use, and from the disposal of distillation sludge. Solvents escape to the subsurface environment mainly as a result of careless handling, accidental spillages, misuse, poor disposal practices and inadequately designed, poorly maintained or badly operated equipment.

Chlorinated solvents are dense liquids that are volatile and of low viscosity, and consequently

Box 13.3 (continued)

are more mobile than water in a porous medium. On infiltrating through the unsaturated zone, solvents leave behind a residual contamination that partitions into soil gas, which subsequently migrates upward and laterally by diffusion. The remaining contaminant mass migrates downwards and through the water table until halted by the base of the aquifer, or by some other intermediate impermeable barrier, as illustrated in Figure 13.3.1. Residual small amounts of solvent are left in the pore spaces through which the solvent body has passed. TCE and PCE are not completely insoluble, and the non-aqueous phase becomes surrounded by groundwater contaminated with dissolved solvent. TCE and PCE degrade extremely slowly,

and some of the degradation products may be more toxic, soluble and mobile than the parent compounds. For example, tetrachloroethane can be progressively de-halogenated, first to trichloroethane, then to dichloroethane, and finally to carcinogenic vinyl chloride.

Refined mineral oils include petrol, aviation fuel, diesel and heating oils. As a group, their physical characteristics are variable, particularly that of viscosity; but all have a density less than water and a heterogeneous composition dominated by pure hydrocarbons. In the context of groundwater, regulation is aimed primarily at taste and odour control. The WHO guide level is 10 µg/l. Sources of contamination include oil storage depots, cross-country

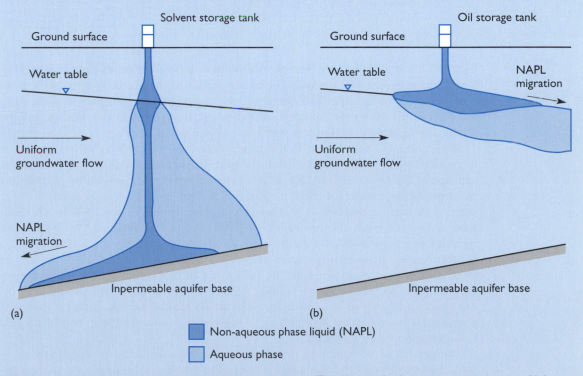

Figure 13.3.1: Behaviour of organic contaminants in groundwater. The chlorinated solvent contaminant shown in (a) has a density greater than that of water and so sinks to the base of the aquifer. Here, transport of the non-aqueous phase is controlled by the slope of the base of the aquifer, while the dissolved aqueous phase moves in the direction of groundwater flow. The hydrocarbon contaminant shown in (b) has a density less than that of water and so floats on the water table. In this case, transport of the non-aqueous phase is controlled by the slope of the water table, while the dissolved aqueous phase moves in the direction of groundwater flow.

Box 13.3 (continued)

oil pipelines, service stations, tanker transport and airfields.

Mineral oils behave in a similar manner to chlorinated solvents, except that, as shown in the figure, by reason of their density, they float on the water table. In this situation, migration is controlled by the water table gradient. Mineral oils are capable of degradation under aerobic conditions, although the rate of degradation is slow.

Box 13.4 Heavy metal pollution of groundwater

The heavy metals of concern in drinking water supplies are, to choose a few, nickel, zinc, lead, copper, mercury, cadmium and chromium. In the form of their reduced species and in acidic waters, heavy metals remain mobile in groundwater; but in soils and aquifers that have a pH buffering capacity, and under oxidising conditions, heavy metals are readily adsorbed or exchanged by clays, oxides and other minerals.

Heavy metal pollution of groundwater therefore becomes a particular threat under extreme conditions such as acidic mine waters, or in leachate beneath landfill sites where there is a high concentration of fatty acids. Other sources of heavy metals include, in general, the metal processing industries, particularly electroplating works with their concentrated acidic electrolytes and other metal surface treatment processes.

Figure 13.3, and define a pollution plume that is about 1300 m long, up to 300 m wide, and as much as 21 m thick. The upper surface of the plume is generally less than 3 m below the water table. The plume is thickest along its longitudinal axis, the principal path of flow from the basins, and is thinnest along its east and west boundaries. The plume appears to be entirely within the upper glacial aquifer.

Differences in chemical quality of water within the plume may reflect the varying types of contamination introduced in the past. In general, groundwater in the southern part of the plume reflects conditions prior to 1948, when extraction of chromium from the plating wastes, before disposal to the basins, was commenced. Since the start of chromium treatment, the maximum observed concentrations in the plume have decreased from about 40 mg/l in 1949 to about 10 mg/l in 1962. The World Health Organisation (WHO) guide value for chromium is 0.05 mg/l. Cadmium concentrations have apparently decreased in some places and increased in others, and peak concentrations do not coincide with those of chromium. These differences are probably due partly to changes in the chemical character of the treated effluent over the years and partly to the influence of hydrogeological factors. A test site near the disposal basins recorded as much as 10 mg/l of cadmium in 1964. The WHO guide value for cadmium is 0.005 mg/l.

The pattern of movement of the plating waste is vertically downwards from the disposal basins, through the unsaturated zone, and into the saturated zone of the upper glacial aquifer. From here, most of the groundwater moves horizontally southwards, with an average velocity of about 0.5 m/day, and discharges to the Massapequa Creek. Solute transport computer modelling indicates that with complete cessation of all discharges, it would take 7–11 years for the plume to move out of the area.

Analysis of cores of aquifer material along the axis of the plume showed that the median concentrations of chromium and cadmium per kilogram of aquifer material are

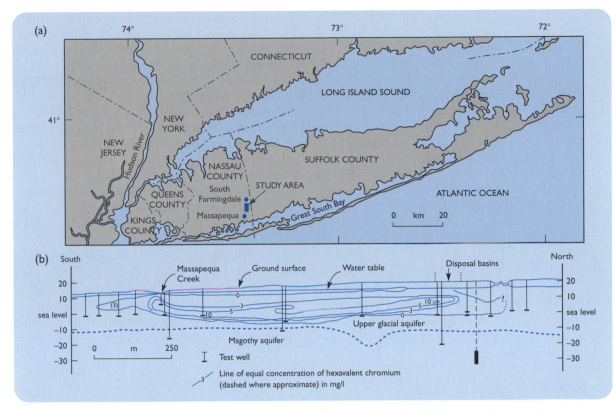

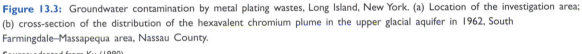

Figure 13.3: Groundwater contamination by metal plating wastes, Long Island, New York. (a) Location of the investigation area; (b) cross-section of the distribution of the hexavalent chromium plume in the upper glacial aquifer in 1962, South Farmingdale–Massapequa area, Nassau County.

Source: adapted from Ku (1980).

7.5 and 1.1 mg, respectively, and the maximum concentrations are 19 and 2.3 mg, respectively. Adsorption occurs on hydrous iron oxide coatings on the aquifer sands. The ability of the aquifer material to adsorb heavy metals complicates the prediction of the movement and concentration of the plume. Furthermore, metals may continue to leach from the aquifer material into the groundwater long after cessation of plating waste discharges, so necessitating continued monitoring of the site.

Nitrate contamination of the Jersey bedrock aquifer

The island of Jersey is the largest of the English Channel Islands. The island setting comprises a plateau, formed largely of Precambrian crystalline rocks, with a steep topographic rise along the coastline. The temperate maritime climate encourages early flowers and vegetables, with intensive agricultural production sustained by large fertiliser applications. During winter and early spring, applications of nitrogen fertiliser to early-cropping potatoes and horticultural crops may exacerbate the problem of nutrient leaching, with estimates of leaching losses of up to 100 kg N/ha expected from Jersey potato crops.

The groundwater resources are affected by both intense exploitation and surface pollution. Mains water supply is principally from surface water storage, but there are large areas which are neither on mains water supply nor on mains sewerage, particularly in the rural north of the island. In these areas, well and borehole supplies meet demands for domestic, agricultural and light industry uses. The main aquifer and isolated perched aquifers occur within a shallow zone of weathering in the bedrock, up to 25 m in depth below the water table surface, with groundwater flow almost entirely dependent on secondary permeability, imparted by dilated fractures.

The fractured bedrock aquifer has a piezometric level which is largely a subdued version of the surface topography. The depth to water is only a few metres beneath much of the island. Well and borehole yields are modest, typically <0.5 l/s. The chemical composition of groundwater is controlled by maritime recharge inputs and water–rock interaction, although the effects of anthropogenic pollution, particularly from nitrate, are in places severe. The combination of intensive agriculture, high livestock densities, shallow groundwater levels and near-surface fissure flow presents a high vulnerability to groundwater contamination by nitrate. In some areas, domestic pollution from septic tank discharges is a further potential hazard.

A 1995 survey of groundwater quality at 46 locations across the island produced the regional distribution of nitrate shown in Figure 13.4 (Green et al., 1997). Elevated nitrate concentrations occur across much of the island and range from undetected to 215 mg/l, with a mean value of 71 mg/l. Of the total, 67% of samples exceed the European Commission maximum admissible concentration for nitrate in drinking water of 50 mg/l. The high nitrate concentrations appear to decrease in a coastal direc-

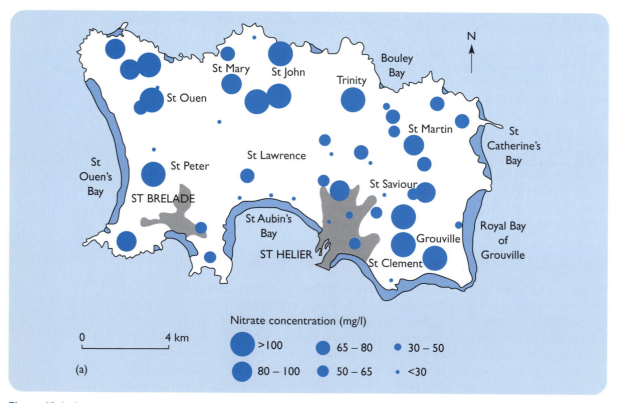

Figure 13.4: Location map of Jersey showing groundwater sampling points and the regional distribution of groundwater nitrate concentrations in the bedrock aquifer in June 1995.

Source: after Green *et al.* (1997).

tion, especially in those samples from the central southern area and in the valley areas around St Saviour. Otherwise, local variations in well and borehole depths and local land use and agricultural practice, combined with the physical heterogeneity of the aquifer, produce no obvious pattern in the distribution of nitrate. It is concluded that the source of dissolved nitrate in the Jersey bedrock aquifer is primarily the intensive agricultural and horticultural practices employed on the island.

Further consideration of the coastal and valley bottom areas would suggest progressive mixing between shallow, rapidly circulating, high-nitrate groundwater with a slowly circulating, more evolved and denitrified water found deeper in the aquifer. Denitrification is conjectured to proceed as a heterotrophic reaction with the oxidation of organic matter followed by autotrophic denitrification involving reduced iron and sulphur species. The likely sources of the latter are ferrous iron- and sulphide-bearing minerals present in the aquifer material. Compared with the contamination of groundwater by nitrate across the island, the process of denitrification is not significant, except for local coastal and valley areas, and is therefore unlikely to provide a long-term natural remediation measure.

Jet fuel clean-up at Heathrow Airport

This case study concerns hydrocarbon contamination of groundwater adjacent to Technical Block L at Heathrow Airport. Heathrow Airport is built on the Taplow Terrace adjacent to the River Thames floodplain. The geology is formed by 4.5 m of coarse clean gravels overlying low-permeability London Clay. The water table is shallow, about 2.5 m from the surface, with groundwater flow southwards beneath the airport towards the River Thames at Shepperton. A leak of jet fuel (kerosene: a light, non-aqueous phase liquid, or LNAPL) occurred from a cracked fuel pipe leading to an engine maintenance facility, the leak having occurred over a number of years. The leak was discovered when fuel was observed floating on drainage water in a manhole north of Technical Block M (Figure 13.5). In response, a large concrete-lined well (well 1), about 1.5 m in diameter, was installed close to the manhole and revealed about 10 cm of kerosene floating on the water table. As a first step in remediating the contaminated site, the leak was traced to the cracked pipe and the fracture repaired.

A detailed site investigation, including the installation of 14 monitoring boreholes, showed that the thickest measured depth of floating kerosene was 0.95 m in borehole 5, with further 'free product' measured in boreholes 1, 11 and 13 and in well 1 (Clark and Sims, 1998). Odour was reported during the drilling of boreholes 3, 9 and 10, indicative of kerosene. No kerosene has been detected in the outlying observation boreholes since monitoring began, including borehole 12.

The basic remediation structures used included large-diameter wells lined with perforated concrete rings about 1.5 m in diameter. Wells 1 and 2, installed close to borehole 1, where a considerable thickness of fuel was shown to be floating on the water table, were used to begin recovery of the floating kerosene. The kerosene was removed by floating oil-skimmer pumps. Surface-mounted centrifugal pumps, installed with their intakes in the two wells, were also used to lower the water table and encourage the kerosene to move towards the recovery wells.

Groundwater levels were monitored regularly, and within two months the cone of depression in the water table produced by the pumping of wells 1 and 2 encompassed the estimated area of the kerosene 'pancake'. Initially, the recovery rate was such that 19 200 litres of kerosene was removed. The recovery rate then dropped substantially, yet the kerosene layer in borehole 11 still remained unaltered, suggesting that wells 1 and 2 were

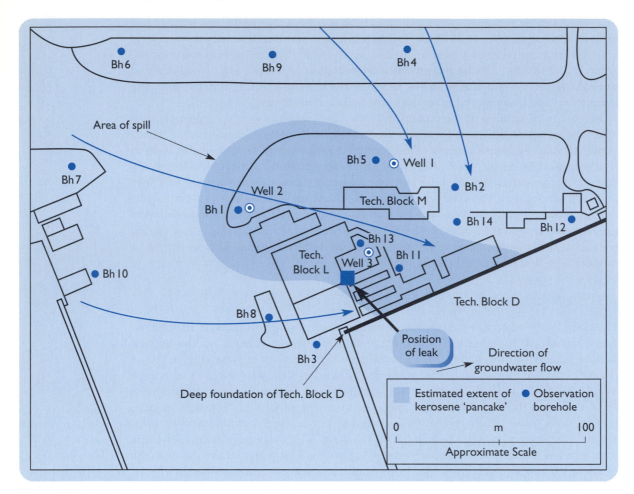

Figure 13.5: Site of jet fuel leak adjacent to Technical Block L at Heathrow Airport showing the estimated extent of the kerosene 'pancake' resting on the gravel aquifer water table. The thickest measured depth of floating kerosene was 0.95 m in borehole 5, with further 'free product' measured in boreholes 1, 11 and 13 and well 1.
Source: after Clark and Sims (1998).

not affecting the southern part of the kerosene 'pancake'. Later, well 3 was installed. It is believed that well 3 tapped a 'pool' of kerosene isolated from the effects of wells 1 and 2 by the foundations of the technical blocks. A further 10 100 litres of kerosene were removed from well 3, making a total recovery of 29 300 litres of kerosene in four years. The removal of the original kerosene 'pancake' is considered to be complete.

Organic solvent contamination at an industrial site

Time spent on assessing the likely distribution of subsurface pollution by methods other than the drilling of expensive boreholes can often be cost-effective. Such methods may include investigations into the locations of chemical use and storage, documented spills, surface cover and local geology, and low-cost reconnaissance. For example, chlorinated organic solvents are volatile compounds (see Box 13.3), and reconnaissance mapping of soil gas solvent concentrations may provide useful information for further site investigations.

The soil gas survey technique has been applied at an urban site in the English Midlands, where chlorinated solvents have been used for many years for cold wipe and soak and vapour degreasing of metal (Bishop *et al*., 1990). Trichloroethene (TCE) use probably started in the 1930s. Solvent use switched from TCE to 1,1,1-trichlorothoethane (TCA) around 1980, although small quantities continued to be used until 1987. In the first half of 1989, 18 600 litres of TCA was used at locations across the site. The only known chlorinated solvent pollution at the site relates to groundwater from an abstraction borehole.

Three bulk storage tanks are located above ground near the northern end of the site (Figure 13.6). Two are disused TCE and oil storage tanks and one is a tank used for TCA storage. Barrels of TCA are filled manually at the tank. The ground in the area of the solvent storage tanks consists of approximately 0.1 m of concrete underlain by 0.2 m of made-ground and 5 m of weathered sandstone.

An initial survey of soil gas solvent concentrations revealed an area of gross soil contamination around the bulk storage tanks. A detailed follow-up survey around the solvent tanks produced the contoured TCE and TCA soil gas concentrations shown in Figure 13.6. Pollution by TCA is more severe than by TCE, reflecting the replacement of TCE with TCA in recent years. Both TCE and TCA soil gas surveys show an apparent pollution plume extending approximately north-west from the storage tanks. The identified pollution plume may have two explanations: first, partitioning of chlorinated solvents from contaminated, possibly perched, groundwater flowing in this direction; and second, direct migration of chlorinated solvent vapours through the unsaturated zone, indicating widespread soil contamination.

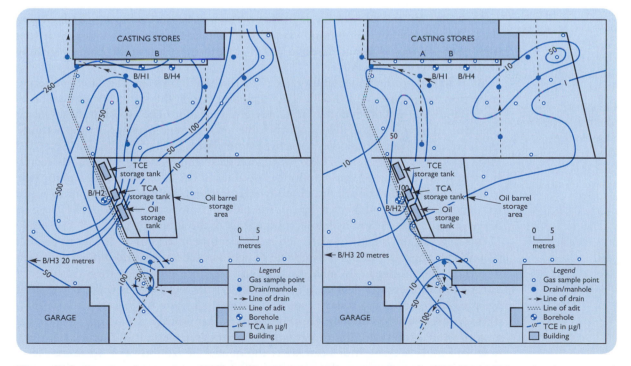

Figure 13.6: Site plans showing disused TCE (trichloroethene) and oil storage tanks and a TCA (1,1,1-trichloroethane) storage tank at an industrial site in the English Midlands. Contoured soil gas concentrations in μg/l of (a) TCA and (b) TCE are shown following a detailed soil gas survey.

Source: after Bishop *et al*. (1990).

In the case of the first explanation, a 5 m thickness of weathered sandstone beneath the hardcore surface is known to have variable amounts of perched groundwater. Analysis of groundwater in investigation borehole 1 indicated mean TCE and TCA concentrations of 523 µg/l and 3890 µg/l, respectively. The two points A and B (see Figure 13.6) analysed for soil gas adjacent to borehole 1 had mean TCE and TCA concentrations of 1 µg/l and 134 µg/l, respectively. High vertical concentration gradients are frequently observed for chlorinated solvent vapour in the unsaturated zone and these concentrations, at about 1 m depth below the surface, could arise from partitioning from polluted groundwater. Equally, the second explanation argues that direct vapour migration in the unsaturated zone may be the origin of the pollution plume. In the highly disturbed ground beneath the industrial site, vapour movement could easily be concentrated along channels originally excavated to locate buried services. The permeable material used to backfill such drains and pipelines has often been known to act as a conduit to laterally diffusing chlorinated solvents.

Groundwater protection policies in industrialised countries

The science of contaminant hydrogeology has largely been driven by both developments in analytical procedures for identifying pollutants and the need to formulate water quality standards and aquifer protection policies. The pace of development also reflects the concerns of individual countries. For example, in the USA during the early 1970s, the public concern for hazardous waste sites and contaminated land led to the introduction in 1980 of the Comprehensive Environmental Response, Compensation and Liability Act: the 'Superfund' legislation. The legislation also embraced concerns about the deterioration of groundwater quality in states such as New Jersey arising from point sources of industrial pollution in situations, typical in the USA, of shallow aquifers with little protection.

To evaluate groundwater vulnerability in the United States, the Environmental Protection Agency uses a methodology designed to permit the systematic evaluation of the groundwater pollution potential for any given location. The system has two major components: first, the designation of mappable units, termed hydrogeological settings; and second, the superposition of a relative rating system having the acronym DRASTIC (Aller *et al.*, 1987). Inherent in each hydrogeological setting are the physical characteristics that affect groundwater pollution potential. The most important mappable factors considered to control the groundwater pollution potential are depth to water (D), net recharge (R), aquifer media (A), soil media (S), topography (slope) (T), impact of the vadose zone (I), and hydraulic conductivity of the aquifer (C).

To safeguard the quality of water intended for human consumption and to protect groundwater against pollution by certain dangerous substances and diffuse pollution by nitrate, the European Community has enacted controlling legislation (Council of European Communities 1980a, 1980b and 1991). In the UK, it is the Environment Agency which, under the Water Resources Act 1991 (Section 84), has the responsibility to protect the quality of groundwater. In promoting a national policy for groundwater protection, the Environment Agency is seeking to influence the decisions of others whose actions can affect the quality of groundwater, for example in response to consultation under planning legislation. The policy is directed towards mapping regional groundwater vulnerability and also defining groundwater source protection zones (National Rivers Authority, 1992). The methodology in each case is based on hydrogeological principles as outlined in Box 13.5.

Box 13.5 Groundwater vulnerability maps and protection zones

In the implementation of national policy, the Environment Agency is producing a series of 53 groundwater vulnerability maps, covering the whole of England and Wales. This approach, as illustrated in Figure 13.5.1, defines vulnerability as a function of the nature of the overlying soil, the presence and nature of any overlying superficial or glacial deposits, the nature of the geological strata forming the aquifer, and the thickness of the saturated zone or thickness of confining beds. The vulnerability maps are being produced from the overlay of information on hydrogeology and soils at a scale of 1:100 000.

The value of regional maps based upon vulnerability classes is their provision of information for planning purposes. Properly used, they can encourage the development of potentially polluting practices in those areas where it will present least concern, allowing a balance of interests to be achieved. The regulation of diffuse pollution can be readily related to aquifer vulnerability zones.

It is recognised that such maps must always be interpreted with caution, since in strict terms, the concept of a 'general vulnerability to a universal pollutant in a typical pollution scenario' is scientifically meaningless. It is more consistent to evaluate vulnerability to pollution by each contaminant or, failing this, by each class of contaminant or by each class of polluting activity. A general limitation of groundwater vulnerability maps is that they provide a regional picture that is insufficiently detailed to demonstrate the actual threat to the groundwater resource at a local scale. The true vulnerability can be established with confidence only through supporting, site-specific ground investigations.

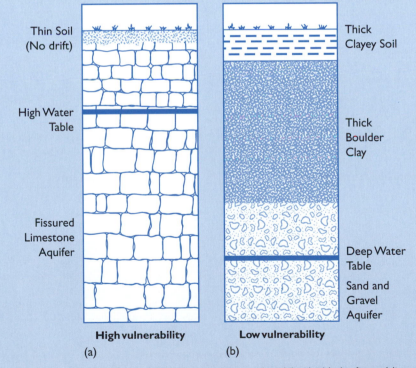

High vulnerability
(a)

Low vulnerability
(b)

Thin Soil (No drift)

High Water Table

Fissured Limestone Aquifer

Thick Clayey Soil

Thick Boulder Clay

Deep Water Table

Sand and Gravel Aquifer

Figure 13.5.1: Illustration of two situations of differing groundwater vulnerability. In (a) the fissured limestone aquifer with a permeable soil cover, no superficial drift deposits and high water table is an example of very high vulnerability, whereas in (b) the sand and gravel aquifer, overlain by low-permeability clay cover and soil, is much less vulnerable.

Source: after National Rivers Authority (1992).

Box 13.5 (continued)

In the definition of groundwater source protection zones, the proximity of a hazardous activity to a point of groundwater abstraction (including springs, wells and boreholes) is one of the most important factors in assessing the risk to an existing groundwater source. In principle, the entire recharge area in the vicinity of the wellfields should be protected, but this is unrealistic on socio-economic grounds. In this situation, a system of zoning of the borehole recharge area, or the protection area, is desirable. In the UK, the Environment Agency recognises three source protection zones. Figure 13.5.2 shows that the orientation, shape and size of the zones are determined by the hydrogeological characteristics of the strata and the direction of groundwater flow. Effort is currently directed towards the use of steady-state groundwater flow computer modelling to define protection zones around sources used for public supply (there are nearly 2000

major sources), other private potable supply (including mineral and bottled water) and in commercial food and drink production.

Zone I, or the inner source protection zone, is located immediately adjacent to the groundwater source and is designed to protect against the effects of human activity which might have an immediate effect upon the source. The area is defined by a 50-day travel time from any point below the water table to the source and as a minimum 50-metre radius from the source. This approach is used in other countries and is based on the time taken for biological contaminants to decay in groundwater. The land immediately adjacent to the source and controlled by the operator of the source is included within this zone.

Zone II, or the outer protection zone, is the area around the source defined by a 400-day travel time, and is based on the requirement to provide delay and attenuation of slowly degrad-

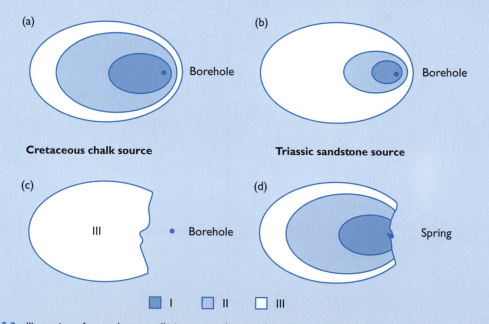

Figure 13.5.2: Illustration of groundwater pollution zones showing the relationship between Zones I, II and III and the groundwater source in four idealised hydrogeological situations: (a) represents a low effective porosity limestone aquifer; (b) a high effective porosity sandstone aquifer; (c) a confined aquifer; and (d) a spring. In reality, the size, shape and relationship of the zones will vary significantly depending on the soil, geology, amount of recharge and volume of water abstracted.

Source: after National Rivers Authority (1992).

Box 13.5 (continued)

ing pollutants. In high groundwater storage aquifers, such as the Permo-Triassic sandstones of the English Midlands, it is necessary, in order to provide adequate attenuation, to define further the outer protection zone to be the larger of either the 400-day travel isochron or the recharge catchment area calculated using 25% of the long-term abstraction rate for the source.

Zone III, or the source catchment, is the remaining catchment area of a groundwater source and is defined as the area needed to support an abstraction from long-term annual groundwater recharge (effective rainfall). For wells and boreholes, the source catchment area is defined by the authorised abstraction rate, while for springs it is defined by the best known value of average annual total discharge. In practice, the size of Zone III will vary from tens to a few thousands of hectares, depending on the volume of groundwater abstraction and the amount of recharge. In areas where the aquifer is confined beneath impermeable cover, the source catchment may be some distance from the actual abstraction.

The UK approach to implementing a groundwater protection policy for the fissured, porous limestone and sandstone that comprise the major aquifers is an adaptation of previous European experience with unconsolidated, porous aquifers (Van Waegeningh, 1985). For example, in the Netherlands, abstraction of drinking water supplies is concentrated within 240 wellfields tapping mainly uniform, horizontally layered aquifers of unconsolidated sands and clays. The zoning system includes a first zone based on a delay time of 50 days to protect against pathogenic bacteria and viruses and rapidly degrading chemicals and extends some 30–150 m from an individual borehole. For the sake of the continuity of water supply in the event of a severe pollution incident requiring remedial action, and in order to exclude public health risks, a delay time of at least 10 years is needed in the next zone. In many cases, even 10 years is not sufficient to guarantee the continuity of safe water supplies, and a protection zone of 25 years is necessary. The 10- and 25-years protection zones, extending to about 800 and 1200 m from the borehole, respectively, constitute the source protection area.

Groundwater pollution and protection in developing countries

Groundwater is extensively used for drinking water supplies in developing countries, especially in smaller towns and rural areas, where it is the cheapest and safest source. Often it is the willingness to pay for cheap and clean water that governs progress in providing better sanitation in poorer countries (see Box 13.6).

In developing countries, groundwater schemes consist of large numbers of boreholes, often drilled on an uncontrolled basis, providing untreated, unmonitored and often unconnected supplies. Shallower dug wells continue to be constructed in some cases. Better-yielding boreholes (10–100 l/s) are quite widely developed in larger towns to provide piped supplies. Even in these cases, raw water monitoring and treatment are often limited and intermittent.

In developing countries, unsewered, ventilated and pour-flush pit latrines provide adequate excreta disposal in rural areas, villages and small towns at a much lower cost than main sewerage systems. Consequently, the possibility of an expansion of excreta disposal into the ground is very real, especially in many Asian countries, where many people are without any form of sanitation.

Box 13.6 Willingness to pay for clean water

It is evident that, where people have the choice, they are prepared to pay for clean water and good sanitation for reasons of health, convenience, privacy and stable property prices. Studies in poor countries in Africa show that urban people already pay about 2% of family income on inadequate water and sanitation services. This is similar to the amount paid for electricity. Nowadays developers are being required to provide packaged sewage treatment services as part of a permit to construct housing in cities in even the poorest countries. A World Bank study showed that less than 10% of sewage connections are actually being made, despite regulations. The answers lie in widening the provision of low-technology sewage provision, separating the regulator from the provider, ensuring performance cash bonds are paid beforehand to stop financial defaulting, and enabling the private sector to bid for projects. In this way, health-debilitating groundwater pollution in the cities of many developing countries can be reduced.

The natural soil profile can be effective in purifying human wastes, including the elimination of faecal microbes, and also in the adsorption, breakdown and removal of many chemicals. However, in some cases, for example where thin soils are developed on aquifer outcrops, there is the risk of direct migration of pathogenic microbes, especially viruses, to adjacent groundwater sources. The inevitable result will be the transmission of waterborne diseases.

A further problem with excreta is the organic nitrogen content, which can cause widespread and persistent problems of nitrate in water, even where dilution and biological reduction processes occur. The problem is exacerbated in arid areas without significant regional aquifer flow to provide dilution.

Another source of pollution risk arises from the use of inorganic fertilisers and pesticides in the effort to secure self-sufficiency in food production. The use of irrigation to provide crop moisture requirements poses the risk of leaching losses of nutrients if not carefully managed, especially from thin, coarse-textured soils. The use of wastewater for irrigation may contribute to groundwater salinity and nitrate concentrations, and possibly micropollutants. Increases in chloride, nitrate and perhaps trace elements will result from excessive land application of sewage effluent and sludge or animal slurry.

In many developing countries, extensive urban areas are unsewered, yet it is these areas where an increasing number of small-scale industries, such as textiles, metal processing, vehicle maintenance and paper manufacture, are located. The quantities of liquid effluent generated by each will generally be discharged to the soil; especially in the absence of specific control measures and the prohibitive cost of treatment processes. Larger industrial plants generating large volumes of process water will commonly have unlined surface impoundments for the handling of liquid effluents.

Protection of groundwater supply sources requires a broad-based pollution control policy comprising:

- minimum separations, depending on the hydrogeological environment, between groundwater supply source and excreta disposal unit for microbiological protection; and
- dilution zones of modified land use to alleviate the impact of polluting land-use activities (Foster, 1985).

The water laws and codes of practice of many countries require a minimum spacing between excreta disposal unit and groundwater supply source of 15 m, under favourable hydrogeological conditions. Some countries have selected a greater distance, for example 200 m in Malawi water law. There is, however, considerable pressure to reduce this permitted spacing to as little as 5 m in some developing countries such as Bangladesh and parts of India and Sri Lanka, often resulting from the lack of space in very densely populated settlements.

This one example of law governing the location of excreta disposal units demonstrates that criteria for groundwater pollution protection is rather arbitrary, based on limited or no technical data. In the future, a more comprehensive, flexible and widely applicable policy to groundwater protection is needed, specifically related to aquifer conditions and human activities (Foster, 1985).

The water quality criteria for the case of groundwater pollution protection in developing countries need questioning. In the case of organic micropollutants, precise limits can be set only when there is adequate medical evidence of the toxicological effects. However, in other cases, the existing WHO recommendations are unnecessarily stringent in the light of the disproportionate cost of attaining such standards in relation to other public health risks. For example, the WHO guide value for nitrate in drinking water supplies in tropical countries is currently 10 mg N/l, but it is argued that concentrations up to 22.6 mg N/l are permissible under most circumstances.

Conclusions

The study of groundwater pollution is a multidisciplinary subject requiring skills in physical and chemical hydrogeology, theoretical and practical experience of pollution incidents, and the understanding of the activities and processes giving rise to the pollution. Knowledge of the surrounding environment, particularly soil type and climatic conditions, is necessary in predicting the vulnerability of any underlying aquifer to groundwater contamination.

In translating these skills and knowledge into policies aimed at protecting groundwater resources, the cost of protecting water supplies must be balanced against other socio-economic factors, and whether the protection policy is to be defined for either a developing or an industrialised country.

Groundwater problems can be solved only by careful analysis of actual ground conditions and infrastructure immediately below a site, followed by accurately positioned monitoring boreholes (or by the employment of rapid survey techniques such as soil gas analysis) to provide a detailed view of the spatial distribution of the pollutant. Even then, any pollution plume emanating from a point source may not be properly understood, particularly in strongly heterogeneous media or in situations where subsurface structures, such as building foundations, complicate the groundwater flow regime and its investigation.

Once the extent of pollution is assessed, clean-up of the site can commence, but again problems of non-ideal conditions and deciding at what point the aquifer is deemed restored arise. In future years, the legacy of past polluting practices will engage regulators and clean-up agencies in considerable lucrative work in locating and dealing with groundwater contamination.

Importantly, and as demonstrated by this chapter, no groundwater pollution investigation or protection policy strategy should be undertaken or formulated without first having a clear understanding of the geology of a region. Geological properties are fundamental in governing the nature of groundwater flow. In particular, the flow mechanism, whether it be intergranular or within fractures, must be firmly established, as well as the aquifer boundaries and groundwater conditions determined by geological structure. Such considerations should be uppermost in the minds of groundwater practitioners and administrators alike.

References

Aller, L., Bennett, T., Lehr, J.H., Petty, R.J. and Hackett, G., (1987) DRASTIC: a standardized system for evaluating groundwater pollution potential using hydrogeologic settings. *NWWA/EPA Series*. EPA/600/2-87/035. Washington, DC.

Bishop, P.K., Burston, M.W., Lerner, D.N. and Eastwood, P.R. (1990) Soil gas surveying of chlorinated solvents in relation to groundwater pollution studies. *Quarterly Journal of Engineering Geology*, **23**, 255–265.

Clark, L. and Sims, P.A. (1998) Investigation and clean-up of jet-fuel contaminated groundwater at Heathrow International Airport, UK. In J. Mather, D. Banks, S. Dumpleton and M. Fermor (eds) *Groundwater Contaminants and their Migration*. Special Publication of the Geological Society of London, **128**, 147–157.

Council of European Communities (1980a) *Directive relating to the quality of water intended for human consumption (80/778/EEC)*. Official Journal of the European Communities, L229, Brussels.

Council of European Communities (1980b) *Directive on the protection of groundwater against pollution caused by certain dangerous substances (80/68/EEC)*. Official Journal of the European Communities, L20, Brussels.

Council of European Communities (1991) *Directive concerning the protection of waters against pollution caused by nitrates from agricultural sources (91/676/EEC)*. Official Journal of the European Communities, L375, Brussels.

Foster, S.S.D. (1985) Groundwater protection in developing countries. In G. Matthess, S.S.D. Foster and A.C. Skinner (eds), *Theoretical Background, Hydrogeology and Practice of Groundwater Protection Zones*. Verlag Heinz Heise, Hanover, 167–200.

Green, A.R., Feast, N.A., Hiscock, K.M. and Dennis, P.F. (1997) Identification of the source and fate of nitrate contamination of the Jersey bedrock aquifer using stable nitrogen isotopes. In N.S. Robins (ed.), *Groundwater Pollution, Aquifer Recharge and Vulnerability*. Special Publication of the Geological Society of London, **130**, 23–35.

Jackson, R.E. (ed.) (1980) *Aquifer Contamination and Protection*. UNESCO, Paris.

Ku, H.F.H. (1980) Ground-water contamination by metal-plating wastes, Long Island, New York, USA. In R.E. Jackson (ed.), *Aquifer Contamination and Protection*. UNESCO, Paris, 310–317.

National Rivers Authority (1992) *Policy and Practice for the Protection of Groundwater*. National Rivers Authority, Bristol.

Van Waegeningh, H.G. (1985) Protection of groundwater quality in porous permeable rocks. In G. Matthess, S.S.D. Foster and A.C. Skinner (eds), *Theoretical Background, Hydrogeology and Practice of Groundwater Protection Zones*. Verlag Heinz Heise, Hanover, 111–121.

Further reading

There are a number of textbooks dealing with the physical and chemical principles of hydrogeology which also contain sections on contaminant hydrogeology. Of these, R.A. Freeze and J.A. Cherry (1979), *Groundwater* (Prentice-Hall, New Jersey), and P.A. Domenico and F.W. Schwartz (1990), *Physical and Chemical Hydrogeology* (Wiley, New York), are comprehensive treatments of the subjects. C.W. Fetter (1999), *Contaminant Hydrogeology* (2nd edn) (Macmillan, New York), deals specifically with groundwater contamination. A more easily understood text for those requiring a rigorous, yet immediately accessible, read is M. Price (1996), *Introducing Groundwater* (2nd edn) (Chapman & Hall, London). A discussion of the theoretical and practical aspects of aquifer contamination and protection, together with a compendium of case studies of groundwater pollution, is contained in a UNESCO publication edited by R.E. Jackson (1980). The International Association of Hydrogeologists publishes two practice volumes on defining groundwater protection zones and mapping groundwater vulnerability. The first, edited by G. Matthess, S.S.D. Foster and A.C. Skinner (1985) is *Theoretical Background, Hydrogeology and Practice of Groundwater Protection Zones*, and the second, edited by J. Vrba and A. Zaporozec (1994), is *Guidebook on Mapping Groundwater Vulnerability*. Both books are published by Verlag Heinz Heise, Hanover, and are recommended for those approaching these topics for the first time.

Also see J. Briscoe (1993), 'When the cup is half full: improving water and sanitation services in the developing world', *Environment*, **35**(4), 6–10, 28–36; and K.M. Hiscock, A.A. Lovett, J.S. Brainard and J.P. Parfitt (1995), 'Groundwater vulnerability assessment: two case studies using GIS methodology', *Quarterly Journal of Engineering Geology*, **28**, 179–194.

Chapter Fourteen

Marine and estuarine pollution

Alastair Grant and Tim Jickells

Editorial introduction

Much of this chapter is an exercise in exploration and discovery, because we now know that the oceans play a vital part in absorbing and emitting atmospheric gases, dissipating energy into the northern and southern latitudes, regulating biogeochemical cycles, and providing vast reservoirs of life in complex food chains (see also Chapter 8). The oceans are probably more important for regulating the biosphere than many realise, hence the need for continued scientific monitoring of the physics and chemistry of ocean–atmosphere relationships.

Yet monitoring marine ecology is fraught with difficulty. When plankton blooms, as in the North Sea case reported in this chapter, it is very tempting to blame human agency. But unless there is a good historical record, or unless we have comprehensive monitoring and screening, it is quite possible that plankton may have been subject to regular (or irregular) blooms, or that changes in the north-east Atlantic climate may be a significant cause of environmental stress. In short, we do not know. Toxicological studies of a chemical cost $10 000 just for measures of bioconcentration and acute toxicity, and $100 000–1 000 000 for a fuller, but still incomplete, assessment. Clearly the toxicity of chemical interactions in food chains will only be partially understood.

Species interactions are so complicated, so rooted in relationships that are not well understood, that only an exhaustive and expensive monitoring programme could unravel all the connections. This is obviously impossible except in the most precious and confined of ecosystems. Ecologists tend to look for indicator or keystone species which provide a measure of conditions favouring or afflicting other species. But sadly few species are truly indicator species, and even fewer are reliable indicators from place to place and over time. So the very basis of ecological monitoring is put in question. Add to that subtle changes in environmental conditions caused by cyclical variations in energy, or unusual additions of human interference, and you will readily appreciate how troublesome ecological prediction is becoming. Finally, we are beginning to realise just how random, or chaotic, population fluctuations actually are. McGarvin (1994) cites a study of barnacles which indicated that the number of adults was determined by the number of larvae that survived their plankton stage and settled onto the rocks. These populations vary widely every year and cannot be predicted with any accuracy without very large databases running for hundreds of years.

This has profound implications, not just for theoretical ecology but also for pollution control policy and international co-ordination. For example, the variability of nationally occurring heavy metals is quite large, but not fully known. It makes no sense to identify threshold levels of anthropogenic inputs of these metals, because the thresholds or critical loads may never be known with any accuracy.

In this chapter, there is continual reference to the first level of scientific uncertainty outlined in Chapter 1, namely lack of historical data. Inputs of pollutants to rivers, from rivers to estuaries, and in concentrated form in currents and sediments, are all highly problematic for predictor modelling. Even the most sophisticated models may fail to cover for these data gaps. The implications for expensive commitments to pollution control schemes are obviously serious.

All this implies that even greater weight will be given to the precautionary principle in the years ahead. This will force pollution control agencies to adopt more rigorously the common burden principle of equivalence of effort, regardless of the 'scientific' justification for differential treatment. It will also place much more reliance on 'up the pipe' pollution prevention at source, utilising techniques such as life-cycle analysis and whole-firm environmental management services. The introductory essay to this section suggests a transition that has barely begun to run its course.

The Rio process, outlined in Chapter 2, has still failed to reach consensus over a Convention on the Oceans. This is partly because there is already some element of international agreement over the Law of the Sea, as summarised in Chapter 8. But the major drawback is that negotiators cannot agree on property rights in the oceans generally. This extends to fish, minerals and responsibilities for shipping. The contentious property right issue also extends to the responsibility for the protection of the air–gas exchanges on the active ocean surface. This point is covered by Michael Meredith and co-authors in Chapter 8, as well as by Tim Lenton (1998). The issue here is more important than it appears, as the climate-modulating effect of this surface layer has potentially great economic advantage. This particular aspect, namely the value to society of a biogeochemical function that has no 'market value', is given coverage in Chapter 4. The promised UN convention would be truly trailblazing if it took this approach into account. The science of ocean regulation of climate is reasonably well advanced. It would be greatly assisted if there were to be a serious effort to safeguard this active interface in the name of moderating the possible long-term effects of climate change. Providing some valuation of this service would also help to place a more complete interpretation of the cost of carbon loading in the post-Kyoto negotiations, as noted in Chapter 7. Pollution that damages the ocean surface may be more expensive than we currently imagine.

Introduction

Pollution of oceans, coastal waters and estuaries is an issue of great public concern. It ranks highest in public opinion surveys, with over 90% of respondents stating that it is a serious or very serious problem. The subject was brought to the forefront of public attention, at least in Europe, by two phenomena which occurred in the summer of 1988, a viral epidemic which killed large numbers of seals and a bloom of toxic algae in Scandinavian waters. More recent events, such as the argument over the disposal of the *Brent Spar* oil platform (beginning in spring 1995) and the wreck of the *Sea Empress* oil tanker at the entrance to Milford Haven (February 1996), have kept these issues in public view. See Chapter 8 for a discussion of the science and politics of *Brent Spar*.

In April 1989, near the Danish island of Anholt, nearly 100 pups of the common or harbour seal were born prematurely. Shortly afterwards, sick adult seals were observed on the Danish and Swedish coasts of the Kattegat, and dead bodies began to be washed up on beaches. The epidemic then spread rapidly round the Danish coast and along the Waddensee, reaching the UK in August, where the first cases were observed in the Wash. By October, dead and dying seals were being found along much of the UK coastline. The immediate cause of the deaths was a virus related to canine distemper and named phocine distemper virus (PDV). More than 16 000 dead common seals were washed up on beaches, and best estimates are that at least half the population of this species died. The grey seal was less severely affected, and only about 200 bodies were reported. The big doleful eyes of seals have always attracted public sympathy. Pictures of large numbers of bodies being removed from beaches were regularly featured on television news, and the events achieved a very high level of public awareness. Fundraising appeals by newspapers in the UK and West Germany raised hundreds of thousands of pounds to save the seals, though whether the tabloid press had any viable proposals to tackle the epidemic was never evident.

On 9 May 1988, the owner of a rainbow trout farm in Gullmarfjorden near Gothenburg on the west coast of Sweden had reported that his fish were showing signs of stress. He observed a yellowish colour to the water, which proved to be due to a small flagellated alga, *Chrysochromulina polylepis*. Over the next three weeks, the algal bloom drifted north and west along the Swedish and Norwegian coasts, reaching the coast off Stavanger by 26 May. By mid-June the bloom had dissipated. The bloom caused deaths of caged salmon and trout at several farms in Norway and Sweden, and divers reported extensive mortality of marine invertebrates. Some wild fish were killed, but in general these were able to avoid the bloom, which was concentrated in the surface layers.

The immediate assumption of many was that these two phenomena were the consequence of years of discharging sewage and other effluents to the North Sea. The bloom of *Chrysochromulina* was attributed to eutrophication, while the PDV epidemic was attributed to the action of PCBs and other chemicals suppressing the immune systems of seals. In neither case is the evidence for a human-caused contribution strong, although the lack of causal proof does emphasise the need for detailed understanding of the complex coastal marine ecosystem and the effects of human activities upon it. Here we review some of these important issues. Many of our examples come from the North Sea. This is a shallow-shelf sea and is largely surrounded by the land masses of eight wealthy and industrialised nations, so it is particularly vulnerable to pollution. It is also probably the most intensively studied area of coastal waters. Our conclusions are, however, of relevance globally.

Contamination and pollution

An important distinction is that between *contamination* and *pollution*. The most commonly used definitions of these terms are attributable to an organisation called GESAMP (a UN group of experts in the field of marine pollution).

Contamination is the presence of elevated concentrations of substances in water, sediments or organisms. Pollution is

> [the] *introduction by man, directly or indirectly, of substances or energy into the marine environment (including estuaries) resulting in deleterious effects such as harm to living resources, hazards to human health, hindrance of marine activities, including fishing, impairing quality for use of sea-water and reduction of amenities.*
>
> (GESAMP, 1982)

Clearly all pollution involves contamination, but the converse is not necessarily true. Using these definitions allows us to divide the process of identifying pollution damage into two separate stages. The first problem is the chemical measurement of the contaminant. For various toxic substances, the concentration may be very low, e.g. one atom in 10^9 to 10^{12} of water and salt. The second problem is to determine whether this contamination is having any deleterious effects on the environment, its biota or potential users. The first of these tasks is the simpler, although even that is not always straightforward. To be able to determine whether contamination with a particular chemical has taken place one needs to be able to define a baseline – the environmental concentrations before anthropogenic (human-derived) inputs began. For synthetic organic compounds like PCBs, artificial radionuclides and a number of other substances, the baseline is zero. Other substances such as heavy metals and oil occur naturally, and considerable effort may be required to define a baseline.

For contaminants in sediments, there are a number of options available. One option is to seek to identify environments which are comparable in all aspects except degree of human influence, then to use the pristine environments as a baseline for the contaminated ones. This is sometimes possible, but in some parts of the world there are very few areas which are untouched by humans, and it is always difficult to be sure that the different environments are truly comparable. Another option, used particularly for heavy metals, is to use concentrations in rocks which are assumed to have been deposited in similar environments. For example, concentrations estimated by geochemists for 'average shale' are often used as a baseline with which to compare estuarine muds. A final possibility is to try to obtain pre-industrial sediments from the area which is being studied by sampling former estuarine channels or areas of reclaimed land. This gives data which are most likely to be comparable with modern samples but requires the assumption that there has been no change in metal concentrations after deposition.

The approach of using primitive sediments from the same environment to provide a baseline can be extended to the construction of entire contamination histories. In a few cases, samples have been collected at known dates in the past and stored (see Figure 14.1), but this is unusual. However, it is often possible to identify sites where sediment accumulation has taken place continuously since before the industrial period. Such sites occur in many estuaries and some distance offshore, where water is deep enough for the influence of waves to be minimal. An undisturbed column of sediment can be extracted from such sites by coring, then divided into sections where each is analysed for concentrations of contaminants. When concentrations become constant below a particular depth, this is assumed to represent the baseline. If the cores can be dated (using radioisotopes, artifacts or pollen) then a dated contamination history can be constructed.

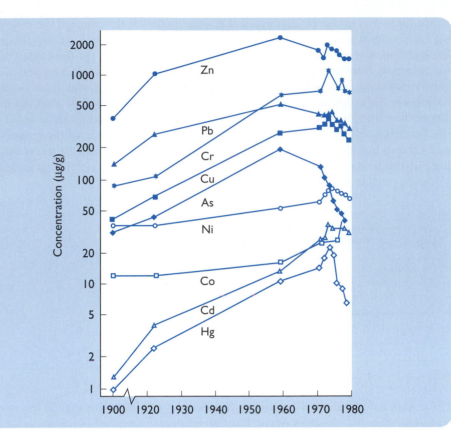

Figure 14.1:
Metal contamination
history of the River
Rhine, based on
analyses of
sediments collected
by the Netherlands
Institute for Soil
Fertility.
Source: Salomons and
de Groot (1978).

Figure 14.2 shows an example of this for the Bristol Channel, UK, showing a marked increase of lead concentrations during the 20th century.

For contaminants in surface waters of the ocean and in biota, it is much more difficult to define baselines. As we shall describe below, there is a substantial input of many contaminants from the atmosphere, and this takes place even at sites well away from population centres. There are therefore no entirely pristine environments which can provide baselines. Consequently, it is difficult to be sure what concentrations of metals or petroleum hydrocarbons in biota or sea water were before they were modified by human actions.

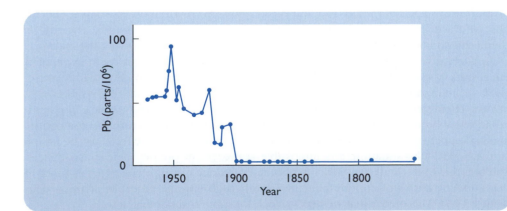

Figure 14.2: Lead
contamination
history of
sediments in the
Bristol Channel.
Based on sediment
core dated using
^{210}Pb chronology,
with natural lead
levels subtracted.
Source: Hamilton and
Clifton (1979).

Another strategy for assessing global contamination is to compare the estimates of the total natural and anthropogenic fluxes into the world oceans. However, quantifying such fluxes is extremely difficult, as illustrated in Table 14.1. In the case of nitrogen we know the river and atmospheric inputs (as nitrate, ammonium and organic nitrogen) to within a factor of three, and we also know that on a global basis both fluxes have been approximately doubled by human activity. However, we do not know the natural flux from N_2 gas fixation to better than an order of magnitude, largely because this process occurs at a low and variable rate and over vast areas of the world's ocean surface, making measurement very difficult. Because of this scale of uncertainty, it is not possible to conclude whether the increase in riverine inputs of nitrogen represents a minor or major perturbation of the total input to the ocean.

Table 14.1:
Estimates of
nitrogen inputs to
the world's oceans
(10^6 tonnes/year).

Input	Flux estimates
N_2 fixation	14–196
Riverine inputs	6–20
Atmospheric inputs	16–60

Source: Cornell *et al*. (1995).

Ways of detecting pollution

The traditional approach to detecting the effects of pollution in the field has been ecological monitoring. Samples of the biota are collected from a large number of places, and concentrations of contaminants are also measured at the same sites. The biotic data are then analysed to see whether there are any signs of deleterious effects at any sites, and if so whether these are correlated with contaminant concentrations. However, the marine environment can be very variable even in the absence of anthropogenic disturbances (see Figure 14.3 and the discussion of eutrophication later in this chapter). A number of statistical methods have been developed which allow the detection of subtle changes in the abundances of species present in a community. These methods can detect changes due to pollution provided one can find uncontaminated control sites that closely match the impacted sites in all respects other than contaminant concentrations. If good control sites are not available, then one can detect the effects only of pollution which is so severe as to eliminate an appreciable proportion of species (Grant and Millward, 1997). Even when an ecological effect is observed, the contamination at any particular site is usually a complex 'cocktail' of chemicals, so it is difficult to establish which contaminants are producing the effects.

An alternative is to carry out toxicity tests in the laboratory, then to try to extrapolate the results of these to the field. This extrapolation procedure is not straightforward, because the marine environment is far more complex than a laboratory glass dish in its physics, chemistry and biology. Contaminants may be complexed by organic compounds in the water or adsorbed by particulate matter. A pollutant may perturb the competitive interaction between species. It is difficult to predict which of these effects may be important in a particular situation, and impossible at present to predict their consequences. The one thing which all the work on laboratory toxicity testing does allow us to say is that concentrations of contaminants are rarely high enough to cause acute toxicity (Figure 14.4). This means that we need to look for subtler 'sub-lethal' effects.

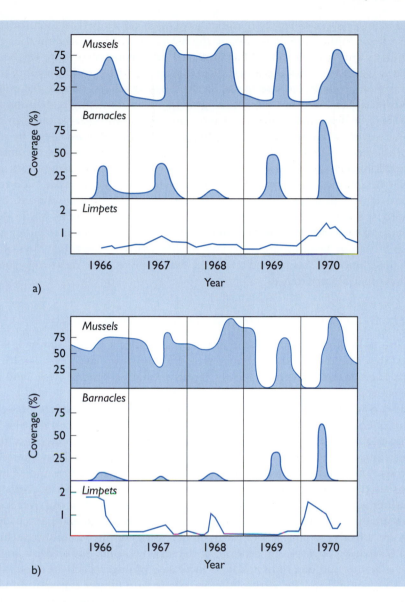

a)

b)

Figure 14.3:
Natural fluctuations
of numbers of
marine organisms
can be large. Data
on percentage
cover of mussels,
barnacles and
limpets at two sites
on the north
Yorkshire coast,
1966–1970.
Source: Lewis (1972).

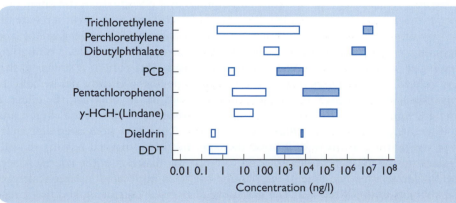

Figure 14.4:
Environmental
concentrations of
most contaminants
(open bars) are
well below those
necessary to cause
acute toxicity (blue
bars). Note the log
scale.
Source: Ernst (1980).

In recent years, a variety of methods have been developed to look for these more subtle effects in the field. Most involve the measurement of biochemical or physiological indices and it is assumed that impaired performance on one of these indices reflects impaired ecological performance. Perhaps the most widely used of these is the effect on the *scope for growth* in the bivalve *Mytilus edulis*, namely the energy which an individual has left over to devote to growth after subtracting respiration and excretion. This involves determining the energy budget of individual animals by measuring feeding and respiration rates and assimilation efficiency. The assay can be used either by collecting animals from different sites or by placing animals with a common origin at sites along a supposed pollution gradient (this can be done using cages attached to buoys when monitoring conditions in open water). The animals are then returned to the laboratory and scope for growth measured in standardised conditions. In uncontaminated sea water, scope for growth is relatively high, but its value declines in response to relatively low concentrations of contaminants because either respiration rate increases or feeding rate decreases. Unlike many of the other physiological and biochemical indices, this has the advantage that it seems to be measuring something of direct ecological importance.

Sources of contaminants to the North Sea

If we are to understand the environmental condition of a marine system it is necessary to first describe the inputs to that system. Only if these have been significantly perturbed is environmental change likely. Inputs to marine systems from land arise from many sources: here we attempt to document the major ones.

Rivers bring material to the seas in both dissolved and particulate phases, and the balance between these two can change during mixing between marine and fresh waters, due to the adsorption or precipitation of dissolved components or the desorption of components from particles under the rapidly changing salinity characteristics of estuaries. The nature of the mixing between fresh and salt waters and of many other processes (including primary production) is further influenced by rivers, since the freshwater is less dense than sea water and will thus tend to float out over it unless turbulent mixing can overcome this stratification. Both dissolved and particulate fluxes in rivers are sensitive to change as a result of increasing inputs within the catchment from direct discharges or increased erosion, or decreasing inputs due to damming removing water, allowing settlement of solids or increased biological activity. Since river transport is episodic, with, for example, more particulate matter transported under high flow conditions, it is hard to obtain reliable estimates of riverine inputs to estuaries.

In Figure 14.5, the increase of nitrate, ammonium and phosphate in the Rhine over recent years is shown as an example of the effect of increasing inputs to river systems. In the case of nitrate, the increase is thought to be the result of increasing use of land for agriculture and more intensive agricultural practices, including deep ploughing and extensive, and relatively inefficient, use of fertilisers. This pattern is similar to that in most other European rivers (e.g. Figure 14.6) and results in current nitrate concentrations considerably higher than they naturally would be. By contrast, the effect of damming is illustrated by the Nile, where the total water flows have been altered and more dramatically the peak discharge reduced. The dams have generated electricity, allowed irrigation and prevented flood damage, but also lowered fertility in soils dependent on the flood-deposited silt and reduced the fisheries of the coastal waters off the Nile delta. The damming of the Danube has been implicated in large-scale ecological changes in the Black Sea. The damming of rivers is increasing and may cause

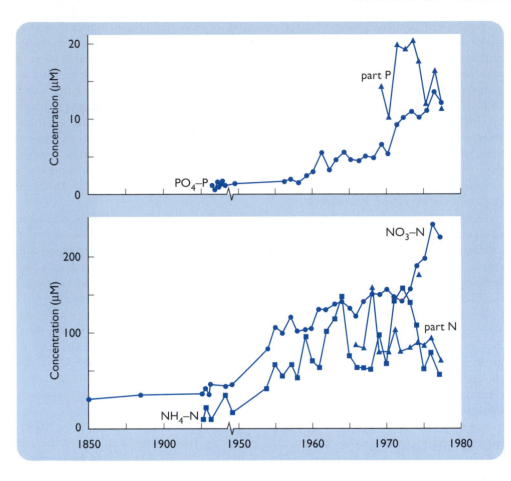

Figure 14.5:
Increase of nitrate
(NO_3), ammonium
(NH_4^+), particulate
nitrogen (part N),
dissolved
phosphorus
(PO_4–P) and
particulate
phophorus (part P)
concentrations in
the River Rhine.
Source: von
Bennekrom and
Salomons (1980).

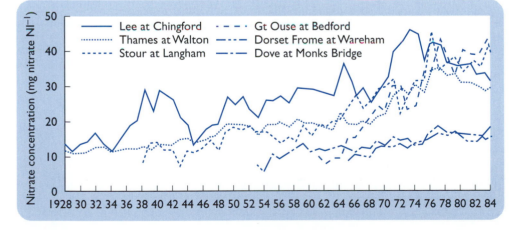

Figure 14.6:
Increase of nitrate
concentrations in
several UK rivers.
Source: Royal Society
(1983).

major perturbation to coastal environments around the world. Damming and other changes in river management can also damage the stability of intertidal ecosystems such as saltmarsh and mangroves.

Atmospheric inputs can bypass the complex removal processes in estuaries and reach the open coastal and oceanic waters direct. The atmosphere can be an efficient transporter of dust and gases to coastal waters since many land-based activities release material to the atmosphere, including dust blown from fields, gases released from agricultural wastes, combustion processes of all kinds and many industrial activities. In addition, there is a large-scale natural cycle transporting material through the atmosphere and contributing to the inputs to coastal waters. In many areas of the world close to major source regions, it is now clear that human activity is discharging material through the atmosphere in amounts comparable with or even exceeding natural fluxes. Atmospheric concentrations vary rapidly on short timescales due to changes in wind direction and rainfall, making it very hard to estimate accurately average concentrations without extensive sampling. In addition, estimating inputs requires a knowledge of rainfall over the sea and/or dry deposition processes at sea. Both are poorly understood and monitored.

As well as these two major inputs, there are a wide variety of other inputs which are of significance in particular areas and for particular compounds. Some of these, such as direct industrial and sewage discharges and the dumping of sewage sludge and dredge spoil, are closely monitored in most industrialised countries, and the amounts of material discharged are relatively well known. Others, such as losses from corrosion *in situ* and coastal erosion and groundwater inputs, are very poorly understood.

Despite the uncertainties involved, estimates of inputs to the North Sea from the various sources are available (Table 14.2) and illustrate several important points. First, despite the considerable attention directed towards them, direct discharges are not of great quantitative significance, though this is not to deny their considerable impact in the local area of dumping. The second point is the considerable importance of atmospheric inputs for many components. These, along with riverine inputs, dominate land-based sources.

For many of the components considered, including nitrogen and lead, emissions have indeed increased dramatically in recent years, though for lead these are now

Table 14.2: Estimates of rates of input of selected elements to the North Sea (tonnes per year).

Input route	Element		
	N^1	*Pb*	*Hg*
Rivers[2]	910 000	1 000	25
Atmosphere[2]	520 000	1 700	6.9
Direct discharge[2]	120 000	160	1.8
Sewage sludge	6 300	77	0.7
Dredge spoil	—[3]	2 700	19
Industrial waste	—[3]	220[4]	0.2[4]
Incineration at sea	—[3]	4.9	0.05
Inputs from transport of waters from offshore	7 705 000[5]		

Notes:
[1] These N fluxes exclude nitrogen gas, because it cannot in general be utilised by algae; they are dominated by nitrate and ammonia.
[2] Maximum estimates.
[3] No data are available, but likely to be small.
[4] Minimum values.
[5] Nelissen and Stefels (1988).

Source: North Sea Task Force (1993).

decreasing even more rapidly as the use of unleaded petrol grows. The magnitudes of the fluxes of many synthetic organic components are poorly known, but, since there are no natural sources of these compounds, their very existence is testimony to anthropogenic sources. However, there are still many elements and chemical species with large natural fluxes and relatively little interference from industrial activity (e.g. aluminium) for which the fluxes still approach their natural levels.

A final important point from Table 14.2 is that, while the riverine and atmospheric inputs of nitrate and ammonia have increased dramatically over the last 50 years or so, the input from offshore is still the major source, though a particularly difficult one to quantify.

In order to determine the effects of recent increases in inputs, it is necessary to know not only the size and form of all the inputs but also their location. Thus again considering nitrogen, the riverine inputs are centred in the southern North Sea, where the big rivers discharge (Figure 14.7). The atmospheric inputs are less focused in this area but are probably still higher in the southern area closest to industrial and urban sources. Thus the extra inputs are concentrated in the southern, shallowest, least well-flushed regions, while the natural input from offshore is concentrated in the northern region. Thus it is no simple task to take input information alone and convert it into an assessment of the environmental condition of the North Sea, or indeed other coastal regions.

Examples of environmental damage by specific chemicals

PCBs (polychlorinated biphenyls) are organic molecules with a structure which includes two linked benzene rings with a varying number of chlorine atoms attached. PCBs are very stable to heat and burning and have been used widely in electrical equipment, paints, plastics and adhesives.

PCBs are slow to be broken down once inside the bodies of organisms. They are also fat-soluble so are readily absorbed into the body and then tend to accumulate in fatty tissues rather than being excreted. In animals, the majority of PCBs in the diet will be absorbed into the body. The majority of energy which an animal derives from its food is used to fuel activity and maintenance, and on average only about 10% of the food is converted into body tissue. If all of the PCBs in the diet are absorbed, an animal will end up with concentrations of PCB in its tissues which are ten times those in its food. Food chains usually have several links in them, so concentrations of PCBs in top carnivores, such as seals and seabirds, can become very high – a phenomenon known as bioaccumulation. This phenomenon occurs with some other organic contaminants such as DDT but not for most metals, since many organisms have physiological mechanisms to regulate them.

PCBs have been implicated as a cause of mass mortalities in seabirds, and in the decline of seal populations in the Baltic. However, the most convincing evidence of an impact by PCBs comes from the Dutch Waddensee. Between 1950 and 1975, numbers of the common seal *Phoca vitulina* in the western part of the Waddensee declined from more than 3000 individuals to fewer than 500. The production of pups by seals in Dutch waters also declined sharply in the same period. A comparison of body burdens of heavy metals and organochlorine compounds between the Dutch seals and those from the more northern parts of the Waddensee showed that only levels of PCBs differed significantly between the two. In an experiment carried out at Texel in the Netherlands, two groups of 12 seals were fed diets of fish containing different levels of PCBs. Seals fed on fish caught in the Waddensee which had high concentrations of PCBs showed reduced reproductive success when compared with the group fed on

Figure 14.7: Map of the North Sea, showing main features of water circulation and major rivers.

Source: adapted from Eisma and Irion in Salomons *et al.* (1988).

fish from the north-east Atlantic. So there seems good evidence that PCBs are having an adverse effect on seal reproduction in the Waddensee. However, in the phocine distemper virus epidemic of 1988, the first seals to be affected were not those in the Waddensee but those in the rather less contaminated Kattegat, suggesting that PCBs were not an important contributory factor.

TBT – tributyltin

If any hard substrate is placed into the sea, it is rapidly colonised by a range of animals and plants commonly labelled 'fouling organisms'. An analogy would be the way in which any piece of open ground on land is rapidly colonised by weeds. If the hard substrate is the hull of a ship, this colonisation by fouling organisms replaces a smooth surface, which has relatively little friction with the water, with a much rougher surface. A well-developed fouling community reduces the maximum speed of a ship and increases fuel consumption considerably. To try to avoid this difficulty, the hulls of ships are usually painted with an anti-fouling paint, which prevents or reduces the settlement of fouling organisms. Traditional anti-fouling paints contained copper, arsenic and other highly toxic substances. In consequence, such paints required considerable safety precautions when being applied. In the 1960s, a new type of anti-fouling paint was introduced, with tributyltin oxide (TBTO, or more usually just TBT) as the active ingredient. These new paints appeared to be of relatively low toxicity and rapidly gained a large share of the world market.

The first sign that these paints were not quite as innocuous as first appeared came in the early 1980s, when evidence was found in the USA that TBT leaching from anti-fouling paints was causing females of the mud snail *Nassarius* to develop male secondary sexual characteristics. Shortly after this, French oyster growers began to find oysters with abnormally thickened shells. The oysters were unsaleable because of their deformed shells, and the effect was traced to TBT.

Some of the most detailed studies of the effect of TBT have been carried out on the dog whelk *Nucella lapillus*. When exposed to TBT, females of the species begin to develop male secondary sexual characteristics, including the growth of a penis and blocking of the oviduct. The combination of characteristics has been given the name *imposex*. This prevents the females laying eggs, and consequently populations decline. *Nucella lapillus* is completely absent around some marinas where there is a high concentration of TBT in water and sediments.

TBT has these effects at minute concentrations in the water. Shell thickening occurs in oysters exposed to 80 ng/l (80 parts in 10^{12}), and imposex in *N. lapillus* is induced by concentrations of only a few nanograms per litre. Because the effect on *N. lapillus* is easily detectable and clearly attributable to TBT, it has been easier to establish cause-and-effect links here than in virtually any other case. There are some suggestions that TBT may have effects on the broader community, but these conclusions are based on correlations between TBT levels and community structure so are subject to the difficulties facing all such studies (see section on detecting pollution above). Since it became clear that TBT was so damaging to marine organisms, the use of TBT in anti-fouling paints is gradually being banned throughout the world. A ban on its use on vessels smaller than 25 m was introduced in the UK in 1987. Fortunately, TBT is less persistent in the environment than many other pesticides, so concentrations in coastal areas have declined. Oysters rapidly resume normal shell growth when no longer exposed to TBT, and dog whelk populations are showing signs of recovery. However, TBT-based paints are still in use on vessels larger than 25 m, and TBT from this source appears to be the cause of imposex in whelks in the central English Channel (North Sea Task Force, 1993). It is not clear how thoroughly the lessons of the TBT story have been learned. One of the substances which has replaced TBT in anti-fouling paints is the triazine herbicide Irgarol. Preliminary studies indicate that environmental concentrations of Irgarol close to marinas are high enough to produce deleterious effects on algal communities (Dahl and Blanck, 1996).

Oil pollution

Oil discharges to the marine environment are often divided into two classes. The first, 'operational discharges', reflects the chronic low-level inputs arising during transport, production, processing and disposal of oil products, and dominates the total amounts of oil discharged, though they attract little attention. Improving pollution control measures and greater attention to minimising discharges is resulting in declining inputs generally from this source. The second source of oil inputs is the occasional large-scale discharges from accidents such as shipwrecks, damage to offshore oil production facilities, broken pipelines, or sabotage such as that during the Iraqi occupation of Kuwait. These occasional and often large-scale discharges attract great public concern and clearly do significant environmental damage. This damage arises because of certain specific features of oil as a contaminant. Oil, like some litter (see later), does not mix with sea water but instead floats and hence cannot readily be diluted by the vast volume of the ocean. After a spill, the oil spreads out as a thin surface film and consequently does major damage only to organisms that encounter this air–sea interface.

Oil itself is a complex mixture of thousands of different organic molecules dominated by aliphatic carbon compounds containing one to 24 carbon atoms, and this complexity makes it very difficult to generalise about the behaviour of oil. The toxicity of many of the individual compounds has been measured in some organisms, but it is difficult to extrapolate data from a few compounds and biological species to a real oil spill in the environment, where the species affected may be different and where the various compounds may act together to increase or reduce the toxicity of the individual compounds. Despite these complications, a few generalisations are possible. Aromatic and low molecular weight compounds are in general more toxic, but these compounds also tend to be those most rapidly lost from spilled oil within a few hours by evaporation and dissolution because they are most volatile and water-soluble. Such losses are, of course, temperature-dependent and will be much less important for oil spills in polar waters. Left alone, the remaining individual compounds in the oil will slowly degrade to CO_2 and water and the other raw materials of oil on timescales ranging from days to centuries. The mechanisms of loss include photo-oxidation, dissolution and bacterial breakdown. Oil is a natural product, and bacteria capable of degrading it are present throughout the marine environment, albeit in low numbers. The capacity of an environment to disperse even large amounts of oil was illustrated by the *Braer* incident in the winter of 1993 off the Shetland Islands. A discharge of 87 000 tonnes of oil was relatively rapidly dispersed due to exceptionally violent weather conditions, with only modest local environmental damage. In March 1989, the tanker *Exxon Valdez* ran aground in Prince William Sound, Alaska, releasing 37 000 tonnes of Alaskan crude oil. Extensive monitoring data are now available, which show that there were ecological effects, but that after one year most of these were small (Wells *et al*., 1995). Thus the problems of oil pollution in the marine environment are short-term, (years) in contrast to those caused by persistent contaminants such as metals or PCBs.

After extensive degradation in the environment, the oil residue is tarry matter which forms tar balls floating on the sea. These are also released by tanker cleaning operations, being a residue of crude oil transportation as well. Tar balls have a residence time on the surface of the ocean of about a year and can cause considerable nuisance when beached in large numbers. However, the observation of many of these tar balls encrusted with fauna such as barnacles is testimony to the limited toxicity of these final residues of oil degradation.

As noted above, the chemical toxicity of oil is relatively low after a matter of a few hours in a warm environment, but the physical properties of the oil to coat and subsequently kill organisms persist as long as there is a coherent oil slick. Thus seabirds and intertidal communities are are risk since these come into contact with the floating oil, whereas subtidal benthic communities are less at risk. The experience of beached oil spills suggests that the effects on intertidal communities can persist for several years, particularly in high-latitude seas, where low temperatures reduce the rate at which the oil degrades.

Experience over the last 30 years with major oil spills has suggested various strategies to minimise their environmental impact. If the slick is well out to sea and away from major bird communities, the most practical and economic action is to leave the natural processes of evaporation, dissolution and physical break-up to dissipate the slick. Nearer to land, however, it is often necessary to consider action to try to minimise the environmental damage from an oil spill. It is possible to use chemical dispersants, which act as powerful surfactants (or detergents) to aid the dissolution of the oil. These have the effect of allowing much more effective dilution of the oil but do expose biological systems, other than those that inhabit the water surface, to the oil. The first large-scale use of oil dispersants during the *Torrey Canyon* incident was a disaster, with the dispersants proving to be more toxic than the oil. Recently, less toxic dispersants have become available, but there is still reluctance to use this approach in many situations, and in addition the use of dispersants on large spills is logistically very difficult. Some of the clean-up activities which followed the *Exxon Valdez* spill involved hosing intertidal areas with hot water, which very effectively killed all marine life in the affected areas. During the *Torrey Canyon* incident, burning the oil was also attempted but then, and in subsequent attempts, this has proved a very inefficient and often counter-productive clean-up procedure. In calm waters, it is possible to mop up the oil onto an absorbent solid or skim it off the surface using specialised equipment; these approaches are often used in harbours for small spills, since the waters are usually calm and equipment can be made available if the eventuality of a spill is planned for.

If a large slick is heading for a shoreline, it is possible to protect the area by using floating booms. These are effective only in calm waters since strong winds or tidal currents can force oil under the booms. In order to protect even a small area of coast, hundreds of metres of boom are needed, which must have been stored near to where it is needed, since it is heavy, cumbersome, slow and expensive to move and deploy. Thus booming is best used across inlets to bays or estuaries, where a relatively small boom can protect a long length of coast. Even then it is necessary to make choices about the coast to protect, since booming can never protect a whole coastline. Booms were used successfully to protect vulnerable saltmarshes after the *Sea Empress* spill. This again illustrates the need for planning in advance for oil spill contingencies.

Coasts can be categorised in terms of their sensitivity to damage from oil, the ease with which they can be cleaned, and other factors such as amenity, fishery or economic value if appropriate. Thus, for example, cliffs and harbour walls are rarely affected by oil because wave patterns tend to keep oil away from them, and any oil that does reach them is readily removed by wave action or remedial cleaning. Sand and gravel beaches are much more readily impacted by oil as the oil tends to move deep into the beach, making it persistent and difficult to remove; these areas may thus attract a higher priority than cliffs. Such beaches have low biological diversity but may have high amenity value, so it is necessary to make value judgements at some stage. Sheltered rocky coasts, mudflats and marsh or mangrove communities are all sensitive to oil, which would be persistent and difficult to remove, so these areas usually rate a high priority for protection. However, the most effective courses of action are those which minimise or avoid spills in the first place. Human error played a major role in the initial grounding of the *Sea Empress*. Even after this had occurred, the amounts of oil spilled would have been small if salvage attempts had been better handled.

Heavy metals

Metals are released to the marine environment from a multitude of sources. Once in the sea, they do not degrade, although they may be removed by burial in sediments. Mercury is liable to bioaccumulate, particularly when incorporated into organic molecules such as methyl mercury. This was shown vividly in the Minamata pollution incident in Japan in the 1960s. A factory manufacturing vinyl chloride discharged large amounts of methyl mercury into the sea. Bioaccumulation led to large concentrations of methyl mercury being present in fish. The Japanese as a nation are rather fond of fish, and Minamata is a fishing community. Fishermen and their families were eating several kilograms of fish per week and received large enough mercury dosages to produce neurotoxic effects. About 2000 people were affected; 43 died during the epidemic and more than 700 were permanently disabled. Discharges of mercury are now tightly regulated.

Cadmium also has a tendency to bioaccumulate, so is placed on most black or red lists of priority pollutants. It is, however, of relatively low toxicity to marine life, and the strictness of most national regulatory regimes is such that damage to the marine environment by cadmium is unlikely to occur. Copper and silver are much more toxic to marine organisms and are more likely to present a threat. Fortunately, the chemical composition of sea water reduces the environmental impact of heavy metals. The relatively high pH of sea water, coupled with its high carbonate content, mean that the solubility of many metals is low. Metals can also be complexed by organic matter in sea water, a process that can greatly modify their bioavailability. Metal concentrations are generally several orders of magnitude lower than saturation, as a result of scavenging by particles and other processes. Consequently, metals discharged to the marine environment are in general not readily bioavailable. High concentrations of metals in biota are observed only in rather anomalous situations, such as estuaries receiving drainage water from metalliferous mines or effluents from smelters (e.g. Restronguet Creek in Cornwall and Sorfjorden in Norway). In the most contaminated sites, there are some signs of a reduction in faunal diversity, but this occurs only at concentrations well in excess of those found in normal industrialised estuaries.

Summary of specific effects

We have given some examples where we can be fairly certain that specific chemicals are having identifiable effects on the biota. Such cases are, however, rather few in number, and the clearest examples of ecological damage are the result of deoxygenation caused by sewage discharges to estuaries. In general, field-based research has not provided us with great ability to predict ecological effects of given concentrations of particular substances. As described earlier in this chapter, it is remarkably difficult to predict effects in the field from laboratory toxicity testing, so we cannot rely on that either, though models of dispersion and organic matter degradation have been used successfully to aid in identifying optimum locations for discharges of sewage and other contaminants. Regulation of discharges must therefore be done in a more arbitrary way. The legal regulation of the environmental release of individual chemicals in the UK and the European Community places a high priority on reducing releases of substances which are toxic, persistent and liable to bioaccumulate. These are those substances which are most likely to cause long-lasting environmental damage, and discharge limits are usually set using uniform emission standards – an assessment of the lowest releases to the environment per unit quantity of product which it would be reasonable to achieve using the 'best available techniques not entailing excessive cost'. In the UK, other substances are regulated using environmental quality standards. Discharge consents are set at levels which will lead to concentrations in the environment that are well below those which have been observed to cause an effect in laboratory tests. This approach takes into account the capacity of the receiving environment to dilute effluents, but precisely because of this it is more complicated to administer than uniform emission standards as consents have to be set on a case-by-case basis. Neither approach regulates atmospheric or other non-point inputs of contaminants, which, as we have seen earlier, can be the most important source for some substances.

Eutrophication

In the previous section the effects of the oxygen consumption arising from the breakdown of organic matter were discussed. However, if the breakdown of organic matter takes place after sufficient dilution so as to allow the rate of oxygen supply to exceed the oxygen consumption, then the organic matter can be broken down to its constituent parts – carbon dioxide and nutrients, including nitrogen (as the oxidised species nitrate or the reduced form ammonium) and phosphate – without deleterious effects on oxygen levels. These nutrients can then be reused by small unicellular free-floating algae called phytoplankton in their growth, thus forming new organic matter in a very efficient recycling process.

The phytoplankton form the basis of the marine food chain, and their rate of primary production can be limited either by the availability of light (as often happens in the turbid waters of estuaries and during winter throughout much of the North Sea) or by the availability of nutrients. If nutrients are the limiting factor, then increasing nutrient levels will increase primary production. At a modest level this can be beneficial in providing increased food supplies, but a large increase in nutrient supply can have several deleterious effects, which are usually referred to as *eutrophication*.

The deleterious effects that are possible fall into two groups: changes in species and increased oxygen consumption in bottom waters. The change of species arises because different algal species are adapted to different nutrient environments, and a change in the nutrient status can favour certain species while disadvantaging others. As an example, some phytoplankton, the diatoms, require silicon to build their skeletons. Silicon is introduced to the marine environment primarily by weathering via rivers. If the nitrogen and phosphate inputs are increased without changing the silicon inputs, then diatoms will be disadvantaged compared with other phytoplankton. The change in phytoplankton species can affect other parts of the marine food web, because certain predator species are adapted to certain types of algal food and changes in species composition can thus affect predators. In addition, some phytoplankton species are apparently avoided as food supplies and some are actually toxic. In particular, certain dinoflagellates can form so-called 'red tides', with massive blooms which release toxins that can be accumulated by shellfish. Such blooms have been reported occasionally from the North Sea but are much more common in tropical waters, where they regularly force the temporary closure of commercial shellfish farming operations. However, a clear link between eutrophication and such red tides has yet to be established.

The second problem that can arise from increased nutrient inputs results when the phytoplankton die and then sink into deeper waters. Here, if there is no light and hence no oxygen production by photosynthesis, the increased stock of dead phytoplankton represents an increased load of organic matter which requires oxygen to break it down. If the rate of oxygen consumption in the deep waters exceeds the rate of oxygen input, oxygen concentrations will fall with subsequent deleterious effects, as described earlier for sewage breakdown in estuaries. Several factors besides increased nutrients can act to exacerbate this situation. First, deeper waters tend to stratify, with warmer, less dense, water forming a stable cap over the deep water and preventing transport of oxygen into the deep water. Thus under warm conditions the effect is increased, and in warmer situations there is both more light to drive photosynthesis in the upper water column and also increased rates of breakdown of organic matter throughout the water column.

There is evidence of eutrophication-induced oxygen depletion in several coastal areas, including the Baltic and Chesapeake Bay on the north-east coast of the USA. In both cases, nutrients have been considerably enriched by human activity, leading to increased algal activity. In addition, in both cases the estuarine system includes deep basins (about 100 m) with exchange with offshore waters restricted by shallow sills at the seaward end. This basin shape allows deep waters to be trapped for long periods, during which time the settling organic matter is decomposed by bacteria and oxygen consumed. In Chesapeake Bay on the eastern seaboard of the USA, winter cooling allows the stratification to break down and deep water to mix with surface waters and replace the anoxic deep water by fresh, well-oxygenated waters, and the anoxia is therefore seasonal. In the Baltic Sea, the winter mixing is inadequate to achieve such reoxygenation and the anoxic conditions can persist for many years until flushed out by exchange with the North Sea. The situation in the Baltic has been monitored for many years, and there is a good time series of measurements documenting the increasing nutrient levels and declining oxygen levels (Figure 14.8).

In the southern North Sea the waters are generally shallow (less than 50 m) and well mixed by tidal action throughout most of the year. Under these conditions bottom waters cannot become trapped and anoxic. In the south-east corner of the southern North Sea, regions of low oxygen have been observed intermittently since 1902. While increasing nutrient concentrations may contribute to this situation, a primary cause is probaly the increased stratification in this area due to freshwater inputs. This stratification naturally traps the bottom waters and allows bacterial oxygen consumption to lower oxygen levels in the absence of a supply of fresh oxygen-rich water from above. This does emphasise the critical role of physical mixing induced by tides or winds and stratification in modulating the effects of contamination.

There has been much debate about whether increasing nutrient inputs to the southern North Sea are pushing the area towards eutrophication. However, there is considerable year-to-year variability in plankton stocks arising from climatic effects, so it is very difficult to identify subtle changes. Fortunately, there is a record of plankton populations in the North Sea spanning more than 40 years. This record is collected

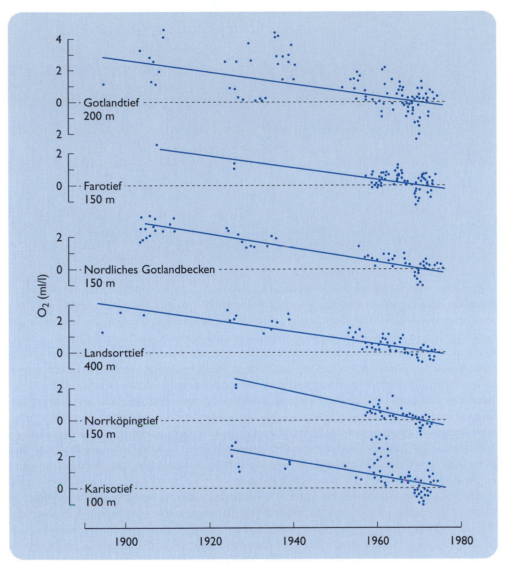

Figure 14.8:
Decreases in oxygen concentrations of the bottom waters of the Baltic. Negative values indicate hydrogen sulphide production.
Source: Fonselius (1982).

using the 'continuous plankton recorder' (CPR), which is towed behind commercial ships. The record documents the considerable year-to-year variation but also suggests long-term changes which occur in all plankton groups throughout the North Sea and the North Atlantic (Figure 14.9). Changes of this magnitude cannot be the result of pollution effects over so large an area and suggest a decrease in plankton rather than the increase which would be expected if eutrophication were occurring. It is now believed that the trend is the result of long-term climatic changes in wind strength and direction. The trend is also seen in some populations of kittiwakes (the only exclusively open-water-feeding bird in this area), suggesting that the decline in their numbers may reflect food shortages rather than pollution effects.

While the trend observed by the CPR is similar throughout the North Sea and North Atlantic, close to the European coast of the North Sea in areas not sampled by the CPR there is evidence of increasing levels of some algae, particularly *Phaeocystis*, a flagellate that releases large amounts of mucus into the water, resulting in the production of foams

Figure 14.9:
Synchronous
long-term changes
in populations of
zooplankton and
kittiwakes shown by
data from the
continuous plankton
recorder.

Source: adapted from
Aebischer *et al.* (1990)
and Dickson *et al.*
(1988).

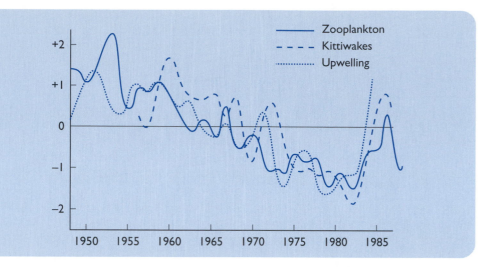

on beaches. These foams can lie several feet thick on the beaches of the Netherlands, Germany and Denmark and fuel considerable public concern over the condition of the marine environment. The circulation of the North Sea acts to confine the inputs of the large European rivers such as the Rhine to the coast (Figure 14.5), and thus it is likely that the increasing algal levels very close to the coast are the result of increasing eutrophication, but this effect appears to be rather local. This problem may not be related simply to increasing inputs of nitrogen and phosphorus but also to the changing balance of nutrients as these nutrients increase and silica does not, as discussed earlier. Exceptional blooms of algae have been reported on occasions throughout the North Sea, sometimes with serious consequences, including the closure of fisheries during red tides and mass fish mortalities during *Chrysochromulina* blooms off the Norwegian coast, as noted earlier. However, the causes of these blooms are not well understood and there is at present no clear evidence to link them to eutrophication.

The North Sea is an environment that is naturally relatively rich in nutrients, at least in winter when production is low. In tropical systems, nutrient levels are very low throughout the year and these systems are hence more sensitive to minor increases in nutrient inputs. Coral reefs are massive and complex ecosystems that grow only in clear, warm, nutrient-poor conditions and provide vital coastal protection and fishery resources. There are many examples of the destruction of coral reefs by increasing sewage and nutrient inputs, with resultant increases in benthos algae overwhelming the corals. In many tropical areas, reef damage by nutrient enrichment is compounded by many other pressures associated with increasing human activity, including overfishing, increased turbidity from shipping activity or change in river catchment, plus crude physical damage from shipwrecks and even use of the reefs as building material. The recovery of these systems after cessation of input is very slow, with the effect still obvious.

Bathing beaches

Seaside resorts are often large centres of population, and this may be increased by several hundred per cent in summer. As noted in our discussion of estuaries, the traditional British approach to sewage disposal on the coast has been to discharge it into the sea with only limited treatment. This has been the practice of many coastal resorts, often involving the discharge of untreated sewage close to low water mark on the most popular beaches, at best after primary screening and perhaps maceration. The sight of children swimming in the sea surrounded by what are euphemistically referred to as 'sewage-derived solids' causes at the very least visual offence, and perhaps infection with sewage-borne diseases. Unless the sewage passes through a fine screen (preferably preceded by maceration), sea outfalls can also be significant sources of plastic litter.

A response to this has been the construction of long sea outfalls. Rather than installing sewage treatment, simply build a longer pipe. The effluent is then diluted by the action of waves and tides and the organic matter can be broken down by bacterial action in a very similar way to the action of a land-based sewage treatment works. The greater distance between the point of discharge and the bathing beaches creates greater potential for dispersal of the effluent. Salt water and sunlight also act as disinfectant. Provided long sea outfalls are properly designed, they avoid the gross fouling of beaches with faecal matter, although this does occur occasionally during periods of onshore winds. Standard methods of assessing sewage contamination which involve counting densities of sewage-derived bacteria in the water show that there is some increase in bacterial populations in bathing waters near long sea outfalls. The question is whether these increased bacterial populations carry any health risks.

There is persistent anecdotal evidence of bathers developing gastro-intestinal and ear infections, apparently as a consequence of sewage contamination. Until recently, the view of the UK regulatory authorities has been that a significant risk of infection occurs only when bathing waters are so contaminated with sewage as to be 'aesthetically revolting'. However, firm scientific evidence on the health risks of sea bathing is sparse. Experimental studies cannot be carried out because of the ethical difficulties in exposing people to something which may be harmful to health. Cohort studies are difficult because of the diverse geographical origins of people using a beach at any one time. The most detailed epidemiological studies have been in North America, but these are based on freshwater bathing beaches, so there are some difficulties in directly applying the results to marine bathing. The main difference between marine and freshwater environments is likely to be in the rate of death of bacteria. Further uncertainty is introduced by the fact that faecal coliforms are being used only as an indicator of the concentrations of human pathogens, as pathogenic organisms are rather more difficult to count. This is particularly true of viruses, which are able to survive for a substantially longer period in sea water than are sewage-derived bacteria. However, recent careful epidemiological studies of the effects of bathing in sewage-contaminated water do now suggest a clear link with the development of various symptoms. The most comprehensive recent study also suggests that the standard tests using faecal coliform abundance in bathing waters as an indicator of risk may be inappropriate, and faecal streptococci may be more appropriate. In Chapter 3, there is a more detailed examination of the institutional politics of managing coastal bathing waters in the UK.

Litter

Litter on land is a widely recognised problem, but the problem of litter at sea has a lower profile because fewer people see it. However, this changes when the litter is washed up on beaches, and it seems safe to assume that the composition of marine litter is similar to that of beached litter. The cleaning of beaches represents a considerable expense to the local community, though it is essential in tourist areas. Some of the beached material also constitutes a significant hazard.

Beach litter has been systematically studied by collecting and identifying all the material on a transect down a beach; the country of origin and the approximate date of manufacture can often be determined. The results of such studies indicate that litter on the beach appears to have three sources – sewage, visitor-dropped litter and material dumped at sea. The sewage-derived material is discussed above. In a series of studies of northern European beaches in the 1970s, 38–50% of the identified material was household and lavatory cleaning bottles, emphasising the importance of ship-derived litter (or the obsessive cleanliness of some beach visitors!). In the 1970s, it was estimated (based on extrapolations from detailed studies on a few ships) that more than 6 million tonnes of waste was dumped annually by ships, though much of this was cargo-related. Recent trends to increased packaging no doubt increase the total amount of dumped wastes and in particular the amount of floatable waste. Much of the dumped material will sink (bottles and tins) or be rapidly broken down (paper and food), but plastics do not degrade rapidly and float, so they cannot be diluted by the full volume of the sea. Ship dumping explains the cosmopolitan origins of the containers washed up on beaches, the majority of which appear to be less than 5 years old. This age implies a relatively rapid removal process, possibly by bleaching or physico-chemical breakdown, though the details of the loss mechanism are unknown.

Some litter is obviously potentially dangerous. For example, pharmaceutical and hospital wastes are occasionally washed up on shore after dumping at sea or loss of cargoes. Other dangerous cargoes, including chemicals and munitions, are quite regularly reported washed up on beaches. While these rare incidents represent an acute hazard, the plastic debris represents a chronic insidious threat to marine life. When the stomach contents of birds from even remote regions of the world are analysed, they often contain plastic debris including pellets a few millimetres in diameter, which are probably precursor material for plastics fabrication and which have been lost at sea or from coastal plastic manufacturing plants. Larger pieces of plastic can entangle birds and marine life, and ultimately kill them. This applies to waste plastics but perhaps more so to lost fishing gear, particularly large drift nets. These nets, often more than 10 km long, will drift around for weeks or more, still catching fish. The entrapped fish lure other fish and marine mammals to their doom in the nets.

The control of marine litter and fishing gear is clearly an international problem, and a series of international agreements have now been implemented to try to control this problem. However, effective enforcement of such agreements at sea is impossible, and compliance must therefore rely on public pressure to ensure that waste storage facilities are provided at sea and waste disposal facilities are provided at ports. Then, as with terrestrial litter, it becomes a matter of public education to persuade people to use them. The fishing nets represent a rather different problem since the construction of robust drift nets, which will inevitably be lost occasionally, means that the problem is unavoidable with this type of fishing and a ban seems the only solution.

Conclusions

There is a lot of evidence for contamination throughout the marine environment and in coastal waters in particular. The evidence for ecological damage is thin on the ground, and limited to rather small areas or to the effects of a small number of persistent chemicals. This contrasts markedly with other sorts of anthropogenic impact in coastal waters. Fisheries have had a major impact on the marine environment, with most fish stocks being at considerably lower levels than their unexploited numbers would be. In the case of the most heavily exploited species, the stocks may be at a point where they are no longer commercially viable as a fishing resource. Fishing can also have other environmental impacts. The whole bed of the North Sea is trawled over on average once a year, which ploughs up the sediment and has an uncertain effect on the benthos. Another anthopogenic impact has been on coastal habitats. Large areas of coastal marshland and mangrove have been reclaimed for agriculture, and estuaries continue to be under threat from proposals to build tidal barrages and marinas. These areas are then lost for ever to nature with uncertain effects, though their importance for bird life and as fish nursery areas is well documented. They are also known to be very important areas for dentrification, the process by which nitrate is broken down to gaseous species and hence returned to the atmoshere. Furthermore, these estuarine and marsh environments represent one of the last truly natural areas in much of northern Europe, and to fail to pass on this legacy intact will seriously diminish the environment of our successors.

References

Aebischer, N.J., Coulson, J.C. and Colebrook, J.M. (1990) Parallel long-term trends across four marine trophic levels and weather. *Nature*, **347**, 751–753.

Cornell, S., Rendell, A. and Jickells, T. (1995) Atmospheric inputs of dissolved organic nitrogen to the oceans. *Nature*, **376**, 243–246.

Dickson, R.R., Kelly, P.M., Colebrook, J.M., Wooster, W.S. and Cushing, D.H. (1988) North winds and production in the eastern North Atlantic. *Journal of Plankton Research*, **10**, 151–169.

Ernst, W. (1980) Effects of pesticides and related organic compounds in the sea. *Helgoländer Meeresuntersuchungen*, **33**, 301–312.

Fonselius, S. (1982) Oxygen and hydrogen sulphide conditions in the Baltic Sea. *Marine Pollution Bulletin*, **12**, 187–194.

GESAMP (1982) *The Review of the Health of the Oceans*. UNESCO, Paris.

Grant, A. and Millward, R.N. (1997) Detecting community responses to pollution. In L.E. Hawkins and S. Hutchinson (eds), Responses of marine organisms to their environment, *Proceedings of the 30th European Marine Biology Symposium*, University of Southampton, Southampton, 201–209.

Hamilton, E.I. and Clifton, R.J. (1979) Isotopic abundances of lead in estuarine sediments, Swansea Bay, Bristol Channel. *Estuarine and Coastal Marine Science*, **8**, 271–278.

Lenton, T. (1998) Gaia and natural selection. *Nature*, **394**, 439–447.

Lewis, J.R. (1972) Problems and approaches to baseline studies in coastal communities. In M. Ruivo (ed.), *Marine Pollution and Sea Life*. Fishing News Books, London.

McGarvin, M. (1994) Precaution, science and the sin of hubris. In T. O'Riordan and J. Cameron (eds), *Interpreting the Precautionary Principle*. Earthscan, London, 69–101.

Nelissen, P. and Stefels, J. (1988) *NIOZ Report 1988–4*. NIOZ, Texel, the Netherlands.

North Sea Task Force (1993) *North Sea Quality Status Report 1993*. Oslo and Paris Commissions, London.

Royal Society (1983) *The Nitrogen Cycle in the UK: a Study Group Report*. The Royal Society, London.

Salomons, W. and de Groot, A.J. (1978) Pollution history of trace metals in sediments as affected by the Rhine River. In W.E. Krumbein (ed.), *Environmental Biogeochemistry and Geomicrobiology, Volume 1. The Aquatic Environment*. Ann Arbor Science, Ann Arbor, MI.

Salomons, W., Bayne, B.L., Duursma, E.K. and Forstner, U. (1988) *Pollution of the North Sea*. Springer-Verlag, Berlin.

von Bennekrom, A.J. and Salomons, W. (1980) Pathways of nutrients and organic matter from land to ocean through rivers. In J.M. Martin *et al.* (eds), *River Inputs to Ocean Systems*. UNEP/UNESCO, Paris.

Wells, P.G., Butler, J.N. and Hughes, J.S. (eds) (1995) *Exxon Valdez Oil Spill: Fate and Effects in Alaskan Waters*. ASTM, Philadelphia.

Further reading

Abel, B.B. and Axiak, V. (1991) *Ecotoxicology and the Marine Environment*. Ellis Horwood Series in Aquaculture and Fisheries Support, London.

Alexander, L.M., Heaven, A., Tennant, A. and Morris, R. (1992) Symptomology of children in contact with sea water contaminated with sewage. *Journal of Epidemiology and Community Health*, **46**, 340–344.

Clark, R.B. (1989) *Marine Pollution* (2nd edn). Oxford Sciences Publication, Oxford.

Coe, J.M. and Rogers, D.B. (eds) (1997) *Marine Debris: Sources, Impacts, and Solutions*. Springer-Verlag, New York.

Dahl, B. and Blanck, H. (1996) Toxic effects of the antifouling agent Irgarol 1051 on periphyton communities in coastal water microcosms. *Marine Pollution Bulletin*, **32**, 342–350.

Dixon, T.R. and Dixon, T.J. (1981) Marine litter surveillance. *Marine Pollution Bulletin*, **18**(68), 303–365.

Kay, D., Fleische, J.M., Salmon, R.L., Jones, F., Wyer, M.D., Godfree, P.F., Zelenauch-Jacquotte, Z. and Share, R. (1994) Predicting likelihood of gastroenteritis from sea bathing: results from randomised exposure. *Lancet*, **344**, 905–909.

Newman, P.S. and Agg, A.R. (1988) *Environmental Protection of the North Sea*. Heinemann, London.

Widdows, J., Donkin, P., Brinsley, M.D., Evans, S.V., Walkeld, P.N., Franklin, A., Law, R.J. and Waldock, M.J. (1995) Scope for growth and contaminant levels in North-Sea mussels *Mytilus edulis. Marine Ecology – Progress Series*, **127**, 131–148.

Wolfe, D. A. (ed.) (1987) Plastics in the sea. *Marine Pollution Bulletin*, **18**(6B), 303–365.

Chapter Fifteen

Urban air pollution and public health

Peter Brimblecombe

Topics covered

- Historical background
- Smoke
- Gaseous pollutants
- Smoke plus fog equals smog
- Health and smog
- Other smoke damage
- Smoke in the modern world
- Photochemical smog
- Effects of photochemical smog
- Other changes and their effects
- Solving the problem: the future

Editorial introduction

Pollution, risk, health effects and energy generation all connect through uncertain science and institutional failure. Pollution is a function of inappropriate property rights, a regulatory regime that is always subservient to capitalism, whether of the state or private kind, and sincere difficulty in proving effect from cause. Despite the lack of some kind of citizens' environmental right, the best arrangement so far is legislation such as the Michigan Environmental Protection Act of 1972 and the South African National Environmental Management Act of 1998. Both statutes permit citizens' groups to take action against a polluter or a regulatory agency in the cause of protecting their health and peace of mind. In the event, such legislation encourages the regulators to be tough and aggressive, rather than spawn a host of citizens' lawsuits.

In the modern world of pollution control, important changes are taking place.

- *Information* is more widely available as a result of regular state-of-the-environment reports nowadays produced by most countries in the northern hemisphere, and through information networks provided by the UN Environment Programme. The creation of the European Environmental Agency by the European Community has helped to standardise the collection and publication of scientific data on environmental quality. This is an important development, which will assist all countries to create independent and competent bureaux of environmental statistics.

● *The definition of environmental harm* is steadily widening to include the integrity of ecosystems and not just human health. Though this is by no means the equivalent of an environmental right, it does provide the regulator with a more ecologically based interpretation of environmental quality, and provides more scope for the precautionary principle to be put into effect.

● *Environmental quality standards*, regulatory assessment levels, *de minimis* risk levels – all these ambient or regional-scale yardsticks of healthy environments – are becoming more commonplace and unified as a basis for determining emissions and waste disposal tolerances. The concept of critical load, namely the threshold value of ecosystem absorption of pollutants to remain in good heart, is also in vogue. These are not easy standards to set. But science has an important role to play in providing advice on what levels to set, even though inevitably there is still much guesswork involved.

● *Arm's length regulatory agencies* and negotiating practices depend very much on the culture of regulation and the actual regulatory guidelines used. In the USA and Germany, the practice is usually more formal and target-driven, with the regulator acting somewhat as an ecological police force. In the UK and in southern Europe, as well as in Scandinavia, a more discretionary approach is taken to environmental standard setting, with the use of the words 'reasonable' or 'practicable' to soften the blow of strict enforcement. In practice, however, most regulatory officials do negotiate around a set of possible actions so long as there are technological or managerial options to best practice, so long as the managers are willing to comply, and so long as the levels of emission reduction are not excessively demanding in too short a time period for an impoverished polluter already in economic difficulties. One can see how, inevitably, there is room for compromise.

● *Up-the-pipe regulation* is a phrase to indicate that firms are increasingly looking to restructure their product lines, creating new products from what was formerly waste and seeking the stimulation of more efficient technology in the reduction of pollution. So the regulator does not seek to control the emission just at the point of discharge. With new environmental management systems such as the European Community eco-audit for industry (and eventually government), the international ISO 14001 and the US risk prevention programme of the USEPA, so management is beginning to incorporate environmental safeguards into the very core of business innovation.

● *Charging for polluting discharges*, for the cost of providing the licence fee, and to encourage the return of used material to proper disposal sites (deposit refund schemes) is becoming more popular. This is partly a reflection of a changing political ideology that favours the customer, rather than the taxpayer, as the source of revenue for services. It is also a reflection of the prevailing mood to reduce the burden of public spending by forcing the consumer to pay via indirect taxation. But, to be fair, it is also a sign of the new mood favouring environmental taxation as an incentive to direct behaviour towards sustainability, and to create revenue for precautionary investment.

All these developments require the reconstruction of institutions, economies and science around a more socially responsive and anticipatory approach to resource management and pollution control. The very nature of these changes also helps to alter the political climate favouring such reforms. Thus, as international action is more and more needed to cope with cross-border transfers of pollutants of all kinds, so countries look to the extended science of participation and communication to determine standards, support enforcement and justify charging for services supplied.

A study of the world's 20 megacities undertaken for the World Health Organisation (WHO) and published by the UN Environment Programme in the journal *Environment* (1994) revealed that each had at least one air pollutant that exceeded the WHO guidelines, 14 had two, and seven had three or more. Suspended particulates are over the limit in 17 of 20 cities, with 12 cities recording double the guideline level. Under winter inversion conditions, lethal combination of SO_x and suspended particulates is widespread in five of the 20 cities. Yet only six of these cities have adequate monitoring networks, whose provision is a real priority if adequate policies are to be tested for their effectiveness. By 2000, there will be 85 cities in the world with more than 3 million people, so the matter of tough regulatory control, linked to the economics and social indicators of improved health, will require capacity building of the utmost urgency. Yet one must bear in mind that over 80% of all air-pollution-related deaths in the developing world are associated with indoor pollution – see Table 15.1 (all figures in thousands).

The total health costs of all air pollution particulates is estimated to be $100 billion. Economic damage to crops due to air pollution ranges from $4.7 billion in Germany to $1.5 billion in Sweden. The car is a major target nowadays, as both the pollutants and the congestion costs may exceed $1 billion per major city per year. This is why action on clean-air technology, limited access to city centres, road pricing, parking charges or banning parking lots, as well as creating more 'village' societised economies in megacities, are now on the cards as serious policy options.

Historical background

Urban air pollution is now an issue of much public interest. In the 1980s it was overshadowed by concern over acid rain, the ozone hole and the greenhouse effect. However, the reaffirmation of the US Clean Air Act by President Bush in 1989, the World Health Organisation's *Healthy Cities* initiative, and the publication of the UK

Table 15.1: Deaths from air pollution (thousands per year).

	Deaths from indoor air pollution		Deaths from outdoor air pollution	Total
	Rural	*Urban*		
India	496	93	84	673
Sub-Saharan Africa	490	32	–	522
China	320	53	70	443
Rest of Asia	363	40	40	443
Latin America	180	113	113	406
Industrialised countries	–	32	147	179
Arab states	–	–	57	57
Total	1849	363	511	2723

Source: UN Development Programme (1998, 71).

government's *This Common Inheritance* and more specifically *Air Quality: Meeting the Challenge*, have reawoken interest in urban air quality. This reinterest stimulated a range of improved urban monitoring sites, such as Britain's Automated Monitoring Network. It has been sustained by a range of toxicological studies which have shown that the interaction of trace particles and gases in the urban air is more complex than was realised, and by the development of new suites of legislation (e.g. EC Directive 96/62/EC on *Ambient air quality monitoring and management*).

The modern concept of air pollution has had to break away from images of smoky chimneys of the past and address something far more subtle. The pollution we see in our cities today is very different from that of the past, so we must trace the way in which it has changed.

Cities have always been polluted, and before there were cities there were polluted huts and houses. In ancient Rome, where wood was burned, Emperor Nero's tutor Seneca complained of the bad effect that smoke had on his health. The Roman courts even dealt with cases where factory smoke annoyed nearby residents.

In London of the 13th century there was a particularly notable transition. The dramatic increase in population caused a fuel crisis, and wood was replaced by coal in some industrial processes such as 'cement' making. The fuel change was so noticeable in terms of the smoke and smell that residents feared for their health, and popular protest led to attempts to restrict the use of coal.

However, by the late 17th century coal was well entrenched in England in domestic as well as industrial use. Later the development of the steam engine and more broadly the Industrial Revolution changed the focus of life by requiring large amounts of labour concentrated around factories. Thus the early part of the 19th century saw a great increase in the population of cities.

This rapid increase in urban population was accompanied by numerous social problems. In particular, the serious health effects of pollution, disease and sanitation were something that urban administrations had never had to cope with on such a scale before. There were very early laws governing the smoke from steam engines in Britain and France, so we know that smoke was a problem which engineers, stokers and officials had to grapple with. In fact, most people were opposed to smoky cities, but smoke was generally seen as a necessary evil. Anderson and Ashby (1981) show how changing views on science, health and social responsibility combined to create Victorian air pollution legislation and action.

Smoke

The concern about air pollution in the 19th century focused on smoke: smoke that soiled clothes, blackened buildings and ruined health by its presence in the urban atmosphere. In fact, smoke has dominated thought about air pollution almost through to the present day. The visibility of smoke, the relative ease of measurement by visual observation, made it easy to study, within the confines of 19th century chemistry. Experts on public health linked air pollution to respiratory illnesses, rickets (as the result of a lack of sunlight) and even moral well-being ('cleanliness is next to godliness'). As we will see, the air pollution issues of the 20th century were to require considerable advance in scientific thinking.

How was this smoke generated? Fuels and their combustion lie at the centre of the air pollution problem. Air pollution does have other sources too, but by and large it is combustion that has traditionally been the most important source. The fuels we use are usually based on carbon combined with small amounts of hydrogen, even though quite exotic fuels such as metals are known for special applications (e.g. as solid rocket fuels).

If we imagine a fuel such as coal or oil, we could write its combustion according to the equation:

$$\text{‘CH’} + 2O_2 \rightarrow CO_2 + 2H_2O$$

coal/oil + oxygen $\rightarrow$ carbon dioxide + water

Now this looks fairly harmless for the urban environment, as carbon dioxide (although a greenhouse gas) is not really poisonous. However, let us imagine that there is not enough oxygen during combustion. The equation might then look more like:

$$\text{‘CH’} + O_2 \rightarrow CO + H_2O$$

coal + oxygen $\rightarrow$ carbon monoxide + water

Now we have produced carbon monoxide. This is a rather poisonous gas which combines with red blood pigment and can kill by asphyxiation at high concentrations. Thus it is hardly a desirable constituent of the urban atmosphere. Or with even less oxygen we might get carbon, or we could simply say smoke:

$$\text{‘CH’} + \tfrac{1}{2}O_2 \rightarrow C + H_2O$$

coal + oxygen $\rightarrow$ ‘smoke’ + water

At low temperatures where there is relatively little oxygen, reactions may cause a rearrangement of atoms that can lead to polycyclic aromatic hydrocarbons (PAHs). Typical of this class of compound is benzo(a)pyrene, B(a)P, a notorious cancer-inducing agent or carcinogen:

$$\text{‘CH’} + \tfrac{1}{2}O_2 \rightarrow B(a)P + H_2O$$

coal + oxygen $\rightarrow$ benzo(a)pyrene + water

Thus, although the combustion of fuels would initially seem a fairly harmless activity, it can produce a range of pollutant carbon compounds. Now the early engineers saw that an excess of oxygen would help to convert all the carbon to carbon dioxide. So they developed a philosophy of consuming smoke by burning it (often known as ‘burning your own smoke’), though this required considerable skill to implement and was often not very successful in practice.

Gaseous pollutants

Although the air pollution problem and the smoke problem have been linked, there were always those who thought there was more to air pollution than just smoke. They were right, because fuels are not burned in oxygen, as is suggested in the equations given above. They are burned in air, which is a mixture of oxygen and nitrogen. In flames molecules may fragment, and even the molecules of air may enter into a series of reactions:

$$O + N_2 \rightarrow NO + N$$

atomic oxygen + nitrogen $\rightarrow$ nitric oxide + atomic nitrogen

$$N + O_2 \rightarrow NO + O$$

atomic nitrogen + oxygen $\rightarrow$ nitric oxide + atomic oxygen

If we add these two reactions we get:

$$N_2 \quad + \; O_2 \quad = \quad 2NO$$

nitrogen + oxygen = nitric oxide

Note that the second of the above reactions produces an oxygen atom, which can go back and re-enter at the first reaction. Once an oxygen atom is formed in a flame it will be regenerated and contribute to a whole chain of reactions.

Nitrogen oxides (i.e. NO_2 + NO, or NO_x) in vehicle exhaust gases are derived in this way. They arise simply because we burn fuels in air rather than just in oxygen. Of course it must be admitted that some fuels contain nitrogen compounds as impurities, so that combustion products of these fuels may contain additional nitrogen oxides.

However, the most common and worrisome impurity in fossil fuels is sulphur. There may be as much as 6% sulphur in some coals, and this is converted to sulphur dioxide on combustion:

$$S \quad + \; O_2 \quad \rightarrow \quad SO_2$$

sulphur + oxygen → sulphur dioxide

There are other impurities in fuels too, but sulphur has been seen as the one most central to the air pollution problems of cities.

If we look at the composition of various fuels (Table 15.2) we see that they contain quite variable amounts of sulphur. In the case of coal we see that the amount of sulphur can be very high. Also, the sulphur concentrations in coal vary geographically. For instance, the coals of the eastern coalfields of the United States are rather higher in sulphur than those in the west. In Europe we find a similar situation, with Eastern European coal often having higher sulphur contents. When concern about pollution by sulphur compounds means that low-sulphur coals become more valued, this can have economic repercussions.

If we look at the list of sulphur contents we see that the highest amounts of sulphur are found in coals, lignites and fuel oils (see Table 15.2). These are fuels that are used in stationary sources such as boilers, furnaces (and traditionally steam engines), domestic chimneys, steam turbines, power stations, etc. As a simple rule, one can thus associate sulphur pollution of the air with stationary sources.

Table 15.2:

Sulphur content of fuels.

Fuel	S (%)
Coal	0.2–7.0
Fuel oils	0.5–4.0
Coke	1.5–2.5
Diesel fuel	0.3–0.9
Petrol	0.1
Kerosene	0.1
Wood	very small
Natural gas	very small[1]

Note:
1 Hydrogen sulphide is often removed from natural gases, but sulphides may also be added as an olfactant.

Smoke too is mainly associated with stationary sources. Naturally, steam trains and boats caused the occasional problem, but it was the stationary source that was most significant.

For many people, sulphur dioxide and smoke came to epitomise the air pollution problems of cities. Smoke and sulphur dioxide are called primary pollutants, because they are formed directly at the pollutant source, as we can see in the equations above. They then enter the atmosphere in this form. So in the case of primary pollutants it can be argued that given pollutants are clearly identified with particular sources. Thus the traditional air pollution problems experienced by cities have been associated with primary pollutants.

Smoke plus fog equals smog

Smoke can be seen, but sulphur dioxide is invisible. However, in the polluted city the combination of these two pollutants became very noticeable in smoke-laden fogs. Some people described these fogs as 'thick enough to spread on bread and butter'. In the first years of the 20th century, an air pollution expert with an enthusiasm for word play named this kind of fog 'smog', i.e., *sm*oke and f*og*, and the word has subsequently been used to describe much urban air pollution, even that not associated with smoke or fogs.

The classical London smog forms under damp conditions when water vapour can condense on smoke particles. Sulphur dioxide can dissolve into this water:

$$SO_2 \quad + H_2O = H^+ \quad + HSO_3^-$$

sulphur dioxide + water = hydrogen ion + bisulphite ion

Traces of metal contaminants catalyse the conversion of dissolved sulphur dioxide to sulphuric acid:

$$2HSO_3^- \quad + O_2 \quad \rightarrow 2H^+ \quad + 2SO_4^{2-}$$

bisulphite ion + oxygen → hydrogen ion + sulphate ion

Sulphuric acid has a great affinity for water, so the droplet tends to absorb more water, it becomes bigger and the fog thickens. It is a strong acid, and during the 1980s the production of sulphuric acid from sulphur dioxide was at the heart of the acid rain problem.

Health and smoke

The strange smells from combustion processes had always caused people to be concerned about the effects these 'vapours' might have on health. By the mid-1600s scientists were beginning to collect evidence of these effects. The higher death rates in London compared with rural areas were sometimes blamed on the smoke from coal. In the areas around metal furnaces local industrial diseases were known and these were often attributed to toxic materials such as antimony, arsenic or mercury within the smoke.

Terrible fogs plagued London at the turn of the 19th century, when Sherlock Holmes and Jack the Ripper paced the streets. Death rates invariably rose in periods of prolonged winter fog: little wonder considering that the fog droplets contained sulphuric acid. The medical experts of Victorian times recognised that the fogs were affecting health, but they, along with others, were not able to legislate smoke out of existence. Even where there was a will, and indeed there were enthusiasts in both Europe and North America who strove for change, the technology was far too naive to achieve really obvious improvements. The improvements that did come about were often due to changes in fuel, in location of industry, in climate, etc., rather than to changes in technology. The political pressures were, in the end, too weak to force profound changes.

In polluted conditions, the respiratory system could not clear itself of the particles that were inhaled. The cilia that normally swept the respiratory passages clear became anaesthetised, and the particles penetrated deeper. Some people susceptible to respiratory diseases became ill. Others were apparently healthier, but the particles could still cause long-term problems. There were toxic trace metals on the surface of some of the tiny soot particles, and others contained compounds such as benzo(a)pyrene, a potent carcinogen. These may have contributed to a high incidence of cancer in smoky cities.

In the 1990s an American scientist, Douglas Dockery, analysed pollution and health records from a number of cities. He showed that fine particles seemed to be the principal cause of death, despite the presence of other pollutants. This discovery has led to small particles being assigned a much greater role in causing ill health. The mechanism is not understood, although some have argued that these very small particles mimic airborne pathogens and induce a kind of chronic inflammation deep in the lung. It is the very finest particles that are most critical in affecting health. Currently, monitoring stations monitor particles of the size we breathe in, i.e. 10 µm in diameter (PM_{10}). However, the USA and Europe are considering legislation that accounts for 2.5 µm particles ($PM_{2.5}$) and some have even raised concern over the ultrafine particles of 0.1 µm diameter that arise from diesel engines ($PM_{0.1}$). It will be hard for monitoring efforts to keep pace with pressures to gather information and regulate smaller and smaller particles (see Box 15.1).

Smoke damage

Smoke not only affected health. Its impact on the urban setting was easy to see. Even today it is possible to notice the black encrustations on older buildings in many large cities. In the past, when there were virtually no effective controls on smoke emissions, the damage done by smoke was even more obvious.

Until the passage of the Clean Air Acts, smoke abatement enthusiasts were anxious to draw attention to the vast costs that were incurred by the presence of smoke in the atmosphere. Clothes were soiled, curtains and hangings blackened and the exteriors of houses spoiled. Much of the cleaning work fell to women, so it is understandable that they had strong views on the undesirability of smoke. It was they who had to wash the white shirts worn by businessmen, who sometimes had to change into a new shirt for the afternoon after the morning's garment had become soiled. Although it was not easy for women to make their views felt, records suggest that some of them saw the presence of smoke in the atmosphere as almost a moral issue. If cleanliness was next to godliness, then smoke which made things so dirty might be seen as evil.

Smoke could also affect the growth of plants. Gradually, we have adopted for agricultural use varieties which are more resistant to air pollution. City gardens too tend to be stocked with more resistant plants. In the past the trees around industrial centres became so blackened that light-coloured butterflies and moths were no longer camouflaged. Melanic (dark) forms became more common because predators could see them less easily. Plants are also very sensitive to sulphur dioxide, and one of the first effects seems to be the inhibition of photosynthesis.

As we have seen, urban air contained sulphuric acid as well as smoke. Sulphuric acid is a powerful corrosive agent. It rusted iron bars and ate away the stone of buildings. Architects sometimes complained of layers of sulphate damage 10 cm thick on calcareous stone. Such building stone was attacked through the reaction

Box 15.1 (a): Air quality guidelines for major air pollutants

Pollutant	– United Kingdom –			WHO	
	Proposed	Standard	Specific Objective	Standard	
Benzene	5 ppb	running annual mean	5 ppb to be achieved by 2005	4.7–7.5×10^{-6} lifetime risk $(\mu g/m^3)^{-1}$	
1,3-Butadiene	1 ppb	running annual mean	1 ppb to be achieved by 2005		
CO	10 ppm	running 8-hour mean	10 ppm to be achieved by 2005	$100\,000\ \mu g/m^3$ $60\,000\ \mu g/m^3$ $30\,000\ \mu g/m^3$ $10\,000\ \mu g/m^3$	15 min 30 min 1 hour 8 hour
Lead	$0.5\ \mu g/m^3$	annual mean	$0.5\ \mu g/m^3$ to be achieved by 2005	$0.5\ \mu g/m^3$	1 year
NO$_2$	104.6 ppb	1-hour mean	104.6 ppb, measured as the 99.9th percentile, to be achieved by 2005	$200\ \mu g/m^3$ 40–$50\ \mu g/m^3$	1 hour 1 year
Ozone	50 ppb	running 8-hour mean	50 ppb, measured as the 97th percentile, to be achieved by 2005	$20\ \mu g/m^3$	8 hour
PM$_{10}$	$50\ \mu g/m^3$	running 24-hour mean	$50\ \mu g/m^3$ measured as the 99th percentile, to be achieved by 2005		
SO$_2$	100 ppb	15-minute mean	100 ppb measured as the 99.9th percentile, to be achieved by 2005	$500\ \mu g/m^3$	10 min

ppm = parts per million; ppb = parts per billion; $\mu g/m^3$ = micrograms per cubic metre

Box 15.1 (b): Proposed European Union limit values for some important air pollutants

	Averaging period	Limit value	Date by which limit value is to be met
Sulphur dioxide			
1. Hourly limit value for the protection of human health	1 hour	350 μg/m^3 not to be exceeded more than 24 times per calendar year	1 January 2005
2. Daily limit value for the protection of human health	24 hours	125 μg/m^3 not to be exceeded more than 3 times per calendar year	1 January 2005
3. Limit value for the protection of ecosystems, to apply away from the immediate vicinity of sources	calendar year and winter (1 October to 31 March)	20 μg/m^3	two years from coming into force of the directive
Nitrogen dioxide and nitric oxide			
1. Hourly limit value for the protection of human health	1 hour	200 μg/m^3 NO$_2$ not to be exceeded more than 8 times per calendar year	1 January 2010
2. Annual limit value for the protection of human health	calendar year	40 μg/m^3 NO$_2$	1 January 2010
3. Annual limit value for the protection of vegetation to apply away from the immediate vicinity of sources	calendar year	30 μg/m^3 NO + NO$_2$	two years from coming into force of the directive
Lead			
Annual limit value for the protection of human health	calendar year	0.5 μg/m^3	1 January 2005
Particulate matter			
1. 24-hour limit value for the protection of human health	24 hours	50 μg/m^3 PM$_{10}$ not to be exceeded more than 25 times per year	1 January 2005
2. Annual limit value for the protection of human health	calendar year	30 μg/m^3 PM$_{10}$	1 January 2005

$$H_2SO_4 \quad + \; CaCO_3 \qquad \rightarrow H_2O \; + \; CO_2 \qquad + \; CaSO_4$$

sulphuric acid + calcium carbonate → water + carbon dioxide + calcium sulphate

This reaction would seem to be rather a good one because it gets rid of the sulphuric acid and converts limestone ($CaCO_3$) into another building material, gypsum ($CaSO_4 \bullet 2H_2O$). However, gypsum is rather soluble and dissolves in rain. The other problem is that gypsum occupies a larger volume than limestone, so the stone almost explodes from within. As sulphur dioxide has decreased within the urban atmosphere, the rate of degradation of building stones in cities seems to have become slower, although it is a complex cumulative process and buildings may be attacked by other air pollutants.

Smoke in the contemporary atmosphere

Smoke and smog are not just problems of the past. There are still many cities where coal is heavily used in poorly controlled furnaces and in domestic fireplaces. This is particularly true in the developing world. Shanghai, for example, uses vast quantities of coal and has great problems in trying to reduce the soot concentrations in its atmosphere. There are a large number of small furnaces to control, and switching the domestic user to less polluting coal-gas takes time.

However, in Western Europe and North America the problems of urban smoke have largely disappeared and sulphur dioxide levels are declining. This is partly due to changes in fuel use, especially domestically, from coal to 'cleaner' fuels such as gas and electricity. In Britain, the Clean Air Acts also contributed something to the reduction of smoke. Where coal is used in developed countries today it tends to be concentrated in large plants, often situated far from cities. These usually have good control on grit and particle emissions, so coal smoke is not such an important problem.

Some of the more positive changes in urban air pollution have not always been so evident outside Europe and North America. In the developing world, there are often great pressures to industrialise, which frequently limits resources available for abatement and control of air pollutants within cities. By the year 2025 as many as 100 of the estimated 135 cities of 4 million people or more will be in developing countries. Poor controls on industrial emissions, especially those that are state-owned, have meant that many of these cities suffer from serious air pollution. China, where a large amount of coal is used, has had special problems with SO_2. Calcutta has a mix of air pollution from industrial and automotive sources. Here a large, though ill-maintained, vehicle fleet often uses diesel fuel and causes a remarkable reduction in the visibility along major roads as the day progresses. The smog of Mexico City has become notorious and seems to have been made worse by an increasingly reactive volatile component of fuels from cars. These hydrocarbons are the catalytic precursors to photochemical smog. Other cities, such as Christchurch in New Zealand, have struggled against the generation of pollution from domestic solid fuels and found themselves confronting serious problems with carbon monoxide and airborne particulate material, despite a high standard of living. Air pollution has often spread well beyond the urban boundaries and wide areas of countryside find themselves suffering from acid rain in China, South-east Asia and South America. The widely reported forest fire smogs of South-east Asia have raised further air pollution concerns in the developing world.

However, even in Europe, where the total smoke loading of the urban atmosphere may have decreased, there have been important changes. Diesel-powered vehicles have been increasingly in evidence, and by no means all of them are large vehicles. Today a substantial number of cars are diesel-powered, taking advantage of the lower fuel costs. Diesel fuels also have the advantage of being unleaded. On the other hand, the droplets of fuel dispersed within the engine may not always burn completely. This means that diesel engines can produce large quantities of smoke (as PM_{10} or finer) if not properly maintained. This smoke now makes a significant contribution to the soiling quality of urban air.

As mentioned earlier, diesel particles are so small that they pose considerable risk to health. In addition, the particles are rich in polyaromatic hydrocarbons (PAHs), which are carcinogens. Fortunately, many PAH compounds show declining concentrations in the atmosphere, but there is some evidence that particularly carcinogenic forms are still found in the atmosphere. PAHs can react and become nitrated or halogenated, which makes these compounds potentially more carcinogenic. The diesel should not take the entire blame for the formation of these substituted organic compounds. Refuse incineration (most especially that involving chlorinated plastics) can be a particularly problematic source of chlorinated compounds. These include chlorinated PAHs, polychlorinated biphenyls (PCBs) and the notorious dioxins. Diesel fuels have also had a high sulphur content, but legislative pressures are lowering that.

Modern winter smogs

Like smoke, sulphur dioxide has been seen as a declining problem in the cities of North America and Europe. The long-term average concentrations have fallen dramatically in many of the larger cities throughout the twentieth century. Some of the decline has been the result of deindustrialisation of the urban core, electrification, the shift away from coal to gas, and legislation. Thus UK cities no longer exceed the limit values defined in *Smoke and sulphur dioxide*, Directive 80/779/EEC, although some cities, such as Belfast, are still smoky enough to exceed guide values.

Older regulations were based on daily and annual average sulphur dioxide concentrations, but recently it has become clear that response to this pollutant within the first few minutes of inhalation and continuing exposure does not increase the effects. This has led to modern regulations accounting for very short exposure. In the UK, the recommendation is that sulphur dioxide should not exceed 100 ppb over a 15-minute averaging period. This standard requires monitoring stations to collect data with a high temporal resolution, and only modern instrumentation can achieve this. It has become clear that the 15-minute standard is widely exceeded at both urban and rural sites in the UK. These exceedances most typically represent exposure of the sites to the plumes of larger sources, such as power stations.

Although areas such as Belfast and Barnsley can also exceed this standard because of the winter use of domestic coal burning, contemporary British cities are more likely to see nitrogen oxide smogs in the winter. These are dominated by high concentrations of NO_2 rather than SO_2. At low temperatures and high NO concentrations, the normally slow reaction

$$2NO + O_2 \rightarrow 2NO_2$$

tends to take place more rapidly, allowing high concentrations of NO_2 to form.

Although diesel fuels can have high concentrations of sulphur, traffic does not appear to be responsible for short-term peaks in sulphur dioxide concentrations, so improvements require emission reduction from large coal-burning plants.

However, sulphur dioxide removal from chimney stacks remains a comparatively costly undertaking. Electricity generators in countries such as Britain, the United States, Poland, etc. have been slow to apply this abatement technology. However, international agreements requiring sulphur dioxide removal from flue gases and shifts in fuel types have reduced sulphur deposition such that acid rain begins to be more the product of nitric than of sulphuric acid. An important fraction of this nitric acid derives from cars, which do not contribute to atmospheric sulphuric acid. In some parts of Europe the sulphur deposit is low enough for farmers to complain about having to add sulphur to soils as fertiliser, when in the past it came free as an air pollutant.

Photochemical smog

The air pollutants that we have been discussing so far usually come from stationary sources. The fuel is predominantly coal. This is very much the traditional type of pollution that has been experienced by cities for as long as coal has been burned; or if we think of smoke alone, we might say for as long as fuel has been burned.

However, the twentieth century has seen the emergence of an entirely new kind of air pollution. This pollution arises particularly when volatile liquid fuels are used, and hence the motor vehicle has been a big contributor. However, most of the actual pollutants which are causing the problems are not themselves emitted by motor vehicles. Rather, they form in the atmosphere. So they are called secondary pollutants – they are formed from the reactions of primary pollutants, such as nitric oxide and unburnt fuel, that come directly from the automobiles. The chemical reactions that produce the secondary pollutants proceed most effectively in sunlight, so the result is called photochemical smog.

Photochemical smog first began to be noticed in Los Angeles during the Second World War (see Table 15.3). It was unique, so initially it was assumed to be similar to

Table 15.3: Comparison of Los Angeles and London smog.

Characteristic	Los Angeles	London
Air temperature	24 to 32°C	–1 to 4°C
Relative humidity	<70%	85% (+ fog)
Type of temperature inversion	Subsidence, at a few thousand metres	Radiation (near ground) at about 100 metres
Wind speed	<3 m/s	Calm
Visibility	<0.8 to 1.6 km	<30 m
Months of most frequent occurrence	Aug.–Sept.	Dec.–Jan.
Major fuels	Petroleum	Largely coal
Principal constituents	O_3, NO, NO_2, CO, organic matter	Particulate matter, CO, S compounds
Type of chemical reaction	Oxidative	Reductive
Time of maximum occurrence	Midday	Early morning
Principal health effects	Temporary eye irritation (PAN)	Bronchial irritation, coughing
Materials damaged	Rubber cracked (O_3)	Iron, concrete corroded (SO_2/smoke)

Source: Andrews et al. (1996).

the air pollution that had been experienced elsewhere. However, when conventional smoke abatement techniques failed to make any impression, it became clear that this pollution must be different, and the experts were initially baffled. It was Haagen-Smit, a biochemist studying vegetation damage, who finally realised that the Los Angeles smog was caused by reactions of automobile exhaust vapours in sunlight.

How does this happen? The reactions are quite complicated, but we can simplify them by substituting a very simple organic molecule such as methane (CH_4) to represent the vehicle emissions:

$$CH_4 + 2O_2 + 2NO + h\upsilon \rightarrow H_2O + HCHO + 2NO_2$$

methane + oxygen + nitric oxide + sunlight → water + formaldehyde + nitrogen dioxide

Nitric oxide is a common pollutant from automobiles. We can see two things taking place in this reaction. First, the automobile hydrocarbon is oxidised to an aldehyde (i.e. a molecule with a CHO group). In the reaction above it is formaldehyde. Aldehydes are eye irritants and, some have argued, also carcinogens. Second, we can see that the nitrogen oxide has oxidised to nitrogen dioxide. This is a brownish gas. It can absorb light and dissociate:

$$NO_2 + h\upsilon \rightarrow O + NO$$

nitrogen dioxide + sunlight → atomic oxygen + nitric oxide

This re-forms the nitric oxide but also gives an isolated and reactive oxygen atom that can react to form ozone:

$$O + O_2 \rightarrow O_3$$

atomic oxygen + molecular oxygen → ozone

It is this ozone that characterises photochemical smog. Note the important fact that ozone, which we regard as such a problem, is not emitted by any major polluter: it is the product of the interaction of a number of pollutants in the atmosphere. The detailed reactions are rather more complicated than illustrated here (see Box 15.2).

The smog found in the Los Angeles basin is very different from what we have previously described as typical of coal-burning cities. There is no fog when Los Angeles smog forms and visibility does not decline to just a few metres as was typical of the London fogs. The Los Angeles smog forms best, of course, on sunny days. London fogs are blown away by a wind, but gentle sea breezes in the Los Angeles basin can hold the pollution in against the mountains and prevent it escaping out to sea. The pollution cannot rise in the atmosphere, because it is trapped by an inversion layer: the air at ground level is cooler than that aloft, thus a cap of warm air prevents the cooler air rising and dispersing the pollutants. A fuller list of the differences between Los Angeles and London-type smogs is given in Table 15.3.

Effects of photochemical smog

Photochemical smog is quite unlike the smoky air typical of cities of old, so we would not expect thick crusts of soot to be growing on buildings. Petrol, unlike coal, contains relatively little sulphur, hence sulphate damage is not so likely to occur either. However, there are plenty of other pollutants that can damage materials.

Ozone is a particularly reactive gas. It attacks the double bonds of organic molecules very readily. Rubber is a polymeric material with many double bonds, so it is attacked very easily by ozone. Rubber exposed to ozone shows cracks, and tyres and

Box 15.2 Reactions in photochemical smog

As seen in the text, reactions involving nitrogen oxides and ozone lie at the heart of photochemical smog:

$$NO_2 + h\upsilon(<310 \text{ nm}) \rightarrow O(^3P) + NO \qquad [R1]$$

$$O(^3P) + O_2 + M \rightarrow O_3 + M \qquad [R2]$$

$$O_3 + NO \rightarrow O_2 + NO_2 \qquad [R3]$$

These may be imagined as represented by a pseudo-equilibrium constant relating the partial pressures of the two nitrogen oxides and ozone:

$$K = [NO].[O_3]/[NO_2] \qquad [R4]$$

If we were to increase NO_2 concentrations (in a way that did not use ozone) then the equilibrium could be maintained by increasing ozone concentrations. This happens in the photochemical smog through the mediation of OH radicals (OH is an important radical present in trace amounts in the atmosphere) in the oxidation of hydrocarbons:

$$OH + CH_4 \rightarrow H_2O + CH_3 \qquad [R5]$$

$$CH_3 + O_2 \rightarrow CH_3O_2 \qquad [R6]$$

$$CH_3O_2 + NO \rightarrow CH_3O + NO_2 \qquad [R7]$$

$$CH_3O + O_2 \rightarrow HCHO + HO_2 \qquad [R8]$$

$$HO_2 + NO \rightarrow NO_2 + OH \qquad [R9]$$

Thus these reactions represent a conversion of NO to NO_2 and of an alkane to an aldehyde. Note that the OH radical is regenerated, so can be thought of as a kind of catalyst. Aldehydes may also undergo attack by OH radicals:

$$CH_3CHO + OH \rightarrow CH_3CO + H_2O \qquad [R10]$$

$$CH_3CO + O_2 \rightarrow CH_3COO_2 \qquad [R11]$$

$$CH_3COO_2 + NO \rightarrow NO_2 + CH_3CO_2 \qquad [R12]$$

$$CH_3CO_2 \rightarrow CH_3 + CO2 \qquad [R13]$$

The methyl radical in [R13] may re-enter [R7]. An important branch to this set of reactions is:

$$CH_3COO_2 + NO_2 \rightarrow CH_3COO_2NO_2 \qquad [R14]$$

leading to the formation of the eye irritant PAN (peroxyacetyl nitrate).

---

windscreen wiper blades are especially vulnerable. Newer synthetic rubbers have the double bonds more protected by other chemical groups, so they are somewhat resistant to damage by ozone. Many pigments and dyes are also attacked by ozone. The usual result of this is that the dye fades. This means that it is important for art galleries in polluted cities to filter their air carefully, particularly where they house collections of paintings using traditional colouring materials, which are especially sensitive.

Nitrogen oxides too are associated with photochemical smogs. They may also damage pigments. It is possible that nitrogen oxides may also increase the rate of damage to building stone, but it is not really clear how this takes place. Some have argued that nitrogen dioxide increases the efficiency of production of sulphuric acid on stone surfaces in those cities that have moderate sulphur dioxide concentrations. Others have suggested that the nitrogen compounds in polluted atmospheres enable micro-organisms to grow more effectively on stone surfaces and enhance the biologically mediated damage.

It is not just materials that are damaged by photochemical smog. Living things suffer too. Plants are especially sensitive to ozone: hence the early observations of damage that led to the research by Haagen-Smit. Ozone damages plants by changing the 'leakiness' of cells to important ions such as potassium. Early symptoms of such injury appear as water-soaked areas on the leaves.

Human health can also be affected by these gases. In general, oxidants such as aldehydes cause eye, nose and throat irritation as well as headaches. Ozone impairs pulmonary function and at high concentrations nitrogen oxides can behave similarly, especially in asthmatics. Eye irritation is a frequent complaint during photochemical smogs. This arises from the presence of a group of chemicals, known as peracyl nitrates (PANs), which form through a reaction of nitrogen oxides and various organic compounds in the smog. The best-known of these eye irritants is peroxyacetyl nitrate, or PAN, which is responsible for the particularly irritant quality of the Los Angeles atmosphere. The reactions that produce PAN are shown in Box 15.2.

Other changes and their effects

Photochemical smog is not the only pollution problem created by vehicles. They are also associated with other pollutants such as lead and benzene. The success of lead tetra-alkyl compounds as anti-knock agents for improving the performance of automotive engines has meant that, in countries with high car use, very large quantities of lead have been mobilised. This lead has been widely dispersed, but particularly large quantities have been deposited in cities and near heavily used roads. Lead is a toxin and has been linked to a number of environmental health problems. Perhaps the most worrying evidence has come from studies which indicate a decline in intelligence among children exposed to quite low concentrations of lead.

Unleaded petrol was introduced in the USA in the 1970s so that catalytic converters could be used on cars. Since that time, unleaded petrol has come to be used more widely. There is evidence that blood lead concentrations have dropped in parallel with the declining automotive source of lead. Nevertheless, a decrease in atmospheric lead may not be sufficient to reduce the risk in children to a satisfactory level. The high ratio of food intake to body weight among children means that they are likely to gain much of their lead burden from sources such as food and water (although some of the lead in these may have come from the atmosphere).

Benzene is a further worrying component of automotive fuels. It occurs naturally in crude oil and is a useful component because it can limit knocking in unleaded petrol (the production process is usually adjusted so that the benzene concentration is about 5%). However, benzene is a potent carcinogen. It appears that more than 10% of the benzene used by society (33 million tonnes per year) ends up being lost to the environment. High concentrations can be found in the air of cities, and these concentrations may increase the number of cancers. Exposure is complicated by the importance of other sources of benzene to humans, such as tobacco smoke.

Toluene is another aromatic compound present in large concentrations in petrol. Evidence would suggest that it is far less likely to be a carcinogen than benzene, but it does have some very undesirable effects. It contributes significantly to the production of ozone and aldehydes in smoggy atmospheres. In addition, it reacts to form a PAN-type compound, peroxybenzoyl nitrate, which is a particularly potent eye irritant.

Solving the problem – the future

A solution to the problem of air pollution would seem relatively simple. We should emit less pollutants. Indeed, that philosophy is inherent in laws such as the UK Clean Air Act 1956. Here the aim was to reduce the amount of smoke by requiring smokeless fuels to be burned in some areas. Later legislation in the UK and elsewhere sought to change the composition of some fuels, requiring them to have low sulphur or low lead contents.

Another solution was to attempt to disperse the pollutants from chimneys more widely through the use of tall stacks. This often has the effect of decreasing the pollutant concentrations locally while increasing them downwind. Consequently, this approach is widely criticised, particularly in relation to the acid rain problem. However, it is hard to imagine that chimneys will vanish from our townscapes, and wisely used they probably represent an appropriate way to lower exposure to pollutants. But on large coal-burning plants, tall chimneys need to be combined with other, more active pollution abatement techniques.

When approaching the question of secondary pollutants we have to be aware that the problem is a good deal more subtle. In some cases stopping the emission of one pollutant can actually increase the secondary photochemical pollutant concentration in the air! Attempts to solve the pollution problems of cities with severe photochemical problems have not been a complete success. Los Angeles typically has great difficulty in meeting the US air quality guidelines.

One way of attacking the problem of secondary pollutants is to eliminate the 'catalyst' for their production. In this case we would wish to remove the hydrocarbons that lie at the heart of the photochemical reactions. These were represented as methane in our illustration (see Box 15.2) but in reality are usually somewhat larger hydrocarbon molecules. This approach was attempted through fitting cars with catalytic converters, which destroyed the hydrocarbons in their exhausts. However, hydrocarbons still escaped by evaporation at filling stations or from hot engines while the cars were stationary. One solution to this is to design fuels of lower volatility so that less evaporates, or to ensure those that do are relatively unreactive (i.e. have a low photochemical ozone).

Such attempts to reduce secondary pollution assume that cities will continue to be filled with hydrocarbon-fuelled vehicles. However, the most obvious way of reducing vehicle emissions is to curtail the use of the private car. This would necessitate the creation of cheap, efficient and convenient public transport systems. Facilities for cyclists and pedestrians would also need to be improved, although a significant reduction in traffic would itself go some way towards making walking and cycling safer, healthier and more attractive. Incentives to encourage car sharing could contribute to reducing traffic, and the use of electric cars would cut down pollution in cities, though the pollution caused by the generation of the electricity cannot be ignored. Moving freight where possible from road to rail (which is generally more fuel-efficient) would also reduce emissions.

Ultimately a lasting solution to the problems of urban air pollution may involve restructuring cities and a change in lifestyle, perhaps with people living nearer their workplaces (or working from home) and buying more locally produced goods, and fuel economy being a priority in the minds of planners. Fuel economy also has advantages, of course, in terms of reducing resource use and curtailing the greenhouse effect.

However, vehicles are not the only source of urban air pollution. Incineration of waste, whether domestic or industrial, can cause considerable air pollution, and modern industrial processes can lead to the emission of a wide range of novel air pollutants. As most city dwellers spend the vast majority of their time indoors, interior pollution can be a significant factor too. For instance, chipboards and insulating foam can release formaldehyde, gas cookers produce nitrogen dioxide, slow-burning wood stoves emit various carcinogens, and cigarette smoking produces a large number of harmful compounds, including benzene and carbon monoxide. So the solution to urban air pollution problems no longer simply revolves around the reduction of smoke. Today, if we really want to reduce human exposure to air pollutants, we need to consider and regulate an enormous range of pollution sources.

References and further reading

For historical and literary background, see Brimblecombe, P. (1987), *The Big Smoke*, Methuen, London; Brimblecombe, P. (1990), Writing on smoke, in H. Bradby (ed.), *Dirty words*, Earthscan, London, 93–114; and Anderson, M. and Ashby, E. (1981), *The Politics of Air Pollution*. Oxford University Press, Oxford.

For air pollutants in the modern world, see Elsom, D.M. (1992), *Atmospheric Pollution*, Blackwell Science, Oxford.

Regarding building damage, see the explanatory report by Cooke, R.U. and Gibbs, G.B. (1995), *Crumbling Heritage?*, National Power plc and PowerGen plc, London.

For the chemistry and photochemistry of smog, see Brimblecombe, P. (1996), *Air Composition and Chemistry*, Cambridge University Press, Cambridge; and Andrews, J.E., Brimblecombe, P., Jickells, T. and Liss, P.S. (1996), *An Introduction to Environmental Chemistry*, Blackwell Science, Oxford.

Looking at contemporary issues in the UK see The Quality of Urban Air Review Group (1993), *Urban Air Quality in the UK*, Department of Environment, London; and Department of Environment (1997), *The United Kingdom National Air Quality Strategy*, London.

The statistics in the editorial introduction came from

UN Environment Programme (1994) Air pollution in megacities. *Environment*, **36**(2), 5–13, 25–38.

UN Development Programme (1998) *Human Development Report 1998*. Oxford University Press, Oxford, 70–71.

Chapter Sixteen

Preventing disease

Robin Haynes

Topics covered

- Clean water and sanitation
- Air pollution and health
- AIDS
- Ultraviolet radiation and skin cancer
- Radiation
- Science and management

Editorial introduction

Health is increasingly being regarded as a holistic and socio-cultural phenomenon rather than a purely medical matter. This means that the four roles of health management – prevention, promotion, cure and rehabilitation – have to be integrated both into so-called 'complementary' medical procedures and into socio-economic development patterns that place the physical and mental well-being of people as first priority. We have already seen in Chapter 6 how there is a close link between health provision at a community level, better civil and economic rights for women, and a lowering of population growth. In this chapter, rural development in certain areas, notably Central and East Africa and parts of South-east Asia, may be critically influenced by the diminishing number of able-bodied young adults, whose siblings are infected by AIDS and by other debilitating and socially influenced diseases.

Health care is also very much a factor of service provision in a manner that is socially acceptable, readily accessible (in distance and money terms) and scientifically sound. Health provision need not require medical specialisation at its primary level: much can also be done through community nursing, preventive health and welfare programmes, health education in schools, and adequate (but modest) funds to ensure satisfactory service. Much can be achieved through self-reliant measures that are interconnected with social and economic development programmes. Sadly, these are rarely in place. For the price of a couple of dozen high-technology fighter aircraft per year, primary health care could be provided to over 90% of the world's peoples with plenty of scope for job training and community employment. Yet after a decade of the international drinking water and sanitation supply effort of various United Nations agencies, over 1.3 billion people still do not have access to drinkable water, and nearly 2.5 billion are unable to enjoy minimal sanitation facilities (UN Development Programme, 1998, 68).

The ideas are there, the practice usually fails. Chemical control of vector-borne diseases such as river blindness (onchocerciasis) and malaria is proving to be inadequate and expensive. In the case of river blindness, biological control via bacteria is being shown to be most successful and culturally accommodating. Mosquito protection netting, combined with environmental improvements of ecologically unnecessary water-pools is also recognised as a better bet than extensive use of DDT or other inexpensive but non-systemic pesticides. Yet malaria claims 100 million clinical cases and at least 5 million deaths every year. In many frontier areas, where forests are being removed, six out of ten young adults are affected.

A matter of much interest to epidemiologists is the relationship between immature, or vulnerable, immune systems and exposure to external agents, such as increased ultraviolet light, carcinogens, and indoor pollutants from fires, furniture and carpets, and mites. At one level, disease is spread amongst individuals with weakened natural defences. At another level, systematic breakdown of immune systems could become genetically inherited and progressive. We simply do not know: hence the concern over microtoxicity, such as is found in organochlorines and organophosphate products linked to fertility suppression and mental depression. Vivian Howard (1997, 192) claims that the additive, or synergistic, effects of the 300–500 chemicals we now have in our bodies since 1950 could have more powerful consequences than the individual effects, taken singly. *The Ecologist* (1998), frequently cited as being extremist and anti-scientific, yet brave enough to publicise 'rethought assumptions', carries a number of articles suggesting that cancer-creating agents are much more widespread than is normally accepted by 'Type I' science of the kind introduced in Chapter 1. The renowned American critic Samuel Epstein (1998, 71) claims that cancer rates are increasing by 6% overall, and by 9% among patients over 75. He looks to carcinogenic pesticides, and to PCB residues, as primary causes, as well as termite-eradication organochlorines (*ibid.*, 76).

Ian Langford and Rosie Day (1998, 18) summarise studies that show increased mortality due to heat stress following very hot days, of the kind that will become more common if global warming takes hold. The estimates are a doubling of such deaths by 2020 and a sevenfold increase by 2050. Malaria incidence may increase by 25% if temperatures continue to rise, so the distribution of malaria-infected mosquitoes widens. Furthermore, incidence of food poisoning may increase by up to 15% over the same period if temperatures rise by 2–3°C.

Epidemiology is a science of both statistical inference and careful analytical judgements. This is a highly politicised realm, so any statement by a medical officer or health specialist is treated with considerable attention. This is a battleground arena for the kinds of science depicted in Box 1.7. Rarely can cause and effect be unambiguously linked. Leukaemia clusters may or may not be attributable to human-created radiation. In the minds of those seeking 'blame' (see Chapter 17), such clusters clearly spell trouble, so scientific 'proof' fits into predetermined frameworks of belief and values. Environmental groups campaigned successfully against lead in petrol and nitrates in water, because formal scientific caution refused to lay blame on causes that were deemed unacceptable and removable. Environmental health and management is adopting the 'civic' approach to science introduced in Chapter 1. This means ensuring that the very best scientific analyses are available, but ensuring a strong element of precaution where there is public alarm or avoidable danger. It also means incorporating the voices of 'lay' viewpoints in scientific advisory bodies, coupled with consensus-building techniques. The environmental health phenomenon is becoming a matter for the integration of ethical and medical values around socially significant themes such as preventive education, access to health services, social justice and poverty, diminishing self-esteem and eating disorders, and reinforcement of holistic approaches to human well-being.

Influences on health

If 'the environment is everything that isn't me', as Albert Einstein said, then the environment is largely responsible for human health and disease. Internal genetic factors certainly affect vulnerability to some diseases, but the environment generally has a stronger influence. Epidemiologists have estimated that 75–80% of all cancers, for example, are caused by environmental conditions, which include diet and smoking. Figure 16.1 summarises the main influences. Since many aspects of our environment are within our own control (unlike our genetic inheritance), much disease is preventable by changing the environment – in theory, at least. There are clear links between the work of the scientist investigating environmental associations with ill health and management of the environment to prevent illness and promote good health. This chapter illustrates some of the connections, and it will also sketch out the differing ethics and responsibilities of the scientist and the manager in areas where the causes of disease have yet to be established.

Clean water and sanitation

Since the early 19th century, death rates in Western industrialised countries have been falling, and they now stand at about half previous levels. The greatest improvements in survival have been in infants and children. Diseases which were rampant such as tuberculosis, cholera, dysentery, typhoid, typhus, smallpox, scarlet fever, measles, whooping cough and diphtheria are now eliminated or controlled. Environmental reforms such as the introduction of clean water supplies, sewerage systems, simple food hygiene, better housing and control of working conditions were responsible for much of the decline in the major endemic infectious diseases. Acute pollution incidents such as inner city smogs (Box 16.1) have largely been eliminated. The other important change was improvements in diet resulting from advances in agriculture, which enhanced

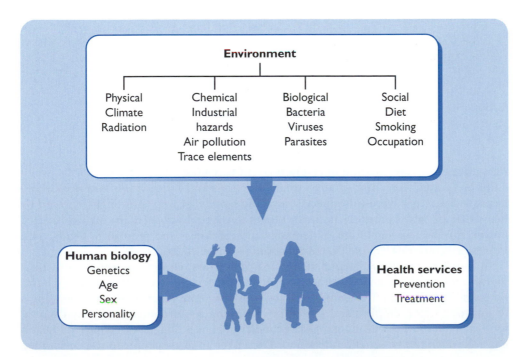

Figure 16.1:
Influences on health.

Box 16.1 Acute effects of air pollution

Most environmental effects are difficult to detect because many years elapse before any symptoms appear. These are called *chronic* effects. Some events, however, produce a short-term *acute* effect. The last great London smog caused a sudden temporary increase in the death rate (see Figure 16.1.1). Chapter 15 gives more details about air pollution.

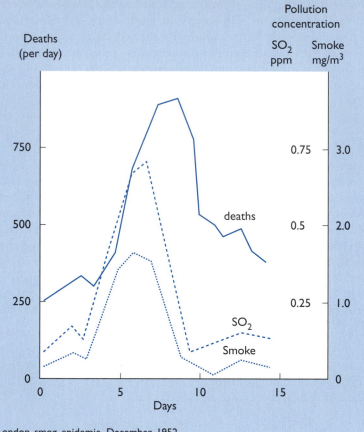

Figure 16.1.1: The London smog epidemic, December 1952.
Source: UK Ministry of Health (1954).

resistance to disease. Medical interventions had very little effect until the 1930s, when sulphonamides were first used to treat infection, and later, when antibiotics and improved vaccines were introduced. Now infectious diseases have been replaced by heart disease and cancer as the main killers in more developed countries. Poverty, however, remains a significant underlying cause of illness and premature death, even in Western countries like the United Kingdom (see Box 16.2).

In much of the Third World, diarrhoea, dysentery, cholera, typhoid, intestinal worms, tuberculosis and respiratory infections still dominate. Many of the gastro-intestinal disorders are transmitted via faeces, so the most effective first intervention is to improve water supplies and sanitation. Drinking water contaminated by excreta, food crops fertilised with human waste, and fish from polluted waters, are all health hazards which can be significantly reduced, but over 1 billion people in the developing

Box 16.2 Social inequalities in health

When the National Health Service was introduced to the United Kingdom in 1948 it was hoped that making the best health services freely available to all would eliminate the health gap between rich and poor. This did not happen. In 1980, the controversial Black Report (DHSS, 1980) revealed that men working in unskilled manual jobs were twice as likely to have a long-term illness than men in professional occupations, and that their babies were twice as likely to die in the first month of life than the babies of professional parents. Since 1980, the inequalities in health between rich and poor in the UK have widened even further. While incomes generally have been rising, particularly at the top end of the scale, the quality of life for the poorest people has deteriorated. For example, it is becoming increasingly difficult for the poorest people to obtain cheap, varied food. Local shops have closed down because of the success of out-of-town supermarkets, so that inner city areas are becoming 'food deserts', putting mothers and children at risk through poor nutrition. Research shows that a woman's diet during pregnancy affects the weight of her baby, and that babies with low birth weights are at greater risk of diseases in later life. The best chance of reducing social inequalities in health is therefore believed to be to implement policies that improve nutrition in mothers on low incomes and eliminate child poverty.

world lack access to safe water (see Figure 16.2) and nearly 2 billion people lack an adequate sewerage system. The World Bank (1993) estimates that the cost of providing adequate water and sanitation services ranges from $15 per person per year for simple rural systems to $200 for urban systems with individual household taps and flush toilets. The benefits to be gained include increased productivity as well as improved health, so cost-effectiveness is extremely high.

The global HIV/AIDS epidemic

The HIV/AIDS epidemic started in the late 1970s and now covers almost the whole inhabited world. Its spread has been both rapid and volatile. HIV (human immunodeficiency virus) is a retrovirus which attacks the cells controlling the immune system, making the body very vulnerable to opportunistic infections. HIV is transmitted

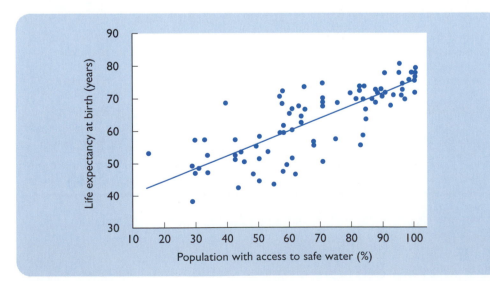

Figure 16.2: Life expectancy at birth and percentage of the population with access to safe water in the countries of the world.

Source: World Bank (1997).

through sexual intercourse, contact with contaminated blood in transfusions or shared hypodermic needles, and from mother to child before birth or through breast milk after birth. AIDS (acquired immune deficiency syndrome) is diagnosed when a person carrying HIV develops an infectious disease which is rare in people whose immune systems are functioning. It was first recognised in San Francisco and New York when a rare cancer (Kaposi's sarcoma) suddenly appeared in apparently healthy young homosexual men. There is a long incubation period between becoming HIV-positive and the onset of AIDS, typically 8–10 years if untreated. AIDS is always fatal.

Up to 1998, 12 million people had died with AIDS. Over 30 million people were living with HIV infection at the end of 1997, that is one in every 100 young adults worldwide, most of whom did not know they were infected. With 16 000 new infections every day, it is estimated that 40 million people will be infected by 2000 (UNAIDS, 1997). Although the onset of AIDS can now be delayed by drug therapy, the only effective way to reduce the devastation ahead is by prevention. Preventive strategies have emerged from studies of the mechanisms of HIV transmission and the social context in which transmission occurs.

The region most crippled by AIDS is sub-Saharan Africa (Table 16.1). In some cities in southern Africa over 40% of pregnant women tested in 1997 were HIV positive, and

Table 16.1:
World distribution of HIV/AIDS in 1997.

Region	Adults and children living with HIV/AIDS	Adults aged 15–49, prevalence rate	Percentage women	Main modes of transmission[1]
Sub-Saharan Africa	20 800 000	7.4%	50%	Hetero
North Africa, Middle East	210 000	0.13%	20%	IDU, Hetero
South and South-east Asia	6 000 000	0.6%	25%	Hetero
East Asia, Pacific	440 000	0.05%	11%	IDU, Hetero, MSM
Latin America	1 300 000	0.5%	19% Hetero	MSM, IDU,
Caribbean	310 000	1.9%	33%	Hetero, MSM
Eastern Europe and Central Asia	150 000	0.07%	25%	IDU, MSM
Western Europe	530 000	0.3%	20%	IDU, MSM
North America	860 000	0.6%	20% Hetero	MSM, IDU
Australia and New Zealand	12 000	0.1%	5%	MSM, IDU
Total	30 600 000	1.0%	41%	

Note:
Hetero: heterosexual transmission; IDU: injecting drug users; MSM: men who have sex with men.
Source: UNAIDS and WHO (1997).

increasing numbers of babies are now born with HIV. Infection through sexual encounters in the towns has spread to destroy family life in the rural villages. The illness of a young adult in a rural community leads to lower crop production and livestock yields, with a consequent reduction in income and nutrition for the family. When whole communities are affected a spiral of decline is set in motion, with cash crops abandoned, food crops barely attended and ravaged by pests, and a generation of orphans to be cared for. By 1998, almost 8 million children had lost their mother or both parents in Africa south of the Sahara. The immediate challenge for the very poorly developed health services is to reduce transmission and provide simple treatments with inexpensive drugs that alleviate the suffering of AIDS-related diseases. Prevention concentrates on programmes to promote later sexual initiation, fewer partners and more condom use. Uganda was one of the first countries to respond with testing centres and a prevention programme, and in 1997 there were signs that infection levels in young adults in that country were beginning to drop.

In other parts of the world, different cultural settings have created variations in the HIV/AIDS epidemic which call for alternative strategies. In Thailand, for example, there was a rapid early increase in HIV infection among drug injecters and commercial sex workers. New infections among sex workers and their clients now appear to be slowing down due to a sustained prevention programme aimed at increasing condom use, discouraging men from visiting sex workers, boosting respect for women and providing better educational and other prospects for young women as alternatives to entering into commercial sex. These, unfortunately, are isolated examples of successful campaigns, but they offer some hope for the future. In most parts of the developing world, the epidemic is still uncontrolled and spreading rapidly.

In the industrialised world, by contrast, new AIDS cases have been falling since 1995, when antiretroviral drugs became available (UNAIDS, 1997). Antiretroviral treatment is expensive, but it is effective in reducing the speed at which HIV-infected people develop AIDS. The fall has been greatest in countries where infection was concentrated in homosexual men, whose HIV rates started to drop some years earlier, probably as the result of safer sex behaviour in the well-educated and well-organised gay community. The danger of using contaminated blood for transfusions (which had infected many young children in orphanages in Romania, for example) has now been reduced by better screening. In spite of these gains, efforts to reduce transmission through drug injection and heterosexual intercourse in disadvantaged sections of society in Western Europe and North America have been less successful. Sharing hypodermic needles is a particular risk for injecting drug abusers. In the United Kingdom, there are local centres where drug users may exchange used needles for new ones, but there are enormous difficulties in enabling socially marginalised people with unstable lives to make regular use of this service.

In parts of the United Kingdom and the United States, tuberculosis has returned to scourge the poor. Tuberculosis is associated with conditions of poverty and is strongly linked to AIDS in the developing world. TB was largely eliminated from Western countries until the HIV epidemic, but now it is spreading again among the homeless, alcoholics and drug abusers. Prisons and shelters for the homeless are ideal breeding grounds, and new, drug-resistant strains of TB have emerged. In the very poorest parts of Britain, such as the deprived overcrowded areas of Liverpool, TB has recently taken hold, sometimes quite independently of AIDS (Elender *et al.*, 1998). In New York, homelessness, drug use, AIDS and TB are inextricably linked, and what is needed is a broad programme to alleviate extreme urban poverty. There is clearly no single AIDS problem with a single solution anywhere in the inhabited world.

Ultraviolet light and skin cancer

Concern over the recent depletion of stratospheric ozone as a result of human activity has drawn attention to the current epidemic of skin cancer in Western countries. Ozone in the stratosphere absorbs UV radiation, and as the protective layer is progressively destroyed by chlorofluorocarbons (CFCs) and other products of human industry more UV radiation is allowed to penetrate to the Earth's surface. This matter is given

Box 16.3 Geographical studies

Some diseases have a pronounced geographical pattern, like stomach cancer, the second most common cancer in the world. Stomach cancer is linked to the consumption of nitrates and nitrites, but this explanation does not account for the two-fold difference in rates between counties in England and Wales (Figure 16.3.1). Geographical studies of disease (also called *ecological studies*)

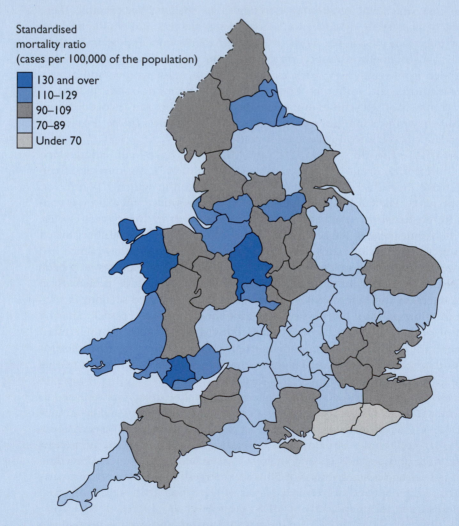

Figure 16.3.1: Incidence of male stomach cancer in England and Wales.

Source: Office of Population Censuses and Surveys (1990).

Box 16.3 (continued)

often attempt to explain such differences by relating disease rates in administrative areas to environmental measures or social indicators from the population census. They can be a very fruitful source of hypotheses about possible risk factors, although they suffer from several drawbacks. The main problem, known as *confounding*, is that it is difficult to control for all other influences on the disease in an ecological study, so these may distort the results. Geographical studies of lung cancer in North America and Britain have found that the areas of highest radon exposure are not the areas with most lung cancer, but their failure to detect a relationship between lung cancer and radon might be because they cannot take smoking variations properly into account. Another potential problem is the *ecological fallacy*, when conclusions about areas are mistakenly applied to individuals. Inner city areas, for example, usually have high mental illness rates and also contain high numbers of students, but it would be wrong to conclude that students are particularly prone to mental illness.

Other features to bear in mind are that statistical correlations are affected by the number and size of the areas being studied (*the modifiable areal unit problem*), and all geographical comparisons are blurred by the migration of people between areas. These problems mean that geographical studies are not as powerful as case–control or prospective cohort studies (see Box 16.4) for investigating individual characteristics which may be risk factors for disease. Increasingly, however, it is becoming clear that the characteristics of neighbourhoods or communities can also affect health. The risk of accidents to young children, for example, is related to the level of poverty or affluence in the neighbourhood as a whole as well as to the characteristics of individual families, and both childhood leukaemia and sudden infant death syndrome are found in communities with a high degree of population mixing. In these circumstances, the conventional methods are insufficient and a new technique, *multi-level modelling*, is required to separate individual and area effects.

extensive treatment in Box 19.7. In countries of the Third World, the main effect of increased exposure to ultraviolet radiation will be an increase in cataracts and in the social and economic consequences of blindness. In Western countries, an increase in skin cancer is predicted.

There are three main types of skin cancer, both of which affect white-skinned people in particular. Non-melanoma skin cancer is common and relatively easily treated. It generally occurs on the most exposed parts of the body (the face, neck, arms and hands) and especially in people who work outdoors. Elderly people are more likely to suffer than younger people, and cumulative exposure to sunlight is thought to be the main cause. Melanoma, on the other hand, is rarer but more serious, with about one in seven cases proving fatal. Melanomas characteristically appear on women's legs and men's trunks, attacking young adults as well as the elderly. Office workers are more at risk of melanoma than outdoor workers, and it is believed that occasional high doses of UV radiation are to blame. Exposure to the sun in childhood is a major risk factor. Ultraviolet radiation seems to have two effects: first to damage the DNA in cells that have been badly sunburnt and second to suppress the natural immune system, which would normally destroy unusual cells. The result is cancerous cells, which are allowed to proliferate.

In most countries with white-skinned populations, the incidence of melanomas has been increasing by about 5% every year since the 1960s, faster than that of any other major cancer. Cancers often have latency periods of 30–40 years, so the main reason has been earlier changes in behaviour in Western societies (more outdoor recreation, more foreign holidays, less clothing cover) which increased exposure to UV radiation, rather than ozone depletion, which is a more recent phenomenon. Queensland,

Australia, where sunshine is abundant and beaches are inviting, has the highest melanoma rates in the world. In Australia, public education campaigns have been used since the mid-1980s to alert the public to the dangers of sunburn and the symptoms of skin cancer. People are advised to wear protective clothing, such as a wide-brimmed hat, and to avoid the midday sun. Sunscreen lotions are helpful in that they reduce the risk of burning, but they may not protect against the immunosuppressive effects of UV radiation. Graphic posters displayed in drugstores have helped the early detection of melanomas before they become invasive, but the incidence rate has still not fallen. The main problem, of course, is that people enjoy sunbathing and a tan is considered to be attractive. Getting people to change their behaviour is a difficult and slow process, as experience with cigarette smoking has proved.

The depletion of ozone in the stratosphere will certainly intensify the problem. Ozone has natural fluctuations, but there is now compelling evidence that significant depletion has been occurring in recent years, particularly over Antarctica. A United Nations Environment Programme report (1994) estimates that if the 10% reduction in ozone experienced at mid to high latitudes over the period 1979–1992 continues for the next 30–40 years and behaviour does not change, there will be 250 000 additional cases of non-melanoma skin cancer every year. For each 1% depletion in stratospheric ozone, cataract incidence is predicted to increase by 0.6–0.8%. These are conservative estimates. Recent research has suggested that the effects of exposure to UV radiation on the immune response system make us less resistant to infectious diseases and possibly increase the risk of some cancers where immune suppression is important (Bentham, 1996). If these links are confirmed, the consequences of no action will be even more damaging to health.

Radiation

Nuclear safety is an issue of great concern. The accident at the Chernobyl nuclear power plant in 1986 caused immediate radiation sickness and early death for those exposed to high levels of radiation at the site and appears to be responsible for regional excesses of thyroid cancer 10 years later. In years to come, up to 7000 additional fatal cancers are expected over Europe as a result of Chernobyl. But setting aside spectacular accidents, what are the dangers of living next to a nuclear plant operating normally? Much attention has been focused on childhood leukaemia as a possible consequence.

Leukaemia is a group of diseases linked to the blood, in which developing white blood cells fail to mature in the bone marrow, immature cells multiply and the body's defence against infection is eventually lost. Acute lymphocytic leukaemia is the commonest form of cancer in children, but it is still a rare disease. The causes of leukaemia are obscure, but exposure to ionising radiation is known to be one. The evidence here is derived from studies of the survivors of the atom bombs dropped on Hiroshima and Nagasaki in 1945 (see Figure 16.3), studies of the effects of exposure to medical X-rays, and studies of people exposed to high doses of radiation through their occupation. When a tele-vision programme in 1983 publicised the existence of a cluster of childhood leukaemias in the vicinity of the Sellafield nuclear reprocessing plant in Cumbria, the link between radioac-tive discharges from Britain's dirtiest nuclear plant and the disease appeared to be self-evident. Subsequent research has uncovered a more complicated story.

For a while, investigations concentrated on searching for excesses of childhood leukaemia around other nuclear plants, and such clusters were found, with the most sig-nificant (five cases) at Sellafield. The two other sites in the world where nuclear waste is reprocessed on a commercial scale (Dounreay in Scotland and La Hague in France) were

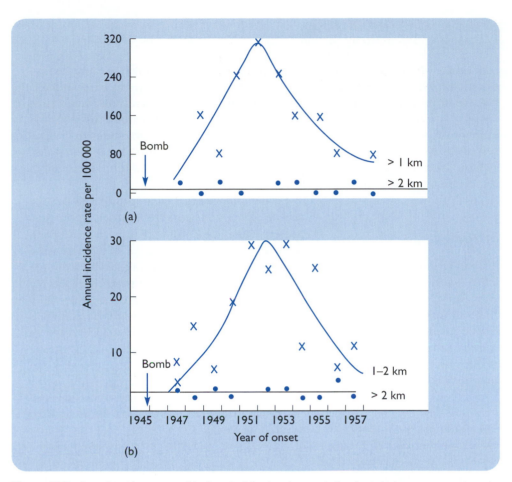

Figure 16.3: Annual incidence rate of leukaemia following the atomic bomb explosion among survivors in Hiroshima. (a) People less than 1 km from the hypocentre; (b) people 1–2 km from the hypocentre, compared with people more than 2 km from the hypocentre at the time of explosion.
Source: Cobb *et al.* (1959).

also the centres of local leukaemia clusters. Circumstantial evidence, however, is not in itself proof of a causal relationship. The missing link in the argument was a demonstration that radiation doses received by children from nuclear plants were sufficient to cause the excess leukaemias. Careful calculations based on what was known about radioactive discharges, pathways of radiation into the human body, and the relationship between the dose and the risk of leukaemia failed to account for the clusters. What, then, could explain the cancers? It is possible that the radiation doses to children playing on the beach were very seriously underestimated, even though the estimates were based on the best scientific models available. Another possibility was that the children's fathers had been exposed to high doses of radioactivity at work in the nuclear plant prior to conception, as was found to be the case in Sellafield, but no supporting evidence could be found in other nuclear plants, or from studies of the children of Japanese atom bomb survivors.

The research team that had identified higher than expected childhood leukaemia rates near all nuclear plants in England and Wales must have been surprised when they investigated places earmarked as potential nuclear sites for the future. These places also had elevated childhood leukaemia rates, even though no nuclear facilities existed

Box 16.4 Epidemiological methods

Epidemiology is the study of the distribution of disease in populations. While clinical doctors investigate disease in individual patients, epidemiologists look for patterns of occurrence in large numbers of people in order to identify the causes of disease and to control health problems. There are three types of epidemiological study: descriptive, analytical and intervention.

Descriptive studies

The incidence of disease may vary with personal characteristics such as age, sex and occupation, it may fluctuate over time (as do epidemics of infectious disease) or it may have a particular geographical distribution. Recognising the pattern might generate new hypotheses about the cause, but it cannot by itself prove a causal association. For example, the incidence of lung cancer in some urban areas of Britain is almost twice that in rural areas, suggesting that ambient air pollution or industrial exposure may be causal factors. Further investigation would need to take the prevalence of cigarette smoking in previous years in both environments into account and to show that in cities those people most exposed to the hypothesised risk factors had the highest rates of lung cancer.

Analytical studies

Hypotheses are usually tested by studies designed for the purpose (see Figure 16.4.1). *Case–control studies* start by identifying a sample of people with the disease (cases) and a sample of people without the disease drawn from the same population (controls). Often the controls are selected so that some characteristics known to be associated with the disease, but not of interest to the study (such as age or sex), are 'matched' with the cases. The past history of both cases and controls is then investigated to test whether the cases were more likely to be exposed to the suspected risk factor than the controls. A case–control study in southwest England has recently established that lung cancer victims are more likely to have been exposed to high domestic radon levels than healthy controls.

Cohort studies start with a sample of healthy people, some of whom are exposed to the suspected cause and some who are not. Both groups are followed over a period of time (often several years) to determine which people develop disease. The long-term effects of methyl isocyanate poisoning are the subject of a cohort study of residents around the pesticide factory in Bhopal, India, which was the scene of a major accident in 1984. Cohort studies take longer and are much more expensive than case–control studies, but they provide better information about the causation of disease and the most direct measurement of the risk of developing disease.

Intervention studies

Randomised controlled trials are experiments to study new methods of disease prevention or treatment as well as disease risks. Subjects are randomly allocated to either an experimental or a control group and only the first group is given treatment. The results are assessed by comparing the outcome of the two situations. A randomised controlled trial was used in a study of the risks of sea bathing, in which healthy volunteers in four UK coastal resorts were randomly divided into those who were asked to immerse themselves in the sea and those who were required to stay on the beach. Health checks later confirmed that sea bathing increases the likelihood of gastroenteritis, even on beaches that meet current EC criteria for safe bathing.

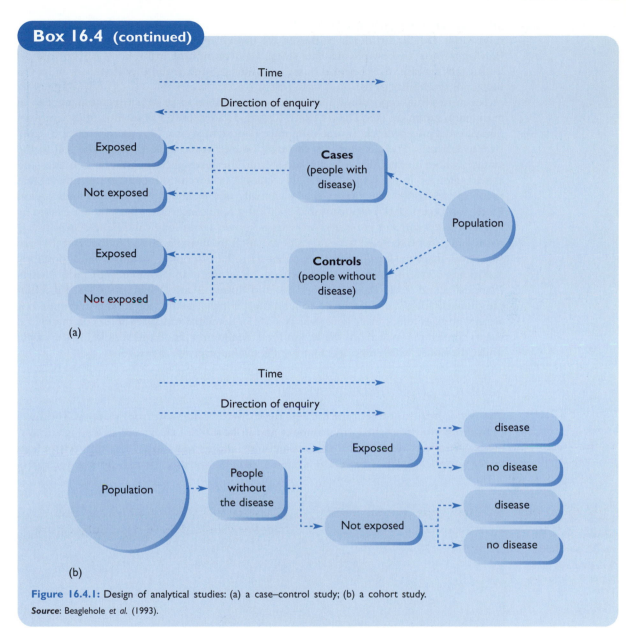

Figure 16.4.1: Design of analytical studies: (a) a case–control study; (b) a cohort study.
Source: Beaglehole *et al.* (1993).

there yet. Perhaps childhood leukaemia is associated not with nuclear plants but with some unknown characteristic of people in the type of place where nuclear establishments are to be found. Remote rural areas, for example, have higher than average leukaemia rates, as do areas with rapidly growing populations, and both features are typical of nuclear sites. New towns built in rural surroundings in the 1950s which quickly grew in population had high childhood leukaemia rates. Perhaps a sudden mixing of population in a previously stable situation might be responsible for spreading the disease, as could occur, for example, if a virus were involved. A more sophisticated theory is that childhood leukaemia results when an insufficiently developed immune response system is challenged by sudden exposure to a range of infections. This might happen in a situation where children are shielded from infections at a very early age, only to be exposed later when infections are imported by migration. This and other hypotheses are under active investigation, but each new finding seems to complicate the story. Environmental influences on childhood leukaemia remain, for the time being, a mystery. Such ambiguity does not help the image, or credibility, of the nuclear industry.

For most of us, exposure to natural sources of radiation far outstrips the danger from man-made radiation (see Figure 16.4). Radon, a radioactive gas produced by uranium decaying deep in bedrock, is much the most important source of ionising radiation exposure to humans. The gas seeps up through fissures in the rock, through the soil to the surface, where it either disperses harmlessly into the atmosphere or collects in homes and other buildings. In the United States, the National Research Council (1998) estimates that radon causes 18 000 lung cancer deaths annually, and in Britain the National Radiological Protection Board (1990) estimates that 2500 extra lung cancer cases per year are caused by radon. These estimates are based on what is known about lung cancer incidence in uranium miners who received large doses of radiation from underground radon, and they are sufficiently alarming for most Western countries to have adopted 'action levels' of domestic radon gas concentrations. In areas with geology likely to produce high radon levels (Minnesota and North Dakota in the USA, and Cornwall and Devon in the UK, for example) householders are encouraged to have their homes tested, even though these are not areas with high lung cancer rates (see Box 16.3). Remedial measures are recommended if radon concentrations exceed the action level. Sealing the floor and ventilating the cavity under the house with a pipe and a fan are the main preventive measures.

Science and management

The major health problems of the less developed world stem from poverty. The links between an impoverished environment and ill health are clear, and defeating the diseases of poverty depends on raising incomes and expectations, together with clean water supplies, proper sewerage systems, decent housing, better diets, safer work-

Figure 16.4:
Sources of average radiation doses to the UK population.

Source: Adapted from Hughes and O'Riordan (1993).

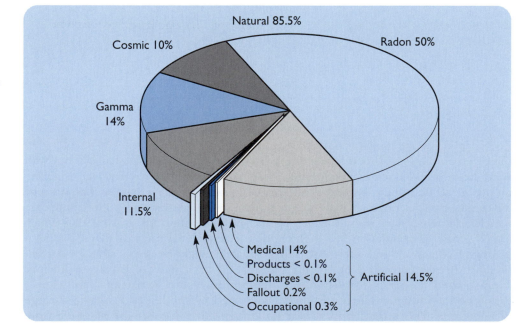

places and effective pest control as well as immunisation programmes and basic health care facilities. In the developed world, the most obvious environmental hazards to health have been eliminated (with the exception of tobacco smoking), and the challenge now is to investigate insidious and rare effects. It may be, for example, that soft water is a risk factor for cardiovascular disease, that nitrates in the water supply could help to produce stomach cancer, that agricultural pesticides induce asthma, that lead released from petrol combustion could affect children's intelligence, or that any one of the new chemical products being manufactured will create unforeseen problems in the future. Hypotheses are easily generated but not so easy to prove.

There are several reasons. First, there are problems of data accuracy. A high proportion of diagnoses are known to be inaccurate, even in countries with advanced health care facilities, but much illness is never detected or reported to a doctor and does not even reach the diagnosis stage. Many diseases have a long latent period between exposure to the environmental agent which initiates the disease and the appearance of symptoms. Figure 16.3 illustrates this in the case of leukaemia. Lung cancer, to take another example, has an average latent period of 20 years. Past smoking habits are known to be the major cause of lung cancer, but for conditions whose cause is unknown it is difficult to identify risk factors operating many years ago. Similarly, the effects of new pollution problems today will not be evident for many years. To add to the difficulties, most diseases are caused not by a single environmental factor but by a combination of many things, with perhaps different combinations triggering different stages of the disease. Confounding variables, such as age, sex, socio-economic circumstances, occupation, smoking, and so on, are all very strongly related to health and must be properly controlled in an epidemiological study (see Box 16.4) so that their effects do not mask the effect being investigated. Then there is the problem of how to use what is known about the effects of high doses of the pollutant to predict the health effects at low doses, a difficulty that besets studies of radiation effects, for example.

Many illnesses are comparatively rare events. A study population of 1 million people might be expected to produce between 10 and 20 acute lymphocytic leukaemias per year, and the infamous 'cluster' of childhood leukaemias near Sellafield consisted of just five cases, so any search for patterns of association is made difficult. Circumstantial evidence, such as the observation that childhood leukaemias occur near a nuclear reprocessing plant, does not prove that the cancers were caused by the plant, since both might be linked to another factor such as a sudden influx of population into a remote area when the plant was built. Other associations could be the result of pure chance, and here the methods of statistics are invaluable. Conventionally, if there is a probability of more than 5% that any result could have been caused by chance, the scientist will reject the evidence as not being sufficiently strong. It is the scientist's responsibility never to claim more than the evidence allows, to be mindful of the complications and to be cautious in interpretation. Scientific scepticism ensures that investigations continue and do not stagnate.

The environmental manager, on the other hand, has a different responsibility: to safeguard public welfare in the light of current knowledge. Part of that responsibility is to make current knowledge more accessible to the general public: to make clearer, for example, the relative risks arising from natural as opposed to human-caused radiation. Public concern needs to be alleviated by more effective communication when scientific investigation has determined that no serious threat exists, but scientific evidence is frequently incomplete and even contradictory. Like the scientist, the manager must err on the side of caution, but for the manager the cautious strategy in cases of doubt is to act as if the hazard were real. Statistical significance is a criterion for accepting or rejecting

Box 16.5 Public campaigns

Woburn in Massachusetts is a town which contains a toxic waste dump. In the 1970s, the local residents were alarmed by cases of leukaemia that kept occurring, and they campaigned for the wells that supplied the town's water to be shut off. Subsequent investigation showed that the well water was contaminated with industrial wastes. The parents of children with leukaemia spent years in litigation against two companies. One company was eventually found guilty of negligently dumping chemicals but the other was absolved. The evidence collected by the activists did, however, spur several official studies and contributed to the momentum of a national movement for better management of toxic wastes.

The Woburn case is typical of many public campaigns where ordinary citizens react against the complacency of companies or government agencies and seek to determine the nature and extent of health risks themselves. Campaigns usually begin with a few activists who note a strange pattern of ill health and contact the local doctor, who is instrumental in providing data on birth abnormalities, skin rashes or whatever condition causes concern. The affected groups mobilise and stir up the media. Local health officials question the evidence, which they claim is inconclusive. The angered citizens bring in their own experts, take legal action and become confrontational. They press for official corroboration of their findings and file for adequate compensation. Campaigners face an uphill struggle, with access to only limited information. Their own consultant scientists may be branded as biased,

and the weight of scientific testimony seems to be loaded against them.

Why is it that evidence that appears so obviously to show a link between a pollution source and cases of illness is not acceptable as proof to the scientist? The scientist requires more than the existence of a 'cluster', because even randomly occurring events sometimes cluster in a particular place just by chance. The scientist looks for confirmation that the suspected exposure happened before the illness appeared, that the incidence of illness correlates with the amount and duration of the exposure, that there is a plausible biological mechanism for the pollution to cause the illness and that the same effect can be observed in similar circumstances elsewhere. From a scientific point of view, evidence assembled to support a predetermined conviction is often one-sided and lacks an adequate comparison with a control group.

Good studies are expensive and may take years to deliver a conclusion, so public frustration with science is understandable. Satisfying strict scientific standards should not be a condition for taking precautionary action. Where doubt exists the real test is whether a reasonable person would expose themselves to the potential hazard. While public campaigns rarely prove that a source of pollution causes ill health, they do expose ignorance, complacency and inaction, and they are a powerful stimulant to provoke well-designed studies and government regulation.

Sources: Crouch and Krollersmith (1992); Brown (1993).

evidence, but the manager is more concerned with public health significance, which is not necessarily the same. Statistical significance is in part dependent on sample size, so a result that is important for disease prevention could be overlooked because the study sample is not large enough for scientifically safe conclusions to be drawn. A real hazard might be so entangled with confounding effects that it is difficult to design the perfect study that isolates one particular relationship; but while the scientist grapples with methodological problems the public health consultant must decide what action to take. The precise effects of low doses of radiation are still open to scientific question, but few would argue with the principle of keeping human exposure to radiation as low as reasonably practicable. Defining what is reasonably practicable is an ethical minefield through which the manager must tread. Even when the scientific evidence is firm, as with the knowledge that fluoride in the water supply protects against dental caries, the question whether to add fluoride to the water supply in fluoride-deficient areas is not easy, for excess fluoride discolours teeth and may cause harmful effects not yet established. There are too many environmental changes occurring and new materials coming into use for each to be investigated, and, in the case of possible long-term effects, it would be unethical to wait until the damage has become obvious before taking action. Environmental management to protect health therefore requires a mixture of science and judgement.

References

Beaglehole, R., Bonita, R. and Kjellstrom, T. (1993) *Basic Epidemiology*. World Health Organisation, Geneva.

Bentham, G. (1996) Association between incidence of non-Hodgkin's lymphoma and solar ultraviolet radiation in England and Wales. *British Medical Journal*, **312**, 1128–1131.

Brown, P. (1993) When the public knows better: popular epidemiology challenges the system. *Environment*, **35**(8), 16–20, 29–41.

Cobb, S., Miller, M. and Wald, N. (1959) On the estimation of the incubation period in malignant disease. *Journal of Chronic Diseases*, **9**, 385–393.

Crouch, J.R. and Krollersmith, J.S. (eds) (1992) *Communities at Risk: Collective Responses to Technological Hazard*. Peter Lang, New York.

Department of Health and Social Security (1980) *Inequalities in Health: Report of a Research Working Group*. DHSS, London.

The Ecologist (1998) Cancer: are the experts lying? *The Ecologist*, **28**(3), whole issue.

Elender, F., Bentham, G. and Langford, I. (1998) Tuberculosis mortality in England and Wales during 1982–92: its association with poverty, ethnicity and AIDS. *Social Science and Medicine*, **46**, 673–681.

Epstein, S. (1998) Winning the war against cancer: are they even fighting it? *The Ecologist*, **28**(2), 69–81.

Howard, V. (1997) Synergistic effects of chemical mixtures: can we rely on traditional toxicology? *The Ecologist*, **27**(5), 192–196.

Hughes, J.S. and O'Riordan, M.C. (1993) *Radiation exposure of the UK population: 1993 review (NRPB R263)*. National Radiological Protection Board (NRPB), Chilton, Oxon.

Hughes, J.S., Shaw, K.B. and O'Riordan, M.C. (1989) *Radiation exposure of the UK population: 1988 review (NRPB R277)*. National Radiological Protection Board (NRPB), Chilton, Oxon.

Langford, I. and Day, R. (1998) Climate change, ozone depletion and health: implications for public health and the tourism industry. In T. O'Riordan (ed.), *Climate change and the tourism industry*. Working Paper PA-98-07, CSERGE, University of East Anglia, Norwich, 17–21.

National Radiological Protection Board (1990) Board statement on radon in homes. *Documents of the NRPB*, **1**. NRPB, Chilton, Oxon.

National Research Council (1988) *Health Risks of Radon and Other Internally Deposited Alpha-emitters: BEIR IV*. National Academy Press, Washington DC.

Office of Population Censuses and Surveys (1990) *Mortality and Geography: a Review in the Mid 1980s*. HMSO, London.

UNAIDS (1997) *Report on the Global HIV/AIDS Epidemic: December 1997*. United Nations and World Health Organisation, Geneva.

United Kingdom Ministry of Health (1954) *Mortality and Morbidity During the London Fog of December 1952*. HMSO, London.

United Nations Development Programme (1998) *Human Development Report 1998.* Oxford University Press, Oxford.

United Nations Environment Programme (1994) *Environmental Effects of Ozone Depletion: 1994 Assessment.* UNEP, Nairobi.

World Bank (1993) *World Development Report 1993: Investing in Health.* Oxford University Press, New York.

World Bank (1997) *World Development Report 1997: the State in a Changing World.* Oxford University Press, New York.

Further reading

Curtis, S. and Taket, A. (1996) *Health and Societies: Changing Perspectives.* Arnold, London.

Davey, B., Gray, A. and Seale, C. (1995) *Health and Disease: a Reader* (2nd edn). Open University Press, Buckingham.

Farmer, R. and Miller, D. (1991) *Lecture Notes on Epidemiology and Public Health Medicine* (3rd edn). Blackwell, Oxford.

Jones, K. and Moon, G. (1987) *Health, Disease and Society: an Introduction to Medical Geography.* Routledge, London.

McMichael, A.J., Haines, A., Slooff, R. and Kovatts, S. (1996) *Climate Change and Human Health.* World Health Organisation, Geneva.

Rodricks, J.V. (1992) *Calculated Risks.* Cambridge University Press, Cambridge.

Chapter Seventeen

Environmental risk management

Simon Gerrard

Topics covered

- Risk assessment
- Risk perception
- Risk communication
- Risk management

Editorial introduction

A central theme in this volume is the repositioning of science. Nowhere is this more necessary than in the area of risk management. Considerable uncertainty exists about the complex interactions within and between the environmental media (air, water and land) and ecosystems. Here, profound uncertainties and indeterminacies create a major challenge to science and scientific method. This has become apparent at a time when the liabilities of economic growth coupled with the seemingly pervasive problems associated with hazardous technologies have become a characteristic of our modern society. We live in a world that is not risk-free: the notion of zero risk has been largely dispelled.

Ulrich Beck (1992) describes our present state as the *risk society* – a society that is defined not by traditional economic and social class characteristics but by the distribution of new forms of technological risks which are harder to perceive directly, may transcend generations and for which no appropriate compensatory mechanisms exist. Developing solutions to these new forms of risk is limited by a heavy reliance upon the very scientific and technological systems that generated the problems in the first place. This realisation became most evident in the wake of the BSE (bovine spongiform encephalopathy) crisis, and in the flare-up of political anxiety over the incursion of genetically modified crops. The protest is essentially about re-democratising the debate so that values other than purely scientific and technological ones can be injected into the deliberating process.

Never has it been more important that individuals or their institutions are clearly responsible and accountable for the direction in which development occurs. It is not solely the traditional 'hard' sciences that are driving this process but the so-called 'softer' sciences too. Economics, psychology, sociology and anthropology rely to greater or lesser degrees upon empirical and analytical techniques. The crucial point here is that the concept of risk is neither entirely abstract nor wholly physical: it is *socially constructed* and so must be *socially resolved* through mediation and negotiation between the parties involved. In the context of risk management, the domination of scientific debate manifests itself as an over-reliance on narrowly focused hazard and risk assessment modelling. This has tended to underplay the importance of other 'non-scientific' factors and thus can provide only a partial perspective of the wider risk problem. As Brian Wynne notes:

> *The social–situational factors, such as the competence and trustworthiness of organizations or individuals, are an essential part of the risk process, hence the risks. Expert risk analyses usually misrepresent the intrinsically open-ended, indeterminate nature of these dimensions, treating them as if they were deterministic, and assuming that technical imprecision can be overcome by statistical rules and techniques of uncertainty analysis. Alternatively, they adopt naively idealistic models that assume that social and organizational behaviour follows dependable laws.*
>
> (Wynne, 1992, 281)

John Stuart Mill recognised that retaining a plurality of viewpoints is necessary for any democracy and identified three reasons why this is advisable (Feyerabend 1987). First, because to deny any view assumes infallibility – a dangerous assumption at any time. Second, because even if a view appears to be wrong, it may still hold a portion of truth. Any one view never holds the whole truth. Third, a point of view that is largely true but not contested becomes a prejudice. It is important to measure up these views against those of others even if the original view is retained. This is because the meaning of any single view cannot be fully understood unless contrasted with other opinions. We shall see that these are all very important considerations for risk management.

As public distrust and disbelief of scientists and regulators have increased, other arenas thrive in their new-found importance as influences in the policy-making debate. The media, the judiciary, citizens' initiative groups and new social movements are emerging as alternative forums in which the variety of perceptions of environment and society can be aired. Peter Rawcliffe (1998, 60–61) shows that a new culture of openness and accessibility to the policy process, and to management generally, has encouraged a more sophisticated pressure group style and structure in terms of their policy expertise and authority, their educational role and their effectiveness in mobilisation. One nascent effect of this opening up of debate is to reinforce the existing rights of individuals and extend these rights to encompass broader aspects of everyday life. In the social sciences, the emergence of these rights is talked about in terms of deliberative participation, as introduced more fully in Chapter 1.

The rather pessimistic notion of the risk society has emerged alongside a more optimistic vision of the future known as *ecological modernisation*. This seeks to connect the technological momentiveness of 'Factor 4' (see Chapter 2) with new approaches to regulation and negotiation of sustainable development. Ecological modernisation requires changes at three distinct levels. First, it requires a shift to eco-efficient technology with a focus on clean production and waste minimisation. This shift is coupled with anticipatory planning processes delivered through respect for the precautionary principle. The more fundamental shift occurs in the mindset of organisational behaviour, which internalises ecological responsibility to the extent that it becomes second nature. In an ecologically modern society, there are no trade-offs between economy and environment, and pollution reduction is seen as a competitive strategy (see Figure 3.1 for a fuller explanation).

Part of the outcome of ecological modernisation have been the growth in influence and positioning of the judgement of scientific experts. As noted in Chapter 1, expert judgement is often more subjective and value-laden than citizens have realised or scientists themselves have admitted. Science and scientific analysis have been presented as rational and therefore logical and reasonable, even though within the scientific community many different and conflicting views and beliefs are held. For example, empiricists argue that it is irrational to retain views which contradict experimental evidence, while theoreticians regard it as irrational to change views at every flicker of evidence. Ultimately, and despite these conflicts, science has been built upon the idea that there is a single right way of doing things and that the world should be made to accept this.

The net result of this approach to scientific analysis has been to disenfranchise the citizen from the decision-making process to the point where, in our so-called democratic society, the citizen is often left bypassed by the development of environmental policy and the management of environmental risk. We have reached a situation where the public are involved in decision making after the science has been completed and, in many cases, when the decision has virtually been reached. Public involvement in decision making in the UK, and in others like it, has sometimes served merely as a legitimation device leaving the public exposed to risks largely without question, or to be branded as irrational opposition protecting self-interest at the expense of the benefit of society. Boxes 17.7 and 17.8 in this chapter illustrate these points using two waste management case studies at different levels of government: domestic waste management planning by UK county councils and radioactive waste disposal by NIREX under the supervision of central government.

But the tide is turning. The Royal Commission on Environmental Pollution, in its influential 21st Report (1998, 100–106), urged all levels of government to explore a range of innovative means of articulating the widest possible array of values at the earliest stage possible in setting standards and developing policy. 'The public', the commission stressed (page 105), 'should be involved in the formulation of strategies, rather than merely being consulted on already drafted proposals'. Indeed, the commission (page 29) went further and proposed the formal linking of scientific assessments for all types of problem analysis and solving with more wide-ranging and 'people friendly' means of articulating social values, including feelings on trust and accountability.

We shall see that this means that the whole notion of risk management will enter the realms of 'regulatory science' and 'deliberative science', and increasingly merge the two, as the nature of discourse around risk moves into the participatory and interdisciplinary realms.

What is risk management?

In general, but not wholly, the development of modern society has regarded the promotion of safety as a good thing. The wish to reduce risks, particularly those associated with human-made technologies such as nuclear power and the motor car, is perhaps part of our underlying desire to promote life as long as possible. Whatever the precise reason, the process of managing risks, and thus increasing safety, involves a balancing act – that between the cost of reducing risk and the benefits arising from the amount of risk reduced. This chapter explores the whole concept of risk and examines some of the key issues associated with the management of risk, including its assessment, characterisation, perception, communication and evaluation. As Box 17.1 shows, these are the basic components of risk management. Box 17.2 provides some basic definitions.

The idea of management implies some form of decision making, increasingly often in an atmosphere of intense conflict between competing parties. Such conflict might be resolved simply by power, where the strongest party wins, imposing its rule and eliminating behaviour contrary to it (see Chapter 3). In this case, there is little or no argument or discussion about alternative solutions to a problem. Conflict can also be resolved through the use of discourse where special groups, including scientists, analyse a problem, develop means of characterising the problem, and provide guidelines for its resolution.

This approach has been the mainstay of the development of environmental policy in the UK, but it suffers from a series of drawbacks. First, it assumes that only scientists and technical specialists are worthy of taking part in the debate. Even though they may not always agree, if consensus can be reached, so the analysis went, then society will benefit accordingly. Second, there are many examples where the adoption of a largely scientific approach, at the expense of the intuitive reactions of individuals and groups within society, has led to disastrous consequences for certain people, or for ecosys-

Box 17.1 The risk management cycle

Figure 17.1.1 shows the interrelationships between the basic components of risk management. Two points should be noted: first, the pervasive nature of risk perception and multi-way risk communication throughout the cycle; and second, the feedback loop provided by evaluation, which informs the hazard identification component. The latter feature enables organisa-tions responsible for managing risks to become self-learning, more responsive and thus anticipatory. The cyclical nature of risk management has now been recognised by the US Presidential/Congressional Committee on Risk Assessment and Risk Management and the UK Parliamentary Office of Science and Technology.

Figure 17.1.1

Box 17.2 Definitions of the stages of risk management

Although no universally agreed risk terminology exists, some of the more common terms are defined below. However, a note of caution is urged as these terms are open to differing interpretations: for instance, one person's risk assessment could be another's risk estimation.

● **Hazard identification** – determines what can go wrong by identifying a set of circumstances.

● **Risk estimation** – predicts how likely it is that a particular set of circumstances will arise.

● **Consequence analysis** – determines subjectively the significance of the outcome of a hazard event.

● **Risk assessment** – combines two-dimensionally the risk estimation and consequence analysis stages.

● **Risk characterisation** – the collation of information from the stages above to produce risk numbers describing adverse effects and their likelihood which can be used in conjunction with other factors to determine a tolerable level of risk.

● **Risk mitigation** – considers how risks can best be avoided, reduced to tolerable levels or controlled.

● **Risk monitoring** – provides feedback about the net effects of risk mitigation.

● **Risk policy evaluation** – this is sometimes confused with risk characterisation, though here it refers to the evaluation of a risk management intervention such as a piece of legislation or a new working practice designed to reduce risk.

tems, and unnecessary human suffering. Examples include the failure to control persistent organochlorine pesticides that led to the death of birds of prey in the 1960s, the delayed lowering of occupational radiation exposure levels, which created widespread disease amongst workers using radionuclides, and the use of agricultural chemicals. In the latter case, the toxicological assessment of herbicides undertaken by the UK government's Pesticides Advisory Council (PAC) repeatedly stated that there was no scientific evidence, based upon laboratory studies, that the substances caused any harm to humans, despite large volumes of anecdotal evidence to the contrary provided by the farm workers themselves. This is the 'false positive' position examined in detail in Chapter 1. Ultimately, in the face of mounting opposition, the PAC was forced to qualify its original statement, acknowledging that there was no harm *providing that the herbicides were produced and used under the correct conditions* (Wynne, 1992, italics added). This apparently minor amendment to the scientific facts is crucial as it introduces the concept of *management* and *operation* into the safety debate – two points that we shall see are as important as the scientific and technical aspects of assessing risks. In all of these cases, either the scientific assessments were flawed or the wider management and operational considerations had not been considered. In rather too many cases both kinds of misjudgement have been made.

An alternative method for resolving conflict lies in the free and open exchange of ideas between interested parties. This is not to say that conflict between parties will thus be avoided, but this approach does recognise that where conflict does occur it can be, and should be, a positive experience involving sophisticated debates through which ideas are transformed into solutions. Increasingly, as people discover that Western ways do not necessarily offer the panacea that they were once promised, they are beginning to realise that the decision-making process over risk covers issues such as how we want to live, to feel, to think. If this is so then decisions about how to manage the environment should not rely wholly on scientific estimations but should include the ethical and moral considerations of other members of society. Box 17.3

Box 17.3 Isolation or integration? The role of risk assessment in risk management

One of the key issues highlighted by the multi-disciplinarity of risk management is whether risk assessment as a scientific process can and should be separated from risk management. The basis of the arguments for and against separation are rooted in fundamental views of the role of science and society.

At the beginning of the 1980s, an increasing level of concern was expressed in the United States that the scientific aspects of risk assessment were being corrupted by extraneous and irrelevant social policy dimensions. In response to this, a high-level study conducted by the US National Academy of Sciences proposed a return to the separation of facts and values in the management of risks. The proposed scheme relied upon a tripartite system of scientific risk assessment, risk assessment policy and risk management. The former dealt with facts, the latter with values, leaving risk assessment policy to liaise between the two. The point to make is that eventually science and policy *must* interact, so science can never be wholly isolated. Furthermore, it is clear that policy issues drive the kinds of science that are conducted, so the assumption that science is conducted in advance of policy is overly simplistic. The bridge between the science and policy, here called risk assessment policy, is constructed from decision rules based either on fact or on value, whichever is chosen being a legitimate matter of policy choice. That science might somehow be conducted in isolation and occasionally deliver objective information which the policy makers can then choose to either accept or reject is not a plausible vision. In considering the nature of quantitative risk assessment, the National Academy of Sciences identified possibly 50 opportunities where scientists may have to make discretionary judgements ranging from the kinds of hazard to study to the identification of most relevant exposure pathways. This is supported by a number of studies that have investigated different teams' efforts to conduct risk assessments for the same process, where, typically, risk estimates can vary by at least one order of magnitude depending on the assumptions made at the outset.

In the UK, similar discomfort has been felt by some scientists concerning the gradual incursion into their domain of social and political dimensions. An illustration of this can be found in the preface to the Royal Society's second risk management volume published nine years after its initial report on the techniques of risk assessment. The second report focused more broadly on risk management and included a chapter on risk perception and communication. The preface to the report contained the explanation that the content should be viewed as a report not of the society's, but of the chapter authors' views. By disguising its inability to address the issue of how best to manage scientific and non-scientific material in terms of not wanting to 'pre-empt the very debate that the Council and the contributors wish to encourage', the society left open the whole issue of the increasingly unhappy marriage of facts and values.

More recently, the US Presidential and Congressional Commission presented a new framework for risk management which emphasises the engagement of stakeholders as active partners. Here risk assessment is conducted iteratively, which allows for new information to be built into the management process at any stage. The framework recognises the need for health and environmental risks to be evaluated in their broader context rather than focused on single chemicals in single media.

The integration of risk assessment into risk management is now widely recognised to be of crucial importance. Moreover, the *procedural* aspects of risk assessment are becoming more important than the *substantive* issues: how risk assessment is conducted is more important than the results of the assessment itself.

Source: Presidential/Congressional Commission on Risk Assessment and Risk Management (1997).

examines the current debate on whether or not risk assessment as a scientific exercise should be isolated from or integrated into risk management. The role of risk assessment in risk management is discussed later in the chapter.

Opening up the decision-making process has two implications. One involves the faithful two-way communication about dangers and tolerance to danger. The other provides a mediative or dispute-resolving mechanism for seeking common ground between interested parties. Environmental risk mediation is becoming more common in Europe and Japan, though the more hierarchical structures of power which operate may take some time to adopt it fully. But in the US and Canada and increasingly elsewhere, environmental mediation generally is already a valuable method of reaching solutions (for a good review see Bingham, 1986). This involves an element of power sharing between parties and hence requires a more open, trusting and pluralistic approach to politics and power. Here is when environmental science in its interdisciplinary mode, as covered in Chapter 1, can come into its own: see, in particular, Chapters 5 and 6 of the Royal Society (1992), Krimsky and Golding (1992) and the more recent Royal Society report on science, policy and risk (Royal Society, 1997).

Understanding hazard and risk

There are no universally accepted definitions of hazard and risk. As a consequence, some confusion can be found within risk literature as terms such as hazard, risk, risk assessment, risk evaluation, and risk analysis are interpreted in different ways. This is in part a function of the relatively embryonic nature of the field, but as we shall see it can be argued that the lack of universal definitions for risk terms reflects the variety of the situations and political systems in which the management of risk occurs. An initial reaction may be to attempt to develop a standardised risk terminology akin to 'risk Esperanto'. However, such efforts have been largely misplaced, and it is now proving more effective to promote mutual understanding by encouraging risk assessors and managers to make explicit the assumptions and interpretations of the risk terms they use.

One of the most common confusions lies in the often synonymous use of the terms 'hazard' and 'risk'. In practice, these form the two endpoints of a continuum from qualitative descriptions of hazard to quantitative estimates of risk. Put at its simplest, the term 'hazard' relates to the property of a substance, or an activity, to cause harm. The property or activity may best be described as a set of defined circumstances. For example, one hazard associated with landfill gas is its flammability at concentrations in air of 5–15%. In confined spaces, this flammability leads to a potentially explosive situation. For this potential to occur a series of individual circumstances must occur: for example, landfill gas must be generated and migrate off-site; it must accumulate in a confined space such as beneath a building; and some ignition source must be present to ignite the gas between the concentration limits specified above. *Understanding the hazard* as a set of defined circumstances, how ever simple or complex, is the basis of developing systems to reduce the potential of the hazard to occur and thus to increase safety. This is the established, scientific method of framing risk problems. However, it suffers from the problem that it is logically impossible to define all the possible circumstances that may occur. For instance, few nuclear engineering experts would have considered feral pigeons as a migration pathway for radioactive contamination, yet this is exactly what has happened at the Sellafield nuclear installation in the UK, where caesium-137 contamination spread by pigeons has caused much public alarm, and no clear consensus as to the possible risk effect.

Defining the set of circumstances allows us to understand the nature and characteristics of the hazard. The next stage is to ask how likely it is that the set of circumstances will occur – to *realise the hazard potential*. Sometimes, but not always, it is possible to attribute a statistical probability to each of the factors within the set of circumstances. In the pigeon example, this would involve knowing about the likelihood that pigeons could become contaminated and the chances that they might enter the human food chain. Ornithologists indicated that the feral pigeons were unlikely to stray more than 10 miles from the plant, as they are very territorial birds. However, preliminary tests showed that eating the breast meat from six contaminated pigeons could cause the annual radiation dose limit for an exposed individual to be reached.

Most definitions of risk use terms such as 'probability', 'likelihood' and 'chance', which imply *the potential for something to occur*. Such an outcome is usually adverse – it is not usual to talk about the risk of benefiting from a particular adverse outcome such as winning the national lottery or passing exams. Risk, then, is the likelihood, or probability, that a particular set of circumstances will occur, resulting in a particular adverse consequence, over a particular time period. It is usually expressed as a frequency – for example, the risk of being struck by lightning is about one chance in 10 million per year. This is based on the average number of people being struck by lightning each year divided by the total population. Clearly, this statistical probability is only an estimate. Some people, for example golfers playing in thunderstorms, will be at a greater risk than others. Traditionally, the time period used in calculating risks is one year, and the most common outcome or consequence of the risk event is human death. This can be seriously misleading, as environmental impairment can be an equally valid outcome.

Box 17.4 illustrates two common methods of *identifying and organising hazards*. If defining sets of circumstances is often very problematic, then the estimation of proba-

Box 17.4 Engineering risk assessment

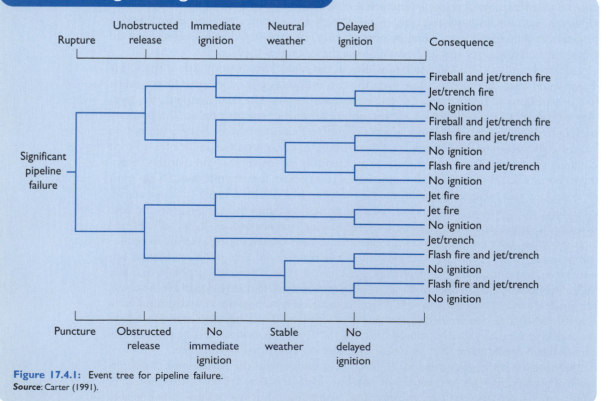

Figure 17.4.1: Event tree for pipeline failure.
Source: Carter (1991).

Box 17.4 (continued)

The logic diagrams shown in Figures 17.4.1 and 17.4.2 have been applied to engineering systems analysis since the 1960s. Each form of analysis attempts to identify the components of a system that may lead to a hazard being realised.

Event trees (Figure 17.4.1) specify a range of possible outcomes of a hazardous event, such as a pipeline rupture. Each outcome is then linked to the main event by a series of circumstances known as an accident sequence. Probabilities of any part of the sequence occurring can be estimated and an overall probability of a particular outcome, say, a flash fire and explosion, can be calculated by multiplying the probabilities throughout the sequence.

Fault trees (Figure 17.4.2) use the reverse process, beginning with the initial event. A 'top-down' analysis is then constructed using a series of formal statements or 'logic gates', which determine the relationship between the individual circumstances within the accident sequence. Such statements will identify whether circumstances are mutually inclusive ('AND' gates), whereby all circumstances must occur for the hazard to be

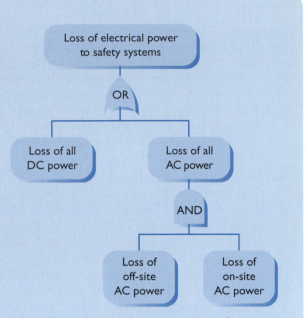

Figure 17.4.2: Simplified fault tree on electrical power. **Source**: Rasmusen (1990).

realised, or whether the circumstances are mutually independent ('OR' gates).

bilities is even more so. First, there is the problem of how to frame the problem such that it is possible to undertake numerical analyses. Any framing must be balanced against the need to provide a sufficient level of comprehensiveness such that the majority of possible sequences are considered. Second, it is important to account for uncertainties within the system. These may be compounded throughout the analysis to the point where the numerical assessment of probability becomes meaningless. Third, it is important to consider the nature of the probability data itself, as the combination of data requires care and sensitivity.

Although this definition of risk implies a particular consequence, say, human death, it is only when the significance of the consequence is evaluated, in terms of societal judgement of the cost of the consequences, that a *risk assessment* is performed (see Box 17.5). Thus, a common definition of risk assessment would be the combination of a frequency estimation – a risk – with some form of consequence analysis – an evaluation of the seriousness of the consequences in question, such as death, environmental damage or passing on a limitation of choice to future generations about the amount of fossil fuels they can burn. This two-dimensional approach is what many people in the risk field understand by the term 'risk assessment'. These two dimensions can be depicted graphically as the *fN* curves shown in Box 17.6.

Extending this definition further, risk management can be seen as a series of inter-linked components with technical risk estimation at the outset, the assessment of that risk, and the development of some kind of policy response to that assessment. Thus, as Boxes 17.1 and 17.2 show, risk management involves much more than just the two-

Box 17.5 Health risk assessment

● *Hazard identification.* The qualitative determination of the kinds of adverse health or ecological effect a substance (or activity) can cause. As agencies in the US have tended to focus on the ability of substances to cause cancer, hazard identification involves mutagenicity and carcinogenicity tests, epidemiological studies, and case–control studies.

● *Dose–response assessment.* The estimation of probability of harm that the exposure levels identified above will cause. This involves the (uncertain) interpretation and scaling of animal carcinogenicity tests to the human case. It is assumed that for carcinogenic substances no threshold dose exists (i.e. the dose–response relationship is linear). For non-carcinogenic substances; thresholds are established using animal, plant or micro-organism studies.

● *Exposure assessment.* A determination of the 'maximally exposed individual' (MEI), normally a hypothetical person derived statistically, or the total population exposed. It involves emissions modelling, fate and transport analysis, uptake analysis, and demographic analysis of the target population.

● *Risk characterisation.* The results of the above steps are collated and risk numbers produced to describe the adverse effects and their likelihood. The estimates include safety factors incorporated from dose–response and exposure uncertainties. Typically in setting standards a factor of 10 is applied at each of the two stages, resulting in a safety factor of 100. Depending on the particular case, the safety standards applied may be more or less based on scientific studies than on expert guesswork.

Box 17.6 Societal risk and *fN* curves

Societal, as opposed to individual, risk involves risks to large numbers of people. Frequency–consequence graphs, or *fN* curves, have been used extensively to depict and compare societal or group risks from different activities. Where the consequences of an event vary over a wide range, this type of approach is particularly useful. When the cumulative frequency of events with impacts over a certain size is plotted, the distribution can be used to indicate levels of acceptable or tolerable risk. However, a number of problems exist with *fN* curves. Attributing levels of tolerable risk is a largely subjective exercise about which there is little agreement. In particular, the question of whether a few large-scale incidents that kill many people are more significant, and thus deserve higher priority, than many small-scale incidents that kill the same number of people remains unresolved.

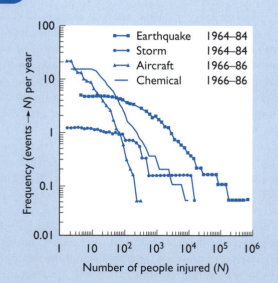

Figure 17.6.1
Source: Fernandes-Russell (1988).

Box 17.6 (continued)

For the policy maker, a reliance on *fN* curves is problematic, because the calculation of fatalities is not always straightforward, as it relies on historical data usually reported in a variety of ways and, by definition, does not include potential risk events which have not yet occurred. For example, there is the difficulty of accounting for delayed deaths, those fatalities that occur long after the event. In certain circumstances, there may be significant under-reporting of fatalities or problems defining accident type. Also, by considering only fatalities one omits to account for fates worse than death.

dimensional probability/consequence assessment of risk. It also encompasses much broader factors such as public perceptions of risk, which may differ significantly from those of the experts, legal considerations regarding the style of regulation, and cost–benefit analyses of various strategies for risk mitigation. These points are developed later in the chapter. Ultimately, the development of environmental policy, and the decision-making process through which policy is derived, faces the challenge of incorporating a wide range of concerns to determine what level of risk will be tolerable in any given set of defined circumstances. These include not only scientific evidence and probability estimates but also a cultural sense of fairness in the treatment of citizens and enterprise, the degree of trust in regulating agencies, and the underlying feelings about how much manipulation of our lives we are prepared to tolerate when we have no means of avoiding certain outcomes, for example the possible consequences from extra ultraviolet radiation from ozone depletion.

The role of risk assessment in managing risks

Traditionally, risk assessment has been the foundation on which many risk management decisions have been based. Risk assessment is one component of risk management and comes in many different forms. At an individual level, assessing risks has always been a daily activity for everyone. At one level we are faced with risk decisions all the time: where to cross the road; what kinds of food to eat; how long to sunbathe for; and so on. We sometimes arrive at answers to these questions without appearing to think very carefully, acting on impulse or driven by habit or custom. At other times we may wish to weigh up more systematically the reported risks against the benefits that will be accrued.

The systematic assessment of risk is used in many different instances. Box 17.4 illustrated some examples drawn from the engineering sector. Engineering risk assessment has been practised for decades and was initially designed to make systems and structures more reliable. At first, these reliability assessments tended to be used internally to the engineering process. However, as time has progressed, these types of assessment have been used more frequently to show external interests that engineering systems will be safe enough. The rapid expansion of health risk assessment (see Box 17.5), which seeks to establish the probability of harm from exposure to certain substances or activities, focuses more intently on acute and chronic impacts to humans. This four-stage approach has been adopted by the relatively new field of ecological risk assessment with some modifications to account for the differences between human and environmental systems. However, efforts to develop the field of ecological risk assessment by the US Environmental Protection Agency have met resistance from some areas and, as purse strings tighten in times of economic recession, so resistance

Continued on page 449

Box 17.7 Community consultation in siting hazardous facilities

Environmental conflicts are often related closely to the many different ways of understanding or 'knowledges' that exist about certain issues. In siting hazardous technologies, community knowledge has often been at odds with more traditional scientific knowledge, not least because the two rarely interact in any meaningful manner. The latter applies 'objective' analytical techniques to problems such as those posed by the mounting levels of waste generated by society. Here scientific knowledge follows the path described in Figure 17.7.1 from quantification to the development of technological solutions, and finally to siting. However, the rise of opposition to this expert-oriented approach has led some local authorities to reassess the way in which waste management plans are developed. Before discussing the novel forms of community involvement, the three stages of community opposition to the development of waste management facilities are described below.

Phase 1: geographical opposition

Community opposition to waste management facilities is often encountered once siting questions have been raised by the planning system. This concern might be characterised as geographical. Expressed as *not in my back yard* (NIMBY), this protest has been regarded by developers, and in some cases local authorities, as selfish, small-minded and an unavoidable nuisance. Through time it has become clear within the risk and planning literature that NIMBY attitudes represent more than this and that many of the issues being raised by public participants held a high proportion of 'common' sense.

The response of the waste management industry to the NIMBY syndrome has been twofold. At first, attempts were made to relocate to alternative sites. When this failed, public relations exercises were initiated, based largely upon one-way risk communication strategies, to educate and inform the public

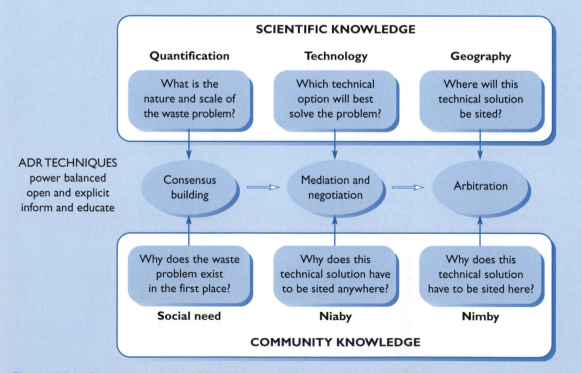

Figure 17.7.1: The potential role for ADR in bridging scientific and community knowledge for waste management planning.

Box 17.7 (continued)

about the scientific and technical aspects of the proposals. Although over time these exercises have become more responsive to public needs, the focus of the majority of communication efforts remains on the allaying of (perceived) irrational anxieties about proposed waste management facilities.

Phase 2: technological opposition

As community groups opposing waste management facilities have grown in number, so has the nature of their opposition. They have responded to the industrial strategy of moving from one community to the next by opposing developments on *technical* rather than *geographical* grounds. United in their campaign against the waste management industry, it was no longer appropriate for the basis of community opposition to be classed as NIMBY. In the UK, the key catalyst in the emergence of more co-ordinated opposition to waste management technologies has been Communities Against Toxics (CATs).

The CATs network developed from a conference organised by Greenpeace UK in 1990. This conference brought together community groups campaigning against a wide range of environmental pollution sources, including landfilling, land and sea incineration, waste importation and transportation, reprocessing, and sewage and industrial waste outfalls to freshwater and marine environments. It was attended by over 200 delegates representing over 20 campaigns from the UK and Ireland, trades unions and local authorities, and one member of parliament. Nine years on, CATs has emerged as a national network aiming to help local groups, to assist new campaigns, break down group isolation, encourage the sharing of information, ideas and experiences, and co-ordinate national activities. CATs acts as an information broker between the social justice movement in the USA (see Box 17.9 below) and the 100 or so groups in its UK membership.

The development of the CATs network, nationally and internationally, has played a sig-

nificant part in the shift in attitude of geographically based NIMBY opposition to the more technically based *not in anybody's back yard* (NIABY) stance. As community groups have interacted, a new consensus knowledge has developed based around answers to the question: '*why* does this technical solution have to be sited anywhere?'

Phase 3: moral opposition

As community groups under the CATs banner have exchanged ideas and information, the focus of debate is developing further. Questions of social need, that is the fundamental need for various waste management technologies, have been raised. These have included a reappraisal of basic assumptions, which include why waste is generated in the first place and *why* waste is being imposed upon certain sectors of society who are as unwilling to tolerate its consequences as other sectors. In this sense the problem possesses very clear social dimensions, overlooked by established scientific knowledge in the initial framing of the waste problem using hazard assessment methodologies for screening problem sites from a technical perspective. Community opposition challenges from a moral standpoint the basis for economic growth through industrial production. Emerging debates about sustainable futures bring this focus into sharp relief.

As Figure 17.7.1 illustrates, a number of novel methods of incorporating community views into waste management planning exist to bridge the divide between scientific and community knowledge. These include waste generation monitoring exercises, often run as competitions, which raise awareness of the nature and range of wastes generated at the household level, structured discussion groups designed to explore possible waste management futures, visioning exercises as a means of developing action plans to deliver specific aims and objectives, and the incorporation of local communities into environmental monitoring for EIA and plant operation.

Box 17.8 Going deep? Radioactive waste disposal in the UK

Since the early 1980s, NIREX, the Nuclear Industry Radioactive Waste Executive, a company responsible for solving the radioactive waste disposal problem, has investigated and proposed a wide range of technical and geographical solutions. Initially, low- and intermediate-level (LLW and ILW) radioactive wastes were to be disposed of separately, LLW in shallow trenches at Elstow in Bedfordshire, and ILW at a deep mine location in north-east England. However, public opposition to these proposals, along with criticism from experts concerning the limited range of options considered, led to a first rethink of options.

In January 1985, it was announced that three additional shallow burial sites would be investigated along with Elstow. The formal announcement was made in the House of Commons, and as NIREX had been prevented from liaising with the local authorities in the three areas, it was no surprise that vociferous opposition developed. This opposition was fostered by the news that planning permission for investigation into the sites was to be granted by special development order (SDO). This removed power over planning decisions from the local authorities and gave it to the House of Commons. Disenfranchised from the planning process, the local authorities in the three areas formed a political coalition against the proposals.

In the run-up to the General Election in May 1987 another policy shift occurred. This time economic reasons were given for a move to combine the disposal of LLW and ILW to a deep repository. There was considerable scepticism about the timing of this policy reversal, not least because the four shallow burial sites were each in constituencies of prominent Conservative members of parliament.

In an attempt to pre-empt opposition to this latest switch in policy, NIREX launched a public consultation exercise to garner views from interest groups and the public at large. However, the history of administrative mismanagement served only to reduce the effectiveness of the exercise. The nine-month consultation period generated over 2500 responses, and, though a small minority of the public invited NIREX to provide a 'short courtship and a large dowry', the vast majority felt that this novel policy of openness by NIREX was merely a 'Trojan horse'.

Following this exercise, two deep disposal sites were identified: Sellafield, in Cumbria, England, and Dounreay on the north coast of Scotland. Both sites had existing nuclear facilities, housing communities with familiarity with radioactive issues and, perhaps more importantly, economic dependence on the nuclear industry. Following vociferous opposition from Scotland about the transport of radioactive wastes, NIREX began intensive site investigations at Sellafield for a deep repository accessed from land stretching out under the Irish Sea. Technically, Sellafield proved to be a challenging site, but given the relatively low levels of public opposition the trade-off between technical suitability and political accessibility was considered to make Sellafield the prime choice.

In 1996, NIREX proposed the creation of an underground rock laboratory which would enable more detailed studies of hygrogeological characteristics of the Sellafield site. Others interpreted this as the thin end of the wedge. Once such a 'laboratory' had been created it could easily be manipulated into a final resting place for radioactive wastes. Cumbria County Council opposed the planning application and a public inquiry was launched. On the same day that John Major announced the date for the General Election in May 1997, John Gummer, then the Secretary of State for the Environment, rejected the proposed rock laboratory. Since then, NIREX has come under intense criticism from organisations such as RWMAC, the government's own radioactive waste advisors, concerning a number of important issues, not least the ill-prepared safety case.

With NIREX's decision not to appeal against the decision, the UK's radioactive waste management policy has come full circle. The recent House of Lords Science and Technology Committee report on radioactive waste disposal (HoL, 1999) recognised the value of a deep repository. However, it was also clear that no solution would go forward without full public support. Securing an appropriate site is now the major challenge for the new millennium, though the quest may prove impossible in the foreseeable future, such is the public antipathy for deep storage.

For a thorough and entertaining discussion of the UK radioactive waste disposal story, see Kemp (1992).

increases. As yet the benefits of being able to model accurately complex ecosystems do not appear to outweigh the considerable costs of achieving this, particularly as less sophisticated hazard assessment techniques are able to highlight the majority of problem sites where some form of clean-up action is required.

Typically, risk assessments are conducted using data from various sources, including expert judgements, accident records, actuarial data for insurance purposes, laboratory tests for mechanical reliability, toxicological tests such as LD_{50} on animals, plants or individual cells thereof, and data from predictive mathematical modelling. It is possible to identify five different levels at which information may be available for risk assessments and risk comparisons made.

- Good, direct statistical evidence either from the historical record or from laboratory studies.

- Direct statistical evidence is not available for the whole process but is available for individual components within the process. Thus the whole process may be disaggregated using fault trees or event trees, and aggregate probabilities for the whole process can be derived.

- No good data are available for the process under consideration, but data are available for a similar process. These can be adapted and extended either directly or indirectly.

- Little or no direct evidence exists, but it is possible to use expert intuition to provide authoritative subjective judgement.

- Little or no direct or indirect evidence exists, and even the experts have trouble producing reliable or comparable subjective judgements.

Too often risk assessors are forced to base assessments on data derived from subjective sources. More often than not, assessments cannot be based upon probabilistic estimates but must rely on expert guesses. Hopefully, as databases become larger and processes better understood it will be possible to undertake risk assessments more reliably. Until then, however, we are restricted to assessments of the likely hazards associated with a particular activity. Identifying the sets of circumstances that may give rise to adverse events is a necessary part of the risk assessment process, but risk managers should be aware that what is often called a risk assessment is in fact nothing more sophisticated than an exercise in hazard identification. Means of getting round this dilemma include assuming worst-case evidence every time, applying sensitivity analysis to the uncertainties involved, experimenting with carefully monitored trials or experiments, adopting the precautionary principle, or introducing various measures of deliberation (Royal Commission on Environmental Pollution, 1998, 52–59).

Comparative risk assessment: the apples and oranges problem

Whilst QRA is normally used at the project level, at the national or global level, it is common to find *comparative risk assessment* (CRA) used to rank environmental problems by their severity so that resources can be targeted effectively and efficiently. CRA provides a mechanism whereby the decision maker can justify the many difficult decisions that have to be made concerning resource prioritisation.

In comparing the relative merit of alternative, and sometimes competing, environmental risk management strategies a clear problem emerges. How is it possible to compare meaningfully alternatives that are not easily comparable? If we are faced with choices between different amounts and qualities of air pollution and water pollution, which should we choose, and perhaps more importantly, how can we justify the decisions that we make? At the risk of being overly simplistic, it could be argued that comparisons can be made between apples and oranges, though less on the basis of technical parameters such as the vitamin content and more on subjective parameters such as taste, flavour and texture. The point here is that simply to give up attempting to make comparisons defeats the purpose of interdisciplinary studies. What is required are new ways of comparing things which from our traditional scientific viewpoints appear incommensurate. This problem is well known to risk analysts and is becoming more well known to environmental policy makers, particularly those in search of the best practicable environmental option (BPEO): see Box 17.14 later in this chapter.

Where risk assessments are possible, it is important that risk managers should realise that a meaningful risk comparison involves more than the accurate estimation of each option in isolation. Given the levels of scientific and technical uncertainty that exist within the development of environmental management policy, it seems appropriate to call for an opening up of decision-making processes to a wider audience. In any case, those responsible for undertaking comparative risk evaluations should be wise to the following:

- comparisons should include as wide a variety of options as possible. It is easier to exclude options after consideration than to include options once the evaluation process is underway;

- comparative assessments must include geographical and technical parameters – a solution that is tolerable in one location may be intolerable elsewhere;

- data limitations may require the extensive use of conservative assumptions, and methods for estimating environmental effects should err on the side of caution;

- value judgements inherent within any assessment technique should be stated explicitly to avoid the danger of misinterpretation.

The rise of risk assessment as a tool for decision makers comes in the face of mounting criticism from industry, which argues that it is too conservative in its assumptions and thus inflames or initiates public fear and leads to unnecessary financial hardship. Environmental and community groups, on the other hand, see the tool as too simplistic and narrow to deal with the complex reality of risk issues. Characterising risk debates is fraught with difficulty as terms are interchanged and possess different meanings to different parties. Whether risk assessment is seen as a device for informing the decision-making process or as the decision-making process in itself is one aspect of risk management that divides opinions. Recent research suggests that procedural matters are now as important as substantive issues: it is more important *how* risks are assessed than the results of the assessments themselves. This point is given full coverage in O'Riordan *et al.* (1997).

Risk characterisation: how safe is safe enough?

The question of achieving sufficient levels of safety is one that has demanded much attention from risk analysts and decision makers. Determining acceptable, or more correctly tolerable, levels of risk has proved to be difficult for many reasons, not least because the answer lies not in scientific and technical assessment but in economic, moral and ethical bases. Thus, decision makers are forced to consider an array of much wider social concerns in order to achieve a tolerable level of risk. The simple fact that different groups within our democratic society adopt vastly different definitions of the term 'tolerable' gives some indication of how complex a task this is. Decision makers have often been accused of trying to avoid criticism by concealing how decisions were reached. This has led to increasing public disquiet about the fairness of risk management, particularly in situations where a local public is asked to bear the brunt of the risks associated with an activity when the benefits are shared more widely. Examples here include radioactive waste disposal and the incineration of industrial wastes, but the siting of new road schemes and high-voltage transmission lines is also becoming a matter of controversy. The unequal distribution of risks and benefits of societal risks are a feature of modern risk management that require urgent attention, to the point that within the risk field the question becomes one of how fair is fair enough? Box 17.9 looks at the issues of social justice emerging in the United States, which have led to the

Box 17.9 Social justice

In the USA, hazard and risk assessment techniques have been linked to the unfair prioritisation of contaminated site clean-up efforts. For example, the US Environmental Protection Agency uses a prioritisation tool known as the hazard ranking system (HRS) for deciding which contaminated sites ought to be cleaned up first. Based on models similar to those outlined in Box 17.4, the HRS combines factors indicative of the hazards associated with contaminated land using a prescribed scoring and weighting system. An internal bias in the mathematical algorithm of the HRS towards the groundwater component of the HRS has been argued to explain why fewer contaminated sites have been identified in large dense urban areas which have a dependence on surface water supplies. As urban centres have a higher than average minority population, concerns have been raised about the social equity issues surrounding the distribution of risk and benefits.

Other studies have shown similarly disturbing trends. For example, a US Greenpeace study found:

● the racial minority population in areas of municipal incinerators is 89% higher than the national average;

● communities where incinerators are proposed have 60% higher racial minority population;

● average income in communities with incinerators is 15% less than the national average;

● in communities where incinerators are proposed, property values are 35% less than the national average.

In response to increasing concerns about the inequitable distribution of risk and benefits in February 1994, an Executive Order was passed designed to address environmental justice in minority and low-income populations. Achieving environmental justice is now part of the mission for each US federal agency. In particular, efforts are being made to (1) promote enforcement of all health and environmental laws, (2) ensure greater public participation, (3) improve research and data collection relating to health and environmental quality amongst minority populations, and (4) identify differential patterns of consumption of natural resources among minority populations, which was intended to apply some level of protection to populations who rely principally on fish or wildlife for subsistence.

In the context of cleaning up contaminated land, research into the unfair distributions of risks and benefits has yet to prove conclusively that minority populations are disadvantaged. Much depends on how largely subjective boundaries are drawn around sites which may include or exclude certain populations.

The temporal aspects of perceptions of environment are important here too. Greenberg and Schneider (see Bullard, 1993) discovered that in areas close to three of the most prominent contaminated sites in the USA, recent residents rated the neighbourhood quality significantly higher than long-term residents. The asymmetry behind these perceptions could be related to decreases in property values, fears of health effects, odours and community disruption caused by the site. Whichever the case, this illustrates that to fully understand the attitudes of residents it is important to reflect on the wider context of the area, both geographically and temporally.

Source: Heinman (1996).

creation of new legislation enacted specifically to deal with unfair risk distributions amongst minority populations.

Clearly there are a vast number of different types of risk that need to be evaluated. Some risks we cope with subconsciously on an everyday basis – where to cross the road, what to eat, which hobby to pursue. Others are evaluated for the benefit of society as a whole. It is on the latter that this chapter will concentrate. But first let us consider briefly how individual risks are judged. Why is it that some people choose to undertake dangerous activities, seeking relatively high levels of risk, whilst others take extensive measures to avoid risk? One suggestion is that, as individuals, we each have an inherent level of risk at which we feel comfortable. Thus, if activities are made safer for us we may strive to undertake them in a more dangerous way, compensating for the increased safety, and so restoring our chosen level of personal risk. An example of this concerns driving habits. The introduction of seat belts was a safety measure designed to reduce the number of people killed in road accidents. The UK Department of Transport weighed up the cost of introducing seat belts against the implied statistical value of the lives that seat belts would save and concluded that it was cost-effective to require car manufacturers to fit seat belts to all new cars. The net result of this was a reduction in the number of deaths to car drivers and passengers. However, one side-effect of this reduction measure was to change the way in which some people drive. Not only do some people drive faster when wearing a seat belt than when not, but they accelerate and brake more sharply to compensate for the increased safety that they perceive. This may be tolerable for those fortunate enough to be contained within the car, but for cyclists and pedestrians there is a net increase in risk. This example illustrates the complexity of managing risks, particularly when dealing with the transfer of risks from one group to another.

Balancing risks, costs and benefits

Having established the means by which risks can be assessed, it is then necessary for the decision maker to offset these judgements against the costs and benefits of mitigation measures. In the example above, the method used by the government was a fairly straightforward risk cost–benefit analysis based on economic efficiency. However, other methods may be used:

- *Risk cost–benefit analysis* is based on the economic efficiency of introducing particular risk reduction measures.

- *Revealed preference* uses the tolerance of current risks as a basis for the introduction of new risks, providing the benefits are similar. This method assumes that the degree to which people expose themselves to any danger reveals their willingness to tolerate the associated risk. This technique is based on actual evidence of behaviour and exposure.

- *Expressed preference* is more direct in that it measures risk by asking people what levels of safety and danger they are prepared to tolerate for certain classes of hazard. This may be done statistically or by comparative measures, for instance by determining how much more or less dangerous a hazard should be in comparison with another hazard. The difficulty here is to select comparable hazards. One cannot compare the risks of involuntary exposure, such as nuclear power, with voluntary risks, such as smoking. This method also assumes that people are well informed and it provides only a snapshot set of preferences at a given time.

- *Multi-objective approach* uses multi-attribute utility theory (MAUT), which evaluates different factors independently, then combines them into some form of overall assessment. This approach is often used when dealing with complex environmental issues, such as deciding on the appropriate balance of waste management options (see Chapter 18).

- *De minimis* – literally 'at the lowest level', assumes that some risks are too trivial to bother with, or that levels of safety are as high as they possibly can be. It seeks to establish thresholds below which risks are deemed acceptable. This suffers from the fallacy that background levels are thought of as 'natural', and therefore are 'normal', even to the extent of being considered moral. It is also couched in terms of discretion, that is, the judgements of regulators as to what level of safety is tolerable are based on the extent of technological development, the economic circumstances of the risk creator and the likely public reaction.

The underlying principle by which risks are judged in the UK is 'reasonable practicability'. The notion of practicability is based primarily on cost grounds with consideration of advances in new technology (see Box 17.10). The concept of reasonableness comes

Box 17.10 Risk regulation and the ALARP principle

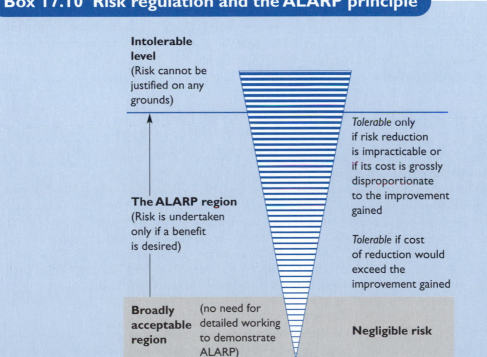

Figure 17.10.1

ALARP, or *as low as reasonably practicable*, is couched in the British discretionary manner of determining safety. The onus of the safety case rests with the creator of the risk, who must satisfy the inspector that the ALARP principle has been followed in the management of industrial activities; that a safety culture is in place; and that action in the event of an accident is judged to be appropriate and fully intelligible to the nearby public. Finally, ALARP is based on a

Box 17.10 (continued)

crude cost–benefit comparison. Even for private concerns, the benefits from a given technology or management approach may be only a tenth of the costs. Yet the adoption of the scheme can be enforced on the bases of tolerance and precaution.

In the UK, the fundamental driving force behind risk assessment and management is the ALARP principle. This principle aims to reduce risks to as low as reasonably practicable and introduces the idea that, as zero risk is unachievable, efforts should be made to reduce risks only where they are entirely intolerable, or where the cost of reducing them further still is reasonable given the extra risk reduction gained. Here, the notion of reasonableness is argued in terms of gross disproportion between the costs and the benefits of risk reduction. Three areas exist: at the top end of the diagram, the level of risk is deemed to be entirely intolerable and risks have to be reduced *whatever the cost* or the activity terminated. At the bottom end of the diagram the risks are deemed to be negligible, where no reduction is necessary and no systematic risk assessment need be undertaken. The middle section, the ALARP area, represents risks which should be reduced if the cost of doing so is reasonable in relation to the activity concerned. The ALARP principle is designed to act as an ever-tightening risk reduction device such that, as new technologies are introduced and novel pollution control systems become available, industrial risks are gradually reduced. This is a classic example of the UK approach to safety management. The aim is to place the burden of safety on the regulated, who must negotiate ALARP for every piece of equipment and process. For its part, the safety regulator will grant a permit only when the conditions of tolerability noted above are satisfied. In addition there is a rule of thumb about risk–benefit calculations. In general, if the risk is relatively localised the benefit to cost ratio should be about 1 to 3. But if the risk is potentially catastrophic, the benefit to cost ratio may rise to as large as 1 to 1000, as is the case with radioactive waste disposal (see also the previous chapter on air pollution studies and public health).

The HSE guidelines are as follows:

- 1 in 1000 as a just tolerable rise for any substantial category of workers for any large part of their working life;

- 1 in 10 000 as the maximum tolerable risk for members of the public for any single non-nuclear plant;

- 1 in 100 000 as the maximum tolerable risk for any member of the public for any new nuclear power station;

- 1 in 1 000 000 as the level of acceptable risk at which no further improvements in safety can be made.

At the point of tolerability (1 in 100 000), risk managers are expected to spend on safety measures amounts up to the point where further expenditure would be grossly disproportionate to the reduction in the risk obtained. There is an implicit comparison of costs and benefits at each of these points in the ALARP chain.

from legal judgements concerning specific cases brought to court. The emphasis in these cases has been and remains establishing if particular actions or inactions are those of a reasonable man or woman: in UK legal parlance, 'the individual on the Clapham omnibus'. Reasonableness is considered in terms of aspects such as negligence and foreseeability.

It is tempting to view the technical assessment of risk as being somehow objective and rational, while the wider evaluation of risk tolerability is filled with value judgements and internal and external biases. In reality, each of the components within the risk management process is subject to value judgements, biases and uncertainties. Risk management, although a very useful process, is not a precise science, nor is it a particularly well-developed art form. Open and honest scientific disagreement is a valuable part of the process of developing risk policies. Proponents see these processes as rational, scientific tools providing accurate assessment and effective management, whilst opponents consider them deeply biased and inherently flawed. However, even if risk assessment and evaluation were somehow wholly objective, they would not be able to answer the wider question of how safe is safe enough, or to put it another way, what level of risk is tolerable?

Risk and tolerability

The notion of risk tolerability as opposed to acceptability emerged from the long-running Sizewell Inquiry into a major new nuclear reactor in East Anglia, England, held between 1983 and 1986. The Inspector, Sir Frank Layfield, coined the term as it imparted more clearly the idea that people judge risks in terms not only of costs but also of benefits (O'Riordan *et al.*, 1987). He argued that if the benefits of a particular risk were considered to outweigh the costs then the risk would be tolerated, but never fully accepted. While scientific and technical assessments enable a clearer understanding of the nature of risks, the question of whether such risks are tolerable is one for wider debate. This debate should include input not only from the scientific community but also from other sections of society – not least the public itself. Clearly, opinions will differ. However, it is vital to realise that, despite the many and varied philosophies and world views supporting these perceptions, each has legitimacy. Facilitating open debate, where differing opinions are not only tolerated but welcome, is the challenge facing risk management today.

Underlying the notion of tolerability are two fundamental principles. The first is that of ensuring that language assists communication and is not a medium of dominance supported by indecipherable jargon. The second is that of process. The procedures through which risks are communicated have to be based on a mutual sense of respect for all positions of the interested parties. This in turn means applying some of the features of the precautionary principle outlined in Chapter 1, namely providing room for error, ensuring that the risk creator has to show that no unreasonable harm will occur, and guaranteeing some form of compensation in the event of unforeseeable circumstances. All these points are given a full airing by the Royal Commission on Environmental Pollution (1998, 51–62).

Psychological and sociological theories of risk perception

The call for increasing levels of public participation within risk management has echoed around the risk literature for over a decade. The most recent exploration of this literature can be found in Marris *et al.* (1997; 1998) and Renn (1998). To satisfy this demand, it is necessary to understand more clearly how lay perceptions of risk, which often differ from scientific and technical estimations, are developed. Initial attempts to understand public perceptions of risk focused on the determination of 'acceptable' levels. As it became clear that no single, standard level of risk could be called acceptable, psychologists began to realise that the nature of risk perceptions was far from simple. A complex web of factors that were thought to influence public perception were identified and tested through psychometric studies, which involved the rating of different hazards either in direct comparison or in terms of different influencing factors such as controllability, familiarity, dread, novelty, and catastrophic

Box 17.11 The assessment of PCB and dioxin emissions from high-temperature incineration

PCBs and dioxins are complex chemicals and notoriously difficult to monitor. They are persistent bioaccumulators which have been shown to cause cancer in mice and rats, though no definitive link to human cancer has yet been proved.

The Centre for Environmental Risk at the University of East Anglia has conducted what is probably the most comprehensive investigation into PCB and dioxin contamination in an area of south Wales close to a high-temperature chemical waste incinerator. The study, commissioned by the Welsh Office, cost almost £500 000 and took 36 months to complete. It involved the use of several laboratories, which each adopted slightly different techniques for determining the levels of PCBs and dioxins in air, grass, soil and duck eggs.

Ultimately, the data showed that there were higher PCB and dioxin levels in the local environment than one would normally expect. When air concentrations were combined with measurements of wind direction, the results suggested that the major source of these emissions was the incinerator.

However, despite the intensity of the monitoring regime the study was not able to draw any conclusions as to how much of the contamination was related to past compared with recent operation. Nor could the study identify which part of the waste disposal process was the major contributor to the contamination. Much effort has been placed upon air emissions from incinerator stacks, but in this case fugitive emissions from other areas of the site might be equally, if not more, responsible for environmental releases. Further still, the study could not determine the significance of the contamination in terms both of the migration of PCBs and dioxins through the environment and, ultimately, of their effect on human health.

This state-of-the-art assessment represented a milestone in environmental monitoring. However, in terms of managing the problem, the lessons that can be drawn from the study are limited. Those responsible for managing the plant have been forced to 'best guess' what form of risk management strategy should be implemented.

Although the nature and extent of the contamination have now been agreed by all parties, the local public remain unconvinced about attempts to clean up the incinerator and manage the risks that that particular industrial facility has imposed on the local community. Despite the recent reassessment of the health effects of dioxin conducted by the US Environmental Protection Agency, until further expensive studies are undertaken to determine human health effects it seems unlikely that residents immediately adjacent to the plant will feel reassured about the situation.

potential. The cognitive psychology approach introduced a whole range of qualitative factors that required some form of evaluation. One method linked risk perceptions to a variety of 'rules of thumb', or heuristics, of which the most obvious one was availability. This suggested that risk perceptions were driven in part by direct and recent experience of a particular event. The psychometric approach used factor analysis to combine certain influences thought to affect risk perception onto two axes, from which a 'hazard map' could be drawn. Factors such as dread, voluntariness, controllability, familiarity and knowledge were assessed using questionnaires. Although this allowed comparisons to be drawn between different groups within the public (typically students), it became obvious that the psychometric approach was painting only part of the overall picture of risk perception. Though widely used, there were a number of disadvantages to the psychometric approach. Studies tended to use small samples, and these were mostly from the United States. Results could provide a view of perception only at the time that they were undertaken. When rating scales were used, these had a framing effect on the answers that respondents could give. More

Box 17.12 The management of genetically modified organisms

We are now in a position to swap genes at will. Designer technology can produce millions of variants of the building blocks of life, so it is now possible to create food that has special properties of disease resistance, colour or long shelf life. In utilising this technology, there is a moral question as to how far it is right to manipulate food so boldly, and whether a future society may ultimately be deprived of purely 'natural' products. It is not easy to incorporate such moral concerns into the review structures that evaluate the risks of novel foods. One method is to include representatives of consumer groups and institutions such as the Church in scientific review panels advising government on the introduction of such substances. This is becoming a more widespread solution, though problems with the (perception of the) ability of such groups to assimilate technical data have frustrated this process to some extent.

The use of three- to four-day consensus conferences in technology assessment is already widespread in Denmark, which has held over 10 such events. The consensus conference model is unusual both in its format and in that lay people have a key role to play in controlling the event. As with citizens' panels, experts are invited to 'testify' in front of a lay panel. In terms of the scientific input, the *content* of these events may not be new. However, the procedural aspects are truly innovative. Not only do consensus conferences open up the debate but they can also broaden and raise knowledge levels for all participants. The success of such events depends upon the cultural context in which they are conducted. Important here is how the results are used: whether reports from such events are simply advisory documents or carry some predefined legal or administrative weight.

Another solution is to label products clearly and comprehensively so that the consumer has a definite choice as to which products to endorse. Labelling is a particularly thorny issue, with much resistance from industrial quarters, which argue that some types of labelling may arouse undue consumer concern. Those in favour of labelling argue that such information provides the basis for making informed judgements about product purchases. A key question here is whether labelling should take the form of a warning notice which would require justification of a safety issue, or whether it should be simply informative, as with much of the present nutritional information.

The British government has decided to place a moratorium on the planting for commercial use of any genetically modified crops for which approval had not been granted by 1998. This allows time for more scientific research on the possibilities of pollination transfer to non-genetically modified crops, and for a deliberative process via a stakeholder forum to be put in place. This is a radical innovation in deliberative environmental risk management.

importantly, there was no easy way to extrapolate from small-scale psychometric studies to the population at large, and therefore it remained difficult to incorporate public perceptions of risk into policy making at a national level.

Public perception of risk is influenced by a wide array of different individual and social concerns, including:

- *the degree of familiarity and control.* Everyday risks such as smoking or driving are usually tolerated. Unfamiliar and uncontrollable dangers such as electromagnetic radiation from overhead power lines, or genetically modified organisms, are resisted;
- *the potential for catastrophe of irreversibility.* The more a danger is perceived to have the potential for causing a major, sudden disaster, or the more a possible hazard could realise a truly worldwide potential in its impact, the more it will be resisted;

● *the degree to and the point at which public opinion is seen to be taken into account.* This affects the willingness to tolerate the outcome of risk management. When regulators show that they require public opinions in order to make judgements about risk and safety, trust in the decision-making process is fostered, particularly if public opinion is sought early in the risk management process;

● *the notion of compensation should anything go wrong.* This is also a vital element in risk tolerance. People need to feel that they will be protected by, say, long-term medical surveillance, in the light of a chemical hazard showing up in a nearby waste disposal facility. This is partly why the US Superfund Program was set up in 1981, namely to levy a charge on chemical production to pay for the clean-up of old sites where the responsible party could not be identified. In practice, Superfund is not achieving all that it might, but this is a result of administrative difficulties and not a fault of the principle itself.

Marris *et al.* (1998) reanalyse the psychometric approach in a detailed study of 13 risks, as summarised in Table 17.1. Here it can be seen that no single characterisation of 'risk' applies to each of the 13 phenomena. But riskiness was more related to unacceptability and potential for environmental harm (both loaded with social meaning) than with fatalities or injuries. In general, the 'personality profiles' of hazards are not universal, and different individuals attribute different characteristics to the same risk issue. However, individuals reacted in a similar way to the aggregates of individuals, suggesting that there is a good degree of coherence in the patterns of risk perception across the spectrum of society.

Partly in response to the failure of psychometric studies, in particular the way they tried to homogenise the population, an anthropological approach to risk perception began to gain credence.

Table 17.1: Mean risk perception scores, with 'risk' defined in five different ways.

	Riskiness	Fatalities	Injuries	Environmental harm	Unacceptability
Nuclear power	4.1	2.4	**2.7**	**3.3**	**3.6**
Ozone depletion	4.0	2.0	2.3	**3.4**	4.0
Sunbathing	3.9	2.1	**3.0**	1.6	**2.9**
Genetic	3.7	1.8	1.9	**2.5**	**3.3**
War	3.6	**3.7**	3.7	**3.7**	4.3
Car driving	**3.5**	**3.7**	3.9	**3.4**	**3.5**
AIDS	3.4	**3.1**	**3.2**	1.8	4.0
Mugging	3.4	**2.5**	**3.0**	1.8	**3.8**
Terrorism	**3.3**	2.7	2.9	2.6	4.1
Home accidents	**3.2**	**2.8**	**3.4**	1.7	**3.2**
Alcoholic drinks	**3.0**	**3.0**	**3.4**	1.7	**3.0**
Food colourings	**2.8**	1.3	1.8	1.5	**2.5**
Microwave ovens	2.0	1.2	1.4	1.4	1.9

Note: The table lists the mean risk perception scores obtained for each of the risk issues, with 'risk' defined as riskiness, fatalities, injuries, environmental harm or unacceptability. Scores shown in normal type indicate risk ratings which were virtually universal, i.e., less than 15% of respondents gave a good rating at one or the other end of the 5-point scale (1 and 2; or 4 and 5). Scores shown in bold indicate scores for which there was more variability.

Source: Marris *et al.* (1997, 306).

Cultural theory and risk perception

Cultural theory was developed by the British anthropologist Mary Douglas (1966), expanded by Douglas and Wildawsky (1983) and further theorised by Thompson *et al.* (1990). Tansey and O'Riordan (1999) provide a fuller analysis. The essence of the cultural theoretic approach is that risks are set in a culture of personal esteem, social identification and blame. Furthermore, people form into groups of common outlook as part of these processes and assign 'symbolic' meaning to events in a social and natural world to create order and coherence. This does not mean that humans are biologically predisposed to communality. Rather, they rely on patterns of habit and socialised re-enforcement of their values and beliefs in order to cope with the world around them. Richard Eiser (1994) provides a very innovative and multidisciplinary psychological theory for all of this. In essence, then, risks are defined, perceived and managed according to principles that are inherent in particular forms of social organisation (Rayner, 1992, 84).

The two most central social frameworks are *market and hierarchy* on the one hand, and *fairness* on the other. According to Douglas and Wildawsky (1983, 125), the former represents the 'centre', the economy of governance, while the latter represents the 'border', or the critical element of the mainstream, the common belief in the corruptibility of bureaucracies, government and business. The 'border' was identified by 'sects' or voluntary associations of like-minded people, who held together through the identification of blame. Organisation along egalitarian lines means debates are framed in terms of pure principle, based primarily on accountability and trust.

Cultural theory was further expanded to create a typology of outlooks; on the basis of these outlooks, underlying meanings or constructs could be solicited as to why certain individuals perceived risks in the kinds of ways depicted in Table 17.1. Douglas (1982) proposed a breakdown on the basis of the degree to which individuals interacted with others in the community, by means of some indicator of bondedness or identity. This dimension she termed *group*. The other spectrum was the extent to which individual norms were prescribed or influenced by external structures of authority. This she termed *grid*. For example, strongly hierarchical structures such as the military, the police, scout troops, etc. would have high grid, while loose organisations such as anti-nuclear protesters would have low grid. Thompson defines these four positions, outlined in Figure 17.1, as *solidarities*, or collectivities of similar mindedness. Along with others, he linked this to the characterisation of nature outlined by Holling and described in the introduction to Chapter 7. The diagram therefore depicts patterns of outlook which frame responses to particular associations of risk, as indicated in Figure 17.1.1.

One should be careful of grid–group characterisations. Marris *et al.* (1998) found that only 14% of a sample of Norwich residents fitted into the framework. Boholm (1996) argues persuasively that the psychometric tests used to determine the solidarities are self-creating, particularly for egalitarians. Douglas (1997) is especially critical of the misuse of the methodology of characterisation, suggesting that the approach deliberately misleads the analyst. More penetrating sociological techniques should be used, using groups and deliberative approaches, as summarised in Chapter 1.

Risk communication

The development of an understanding of risk perceptions has enabled risk managers to reflect on how best to manage conflicts between experts and the public. Initially,

Figure 17.1: The four 'ways' of life. Cultural theory applies to the relationship between patterns of belief and values, and solidarities in social settings and networks. The principle is that people in particular social groupings and frameworks of social control will have similar outlooks on risk-related themes. Research indicates that the methodology of proving this is tricky, and that, at best, the framework outlined here is supportive rather than prescriptive.

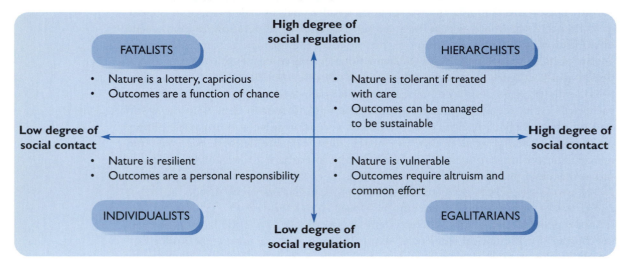

disparities between the scientific and technical estimation of risk and public perceptions of risk were thought to be merely a question of education – that risk assessors were not getting the right message across. Thus, much effort was spent attempting to provide clear advice as to effective ways to make the public understand the science. Risk comparisons have been a favourite tool. This can be useful if presented in a meaningful way, using risks that are strictly comparable. But all too often rather spurious comparisons were made, comparing the risks of nuclear power with crossing the road or smoking, risks which possess inherently different dimensions. Not surprisingly, most of these efforts failed and appeared to reinforce the view that the factors that influenced public perception of risk lay in the statistical likelihood of death.

More recent risk communication efforts have tended to focus on the necessity to address the fact that the public often see the world in a different way from the scientific and technical community – that is not to say that scientific and technical people are not part of the public, but that a whole array of alternative rationalities exist. None can be wholly right, none wholly wrong, but they exist and, in our democratic society, they have a right to be considered in the decision-making process. So, risk communication has become a much more multi-way process that is designed to make risk management more accessible to everyone. It attempts to promote mutual understanding, if not consensus, and, whilst nobody would argue that emphasising risk communication is a one-stop solution for reducing conflict, it does offer an opportunity for people to become part of the solution and to gain a greater insight into how decisions are made, even if they do not necessarily agree with the outcome of the decision itself.

Risk communication is a subset of general communication theory, which is why many of the practical lessons relate to existing fields such as public relations and management science. Many of the arguments for greater quality and sophistication of communication are being advanced by other fields such as Local Agenda 21 (see Local Government Management Board, 1996).

Possibly because of the tensions that exist between the different social science disciplines, efforts to combine risk perception and communication research have been rather limited. Kasperson and his colleagues (1988) developed an integrated model relating to the social amplification of risk, which looked at why some minor risks (as judged by technical experts) lead to significant public concerns while others, judged as relatively high-risk by experts, are not perceived by the public to be causes for concern. Using a source–signal–receiver type structure, the model attempted systematically to link the psychological, sociological and cultural aspects of risk perception and risk-related behaviour.

More recently, researchers at Carnegie Mellon University have developed the mental models approach, which is based on four stages that attempt to link the communication activity with an audience's needs. This links directly the type of information needed to make an informed decision with an understanding of the audience's underlying belief system. The four-stage process begins with an elicitation of people's accurate and inaccurate beliefs about a hazard. Once an understanding of these underlying beliefs has been established, a communications strategy is planned based on what people need to know to make an informed decision. The development of the communications strategy involves iterative testing of successive versions of material using a variety of different analytical techniques.

Barriers to integrated risk management – substantive and procedural issues

Efforts to develop more integrated approaches to risk management are increasing. In particular, there is a need for systematic evaluation through which it becomes possible to learn from experience and understand what processes promote effective risk management. This matter is studied in detail by the Royal Commission on Environmental Pollution (1998, 51–62). However, the movement towards a more comprehensive approach to environmental risk management is still inhibited by three key impediments:

- *poor frameworks for analysis* impede the kind of comprehensiveness and decision making necessary to make informal judgements in the modern age. This is where multi-attribute analysis requires more attention. This method accepts that there are various non-comparable dimensions to any given valuation, and elicits a merit ordering of ranked priorities through deliberative processes;

- *inadequate regulatory principles* allow too much discretion at the door of the safety official and tend to lead to discrepancies in levels of safety provided for different groups in society according to their environmental circumstances and income levels – i.e. the poor get higher levels of exposure whereas the wealthy are able to avoid industrial hazards. This is now being addressed through more transparent regulatory arrangements;

- *insufficient consultation procedures* restrict the different interests that have a legitimate role to play in the final determination of risk. This has improved recently with regard to chemical plants and local communities but remains a problem for more general consultation and involvement in the assessment of risks and the development of regulation. The creation of local community forums is one method of building trust.

These failures relate to both substantive and procedural issues. Much of the efforts expended to improve the situation have been directed at the former. Only now, following issues such as the *Brent Spar* and BSE/CJD, has the pendulum begun to shift towards a focus on the procedural aspects of the decision-making process (see Boxes 17.13 and 17.14).

Box 17.13 Trust, confidence and the management of agri-food risks

In June 1997, the UK Environment Agency produced a press release outlining the results of trial burns it had conducted on cattle infected with bovine spongiform encephalopathy (BSE) at two coal-fired power stations. The press release reported that:

The risk of human infection resulting from burning cattle cull wastes in power stations would be negligible. A detailed risk assessment, carried out by the Agency, based on test rig trial burning of meat and bonemeal (MBM) and tallow from cattle slaughtered under the Government's Over Thirty Month Scheme (OTMS) shows that the risk of an individual contracting Creutzfeld Jakob Disease (CJD) would be as low as one in thirty thousand million. This is three thousand times less than the risk of being struck by lightning.

The press release went on to confirm that, though the risks were judged to be negligible, the agency would conduct widespread public consultation before a decision was reached, an activity that might make some people wonder about the validity of the risk assessment. Public trust in regulatory agencies may not be served by promising continual widespread consultation over issues that appear not to be problematic to the agency itself. This approach may simply increase community scepticism of the risk assessment process.

In the UK, it is widely assumed that public confidence in food safety is at an all-time low. However, as beef prices have dropped so beef sales have increased. This might lead one to assume that confidence comes merely at a price.

An interesting paradox is emerging. While the intention is to promote public confidence, the separation of regulatory functions between agriculture and food production via the new Food Standards Agency may hinder the development of the holistic picture that developing sensible agri-food policies requires. As James (1997) states:

International experience demonstrates that there is a need for instituting substantial structural and cultural change in Government before public opinion begins to shift. The introduction of a high quality authority which manifestly takes public health and consumer protection as its first priority and has investigative and executive powers to rectify problems is a fundamental component of the series of developments needed to build confidence.

Source: James (1997).

Box 17.14 Ecological modernity: BPEO and recommissioning offshore oil and gas platforms

In recent years, the offshore industry has begun to recognise its embeddedness within the social *and* natural environments. The *Brent Spar* oil storage buoy became a symbol of this embeddedness as a consequence of the conflict which occurred over its proposed disposal to the North Atlantic seabed in 1994. Shell commissioned 30 separate studies in its evaluation of the best practicable environmental option (BPEO) for platform disposal. Four options were identified: disposal to land, sinking the buoy at its existing location, sinking the buoy in UK waters, and breaking up the buoy in the North Atlantic. Finally, on the grounds of cost and low environmental impact, Shell applied to the UK Department of Trade and Industry for a licence to sink the *Brent Spar* in UK waters. This application was granted, and subsequently vigorously defended by the UK government, as the BPEO.

Days before the disposal licence was issued, Greenpeace occupied the platform and orchestrated a carefully planned media campaign asking for a consumer boycott of Shell petrol stations. Soon afterwards, European governments were lodging protests with the UK, which remained unmoved (in part because the protests

Box 17.14 (continued)

were received outside the 60-day formal objection period, which had elapsed at the end of April). In Germany, Shell petrol sales were 20% below average, 200 stations were threatened with attacks, and 50 stations were vandalised including two which were firebombed (Löfstedt and Renn, 1997). In the face of this economic and public pressure, on 20 June Shell announced it was abandoning its plans to sink the *Brent Spar*. Shell then sought to regenerate its lost credibility, notably with a major advertising campaign that announced 'We will change'.

The offshore industry is now challenged to change both its internal management of technical innovation and its relationship with other stakeholders. This challenge represents a shift to a more ecologically modern operation. There are 6500 platforms in 53 different countries located in water depths from less than 30 m, to over 860 m. Up to 1995, more than 900 platforms had been removed from the Gulf of Mexico, but fewer than 3% of these were in water depths greater than 75 m. In shallower waters, the structures are primarily steel, but in deeper waters concrete gravity structures are more common. Managing these structures as their usefulness declines has proved to be a challenging problem.

The language describing changes to the fate of offshore installations is telling. Initially, the industry spoke of 'abandonment', then of decommissioning, when the potential for recycling was recognised. More recently, following the initiative of a consortium of local authorities and industrial concerns in Great Yarmouth, the potential for reuse and *recommissioning* has come to the fore. The possible fate of offshore oil and gas platforms is ascending the waste management hierarchy, away from disposal and towards recovery, reuse and recycling. In essence, what was once considered waste is now considered a mature asset to be managed appropriately.

This example is indicative of a number of changes that have occurred in the latter half of the 20th century which have led to business reassessing the nature and quality of its contribution to social welfare. These changes generate two forms of challenge: the first challenge is internal and involves adjusting operational processes on the basis of eco-efficiency; the second challenge is larger and more strategic. It involves changes to the relationships between the business and wider society. These areas clearly interact; the second challenge involves changing the wider framework within which business operates. This will then influence internal operational processes such as the development of waste minimisation schemes. Where regulations establish new limits, they do not always prescribe how these limits should be achieved. The implementation of regulations may provide an opportunity for businesses to apply technical innovation to reducing the costs of compliance.

The need for businesses and other stakeholders to address this change in relationships is pressing. As the *Brent Spar* case demonstrated, a company may operate within all the formal prescribed regulations and yet can still incur the wrath of the NGO sector. Recent research from risk management and other fields indicates that the 'wrath' originates less from substantive issues and more from procedural concerns. In this and many other cases, the *process* of communication between key stakeholders is sub-optimal. For instance, stakeholders generally rely on the media to represent their positions, and this tends to distort debate, not least because of the media's own agenda. Through attention to the processes of communication, it should be possible to engender more productive negotiation over issues of conflict. Whilst it is impossible and undesirable to remove conflict altogether, there is a pressing need to raise the quality and sophistication of debate.

In 1996, a Recommissioning Partnership was established in Great Yarmouth, UK. From the beginning of the Great Yarmouth initiative, the emphasis has been upon building a mature partnership between key businesses, SMEs and potential customers and co-operating for the commercial benefit of all involved. For the initiative to be considered sustainable, a more fundamental review must occur which includes

Box 17.14 (continued)

local and national government, environmental NGOs and the local communities. This would set an example nationally and internationally of how to work strategically towards a more sustainable future.

Already there are signs that some major oil and gas production companies are moving towards a focus on more diverse energy sources. In the longer term, this kind of partnership approach will be well positioned to tackle energy management, moving from non-renewable to renewable energy production as well as considering important issues such as appropriate energy usage.

The need for systematic evaluation

Returning to Box 17.1, the weakest link in the risk management cycle is policy evaluation, yet this has the greatest capacity to reinforce the potential of risk management as a crucial tool in modern environmental management. Box 17.15 illustrates one example of how organisations are gradually waking up to the importance of systematic evaluation. There are many barriers to this, not least the discovery that one's efforts have been wasted. However, the advantages of systematic evaluation outweigh these frustrations. Evaluating the performance of a particular policy or programme intervention is becoming widely recognised as a crucial aspect of effective management, yet it is an aspect of management that is often overlooked. Far too often, risk management policies are developed in an uninformed manner that stems from a lack of comprehensive evaluation. Such policies are the ones that are most likely to be ineffectual. Recognising the weaknesses as well as the strengths of policy decisions is an important step towards increasing overall effectiveness.

Planning an evaluation should be linked closely to policy design, though in practice many evaluations are conducted in an *ad hoc* manner with little consideration of underlying assumptions or of potential variables that may well influence specific changes if appropriately identified in good time. Such 'quick and dirty' evaluations are unlikely to help the organisation to improve its longer-term performance and are usually associated with short-term justifications to maintain the *status quo*. More systematic evaluations involve reassessments of risk management fundamentals which are linked closely to deep-rooted dimensions such as organisational culture. Establishing evaluation as a routine management activity cannot happen overnight. It involves a substantial commitment to change, which can be costly and painful in the short term but a small price to pay for long-term success. Chapters 1 and 3 provide the basis for justifying deliberation and more accountable policy networks to assist this process.

Conclusions

Over-reliance on the universality of scientific laws has led us to reject anomalies and idiosyncrasies that do not conform to the restrictive rationality of our scientific system. Yet rejection of them does not mean that they cease to exist and, as has been shown on many occasions, these idiosyncrasies can provide key information in the process of managing environmental risks. The process by which risk assessments are carried out is at least as influential as the results themselves. Paying greater attention to scientific proceduralism may result in the development of processes that permit the full integration of robust scientific and technical analyses, effective communication and stakeholder participation.

Box 17.15 Systematic evaluation of risk management interventions

The UK Health and Safety Executive (HSE) has responsibility for implementing, enforcing and monitoring health and safety legislation. Traditionally, inspections at commercial and industrial premises have played a central part in discharging these responsibilities. However, as the industrial and commercial base of the UK has changed, in particular with the increase in number of small and medium-sized enterprises (SMEs) and the changing nature and range of potential hazards, the HSE has adopted additional forms of risk communication, including the use of mailshots and seminars. However, little is known about the effectiveness of these techniques in changing business attitudes towards health and safety matters.

The evaluation programme developed in this context focused on the changing patterns of three factors deemed to be important in the health and safety practice of SMEs: the relevant knowledge of industry personnel, the arrangements made by managers to implement improvements, and the physical precautions introduced by the company. Evaluation data were collected in a structured format during inspectors' visits to SMEs that had either received a mailshot or attended a seminar about a health and safety issue, and a control group of similar industries that had not been contacted. The health and safety issues were mainly specific topics such as electrical safety and the manual lifting of heavy objects. The evaluation programme compared the changes in knowledge, arrangements and precautions noted by inspectors in both groups to establish whether the contact techniques could be identified as the primary mechanism for stimulating change. Understanding the role of contact techniques in effecting change allows the HSE to decide whether to promote or reduce the usage of such techniques in their intervention strategies for SMEs. (Rakel *et al.*, 1998).

Developing evaluation methodologies requires specific aims and objectives for the communication to be developed prior to the communication exercise. This is crucial for any subsequent measurement of effectiveness, as the intended effect needs to be considered before the communication exercise takes place. Therefore, evaluation methodologies should be designed as an integral component of the communication programme rather than as an additional component once the communication has been undertaken. Strong evidence emerged that seminars were having an effect in stimulating change and that many such changes are, on the whole, being attributed to the work of the regulator. However, mailshots did not appear to stimulate change significantly above background level.

One of the central problems for evaluation programmes is the lack of empirical data and experience with evaluation exercises in general. Despite the importance of developing feedback mechanisms through which experiences can be translated into development, evaluation research in the area of risk management is not applied consistently.

Developing more systematic approaches to evaluation is an important step in helping to learn from previous experiences: to build upon the foundations of good practice and avoid repeating mistakes. In time the results of systematic evaluations can be used to inform future policy making, which may help to create more effective risk reduction strategies. Hopefully, as risk management evaluation programmes gain credence, greater knowledge will be gained in developing evaluation processes and mechanisms for further improving the effectiveness of risk management policies.

In the UK and elsewhere, the not in my back yard (NIMBY) phenomenon is being gradually replaced by a more informed and more active local public (see Box 17.7). Reactive campaigning is switching to proactive education. Emphasis is moving from technical and geographical considerations to fundamental questions of need. In some cases, this may lead to increasing tolerance of industrial activities; in others, it may result in widespread resentment. Conflict between interest groups within the decision-making process is inevitable. The task is to create a positive atmosphere for conflict at the right time in the process such that, where at all possible, consensus can be reached based around those arguments that are most robust.

Risk management is moving beyond psychometrics and cultural theory to reflect more closely the need to deal with scientific proceduralism. Traditional decision-making processes must be overhauled and become more inclusive. Adopting an integrated approach to risk management encourages, even demands, communication between groups, which in turn promotes mutual understanding between, and mutual respect of, parties involved in decision making. Collaboration does not need a shared ideology. It opens up the decision-making process, making explicit the method by which decisions are reached. This does not guarantee satisfaction for all about the outcome of the decisions but will allow interest groups to participate and understand how decisions are made, and give the decision makers some justification for the fruits of their labour.

References

Beck, U. (1992) *Risk Society*: *Towards a New Modernity*. Sage, London.

Bingham, G. (1986) *Resolving Environmental Disputes*. Conservation Foundation, Washington DC.

Boholm, A. (1996) Risk perception and social anthropology: critique of cultural theory. *Ethnos*, **61**, 64–84.

Bullard, P. (1993) *Environmental Justice*. Beacon Hill Press, Boston, MA.

Carter, D. (1991) Aspects of risk assessment for a hazardous pipeline containing flammable substances. *Journal of Loss Prevention Process Industries*, **4**, 60–71.

Douglas, M. (1966) *Purity and Danger: an Analysis of the Conceptions of Pollution and Taboo*. Routledge & Kegan Paul, London.

Douglas, M. (1982) *Essays on the Sociology of Perception*. Routledge & Kegan Paul, London.

Douglas, M. (1992) *Risk and Blame: Essays in Cultural Theory*. Routledge, London.

Douglas, M. (1997) The depolitisation of risk. In R.J. Ellis and M. Thompson (eds), *Culture Matters: Essays in Honour of Aaron Wildavsky*. Westview Press, Boulder, CO.

Douglas, M. and Wildawsky, A. (1983) *Risk and Culture: an Essay on the Selection of Technological and Environmental Dangers*. University of California Press, Berkeley, CA, and London.

Eiser, R. (1994) *Attitudes, Chaos and the Connectivist Mind*. University of Exeter Press, Exeter.

Fernandez-Russel, D. (1988) *Societal risk estimates from historical data from the UK and worldwide events*. Centre for Environmental Risk, University of East Anglia, Norwich.

Feyerabend, P. (1987) *Farewell to Reason*. Verso, London.

Greenpeace (1998) *The Turning of the Spar*. Greenpeace, London.

Heinman, M. (1996) Waste management and risk assessment: environmental discrimination through regulation. *Urban Geography*, **17**(5), 400–418.

House of Lords Select Committee on Science and Technology (1999) *Third Report. Management of Nuclear Waste*. **HL41**. The Stationery Office, London.

James, P. (1997) *Food Standards Agency: an Interim Report*. Consumers Association, London.

Kasperson, R.E., Renn, O., Slovic, P., Brown, H.S., Emel, J., Goble, R., Kasperson, J.X. and Ratick, S. (1988) The social amplification of risk: a conceptual framework. *Risk Analysis*, **8**(2), 177–187.

Kemp, R. (1992) *The Politics of Radioactive Waste Disposal*. Manchester University Press, Manchester.

Krimsky, S. and Golding, D. (eds) (1992) *Social Theories of Risk*. Praeger, New York.

Local Government Management Board (1996) *Innovation in Public Participation*. Local Government Management Board, London.

Löfstedt, R. and Renn, O. (1997) The Brent Spar controversy: an example of risk communication gone wrong. *Risk Analysis*, **17**(2), 131–136.

Marris, C., Langford, I., Saunderson, T. and O'Riordan, T. (1997) Exploring the 'psychometric paradigm': comparisons between individual and aggregate analysis. *Risk Analysis*, **17**(3), 303–312.

Marris, C., Langford, I., Saunderson, T. and O'Riordan, T. (1998) A quantitative test of the cultural theory of risk perception: comparison with the psychometric paradigm. *Risk Analysis*, **18**(2), 181–197.

O'Riordan, T., Kemp, R. and Purdue, H.M. (1987) *Sizewell B: an Anatomy of the Inquiry*. Macmillan, London.

O'Riordan, T., Marris, C. and Langford, I. (1997) Images of science underlying public perceptions of risk. In Royal Society, *Science, Policy and Risk*. The Royal Society, London, 13–30.

Rasmussen, N.C. (1990) The application of probalistic risk techniques. In T.S. Glickman (ed.), *Readings in risk*. Johns Hopkins University Press, Baltimore, MD, 151–170.

Rawcliffe, P. (1998) *Environmental Pressure Groups in Transition*. Manchester University Press, Manchester.

Rayner, S. (1992) Culture theory and risk analysis. In S. Krimsky and D. Golding (eds), *Social Theories of Risk*. Praeger, New York.

Renn, O. (1998) Three decades of risk research: accomplishments and new challenges. *Journal of Risk Research*, **1**(1), 49–72.

Royal Commission on Environmental Pollution (1998) *Setting Environmental Standards. Twenty-First Report*. The Stationery Office, London.

Tansey, J.D. and O'Riordan, T. (1999) Risk perceptions and cultural theory: a review. *Journal of Risk and Health*, **1**(1) 71–90.

The Royal Society (1992) Risk: Analysis, Perception, and Management. Report of a Royal Society Study Group, The Royal Society, London.

The Royal Society (1997) *Science, Policy and Risk*. The Royal Society, London.

Thompson, M., Ellis, R. and Wildawsky, A. (1990) *Cultural Theory*. Westview Press, Boulder, Co.

Wynne, B. (1992) Introduction. In U. Beck (ed.), *Risk Society: Towards a New Modernity*. Sage, London.

Further reading

Bromley, D.B. and Segerson, K. (eds) (1992) *The Social Response to Environmental Risk. Policy in an Age of Uncertainty*. Kluwer, Dordrecht.

Glickman, T.S. and Gough, M. (eds) (1990) *Readings in Risk*. Johns Hopkins University Press, Washington DC.

Jungermann, H., Kasperson, R.E. and Weidemann, P. (eds) (1991) *Risk Communication*. KFA Research Centre, Jülich, Germany.

National Research Council (1989) Improving *Risk Communication*. National Academy Press, Washington DC.

O'Riordan, T. and Jordan, A. (1995) The precautionary principle in contemporary environmental politics. *Environmental Values*, **4**, 191–212.

Presidential/Congressional Commission on Risk Assessment and Risk Management (1997) *Framework for Environmental and Health Risk Management,* Volume 1. National Academy of Sciences, Washington DC.

Rakel, H. Gerrard, S., Piggott, G. and Crick, G. (1998) 'Evaluating contact techniques: assessing the impact of a regulator's intervention on the health & safety performance of small and mediium sized businesses'. *Journal of Safety Research*, **29**(4) 235–47.

Shrader-Freschette, K.R. (1991) *Risk and Rationality: Philosophical Foundations for Populist Reforms.* University of California Press, Berkeley, CA.

Stern, P.C. and Fineberg, H.V. (eds) (1996) *Understanding Risk: Informing Decisions in a Democratic Society.* National Academy Press, Washington DC.

The two Royal Society reports cited in the references provide probably the most comprehensive statements of the natural and engineering science and the social science aspects of risk management, respectively. For the US reader, Shrader-Freschette (1991), Bromley and Segerson (1992) and Stern and Fineberg (1996) provide a primarily sociological–political interpretation of risk tolerance and response. The Glickman and Gough reader and the Krimsky and Golding volume take that material one stage further and provide a thorough examination of the social and cultural context of risk. The latter presents some insightful background material on each author which illustrates the many and varied specialisms that comprise the risk field. The two volumes on risk communication by Jungermann *et al.* (1991, published by KFA Jülich) and the National Research Council (1989) provide excellent coverage of the changing interpretations of this important topic. Both contain good examples from North America and Europe.

Chapter Eighteen

Waste management

Jane C. Powell and Amelia Craighill

Topics covered

- Waste generation and collection
- Waste management hierarchy
- Waste regulation and policy
- Use of economic incentives
- UK landfill tax

Editorial introduction

The discussion on ecological footprints in Chapter 2 turns primarily on waste and energy usage. Overall, for every tonne of consumer product, 11 tonnes of waste materials is generated. It is no wonder that wastes in general have become the prime source of attention amongst regulators, economists bent on ecotaxation and tradable permits, and risk assessors. This chapter is connected to the lively debate on rucksacks (Chapter 2), on economic instruments (Chapter 4) and on risk management (previous chapter).

The aim is to kick start a huge effort at waste minimisation, beginning with reuse of existing materials in areas such as aggregates, construction materials, end-of-life vehicles and 'white' electrical goods. Regulations are not enough, as is being amply demonstrated by the failure to agree on an EC Landfill Directive after six years of negotiations. The directive is geared to close off the landfill 'sink' as far as it is technically and economically practicable to do so. Thus the current UK practice of co-disposing municipal and hazardous waste will be banned by 2002 (ENDS, 1998b, 21). So too will the disposal of biodegradables, down to 35% of current practice by 2016/2020, compared with the 80% allowed in municipal waste streams at present. To achieve this, it is possible that some form of tradable permit, of the kind outlined for carbon dioxide emissions in Box 7.11, will be instituted among the waste disposal operators.

The UK landfill tax is a landmark in environmental management. As this chapter discusses, although the actual tax bears little relationship to any notion of an environmental footprint, it does have the advantage of simplicity of collection. For advocates of economic instruments, such an approach, though transgressing the niceties of neoclassicism, is worth pursuing (see Atkinson *et al.*, 1997). The Environmental Audit Committee (1998) of the House of Commons recommended that the Treasury take the hint from its advocacy of the landfill tax and establish a Green Tax Commission to look at all aspects of ecotaxation (see also O'Riordan, 1997). Currently, the UK government is looking at an aggregates tax (ENDS, 1998a, 17–20). Table 18.1 indicates the kinds of prices being contemplated for such a tax, though these figures do not take into account the more subtle ecological and social consequences of extracting and transporting materials. This is why the minister concerned is prepared to consider a broad-brush approach to the determination of the tax, as in the landfill case. Economists are recognising that the purity of the analysis is less important than the practical feasibilities of collecting the tax, and the political realities of setting an initially low levy, then subsequently raising it as the effects of the tax bite.

Ecotaxation is not just an economic instrument. It is also a sustainability objective. So much depends on who collects the revenue, and how that money is spent. The UK Treasury is very anxious to avoid the precedent of earmarking tax revisions for predetermined purposes. This practice is called hypothecation and is still officially frowned on. But as the tax regime steadily shifts away from income and labour to consumption and non-sustainable 'footprints', so the political justification of such new revenue will have to turn to hypothecation. Fertiliser tax monies could go to support sustainable agriculture, carbon taxes for mobility charities to cut the individual use of the car, and aggregates tax for local job training in restoration and local crafts from recycled materials.

All this is still a little distance off. What is of interest is the rapidity through which clean technology, economic incentives, business sustainability strategies and community participation in waste minimisation futures are beginning to entwine around LA21 and urban sustainability initiatives in general. Furthermore, the strictures of the Kyoto Agreement (see Chapter 7) will inevitably encourage governments and industry to look to waste management, as a basis for greenhouse gas reduction, notably methane. Energy for waste schemes, including methane-generated power, will receive subsidies through direct grant or the application of the Clean Development Mechanism. Dealing with waste is a holistic environmental science, not just a technical exercise of remediation and relooping.

All forms of human activity result in the generation of waste which can harm the environment. However, careful management of the waste can limit the damage done

Table 18.1: Environmental costs of aggregate extraction and transport per tonne.

Source	Local environmental costs		Transport-linked costs (noise, pollution)	Average total costs
	Range	Average		
Hard rock quarries	0.03–2.90	2.62	0.33	2.95
Sand and gravel pits[1]	3.11–34.72	9.00	0.04	9.04
Coastal super quarries		0.18	0.45	0.63
Average primary aggregates		*4.77*	*0.22*	*4.99*
Recycling sites		1.06	0.03	1.09

Note:
1 The sand and gravel costs could be as much as £16.00 per tonne, depending on future location and disturbance valuations.

Source: ENDS (1998, 17).

to the environment and conserve scarce resources. To this end, the 1992 Earth Summit in Rio produced several Agenda 21 objectives for the sustainable management of waste, including minimising wastes, maximising reuse and recycling, promoting environmentally sound waste disposal and treatment techniques, and extending waste services to more people. These areas were seen to be interrelated and therefore an integrated plan was sought for the sustainable management of wastes. In this chapter, we outline the various waste management options available and the debate concerning their relative merits. We examine a number of regulatory and economic instruments for waste and comment on the future direction of waste management policy.

Waste generation

There are many different types of waste, which are usually identified according to their source: for example, household waste, industrial waste and sewage sludge (Table 18.2). These waste streams vary considerably in their composition, particularly between developed and developing countries (Table 18.3). For example, municipal waste in developing countries contains greater quantities of biodegradable kitchen waste because food is generally purchased unprocessed. In northern Europe, North America and other developed countries, greater quantities of processed food are purchased, so food waste comes from the food industry rather than individual householders.

Waste collection

In developed countries, household waste is usually collected on a weekly basis from the household or a nearby collection point. Sometimes the collection of separated

Table 18.2: Quantities of different types of waste for selected countries (thousand tonnes).

	Year	Construction/ demolition	Sewage sludge	Scrapped motor vehicles	Packages	Municipal waste (1992)
Australia	1992	1 568	60 000	271	914	12 000
Canada	1992	11 000	7 450	1 000	10 500	18 800
Czech Rep.	1987	2 677	2 750	—[1]	—	—
Denmark	1993	2 374	192	—	—	2 377
France	1992	25 000	865	—	6 900	27 000
Germany (W)	1992	121 892	2 630	—	8 000	21 615[2]
Hungary	1992	—	30 000	—	500	4 000
Italy	1991	34 374	3 428	1 400	—	20 033
Japan	1991	58 431	169 693	—	—	50 767
Spain	1990	22 000	10 000	—	—	14 256
USA	1986	34 692	11 454	—	64 000	187 790

Notes:
1 Not available.
2 1990.

Source: OECD (1993).

Table 18.3: Generation and composition of municipal waste for selected countries.

	Municipal waste (kg/capita) (1997)	Waste composition (%) (1993)					
		Paper and paperboard	Food and garden waste	Plastics	Glass	Metals	Textiles
USA	730	38	23	9	7	8	16
Canada	630	28	34	11	7	8	13
France	560	30	25	10	12	6	17
Denmark	520	30	37	7	6	3	17
Japan	410	46	26	9	7	8	12
Turkey	390	6	64	3	2	1	24
Spain	370	21	44	11	7	4	13
Iran	324	8	74	5	3	1	2
Mexico	320	14	52	4	6	3	20
Greece	310	20	49	9	5	5	13
Poland	290	10	38	10	12	8	23
China	285	3	60	4	1	0	2

Sources: OECD (1993; 1997).

recyclable materials also takes place. Industrial and commercial waste is collected by waste disposal companies, usually in skips or large wheeled bins. Hazardous wastes need special collection and disposal. In developing countries, the effectiveness of waste collection varies considerably. It has been estimated that by the end of the 20th century half the urban population in developing countries will still be without adequate waste disposal services. One of the objectives of Agenda 21 is the provision of a safe waste collection and disposal service to all people. Particular priority should be given to the extension of waste management services to the urban poor, especially those in 'illegal' settlements. At present, it is thought that approximately 5.2 million people, including 4 million children under five years of age, die globally each year from waste-related diseases. Some of the problems with waste collection in poor urban areas are due to poor infrastructure plus the difficulty of access in overcrowded conditions. The problems are exacerbated by the need, in many hot countries, for a daily collection due to the rapid deterioration of the waste, which often has a high biodegradable content.

Waste management

Boxes in this chapter give details of the technological aspects of various waste management techniques. Alongside these technical factors, financial costs play a big part in deciding a country's waste management strategy. However, there are also important environmental and social costs and benefits which, to an increasing extent, are being taken into account. Many countries encourage a hierarchy of waste management, which generally favours waste minimisation, reuse and recycling over landfill and incineration without energy recovery (see Box 18.1). Following waste prevention or minimisation (see Box 18.2), the waste hierarchy promotes recovery, which includes the reuse and recycling of materials in addition to the recovery of energy (see Boxes

Box 18.1 European Commission waste management hierarchy

- ● Prevention
  - – promotion of clean technologies and products
  - – reduction of the hazardous nature of wastes
  - – establishment of technical standards

- ● Recovery
  - – promotion of reuse and recycling
  - – preference should be given to materials recycling over energy recovery operations

- ● Final disposl
  - – care should be taken to avoid incineration without energy recovery and uncontrolled landfill sites

Source: adapted from the EC Review of Community Waste Management Strategy (EC, 1996).

Box 18.2 Waste prevention

Methods of reducing waste at source range from home composting to improving the efficiency of industrial process. Some industries have found they can profitably reduce their waste by good housekeeping methods such as improved levels of maintenance and by recycling by-products within the factory. Waste minimisation can also be achieved by installing less polluting technology. Another aspect to be considered is reducing the harmful nature of the waste. An industry may change its raw materials or processes so that the waste produced is less damaging to the environment, for example the reduction of CFC refrigerants. The design of products is also important for overall waste prevention and for improving recyclability.

18.3 and 18.4). There has been considerable debate as to the relative values of materials and energy recovery from waste, and the position of composting in the hierarchy (see Box 18.5).

The recovery of materials can save large quantities of energy, but for some materials there are reprocessing and marketing problems. Recycling schemes are often judged in terms of their recovery rate and whether they make a profit. However, the financial costs of recycling schemes may not reflect all the environmental and social costs and benefits. The benefits of recycling are well known: conservation of resources; savings in landfill space and emissions associated with landfill and incineration; energy savings in manufacturing and in the transport of waste. Nevertheless, post-consumer recycling also has environmental and social costs. Materials for recycling need to be collected, sorted and transported to often distant processors. Transport can involve health and safety aspects and environmental pollution. In addition, there are social costs associated with source-separating recyclables and storing them at home, plus the noise and litter that may accompany neighbourhood recycling centres. Social benefits include the 'feel good' factor arising from participation in recycling activities, and the educational value of 'saving' resources.

Although the disposal of waste to landfill is the least favoured method in the waste hierarchy, it remains the main disposal method in most countries (see Box 18.6). For example, an estimated 85% of controlled waste is disposed to landfill in the UK, Greece, Portugal and Finland, 60–70% in Spain and 95% in Ireland, and in the United States 70% of municipal waste is thought to follow this route. Even in countries where alternative methods of disposal predominate, there is a need for landfill sites to

Box 18.3 Materials recovery

There are two levels of materials recovery: reuse and recycling. Reuse entails a product or packaging being used more than once in the same system, for example returnable glass milk bottles in the UK. However, a reusable container has to be stronger and heavier than a standard container to withstand the extra use, resulting in additional use of resources and transport costs. It therefore usually has to be used several times before it is more environmentally beneficial than a single-use container. Recycling is the reuse of materials to make similar new products or something different. Most plastics are downgraded to a less demanding role; for example, PVC bottles are recycled into garden furniture or insulation material for clothing.

In developed countries, the recycling of household waste is well established, if at a relatively low level. Recyclable materials are either collected from the household (kerbside collection) or are taken by householders to community recycling centres (bring recycling). From there, they are taken to a materials recycling facility (MRF) and on to reprocessors. Recycling levels are sensitive to the market value of the materials recovered and the cost of the process. In developing countries, the recovery of recyclable materials is an important source of income for many poor people. Recycling is often highly organised, with high-quality materials being collected from individual households while low-value materials are 'picked' from landfill sites. However, recovered materials are often made into low-quality products, thus adding to the perception of those in authority that recycling is not significant to the country's economy.

Box 18.4 Energy recovery from waste

The energy value of waste varies with its composition, with paper and plastics having a high energy value relative to bigdegradable waste. The two main methods of recovering energy from waste are landfill gas (Box 18.6) and combustion. Large-scale incineration of unsorted waste with energy recovery predominates, but alternatively the waste may be pre-sorted to remove recyclable and non-combustible materials, and the resulting 'refuse-derived fuel' (RDF) burned in smaller boilers. From both systems, energy can be recovered as heat, electricity, or combined heat and power (CHP). Heat recovery requires customers to be located close to the waste facility to limit the cost of installing pipelines and the level of heat loss.

Although the generation of electricity is less efficient than heat recovery, it can be exported directly to the national electricity supply grid. CHP improves the efficiency but again requires a pipe network. In many countries, gas-cleaning systems used to control incinerator emissions, such as particulates and acid gases, have to meet high pollution control standards required by national and European Community legislation. The main benefit of recovering energy from waste is that the recovered energy displaces energy generated from other fuels. Where the displaced energy is generated from fossil fuels, large quantities of carbon dioxide emissions are also displaced.

dispose of residues such as incinerator ash. The underlying reason for the popularity of landfill is its relative cheapness, particularly in countries where unregulated landfill is permitted. In recent years, however, there has been increasing awareness of the pollution problems that can arise from waste 'dumps', and in some countries stringent regulations have been introduced, resulting in sophisticated landfill sites with leachate collection systems and landfill gas recovery. There has been a subsequent increase in cost, but generally landfill remains financially cheaper than other disposal options.

Box 18.5 Aerobic digestion (composting)

Aerobic digestion, the decomposition of biodegradable waste, takes place on different scales, from garden compost heaps to centralised treatment plants. The biodegradable material is from either source-separated household waste or industrial and agricultural waste streams. Purely organic waste streams can produce a high-quality product, but mixed wastes need treatment to remove the non-organic waste fractions prior to composting, and the result may still be a low-grade product. To obtain a high level of decomposition, oxygen must be constantly available. This is achieved by the use of forced air, or a 'window' system where the piles of material are regularly turned to expose them to the air.

Box 18.6 Landfill

When organic waste decomposes anaerobically (without oxygen) in a landfill site it produces landfill gas and a liquid (leachate). Both emissions can be polluting if they escape into the environment. Landfill gas consists mainly of carbon dioxide and methane, both major greenhouse gases. It is considered to be one of the main sources of methane, producing 7–20% of all global anthropogenic methane emissions (Thomeloe, 1991). Migrating landfill gas can also become a fire, explosion or asphyxiation hazard. To mitigate these problems, the gas can be flared or used as an energy source, though gas collection systems are considered to be only 40–75% efficient. The amount of energy recovered depends on the waste composition and the environmental conditions of the site. On average, 4 MJ of electricity can be generated per tonne of municipal waste in developed countries.

In the past, it was not considered necessary to control leachate, and a system of 'dilute and disperse' was and still is operated in many countries. Consequently, leachate poses a serious risk to groundwater; an estimated 40 000 landfill sites in the United States may be contaminating groundwater. In some countries, new landfill sites are operated on a dry containment basis with multiple liners and leachate collection and treatment systems. This reduces the rate of waste decomposition, so the landfill sites involve the storage of waste rather than being a form of waste treatment. There is an alternative view that landfill sites should be regarded as 'bioreactors' and the decomposition accelerated by recirculating the leachate. The majority of the gas and leachate production would then take place at the beginning of the landfill site's life, when the gas and leachate collection systems are operating effectively (White *et al.*, 1995).

Most often, a combination of waste management techniques is employed in order to achieve a satisfactory balance of economic cost and environmental impact. This is known as *integrated waste management* and may involve, for example, recycling of non-combustible components of household waste followed by incineration, with energy recovery of the remaining waste stream and landfill disposal of incinerator residues.

Assessment techniques

In developed countries, there has been a move towards determining the *best practicable environmental option* (BPEO) for waste. This takes a cross-media and long-term approach. All waste management options have environmental costs and benefits, which vary with the circumstances under which they operate and the geographical area where they are located. In order to determine these costs and

Box 18.7 Life-cycle assessment

Life-cycle assessment (LCA) is used to quantify the environmental inputs and outputs of a product or process, from the mining of raw materials, through production, distribution, use and reuse or recycling, to final disposal. There are two main stages of LCA: inventory analysis and impact assessment. Inventory analysis involves the quantification of environmental inputs and outputs throughout a product or process's lifetime. The impacts are related to a function unit such as a container for a litre of milk or 1000 tonnes of household waste. The result of the inventory analysis is a list of pollutants that may have an impact on the environment. They are usually in non-comparable units.

The purpose of the impact assessment is to aggregate and evaluate the potential environmental impacts identified in the inventory, such as global warming and eutrophication. The impacts are aggregated on the basis of equivalency factors (e.g. global warming potential and ozone depletion potential). Unless the outcome is obvious, relative values or weights are then attributed to the impacts according to their relative importance. The various 'valuation' methodologies that are used can relate to goals or costs, or may be estimated by a panel of experts or a cross-section of stakeholders. A set of valuation factors that is widely acceptable has not yet been established, and there is controversy about the different valuation methodologies that exist.

benefits, a method is required to calculate the environmental consequences. In the UK, a programme of research has been carried out by the government, examining the way in which *life-cycle assessment* (LCA) can be used to identify the BPEO for waste (Box 18.7).

Once the environmental impacts have been determined, they may be incorporated into the decision-making process by assigning weights to the impacts in order to reflect their relative importance. Those options lower down the hierarchy are usually less expensive than those further up, which leads to the difficulty of balancing financial with non-financial costs and benefits. One method, the economic damage approach, overcomes this by applying monetary values, which allows the comparison of impacts on a common scale and their comparison with financial costs. Economic valuation methodologies are concerned with estimating the value that individuals place on non-market goods and services. A 'value' can be revealed by a consumer's behaviour derived from their 'willingness to pay' or 'to accept compensation', or by the use of dose–response relationships and replacement costs (Turner *et al.*, 1994). Economic damage values are available for a number of environmental and social impacts, including gaseous emissions, road congestion and casualties from road traffic accidents. An alternative approach, multi-criteria evaluation, can be used to appraise alternative waste management options against a series of criteria (Powell, 1996). A scoring approach is used to obtain weightings, which are then applied to each impact, and the weighted impacts are added. Both monetary and non-monetary values can be assessed in addition to non-numeric evaluations. Sensitivity analysis can be used to determine the effect of varying the weights on the outcome of the analysis.

Waste regulation and policy

The EC waste management strategy

In 1996, the European Commission adopted a waste management strategy which confirmed the waste management hierarchy (see Box 18.1) but also stressed the need for a flexible approach, taking into account environmental, economic and social impacts. Waste prevention is held as being the preferable option, and manufacturers are encouraged to set waste reduction targets. Reuse of materials or energy recovery should generally be favoured over disposal, although it is stated that the recovery of energy from waste should not be to the detriment of waste reduction or materials recycling. A number of policies and associated strategies have been brought into play which reflect or reinforce the hierarchy in some way, several of which are considered here. Finally, the Commission seeks to promote the application of economic instruments in the realm of waste management.

The Packaging Directive

The EC Packaging and Packaging Waste Directive of 1994 aims to reduce the overall impact of packaging on the environment, along with removing obstacles to trade. Member states are required to establish systems of return, collection and materials recovery to enable them to meet the targets set in the directive. The operating costs of this should be borne at least in part by companies in the packaging chain. The overall targets are to recover 50% and recycle 25% of packaging by 2001, with a minimum 15% recycling rate for individual materials (glass, aluminium, steel, paper and cardboard, and plastics, with the addition of wood from 2000). The Packaging Directive should result in less packaging going to landfill as the capacity for collection and processing increases.

The implementation of the regulations in the UK (Producer Responsibility Obligations (Packaging Waste) Regulations 1997) has led to the introduction of industry-based compliance schemes to help waste producers to meet their obligations. The obligation has been shared among the different actors in the packaging chain, namely the manufacturer, converter, packer and seller, and therefore embodies the principle of producer responsibility by encouraging the consideration of waste management impacts at the product's design and conception stage. It was estimated that about 30% of the 9 million tonnes of packing used in 1996 in the UK was recovered, most of which was recycled, with a small amount of paper and plastics being incinerated with energy recovery. However, the regulations have so far had a limited effect on the levels of recycling, with only paper and steel likely to achieve the 2001 target. Reprocessing capacity needs to be improved, as does the market for reprocessed materials, if the targets are to be met.

Directive on end-of-life vehicles

Worldwide, about 30 million cars are scrapped each year, with 9 million scrapped in the United States, 8.1 million in Europe and 5 million in Japan. Parts are often recovered for reuse and oil removed for safe disposal, before the remainder is shredded and ferrous metals are removed for recycling. The remaining residue, often contaminated with heavy metals and PCBs, constitutes around 10% of Europe's hazardous waste arisings. Therefore end-of-life vehicles (ELVs) have become a target for tighter regulation, with the development of voluntary recycling agreements in the UK, the Netherlands and France, and laws in Germany, Sweden and Japan. Car companies have also established groups to recover ELVs.

In 1997, the European Commission's Draft Directive on ELVs was finalised. It requires member states to achieve a rate of recovery for ELVs of 85% and a minimum recycling rate of 80% by January 2005, rising to 95% and 85% respectively by January 2015. Currently, the recovery rate in the UK stands at 75%. The annual cost of meeting the directive is estimated to be £360–525 million, half of which would be for the reimbursement by the manufacturer of the last owners of vehicles with a negative market value. The directive will also require dismantling sites to be licensed, whereas in the UK less than half the sites are currently subject to waste management licensing, and many are operating illegally. In this way, the directive will place the responsibility for managing ELVs on the car manufacturing industry, thereby internalising some of the external costs of disposal.

The Landfill Directive

The European Commission's Draft Directive on the Landfilling of Waste, announced in March 1997, aims at reducing the amount of biodegradable and other types of waste disposed in landfills. Biodegradable waste currently accounts for 20–50% of waste disposed of in European landfills, and the push to reduce methane emanating from landfills comes as part of the EC's climate change policy. The proposals would oblige member states to reduce the proportion of biodegradable waste being landfilled to the following targets (with an additional four years allowed for countries, such as the UK, which currently landfill more than 85% of their municipal waste:

- 75% of the total biodegradable waste produced in 1995 by 2006 (2010);
- 50% by 2009 (2013); and
- 35% by 2016 (2020).

The proposal also recommends the pre-treatment of all waste prior to landfill in order to reduce the volume and hazardous nature of the waste, and a ban on the dumping of used tyres two years after the directive's introduction, and shredded tyres five years after. The directive would also prohibit the practice of 'co-disposal', whereby liquid hazardous waste is disposed of in trenches cut into landfill sites. The proposed ban has met with opposition from the UK waste industry, which maintains that this practice is preferable to hazardous waste-only sites, where the material retains its hazardous nature indefinitely. When implemented, the directive may initiate some dramatic changes in the way that municipal waste is managed in the UK; for example, tradable permits for landfilling are being considered. A significant increase in the incineration and/or composting of waste may be necessary if the targets are to be met. The directive is also likely to increase market opportunities in industrial waste treatment techniques, materials recycling and recovery capacity.

The Basle Convention

In Basle in 1989, an international convention held by the United Nations Environment Programme established a consensus on the international movement of hazardous wastes. It was agreed that a principle of self-sufficiency (or 'subsidiarity') be established, whereby waste should be transported to another country for disposal only when it was less environmentally damaging to do so than to dispose of it at home. Where waste is transported, the receiving country should have prior notification and be supplied with adequate information concerning the waste composition and volume.

Economic instruments

Waste management policy has traditionally been secured by the use of regulatory standards, with the threat of a penalty if these are not met. This has resulted in some waste disposal options being priced at levels that do not take into consideration environmental costs and benefits. These costs and benefits are borne by society in general, and are not accounted for in the decisions made about waste. For example, the price of landfill does not include the external costs associated with the global warming potential of methane emissions. Social costs and benefits, known as 'externalities', include emissions to air and water, and the displacement benefits from recovering energy from waste (see Box 18.4).

If all waste management options reflected their true social cost, then market forces could achieve the optimal mix of waste management options. However, due to market and information failures, this is not the case. The overall result is that levels of waste minimisation and recycling are too low and the level of disposal is too high. Several countries have recognised the limitations of the regulatory approach and have introduced market-based instruments to influence behaviour in the direction of sustainable waste management. There are a range of economic instruments, including raw material and product charges, deposit–refund systems, waste collection and disposal charges, tradable recycling targets, taxes and subsidies.

Variable charging

Householders generally pay an average fee for the collection and disposal of their waste, which is unrelated to the quantity of waste they produce. The fee is often included in a general local tax and not separately identified. Variable charging relates the fee paid to the amount of waste produced, excluding materials separated for recycling. This is also known as pay by the bag, volume-based fees or unit pricing. A variety of methods are used, the most common being the purchase of distinctive refuse bags by the householder, the cost of which includes the full cost of collection and disposal. Alternatively, stickers can be purchased for placing on ordinary refuse bags.

The countries that have already introduced this system (approximately 15) have experienced increased recycling and decreases in disposal by up to 50%, reducing landfill demand, rewarding recyclers and fostering long-term changes in consumer behaviour. The disadvantages are a possible increase in fly-tipping and waste burning and the extensive education that is required. It can also be considered unfair for low-income families and those with several children. The system is less effective in larger apartments and rural areas, and is generally expensive to administer.

Deposit–refund schemes

This is essentially a combination of a tax and a subsidy that may be market-generated or imposed by law. A deposit is added to the price of a product such as a can or bottle of drink, which is refunded to the consumer when they return the packaging. The container can be either refilled or recycled, depending on the material and the contents.

This system, which applies mainly to beverage containers, has been introduced in several North American states and in some northern European countries (Table 18.4). Experience has shown that the management of deposit–refund schemes is best carried out by beverage producers, and the containers returned by the customers to the retailers. US beverage container systems have return rates of 72–98% but have led to relatively small reductions in waste and are expensive to operate.

The benefits of deposit–refund systems are a reduction in the cost of waste collection and disposal, reduced litter, and reduced energy and material resources used in container production. The disadvantages are an increase in storage and handling costs and possible inconvenience to the householder. Evidence suggests that return rates are not very sensitive to the size of deposit, more important being the number and convenience of return points, although a greater number of return points means higher system costs for handling, storage and transport. In the UK, where there is no monetary reward, about 20% of glass bottles are returned to bottle banks, but 95% of milk bottles are returned. These are collected from the doorstep by the person delivering the milk, the most important factor here being convenience. Deposit–refund systems can also be used to target the safe disposal of hazardous household waste such as batteries.

Recycling credits

Recycling credits are not a subsidy, but they do correct a market failure of the waste management system. Recycling household waste obviously removes waste from the disposal stream, saving a proportion of the cost of waste collection and disposal. The aim of recycling credits is to pass on these financial savings to local authorities and third parties that operate the recycling schemes. In the UK, the Recycling Credits Scheme was introduced in 1992 under the Environmental Protection Act (1990) to transfer the financial savings to the organisations which collect and sort the waste for recycling. The payments are made by waste collection and disposal authorities to the recycling operators such as local authorities, private organisations and community groups. The rates are updated annually and include an allowance for the UK landfill tax.

Landfill tax

A landfill tax is a disposal charge that internalises the external costs of landfill, raises revenue and reduces the amount of waste going to landfill by encouraging more

Table 18.4: Examples of deposit–refund schemes.

Country	Initiator	Target	Deposit amount (£)	Return rate (%)
Germany	Beverage producer	PET	0.17	96
Netherlands	Beverage producer	PET	0.03–0.32	50–90
Norway	Beverage producer	Glass	0.10–0.20	98
Sweden	Beverage producer	Glass	0.05–0.23	80–98
UK	Beverage producer	Glass	0.05–0.12	90
USA – California	Beverage producer	All types	0.01–0.02	69

Source: ERL (1991).

recycling and/or waste minimisation. Six European Union countries have landfill taxes: Denmark, France, Germany, Belgium, the United Kingdom and the Netherlands, in which the tax rate varies between £1 and £20.67 per tonne of waste (Table 18.5). In some countries, the tax revenue is hypothecated (earmarked for a purpose directly associated with waste management). For example, a landfill tax in France, levied on about 6500 landfill sites, is collected by the French Environmental Protection Agency and used to finance the Modernisation Fund for Waste Management. This funds the development of new waste treatment technologies, the abolition of illegal sites and contaminated land restoration (Fernandez and Tuddenham, 1995). Landfill tax revenue is not hypothecated in Denmark, the Netherlands, Germany or Belgium.

The UK landfill tax came into effect in 1996 at rates of £7 and £2 per tonne for standard and inert waste, respectively; the standard rate was subsequently raised to £10 from April 1999. Many companies have increased waste minimisation and recycling activities because of the tax, although the effect on householders has been limited. The landfill tax may also have caused an increase in the level of fly-tipping, and the abuse of civic amenity sites by small traders. On a more positive note, many waste disposal companies are diversifying into recycling facilities. The tax has also encouraged the generation of much-needed data on waste arisings in the UK.

One of the most significant effects of the UK landfill tax has been the diversion of an estimated 18–21 million tonnes of inert waste (of 42 million tonnes) to tax-exempt uses such as landscaping, golf courses and foundations of new buildings (Hogg, 1997). Landfill operators who previously used inert material for site engineering or ran inert landfill sites have had difficulty obtaining the materials, and in some cases operators have had to buy primary aggregates instead. From October 1999, inert waste used in the restoration of sites will be exempt from the tax. However, the UK government is also considering a tax on the extraction of primary aggregates.

Although the UK government does not embrace hypothecation, the acceptability of the tax was increased by using the revenue to finance environmental protection through the Environmental Body Tax Credit Scheme. Landfill operators receive a tax rebate of 90% of their contribution to an environmental body. The money must be used for one of several approved purposes, including reclamation of contaminated land and research into sustainable waste management. This scheme appears to have

Table 18.5:
Landfill taxes in Europe.

Country	Waste type	Cost (£/tonne)
Denmark	All	£20.67
France	Municipal	£2.50
	Industrial, hazardous	£5–8
Germany	Industrial, hazardous	£10–41
Belgium (Flanders)	Municipal	£1–3
	Industrial, hazardous	£0.60–7
Netherlands	All	£10.50
UK	Inert	£2
	All other	£7 (rising to £10)

Source: UK Waste (1995).

been extremely successful in channelling funds into environmental improvements (see Box 18.8). The landfill tax is seen by the UK government as a potentially cost-effective way of meeting policy targets such as those required by EC directives on waste. Box 18.8 provides some examples of the use of the earmarked revenue.

Economic instruments such as the deposit–refund system can be highly efficient. However, wider problems of hazardous wastes and contaminated sites have proved to be particularly complex and not amenable to simple, single-instrument, policy solution. US experience with Superfund highlights the conflicts and difficulties involved when a direct regulation and liability-based approach has been used (Turner *et al.*, 1996). Furthermore, the efficiency of economic instruments can be reduced by conflicting policy objectives. A consistency of approach needs to be maintained, for example between the EC waste management strategy and the economic instruments deployed.

The future: sustainable integrated waste management

In some countries, it is now recognised that the waste management hierarchy should be applied with some flexibility to take into consideration environmental, economic and social costs. It is understood that the best practicable environmental option (BPEO) will vary for individual waste streams and local circumstances. However, the European Commission does consider that waste prevention should remain the first

Box 18.8 Environmental bodies

By the end of January 1998, over 600 enivornmental bodies had enrolled, undertaking over 2000 projects with a cumulative value of £200 million (Sills, 1998). Examples of donations to environmental bodies made under the Environmental Body Tax Credit Scheme are given in Table 18.8.1.

Table 18.8.1: Donations under Environmental Body Tax Credit Scheme.

Donating company	Receiving body	Amount	Purpose
UK Waste	Groundwork	£220 000	School waste audits and local litter surveys
Biffa	National Recycling Forum	£14 000	Survey on employment characteristics in the recycling industry
Cory Environmental	Cory Environmental Trust	£2.5 million	Includes refurbishment of historic buildings
Durham County Waste Management	County Durham Environmental Trust	£800 000	Research into sustainable waste management
Global Environmental (Yorkshire Environmental)	Construction Industry Research and Information Association (CIRIA)	£20 000	Research on landfill engineering and recycling of construction materials
Grundon Group	Gloucestershire Wildlife	£100 000	Financing of appointments

Source: anonymous (The Waste Manager), 1997–1998.

priority. In some countries, integrated waste management is promoted, accepting that there can be several complementary management options for one waste stream. There is also a gradual shifting of responsibility for waste onto those further up the life cycle rather than just the final disposers of the waste. Initiatives such as the Packaging Directive and the Directive on End-of-Life Vehicles stress the liability of manufacturers, retailers and even designers. This is a welcome change, as it is only by looking for solutions across the whole life cycle of materials, rather than just at the point of disposal, that waste management can become truly sustainable.

References

Anonymous (1997–1998) Various issues of *The Waste Manager*, March to December.

Atkinson, G., Maddison, D. and Pearce, D.W. (1997) *Measuring Sustainable Development*. Edward Elgar, Cheltenham.

ENDS (1998a) Towards an environmental tax on aggregates. *ENDS*, **280**, 17–20.

ENDS (1998b) Implementing the landfill directive: a turning point for UK waste management. *ENDS*, **280**, 21–25.

Environmental Audit Committee (1998) *Environmental Tax Reform*. The Stationery Office, London.

Environmental Resources Ltd (ERL) (1991) *Economic Instruments and Recovery of Resources from Waste*. HMSO, London.

European Commission (EC) (1996) Review of Community waste management strategy. Supplement to *Europe Environment*, **483** (10 September).

Fernandez, V. and Tuddenham, M. (1995) The landfill tax in France. In R. Gales, S. Barg and A. Gillies (eds), *Green Budget Reform: an International Casebook of Leading Practices*. Earthscan, London.

Hogg, D. (1997) The effectiveness of the UK landfill tax–early indicators. Paper presented at the IBC Conference on Environmental Economic Instruments, London.

OECD (1993) *Environmental Data, Compendium 1993*. OECD, Paris.

OECD (1997) *Environmental Data, Compendium 1997*. OECD, Paris.

O'Riordan, T. (ed.) (1997) *Ecotaxation*. Earthscan, London.

Powell, I.C. (1996) The evaluation of waste management options. *Waste Management and Research*, **14**, 515–526.

Sills, R. (1998) How to get the most from the landfill tax. Paper presented at the Construction Industry Environmental Forum, *Construction Waste: Effects of the Landfill Tax*. London.

Turner, R.K., Pearce, D.W. and Bateman, I. (1994) *Environmental Economics*. Harvester Wheatsheaf, London.

Turner, R.K., Powell, J.C. and Craighill, A. (1996) *Green taxes, waste management and political economy*. Working Paper WM 96–03, CSERGE, University of East Anglia and University College London.

Thorneloe, T. (1991) *Proceedings of the Third International Landfill Symposium, Sardinia*. US EPA's Global Change Programme, USEPA, Washington DC.

UK Waste (1995) *Response to the Government's Consultation Document on Landfill Tax*. UK Waste Management Ltd, Bucks.

White, P.R., Franke, M. and Hindle, P. (1995) *Integrated Solid Waste Management, a Lifecycle Inventory*. Blackie Academic & Professional, London.

Further reading

Brisson, I. (1993) Packaging waste and the environment, economics and policy. *Resources, Conservation and Recycling*, **8**, 183–292.

Craighill, A.L. and Powell, LC. (1996) Lifecycle assessment and economic valuation of recycling: a case study. *Resources, Conservation and Recycling*, **17**(2), 75–96.

Department of the Environment (1995) *Making Waste Work: a Strategy for Sustainable Waste Management in England and Wales*. HMSO, London.

Gale, R., Barg, S. and Gillies, A. (eds) (1995) *Green Budget Reform: an International Casebook of Leading Practices*. Earthscan, London.

Gandy, M. (1994) *Recycling and the Politics of Urban Waste*. Earthscan, London.

HM Government (1991) *Recycling*. Waste Management Paper No. 28, HMSO, London.

Pearce, D.W. and Turner, R.K. (1993) Market-based approaches to solid waste management. *Resources, Conservation and Recycling*, **8**, 63–90.

Powell, J.C. and Craighill, A.L. (1997) The UK landfill tax. In T. O'Riordan (ed.), *Ecotaxation*. Earthscan, London, 304–320.

Turner, R.K. (1995) Waste management. In H. Folmer, H.L. Gabel and H. Opschoor (eds), *Principles of Environmental and Resource Economics: a Guide for Students and Decision-makers*. Edward Elgar, Aldershot.

Waite, R. (1995) *Household Waste Recycling*. Earthscan, London.

Chapter Nineteen

Managing the global commons

Timothy O'Riordan and Andrew Jordan

We might . . . conclude that some severe problems result not from the evil of people, but from their helplessness as individuals. This is not to say that there aren't callous, even malicious, noise and waste and vandalism . . . . But some is unwitting; some offers little choice; and some results from the magnification of small incentives into massive results.   *(Thomas Schelling, 1972, 90)*

The bonds of words are too weak to bridle men's ambition, avarice, anger and other passions, without the fear of some coercive power.
(Thomas Hobbes, 1651, 89)

Topics covered

- Managing the commons
- International environmental law
- Common heritage of humankind
- Global commons ethics
- Regimes and compliance
- Ozone chemistry and politics
- Effectiveness of international environmental agreements
- Education for sustainability

Editorial introduction

This chapter covers a mosaic of themes. The central idea is that the commons is both spaceless and timeless. It represents the continuity of life and evolution. It is not just an institutional domain for laws, management and responsibility. It is the essence of our humanness in a world that can reveal itself only through our stewarding and trusteeship-based consciousness as partners in the cosmic game. Commons is shared ownership, the companionship of partnership and the caring for the 'beyond self'. This is why the chapter begins with the concept of collective responsibility and ends with education for citizenship. Both are enveloping and continuous processes. They make us all what we should become if the sustainability ideal is ever to be attained.

Global change poses an enormous challenge for environmental science. We have already touched on the great complexity, indeterminacy and sheer incomprehensibility of many of the processes involved. In essence, we are trying to fit two unfathomable relationships together. One is the evolving planet; the other is the human occupation of that planet. Neither exists outside the other, and each responds to the other in inexplicable ways. Yet somehow we have to make sense of this interrelationship if humanity is to survive as a civilised species.

One way forward is to amalgamate the local with the global. This is not going to be easy, for one defines the other. Global processes are, in essence, the outcome of millions of local actions, and local behaviour is, in part, dictated by globalising events. The major international conventions will succeed in their effectiveness only if they are understood locally. And people will act locally only if they sense, anarchically, that others are doing the same, for similar motives, the world over. That is global citizenship.

Another way forward is through the appropriate development of Local Agenda 21 and the localisation of the Rio process (see O'Riordan and Voisey, 1998, 229–282). LA21 is proving to be a tortuous process as it is locked up in central–local government relations, policy responsibilities and political controls over local spending. The paradox is the need locally to empower, yet centrally to contain overall expenditures. The result is a flood of expectations dashing against the restraining walls of public sector borrowing limitations and politically motivated curtailment of genuine local democracy. It is no good training for citizenship only to find citizenship denied just as empowerment is promoted. Such is the current dilemma of LA21 almost the world over.

The rise of innovative participatory techniques such as visioning, empowering and mobilising the vulnerable clashes with a local institutional setting that is still unable, or unwilling, to experiment. In many cases, the inability to progress imaginatively with the mix of formal and informal economic relationships remains the major stumbling block. Tax controls, social benefit regulations, employment requirements and narrow interpretations of labour regulations all serve to thwart inventiveness and adaptability at the micro-level of civic survival. This is understandable: bureaucracies hate to make exceptions and cannot easily cope with the myriad of subtle differences that exist on an *ad-hominem* scale. Yet somehow the mix of social security, income flexibility, ecotaxation reform and local sustainability charities of the kind initiated by the landfill tax in the UK will begin to appear in exciting configurations at the local scale. As education emancipates the citizen in each individual, so these opportunities will further flourish. The commons is such a strange phenomenon. No wonder we cannot manage it properly. Maybe we should simply let it manage itself through our role as stewards. Civic science should be able to flourish in this atmosphere, embracing the social, life and Earth sciences to the humanities, the media and the Church. There is still plenty for the young environmental scientist to get excited about, if given the tools to serve the Earth.

On managing the commons

We have already looked at the changing interpretation of the commons in Chapter 1, at the nature of environmental politics in Chapter 3, and at the interconnectedness of the natural and social fabric in Chapter 6. The notion of the commons will forever remain central to environmental science. It is a concept that binds humanity to the planet and to its current and future generations. Caring for the commons is an act of individual stewardship and collective trusteeship. It is the very essence of being 'whole', the fundamental basis of interdisciplinarity.

The original 'commons' were neither freely accessible nor messily abused. Throughout history, there have been lands and waters which provide resources of fertility, food, fuel or timber and which are not privately appropriated but exist for people to use according to certain unwritten rules. The medieval village 'common' is but a fraction of the immense range of commons resources that still offer sustenance for many millions of people. Prakesh and Gupta (1997, 61) look at the social, ethical and religious rationales for altruism towards ecosystems and future generations. In their view, there are lasting examples of voluntaristic institutional arrangements to safeguard commons resources. In many ways, it is the rise of imperialism, mercantilism, industrialisation and consumerism that has undermined these mechanisms – though in the process, providing many advances for human well-being.

Commons provide a special quality of open-access property right. Permissions of use are tempered by cultural, kinship or village membership, so there is no 'open house'. Licences to use are also circumscribed by clearly defined custom and obligation to other users and to the well-being of the commons themselves. Anyone who abuses these privileges, in true commons, is subject to local condemnation and the odium of fellow commons users. For such people, the presumption of being helped in time of need no longer holds. In a society that depends on sharing during periods of scarcity or pain, that is a far greater penalty than incarceration.

In modern times, however, commons are destroyed, neglected, eroded or enclosed. For common property resources to suffer any or all of those fates, six quite specific conditions have to apply.

- There is free access, without impediment, to anyone and everyone.

- The assimilative capacity or critical load of the commons is finite and cannot be increased or removed.

- Individuals who enter the commons act in isolation from their neighbours: there are no rules of communal obligation and reciprocal support.

- Individuals are unaware of the side-effects of their use on others, partly because such effects may simply not be known, but partly because they do not take the trouble to find out since they see themselves as being in a state of competition, not co-operation.

- Co-operation will work only when individual users know that *everyone* will do so, but unless there is an agreed management regime with a police force, in a 'pure' commons there will be some who prefer to deceive for advantage. This causes others to do likewise, so deceit and false reporting become widespread.

● Individuals cannot change any of this on their own: their marginal influence is too small, and the gains from citizens' action are very modest given the social and practical costs of taking such action in a pure and unregulated commons.

Managing commons means controlling access, defining property rights, establishing codes of procedure that are accepted by all, enforcing compliance, and ensuring that the whole management process operates on the practice of accountable consent.

So far we have been discussing commons in a general manner. This chapter begins by looking at two special sets of commons, both of which involve intergovernmental co-operation or international agreement. That is, their management cannot depend on the goodwill of a national democracy. There must be some sort of internationally agreed rules that create a supranational democracy, in order that the global interest is protected. These two kinds of commons are usually referred to as *international commons* and *shared resources*.

International commons are physical or biological systems that lie outside the jurisdiction of any one country but whose life-maintaining services are valued by society as a whole. Examples include the biosphere generally and in particular the stratospheric ozone layer, the atmosphere, the diversity of natural species and their habitats on land and in water, deep sea minerals, Antarctica, the electromagnetic spectrum, outer space, and the biogeochemically active surface of the oceans. In some cases, technological advance has created an interest in a commons where before no effective management regime could be put in place. Outer space, the geostationary orbit and the electromagnetic spectrum are obvious illustrations. In other instances, such as Antarctica, some former commons have been effectively nationalised, though subject to international rules. The coastal zone is a case in point where, for the most part, nation states have management rights over waters and minerals up to 200 miles offshore under the principle of exclusive economic zones (EEZs) (Birnie, 1992).

For international commons, the most cited forms of governance include:

● *world government*: in practice promoted by United Nations institutions.

● *extended national responsibility*: by widening the basis of common property rights to some sort of mutual obligation acquired by treaty or convention.

● *restricted common property*: by imposing conditions of use on nation states through contractual obligations, bilateral agreements or international law.

● *shared resources*: physical or biological systems that extend across an international border, but which are geographically contained. Obvious examples include rivers, lakes or groundwaters that cross national boundaries, oil and gas reserves that underlie two or more nation states, and migratory birds, fish or mammals (e.g. whales) that feed, breed or simply roost in one country while *en route* from another nation to a third national jurisdiction. The other major illustration is the regional sea, such as the North Sea and the Mediterranean.

In the absence of a genuine world government, international environmental politics is said to take place in the context of *anarchy*. By anarchy, international relations theorists do not mean constant war and strife. Rather, they mean the absence of a strong, centralised government above the state – a 'Leviathan' to use a term coined by the great English political theorist Thomas Hobbes – capable of holding the ring between competing state objectives. According to Hobbes, contracts which are not enforced by the sword will fail because 'covenants without swords are but vain breath'. In the modern world, the common management solution to treating shared resources is the joint national agency, either by co-ordination or by some sort of multinational regime (Rittberger, 1993; Vogler, 1995). Many of these arrangements are discussed by Birnie and Boyle (1992) in a most useful survey of international environmental law.

A third set of international commons problems is associated with the transfer of pollution or dangerous substances across a border from one country to another. This is essentially a subset of shared resource management, for it includes such matters as acid rain, nuclear explosions, and chemical contamination of international waterways. But *transboundary externalities*, as they are called, do give rise to a number of vitally important principles for international environmental law.

● *Rights of sovereign action across a border.* When a neighbouring country causes environmental damage to its partner, what rights does the latter have over the former to stop the damage from continuing? Sulphur from the dirty coal-burning power stations in Ohio helps to damage the sugar maple industry of Quebec, just as toxic emissions from iron-smelting plants of the Kola Peninsula in north Russia adversely affect the sensitive ecosystems of Finland. What rights have the Canadians or the Finns to take preventive action against their neighbours? (See Box 19.1.)

Box 19.1 Regime and enterprise: Novilsk Nickel in Russia and damage to Finland

Novilsk Nickel is one of the world's biggest metal smelters, producing 27% of the world's nickel, 22% of global platinum, 62% of palladium and 19% of cobalt. It is by far the biggest smelter in the old Soviet Union, though since 1990 it has gone through a slow process of privatisation. The company is a conglomerate of production and research facilities, but the issue in particular contention is the 180 000 tonnes per year nickel smelter on the Kola Peninsula, near Finland. The plant is a major source of air pollution for the whole of northern Scandinavia, emitting over 250 000 tonnes of pollutants annually, of which 95% are SO_2. On the Kola Peninsula, over 8000 ha of land is beyond repair; in northern Siberia, over 400 000 ha of forest is affected. In all northern Arctic Russian cities, children under 14 are widely afflicted with respiratory ailments. About 20% of all depositions in Finland are from the Russian plants, 12% in Norway and 5% in Sweden (see Table 19.1.1).

This is a classic case of transboundary pollution. The normal route in law would be to force the Russian plant to reduce its emissions, or pay a fine equivalent to the losses to habitat and health. But in post-Soviet Russia, there is no money for such measures, no willingness to abide by international law, and no clear property right in the slowly privatising company. The Scandinavian countries are trying to establish a joint implementation (JI) project through which they finance and transfer the technology of clean-up. All attempts have failed, even though a guarantee of $300 million for a JI scheme was on offer. It appears that the complex of plants has no formal ownership, entrepreneurs are interested only in below-the-counter hard currency sales, and enforcement of post-Soviet air pollution laws is non-existent. Bribery is rife, regulators are very susceptible to kickbacks, and an official investigation ordered by the Russian Duma (parliament) resulted in the resignation of the company's president and the reorganisation of its board of directors. The Duma agreed to write off $1.8 billion in unpaid taxes to assist the plant in its own reconstruction. But so far the levels of pollution remain too high, though technically within international law, and all JI deals have collapsed. The uncertain property right, the inadequate chain of responsibility, the weak

Box 19.1 (continued)

regulatory structure and the inadequate status of international law retain a dangerous and techni- cally criminal regime in place.

Source: Kotov and Nikitina (1998, 549–574).

Table 19.1.1: Emissions and transborder fluxes of SO_2 in 1992 (thousand tonnes).

Source	Emissions	Depositions		
		Finland	*Norway*	*Sweden*
European Russia	4.400	30 (20.9)	11 (11.9)	6 (4.9)
Murmansk region	5.4	19 (13.2)	10 (10.8)	4 (3.3)

● *Buying pollution rights*. The 'polluter pays' principle suggests that those who cause damage should stop the source of the damage or pay for so doing. But in practice, where a poverty-stricken neighbour has neither the resources nor the political incentive to clean up, it may well pay a richer neighbour to buy out the pollution in the form of aid assistance via political agreements or economic restructuring payments. Where this cost of clean-up in the polluted country is greater than the cost of removing the source of pollution, or environmental threat, then it may well be more sensible to invest in technological rehabilitation or process substitutes. In theory, such agreements could be negotiated under the joint implementation provisions of the UN Framework Convention on Climate Change (see Chapter 7), but in practice they are politically and legally complex and are proving difficult to implement.

All this takes us into the field of international environmental law and diplomacy, and the changing role of science, economics, politics and ethics in what are probably the most critical and ill-developed areas of applied environmental management.

International environmental law

International law generally requires an enormous amount of trust, goodwill and collective determination to work. Essentially, any agreement involving two or more countries, in effect, requires that both agree to co-operate, or alternatively that the international community is prepared to enforce this agreement by trade sanction or diplomatic pressure. Ideally, there must be inter-party consensus but within a framework of compliance.

Such law is built on the principles of commons-type agreement. In practice, a nation state that violates an international agreement can be censured, but not incarcerated. Sanctions on trade or particular patterns of assistance can be applied, but, in essence, sanctions are usually designed to be busted. So the only real sanction is that of moral commitment to a collective agreement. Most democratic states abide by the rule of moral norms, as that is the only fundamental rule that works for managing global commons. This is why the institutional frameworks for global environmental agreements are so important.

International law is therefore a set of principles, obligations and rules that bind international behaviour. Through various formal or informal arrangements, the freedom to act as a sovereign state must be circumscribed by an obligation to respect the legitimate interests of other sovereign states. For this to have credibility, three conditions should apply.

- *Mutual advantage*. States have to accept that by co-operating and through compliance, everyone will be better off than if anyone fails to play by the agreed rules. So states are to embrace a shared sovereign interest in collective agreements.

- *Credible threat.* States have to be persuaded that non-compliance is not in their national interest. This requires supportive but believable science of the kind outlined in Chapter 1, backed by an independent scientific and technological review mechanism.

- *Credible enforcement.* Equally, states must know that non-compliance will result in differential penalty. By some means or another, there has to be enforced reporting or behaviour, of action to meet agreed objectives, and of failure to achieve what is promised, by some sort of independently validated verification procedure, in order to provide an acceptable basis for international regulation and enforcement. But even well-established international organisations like the European Union (EU) have notoriously weak powers of enforcement in the area of environmental policy compared with sovereign states and suffer serious implementation problems as a direct result (see Box 19.2).

Box 19.2 Implementation problems in the European Union

The European Union (EU) is one of the most fully developed international organisations, but even it struggles to get its policies put into effect. It has an executive (the Commission), a directly elected parliament, a court whose rulings are binding on states, a quasi-constitution and numerous other supranational agencies and committees. Unlike international law, EU law is directly applicable in member states, requiring no further review or ratification. Significantly, EU law provides citizens with rights and obligations and can be directly enforced through action in national courts.

The EU has been created by a group of 15 like-minded states which believe their common problems can be more effectively addressed if they work together. But while realising the benefits of sustained collaboration, they remain, like all states, extremely wary of surrendering important aspects of their autonomy – or sovereignty – to a supranational entity. As a result, the EU displays elements of both intergovernmentalism and supranationalism, although the founders of the union strongly believed it would evolve into a single federal structure. In some policy areas

like the environment, the Maastricht (1993) and Amsterdam (1997) Treaties allow the union to operate like a quasi-federal state; while in other areas like defence and foreign policy, the EU enjoys much less competence and decisions are reached after intergovernmental bargaining following the doctrine of unanimity in international organisations, thereby preserving the principle of state sovereignty. This is why the union finds itself situated in a *terra incognita*, being neither a federation nor an international organisation.

At its founding in 1957, the EU had no environmental policy. The environment was not even mentioned in the Treaty of Rome, which set down in Article 2 the promotion of 'a harmonious development of economic activities' as the primary objective of the then European Economic Community (EEC). Throughout the 1980s, however, EEC environmental policy underwent a relatively rapid and profound transformation. By 1987, the EU had adopted over 200 pieces of environmental legislation and four action programmes of steadily increasing complexity and scope. Subsequent amendments to the founding treaties, namely the Single

Box 19.2 (continued)

European Act (SEA) (1987) and the Maastricht Treaty on European Union (TEU) (1993), formalised and made more legitimate the EU's involvement, placing environmental protection on a firm legal footing and enunciating a set of guiding principles which have enabled the union to legislate in new areas. There are now over 500 legislative items relating to environmental protection, and until recently environment was one of the fastest-growing areas of EU activity.

For environmental policy, a fairly potent legal regime now exists in the European Union. It is supported by a council of environmental ministers, who can agree by a pattern of qualified majority voting, so that no individual nation has veto power. This tends to create binding consensus agreements that produce more powerful legislation than many of the member states would think out or enact on their own. The council of environmental ministers acts like a super-ministry of environmental affairs. This is the secret to the plethora of environmental legislation that floods across the union in a binding and increasingly comprehensive manner. See Jordan (1998) for a fuller analysis.

The unwillingness of member states to surrender power to supranational institutions helps to explain why EU environmental policy experiences serious implementation problems, although admittedly not nearly as bad as is sometimes found under international environmental laws. A recent report by the EU's own Court of Auditors (OJ C245\1, 23-9-92) concluded that environmental directives were 'being implemented slowly' and pointed to a 'significant gap between the set of rules in force and their actual application.' Unlike a federal state like the USA, the EU lacks all relatively integrated political culture and public opinion, political parties operating at both levels of governance and a high degree of economic and cultural homogeneity. Although in principle, EU law has direct effect, the union has a relatively small and largely non-executant centralised bureaucracy. Therefore, in the very places that EU environmental policy needs to 'bite' – factories, river banks and beaches – the Commission, whose responsibility it is to ensure that EU law is properly applied, has little or no presence or direct influence. In the past, it has often preferred not to stir up trouble with non-compliant states in case it endangers the wider integrationist project. The union now has a European Environment Agency, but unlike its administrative cousin in the USA, this has no formal enforcement powers and is restricted to collecting information and to generating indicators of the changing state of the environment.

International environmental law is normally created by means of legally binding *treaties*, *conventions* or *protocols*, sometimes referred to as hard law, or by *custom* and *framework conventions*, based on more ambiguous wording and hence more feasible interpretation. This is sometimes termed soft law. The distinction between hard and soft law is a dubious one, even though the two concepts sit side by side in the literature (Birnie, 1992, 52–54). Hard law in international terms is probably best seen as regulations or enforceable targets of performance bound by treaty. These impose mandatory obligations on states, which must subsequently enforce their commitments by national laws or regulations. An obvious example is the control of ozone-depleting substances under the Montreal Protocol. Here specific chemical formulations are being phased out over a clearly defined time period. Controls preventing parties from trading with non-parties in goods containing ozone-depleting chemicals provide a powerful incentive to join the regime. Poorer countries receive financial and technical assistance from the Global Environment Facility (GEF) so they can leapfrog ozone-depleting technologies (Jordan, 1994).

Soft law is, for some, a contradiction in terms. Law means 'hard' rules and obligations which are entered into or abided by, irrespective of original willingness or dissent. So there can be no 'true' law which is so vague as to allow excessively flexible interpretation or endless delay in implementation without making a mockery of the notion of legal agreement. Others, however, see in the carefully written ambiguity of modern international environmental agreements an element of calculated persuasion. Some lawyers refer to this as *customary law*, that is support for broad agreements that do not require a formal legal instrument. This allows signatories to accept the principle of sovereign consent but also permits them to vary their interpretations of genuinely ambiguous terms such as sustainable development, precautionary principle, cost-effectiveness and critical load analysis.

So soft law is an important aspect of international environmental diplomacy. Arguably it is of the greatest significance in the evolution of international environmental agreements aimed at managing the various global commons. One can say this for four reasons.

- *Scientific doubt*. Where the science is supple or genuinely indeterminate, and where there need to be regular updates on the state of knowledge, then phased commitments to tighter and tighter obligations (or the inverse, if the earlier scientific assessments were too pessimistic) are more likely to result in compliance.

- *Freedom of action*. One of the great achievements of the EU is to promote the twin principles of shared responsibility and burden sharing. In general, when a directive is agreed, the European Commission leaves it up to each member state to determine the precise pathways of implementation. In global matters, however, the EU negotiates as a bloc, commits all its member states to a predetermined outcome, then shares out the implementation arrangements through individual member state agreements (see Sbragia and Damro, 1999). This process is obviously subject to much horse-trading, but at least the less well-off countries are assisted to meet a common objective. This gives the EU considerable prestige in international negotiations.

- *Social learning*. Governments often learn by experiment and also from pursuing a course of action forced upon them by international commitment. This process of adaptation and reinterpretation of an earlier negotiating position is sometimes referred to as social learning. It is the collectivity of devices ranging from scientific discovery, popular alarm, international agreement and supporting institutional change that is the evolutionary element of governance. Soft law can be enormously influential in promoting this, at times, painful adjustment.

- *Face saving*. At the start of difficult negotiations the parties may be implacably opposed to certain principles. Eventually, for agreement to be reached, wording has to be found that permits diplomats and their political masters to save face yet acquiesce to the outcome. Usually the wording agreed is a combining together of different aspects of the same principle. For example, Article 2 of the Maastricht Treaty of the European Union listed 'harmonious and balanced development of economic activities [and] sustainable and non-inflationary growth respecting the environment' as one of the goals of joint action. This amalgam of environmentally sustainable development and continuous economic growth was unsatisfactory and was amended, at the 1997 Amsterdam European Council, to read 'harmonious, balanced and sustainable development of economic activities'. The Climate Convention talks of 'common but differentiated responsibility' in reducing global greenhouse gas emissions. This could mean that every country must play its part, but some should do so much more than others. But it could also be interpreted as a matter of capability or practicality rather than moral principle or burden of contribution to atmospheric warming. For some time to come, different countries will respond to this wording as a cop-out or a diplomatic lever, depending on the political and economic responsibilities involved in greenhouse gas reduction. The significance of this is discussed in Chapter 7.

Common heritage and global legal ethics

Within the notion of the international commons is a specific concept applying to the safeguarding of life support and celestial processes which transcend the infinitesimally brief tenure of each of the human species. This is the doctrine of the *common heritage for humankind* (for a discussion see Brown Weiss, 1989, 48–49). The idea emerged in the 1930s and became incorporated in law in the 1950s. It has five principal elements, namely non-ownership of the heritage; shared management; shared benefits; inclusive use for peaceful purposes; and conservation for humanity, including future generations. The significance of this doctrine lies in its obligation to current generations to act as trustees of the natural and human heritage so as to enhance the biological and spiritual life of its descendants.

The common heritage doctrine combines ethics with the law in a manner that is by no means agreed amongst international jurists. One view holds that no agreement can operate without the consent of national legislatures as part of formal ratification of treaties signed by ministers or heads of state. This is a conservative, but formally correct, view that representatives of public opinion should bind nation states to shared action. The radical, or unconformist, view holds that the law should have a moral or exhortatory character leading nations into uncharted waters on the basis of stewardship (wise management utilising best practice) and trusteeship (handing on to future generations the legacy of what was given to the present generation). For example, Allott (1990) believes 'when law is not merely the multilaterization of individual purposes but the universalization of social purpose, it transcends self interest and becomes self creating of a society.' This body of thought believes that the International Court of Justice at The Hague in the Netherlands should have the power to declare general principles of natural justice for the sustainability of the Earth, as accepted by the consciences of concerned people. This should in turn set the stage for universal declarations of basic principles, such as fundamental environmental rights of 'critical' natural functions and of sustainable development.

The problem lies both in the concept and in the practice of ethical law and common heritage issues. If there is no consensus at state level, then all the declarations in the world will not help, for there will be no collective backing for enforcement and punitive sanction for wrongdoing. Soft law practices can help by creating statements or declarations for international consensus. In the environmental field, there are the declarations of principles arising from both the Stockholm and Rio conferences (see Box 19.3). To date, these are exhortatory only, with moral and some political persuasive support. The fact that they exist allows pressure groups to exert influence on future negotiations, places governments on the spot when new treaties come up for signature, and allows legislatures some room for critical surveillance.

Edith Brown Weiss (1989, 148) believes this matter to be so important that it should not be left to the existing UN machinery. She advocates the creation of a Commission

Box 19.3 The Rio principles of international environmental law

World leaders signed a statement of non-binding principles to guide the transition to sustainable development at the Rio summit in 1992. Some were already well-established customary norms of international law, whereas others were newer and took longer to negotiate. The final form of words reflects a series of compromises between industrialised and industrialising country concerns, and a balance between ecocentric and anthropocentric world views. Their significance lies in their broad legal and ethical basis, a framework for compliance and 1990 activism. The principles include:

- *Principle 1*: Human beings are 'at the centre of concerns for sustainable development' and are 'entitled to a healthy and productive life in harmony with nature'.

- *Principle 2*: States 'have the sovereign right to exploit their own resources' but a 'responsibility to ensure that activities within their jurisdiction ... do not cause damage to the environment of other states or of areas beyond the limits of national jurisdiction'.

- *Principle 3*: The right to development must be fulfilled so as to 'equitably' meet the development and environment needs of present and future generations.

- *Principle 4*: In order to achieve sustainable development, environmental protection shall constitute an integral part of the development process.

- *Principle 7*: in view of their different contributions to global environmental change, states have 'common but differentiated responsibilities' in the international pursuit of sustainable development

- *Principle 11*: All states should enact 'effective environmental legislation' which reflects 'the environmental and developmental context' to which it applies.

- *Principles 15 and 16*: Endorse the precautionary and polluter pays principles, respectively.

Views differ sharply on the long-term importance of these principles. Some lawyers believe they offer a sound basis for a more systematic development of international law (Sands, 1997). Others regard them as a vague wish list of 'high' aspirations, which all states can agree upon but do not actually bind anyone to anything.

on the Future of the Planet. This might be an amalgam of intergovernmental and non-governmental organisations. It would have to be backed by a UN Resolution on Human and Environmental Rights to give it moral force. Brown Weiss suggests that such a body should have six functions:

- *symbolic*, emphasising global commitment to trusteeship;

- *cautionary*, warning of impending threats to any aspect of common heritage;

- *catalytic*, creating new ideas by bringing together interests that may not have to force recognised collective self-interest;

- *advisory*, providing technical and scientific assistance to countries providing heritage as an act of trusteeship;

- *scientific*, investigating research issues of an interdisciplinary nature to ensure that heritage is properly evaluated and given political support; and

- *educational*, providing the necessary information and awareness to promote the cause among the public generally.

She also argues that the commission would need some form of troubleshooting support arm in the form of an ombudsman's office. This would take grievances from NGOs or even the general public where evidence of heritage damage would be shown to be happening or about to occur as a result of an existing or proposed course of action. The office, which might be regionalised on a continental basis, would also have powers to investigate complaints and would be responsible for the scientific and advisory roles suggested above.

Given the general hostility to the UN, and its pervasive incapacity to reform itself even when the budget chips are down, it is highly unlikely that anything remotely resembling the suggestions of Brown Weiss will bear fruit. We noted the problems with the nearest equivalent body, the UN Commission on Sustainable Development, in Chapter 2. Even a modest organisation cannot curry influence on the pragmatics of sustainable development, mostly because it is under-resourced and ill-supported. So there is little hope of anything ethically more radical drifting out of the mist.

Yet the issue involved is important for managing the global commons, which is why this theme is given prominence in this final chapter. What lies behind Brown Weiss's radical notions is a concept of international law and organisation that promotes a broader cause of trusteeship for the planet than would happen if everything were left to politically driven UN organisations and states. Equally important from her vision is a principle of property right that embraces both a private and a public responsibility embodied in the principle of trusteeship and companionship of an evolutionary journey that is life on Earth. Philosophers have coined two terms to espouse this very special application of property obligation. One is *deep ecology*, namely a belief in the solidarity between humanity and the biosphere that is symbiotic, universal and enduring (Box 19.4). The other is *transpersonal ecology*, or the attempt to lose the individual and identifiable sense of being in a holistic engulfment with the biosphere (Box 19.5).

We enter the realm of global commons ethics, namely the underlying moral outlook that binds us to this planet. Here we see the lands of instrumentalist and normative

Box 19.4 Deep ecology

The philosophy of deep ecology was coined by the Norwegian Arne Naess (Naess, 1989) to reflect ever more searching investigation about the true nature of humans to the natural world. The 'deep' of deep ecology means asking deeper and more profound questions. Deep ecology is the ultimate in ecocentrism, the striving for equality with Gaia or the biosphere as part of organic totality with all creation, past, present and future. Deep ecology preaches egalitarianism, non-violence, animism and a non-dualistic appreciative attitude to the whole of the ecosphere. It encourages compassion and benevolence to non-human beings, accords them rights to coexistence with humans and calls on people to forgo mindless consumerism. Deep ecologists believe that materialistic humanity is a slave to gratification, a passive actor in collective global destruction.

The philosophy has paradoxically been taken up by a growing number of ecoactivists, notably Earth First!, Sea Shepherd, Reclaim the Streets and the Environmental Liberation Front. Each of these groupings is shadowy, operating in cells of activism, forming, fragmenting and recreating as needs arise. They are the object of much intelligence surveillance and continue to grow in number and activity. How far they will succeed in changing the order of things depends enormously on the culture of acceptance of ecological protectionism by illegal means. The irony is that all these groups accept that violence and destruction of objects or processes that they believe are helping to destroy the planet are acceptable practices.

Box 19.5 Transpersonal ecology

Transpersonal ecology is the brainchild of an Australian philosopher, Warwick Fox (1992). He was dissatisfied with the formality of deep ecology and looked instead for a more transcendentalist philosophy (see Box 3.1). This, he reasoned, should regard all objects in the world as part of a continuous and connected whole with innate sameness, or utter interconnectedness. He feels that the proper state is that of pure consciousness, that is, the experience of totality, of complete undifferentiatedness. Once we make distinctions in the world around us, by separating object from object, and emotional reaction from emotional reaction, then we lapse back into the realm of conscious contents. Thus, he claims, discreteness comes from mechanistic science, continuity reaches us from the 'new interactive sciences', and identity (or sameness) from mystical tradition. Transpersonal ecology is the middle way between the extended science of interdisciplinarity as outlined in the Introduction, and the totality of experience that comes from transcendentalism: the sense of utter nothingness, of complete loss of personal separateness, of spiritual engulfment in the comprehensiveness of creative evolution. This emancipated state, he believes, is the basis for education for sustainability. It is also the state most likely to result in revelation as an essential part of community empowerment, interdisciplinarity generally, and in the transition of local collective awareness towards sustainability, as introduced in Chapters 1 and 2. This is why Gaian citizenship may become a central theme in future curricula.

science outlined in Chapter 1 grappling with a new notion, namely that of *Gaia*. The purpose of introducing the Gaia concept more formally at this late stage of the text is to emphasise the persisting discomfort the scientific community has about Gaia. For the most part, the whole idea is discredited, figuring very variably in curricula, mostly dependent on the preferences of teachers. On the other hand, the Gaian framework is increasingly being found to work at the level of physical–chemical and biological interrelationships, even though it is not yet possible to prove conclusively that these processes actually operate in a coupled system.

Both terms introduce the notion of *biosphere*, the interacting totality that is life on Earth. According to Lovelock, the biosphere is not just a set of molecules operating in accord with random and mysterious processes. The biosphere has properties that seem to enable it to maintain the conditions that keep life alive on an Earth that it manipulates to be habitable.

In a valuable review of Gaian science, Tim Lenton (1998, 442) summarises the evidence for a combination of non-selective and selective feedbacks operating in abiotic (physical) and biotic (ecological) ways. He cites the case of oceanic plankton, which appear to switch on and off their capacity for producing dimethyl sulphate gas, the basis of marine clouds via aerosols formed in the mid-troposphere. As clouds form, so temperatures fall, and DMS production is reduced. Furthermore, a thermocline of $-10°C$ might limit the formation of plankton, so that this interaction is controlled within set limits. Subsequent research suggests that nitrogen levels also influence DSP formation, and that ammonia production is affected by sulphate scavenging. Warts and all, these amazing scientific discoveries suggest that 'environmental regulation can emerge at levels from the individual to the global' (Lenton, 1998, 446). The Earth may well be a truly amazing place. The more we use science to reveal this truth, the more the ethics of co-responsibility kick in. Keen readers might also like to consult Harrison and Lenton (1998).

Lovelock's ideas are embodied in a thesis known as the *Gaia hypothesis*. The term stems from the Greek word for the goddess of the Earth, who was responsible for the nurturing powers of nature and the evil of pestilence and disease. The Greeks believed that failure to respond to the strictures of Gaia (what geographers would term environmental determinism) would result in earthly retribution in the form of famine, disease and ultimately the withdrawal of life support. Lovelock does not personalise Gaia: for him it is a scientific concept that can be shown to be a glorified version of the physiography of the human organism (Box 19.6). Just as any organism has powers of survival and adaptation to stress or threat, so the biosphere has the capacity to alter its biological and chemical relationships to maintain temperature, water in all its forms and the great chemical fluxes of oxygen, carbon, sulphur and nitrogen, as well as physical processes such as erosion, accretion and sequestration in a way that keeps the Earth in approximate homeostasis.

The Gaia hypothesis borders on the mystical, the imagining of the mind beyond the normal scientific method. Deep ecology and transpersonal ecology attempt to provide

Box 19.6 The Gaia concept

The Gaia hypothesis is the product of an unusual mind – the non-establishment scientist. Jim Lovelock is a geochemist and an inventor who can earn his keep by designing and patenting an imaginative piece of machinery. But his true love is 'real science' – the 'wondering about how the world works and the design of simple experiments to test the theories that thus come to mind' (Lovelock, 1992, 15). Real science, like its companion artistic creativity, is best done quietly and inexpensively. He regrets the plethora of specialised environmental scientists, 'the brightly polished doctors of philosophy', who are most unlikely to achieve anything other than a secure and comfortable employment. Gaia is holistic, not reductionist; integrative, not separatist; imaginative, not pedantic; open-ended, not closed.

Lovelock believes Gaia is a form of planetary medicine similar to the folk medicine of the pre-scientific era. Keeping a healthy planet with an equitable climate is as complicated as maintaining a healthy body free of disease. The Gaian idea is to design a mode of existence to let the Earth live in a healthy fashion – where possible, to let nature be the guide and the gyroscope of homeostasis. To ignore Gaia may mean that humans will have to live for ever in a form of environmental conscription – looking after the planet as if it were on a kidney machine, functioning but always circumscribed. Caring for Gaia means keeping a planetary healthy life –

tending to the small things that accumulate across billions of thoughtful people into collective planetary health. This is Gaian anarchy. We behave appropriately because we just know others are doing so. Gaianism is the transpersonal lifestyle of a sustained world.

Schneider and Boston (1991) conclude that the Gaia notion is really five different ideas:

- Influential Gaia simply states that biota have influence over the abiotic world, adjusting temperature and atmospheric chemistry.

- Co-evolutionary Gaia believes that biotic and abiotic activities interact in a synergistic manner, creating and sustaining the biosphere.

- Homeostatic Gaia asserts that the co-evolutionary role is to maintain dynamic equilibrium of the biosphere, despite external or internal stresses.

- Teleological Gaia holds that the atmosphere is kept in homeostasis not only *by* the biosphere but also *for* the biosphere.

- Optimising Gaia concludes that the biosphere operates in such a manner as to create conditions that are optimal for biological survival.

These are listed in order of interconnectedness. The first two are relatively weak notions. The latter three suggest a purpose in life on Earth, namely to maintain itself in near-optimal conditions. Take your pick.

Sources: Lovelock (1992); Schneider and Boston (1991).

a philosophical basis for sustainability that connects equality of status, equality of opportunity, non-violence and inner conscience to the more descriptive aspects of sustainable development. The aim is to create a philosophical, and in its literal sense a political, foundation for the transition and attainment of sustainability. By 'political', one is referring to a set of rules, customs and accepted practices that allow countries to overcome dissent and reach consensus by reasoned argument and legitimate means of resolving conflict. These are the crucial areas where the fusing of international law and politics remains in great difficulty. For all its controversy, Gaia provides a valuable ethical framework for such co-operative behaviour, as indeed it should provide a model for a more constructive civic education for the sustainability age.

Regime theory and sustainability

A regime is a set of principles, norms, rules and decision-making procedures around which those who have bonded themselves accept a common purpose, and thus a common process of acting and analysing. It is this *soft law* of collective commitment that gives any regime its form. For this to occur there has to be agreement over problems, jointly determined solutions, and shared knowledge for subsequent analysis and prediction.

Regimes are international *institutions*. They need to be differentiated from international *organisations* like the United Nations, which have a tangible presence in the form of offices, personnel and other resources. Regimes converge action around agreed principles and objectives and are based on a combination of international law, credible science and acceptable political custom. They include informal codes of conduct, such as soft laws, and formal statements of law such as conventions and protocols. Regimes are effective only if the parties involved recognise that their self-interest is best served by co-operation, mutual enforcement and collective action. The scope for *free-riding,* namely letting others conform while avoiding responsibility and hence benefiting from the commitment of others, has to be removed by some form of commonly agreed mutual coercion.

The question of why regimes appear to succeed in some areas but not others is currently being investigated by scholars of international relations (Box 19.7). There are three main theoretical standpoints: realism, liberal institutionalism and epistemic communities.

Realism

Realists believe that states are the primary actors on the international stage. The role of international organisations and other non-governmental groups is incidental and very much secondary to that of states. The assumptions of realism are strikingly similar to those of mainstream economics, states being regarded as self-contained, hierarchical units with clearly specified goals. Anarchy can only ever be mitigated, never transcended. Realism suggests that, in the absence of world government, all agreements will be weak and prone to free-riding. Regimes are regarded as the outcome first and foremost of conflict between states operating by rational choice principles and bent on maximising their individual gain (or utility). Regimes are 'successful' when they are created and maintained by a single power, or a small number of states, with the ability to influence or coerce the weaker members into their definition of acceptable action. Otherwise any resultant regime will be weak and unstable (see Paterson, 1996, for a good review).

Box 19.7 Sequential diplomacy: mending the ozone hole

As is now widely known, ozone is a molecular form of oxygen composed of three atoms created by the fusing energy of sunlight. In the outer layers of the atmosphere, some 50 km above the Earth's surface, oxygen molecules seeping out of the stratosphere are constantly converted into ozone. The ozone molecules are subsequently destroyed by the catalytic activities of minute concentrations of certain gases, in parts per billion or even parts per trillion (10^{-9} or 10^{-12}). These gases can 'steal' an oxygen atom to destroy ozone. One such gas is nitric oxide (NO); its reaction with ozone is shown here:

$$O_3 + NO \rightarrow O_2 + NO_2$$

$$O + NO_2 \rightarrow O_2 + NO$$

Chlorine does a similar job, as follows:

$$O_3 + Cl \rightarrow O_2 + ClO$$

$$O + ClO \rightarrow O_2 + Cl$$

The catalytic nature of this process ensures that active chlorine radicals can remove ozone many times over; possibly as many as 100 000 molecules of ozone per free atom may be removed. For the most part, these free trace atoms are locked up in 'reservoirs' of inactivity. The chlorine rests as hydrogen chloride, and the nitrogen as chlorine nitrate.

All this is fine as long as the dynamic equilibrium is not disturbed. However, the source gases of chlorine and nitrogen atoms are being increased by human-induced activities. NO_x or various nitrogen oxides are created from fossil fuel burning, automobile exhausts, ammonia from manure and artificial fertiliser. Chlorine compounds are increasingly found in many chemical products, from pesticides to solvents and from bleach to plastics. Methane (CH_4) also plays its part as an ozone destroyer, and that too is on the increase. However, the most prominent of the new source gases is the chlorofluorocarbon series, a group of chemical compounds combining carbon, chlorine, fluorine and hydrogen. These gases were developed because they appeared to be inert and non-flammable and therefore could be combined with active agents for safety and security. In the stratosphere, the intense sunlight strips away a chlorine atom, thereby adding to the stock of chlorine atoms. The length of time a CFC molecule remains active and the number of chlorine atoms it contains produces an index known as the ozone depleting potential (ODP): a ratio compared with CFC-11 ($CFCl_3$), the most common of the CFC tribe. However, if sufficient quantities are released, then even a low ODP may signal danger. Two examples illustrate this: methyl chloroform is a widely used dry cleaning agent with an output exceeding 500 000 tonnes annually yet has an ODP of 0.1 to 0.2. The hydrogen fluorocarbons, now being produced as substitutes for CFCs, are greenhouse gases as well, contributing about 8–10% of climate warming.

The polar regions, especially the Antarctic atmosphere, become very stable during the sunless winter months. Luminescent clouds of ice particles form in the intensely cold and stable atmospheres. On these ice particles chemical processes, as yet imperfectly understood, prepare the trace gases for liberation when energised by the first rays of spring sunshine. This is why the polar ozone 'holes' are so prominent in the spring, though the feature is much more marked over the Antarctic because of the scale of the polar ice cap. Recent estimates suggest that the temporary loss of ozone in the Antarctic may be as high as 60% and about 15–20% in the Arctic. Overall, the stratospheric ozone losses may be 1 or 2% annually; even if all the additional trace gases were to be stabilised at pre-1990 levels, it would take over 70 years for net ozone loss to cease.

The removal of stratospheric ozone exposes humans and plants to higher levels of the more intense ultraviolet solar radiation. Curiously, actual levels of UV-B, the band of greatest concern, have not been measured as increasing in the mid-latitudes yet. This has led some to speculate that the whole CFC scare is a scientists'

Box 19.7 (continued)

nightmare concocted to generate more research cash (see Kenny, 1994). The problem is outlined in Chapter 1: measurement needs to be accurate, long-term and very reliable before any trend in such a micro-scale can be recorded with confidence. Impatience is not a scientific virtue. Yet for political and very significant commercial reasons, it pays the big chemical companies to listen to, and to finance, scientists who are sceptical of the evidence. It is only since 1985 that instruments of sufficient accuracy have been designed to detect the changes in UV-B in the less polluted areas such as the southern hemisphere continents and Antarctica itself. Meanwhile, the chemical manufacturers are making handsome profits out of new refrigeration technology using HCFCs and other lower-ODP compounds. This is despite the fact that Greenpeace has championed a safe propane–butane refrigerator that is both inexpensive and entirely ozone-friendly.

The sequential diplomacy concerns the negotiations surrounding the 1987 Montreal Protocol on Substances that Deplete the Ozone Layer. Originally it committed every signatory state to reduce its use of certain CFCs by 50% of their level of use in 1986 by 1999. In 1990, the London Amendments added more CFCs, halons and other substances, and widened the net of signatory nations to 80. In 1992, the World Meteorological Organisation reported that the 1991–92 ozone 'winter' could be classified 'among those with the most negative deviation of systematic ozone observations'. Estimates of skin cancer from increased exposure to UV-B alarmed many, as did reports of eye cataracts in humans and animals increasing in mountainous areas, where thin clear air added to the radiation burden. Counterbalancing these scientific discoveries was the political and economic power of multinational chemical companies and nation states still dependent on the manufacture of high-ODP compounds now in the firing line. Sequential environmental diplomacy is a cat-and-mouse game operating at many levels. The science is

only one element; uncertainty is mercilessly exploited, as are the claims of environmental organisations that environmentally friendly substitutes can be manufactured profitably.

In November 1992, the fourth meeting of the Montreal Protocol agreed to phase out CFCs by 1996 with a 75% reduction by 1994. However, 'essential uses' where no practicable substitutes are available can continue, as can CFC output in the developing world until 2006, so long as production is designed to meet 'basic domestic needs'. The scope for black-market racketeering is considerable, though small-scale manufacture of CFCs is not particularly profitable.

Nevertheless, in a space of six years, environmental diplomacy has all but eliminated CFCs, halons, methyl chloroform and carbon tetrachloride. Yet the influence of the HCFC producers should not be underestimated. They managed to delay a phase-out to 2030, to ensure that they would recoup the profits from their expensive research and development procedures. Rowlands (1995, 27) points out that the US chemical company Du Pont sent a delegation of seven people to the negotiations, a number that exceeded the delegations of all but six countries. The delayed phase-out of HCFCs is in line with the demand by US business interests that air-conditioning units in buildings be allowed to run their full 40-year working lives.

Methyl bromide is also a known ozone-depleting substance. Unlike the CFC group, it is used as a fungicide and is widely deployed in the Third World. Israel is a prominent manufacturer. Poorer countries successfully argued for a retention of the use of this substance, with only a modest phase-out subject to more scientific evidence. Sequential diplomacy will undoubtedly target this product as the science of ozone depletion improves. Poorer nations are seeking exceptions or grant aid from the Global Environment Facility to see them through the transition. In the ozone debate, as in many others, environmental science becomes enmeshed in the whirlpool of political economics. Meanwhile, the

Box 19.7 (continued)

ozone hole continues to deepen, despite the sequential diplomacy reviewed here.

Figures 19.7.1 and 19.7.2 illustrate the latest findings from the Antarctic stratosphere. Figure 19.7.1 shows the depth of the 'hole' in the October spring. Figure 19.7.2 reveals the persistent decline in stratospheric ozone levels at the Halley observation site.

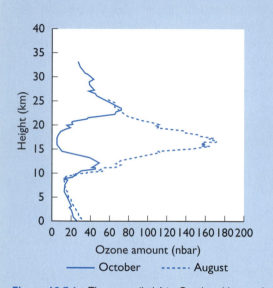

Figure 19.7.1: The ozone 'hole' in October (the southern spring) (left plot) as opposed to August (right plot).
Source: Shanklin (1998, 4).

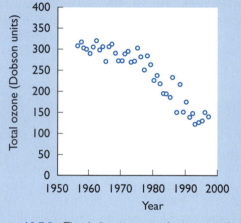

Figure 19.7.2: The decline in stratospheric ozone during October at Halley.
Source: Shanklin (1998, 5).

With the end of the Cold War, the growing dominance of the 'Group of 77' developing countries (symbolised by their attempts to 'greenmail' the industrialised world on matters like ozone and climate change) (Jordan, 1994; Williams, 1997) and the important roles performed by non-state actors such as scientists and environmental groups, the realist perspective is now regarded as an out-of-date interpretation of international environmental action. In some senses, too, the USA has not assumed a dominant role in international environmental negotiations, so its obvious hegemony has not been so apparent. It is interesting that the Rio conference achieved a Framework Convention on Biodiversity without US support, and set targets for greenhouse gas reduction against US opposition. It is even more interesting to note that the Clinton administration not only signed the Biodiversity Convention but also agreed to the stabilisation of US CO_2 emissions at 1990 levels by 2000, when it specifically excluded such targets from the Framework Convention. Furthermore, it acquiesced to the non-inclusion of developing countries in the Kyoto Agreement (see Chapter 7). Yet it pushed hard for carbon trades, mostly to get its domestic implementation off the hook. One could say this was institutional realism. More likely it was political pragmatism. As pointed out in Chapter 7, the USA is governed in such a way as to be unable to deliver greenhouse gas reduction by federal action alone. For all its faults, carbon trading, suitably policed, may be the only way forward if real gains are to be made (von Moltke, 1996, 330–331). These are signs of the significance of global commons regimes. So some other theory is required.

Liberal institutionalism

Liberal institutionalists are much more optimistic about the potential for co-operation between states, brought about by regimes. Unlike realists, liberal institutionalists believe regimes are independent of states, which have the power to set the international agenda and states' perceptions of their interests in ways which facilitate co-operation. Regimes facilitate co-operation by building trust between parties, transferring information and resources, and highlighting free-riding. None of this should be taken to mean that regimes 'supersede or overshadow states': 'to be effective they must create networks over, round and within states that generate the means and the incentives for co-operation' (Haas *et al.*, 1993, 34). The importance of building trust over time is an important lesson learned in the development of the ozone regime (Box 19.7). This perspective requires the kinds of conditions that combine to create more effective compliance for multilateral environmental agreements, as discussed in the section that follows.

Epistemic communities

This theory starts from the claim that states are more 'uncertainty reducers' than they are well-informed 'utility maximisers'. Lack of sound evidence and the recognition that a regime, no matter how imperfect, will create the soft law conditions for civic responsiveness, and hence a collectively induced revelatory condition, become the very basis for this interpretation. The emergence of sudden unexpected crises such as the appearance of the ozone hole prompts states to turn to experts for advice. In his theory of epistemic learning, Peter Haas (1990) highlights the important role performed by international networks of expertise which share knowledge and undertake rigorous peer review to establish and maintain credibility. This aspect was outlined in Chapters 1, 2, 7 and 17. The point is to establish mutual agreement among policy advisors regarding appropriate political action. Much depends on scientific consensus and a reasonable convergence of judgement as to likely outcomes of the policy measures, for example environmental taxes and their influence on jobs and income distribution, or pollutant removal as a precautionary measure and the benefit for specific ecosystems. The points raised in Chapter 1 are relevant here. Science needs both credibility and a broadly based political constituency before epistemic communities can prove successful. The Intergovernmental Panel on Climate Change is a good example of the epistemic science basis for regime formation (see Chapters 1 and 7). Haas argues persuasively that particularly cohesive communities share certain beliefs about the organisation of society which they try to have established in policy. He suggests that an epistemic community of atmospheric scientists was instrumental in the adoption of the Montreal Protocol (see Box 19.7).

In the contemporary world, none of these theories is ideal for identifying negotiating regimes in environmental problem solving. Linkage of environmental issues to trade negotiations, to regional conflict and security, and to the migration of displaced persons means that much of modern regime development is exploratory and revelatory. Indeed, there is a good argument that bargaining over global commons issues is proving to be genuinely innovative and unique, exposing politicians to very different perceptions and placing them face to face with constellations of interests not encountered before. So negotiating regimes have to start small and evolve creatively within larger but supporting international constituencies.

All this puts considerable emphasis on constructive and coherent communication between the parties and demands special skills of 'translation into the vernacular' for *knowledge brokers* or *policy advocates*. This is a relatively new breed of scientist, trained in science and knowledgeable of the complexities and shortfalls of the trade, yet capable of summarising it in a faithful and intelligible manner. Such individuals are becoming half scientists and half politicians, with great media talent, able to explain and emphasise according to the media needs of simple language, evocative illustration and memorable sound-bites.

When and why are regimes effective?

There are around 120 multilateral environmental agreements in existence. For these regimes to be effective they have to have a beneficial impact on the state of the world. This requires states, which are the main parties, to put in place implementation provisions at the national level and ensure that they function effectively. Whether and to what extent regimes actually achieve beneficial outcomes is a matter of great debate among scholars of international relations (Levy, 1996). According to realists, regimes *per se* simply codify what states would have done anyway: they have no independent influence upon a world which, as discussed above, remains fundamentally anarchic. Liberal institutionalists, on the other hand, believe that regimes do have some independent influence, although they disagree about which particular factors determine their effectiveness.

Owen Greene (1996) divides the key factors according to their position relative to the regime. Of those 'external' to the regime, he lists:

- the distribution of *interests* (is there a sufficiently large group of states in support of a particular course of action?);
- the distribution of *power* (is there a powerful state capable of setting the agenda and dragging the rest behind it?);
- the nature of the *issue* (some problems are intrinsically easier to solve than others);
- *knowledge* (can signatories detect non-compliance?).

Internal factors include:

- the *design* of the regime;
- the structure and role of *international* bodies;
- the availability of *side payment mechanisms* (e.g. the GEF) to widen the scope of participation;
- the existence of *dispute settlement* and *verification procedures*.

Thus, the 'success' of the Montreal Protocol is commonly ascribed to the presence of a powerful alliance of Northern states; the availability of relatively simple technical fixes (less ozone-depleting alternatives like HFCs); an international ozone fund to help poor countries; and the pioneering leadership of individuals like the former director of UNEP, Mostafa Tolba.

All this assumes there is a widely agreed definition of 'effectiveness'. At present, this is not the case. This is partly a matter of disciplinary perspective and theoretical standpoint. Lawyers, for instance, approach the question very differently from political scientists. They tend to look at rules and compliance. The political scientist John Vogler (1995) suggests that effectiveness can be measured in at least four ways, although there are several alternative typologies.

- *Legal status.* Are the rules of the regime properly ratified and in force nationally? Are they sufficiently legally binding?

- *Authority.* To what extent does the regime transfer authority to transnational organisations? Following Hobbes, supporters of this perspective claim international government is preferable to the (ultimately destructive) competition between states;

- *Behaviour.* To what extent does the regime actually alter the behaviour of states? Does it lead them to act in ways that are more environmentally sustainable?

- *Problem solving.* Does the regime actually ameliorate the problem it was initially designed to address?

Vogler concludes that the last two are the most important but also the most difficult to establish. Notwithstanding these definitional problems, he concludes that the international whaling regime has performed poorly over the years, the Antarctic regime has made progress but still has many shortcomings, while the ozone regime has been the most effective. Discovering what works and why is a matter of deep concern to policy makers.

David Victor and his colleagues (1998) have completed a 10-year analysis of the effectiveness and implementation of international environmental commitments. They looked at formal treaties and binding agreements, the two main outputs of such commitments. They define implementation as the myriad acts of government to meet a regime objective, as well as the supporters of campaigning behaviour of non-governmental organisations in seeking to ensure compliance. For effectiveness, they look to 'the extent to which the accord causes changes in the behaviour of targets that further the goals of the accord' (Victor *et al.*, 1998, 6). This is a rich notion that covers alterations in outlook, formation of otherwise unconnected coalitions, educational and media-related initiatives to raise awareness, and the scope for increasing civic activism and revolutionary behaviour of the sort introduced in Chapter 1. Effectiveness, therefore, differs from compliance, the legal definition of obeying the letter of an agreement. Compliance may be good, but the agreement may still be weak or misguided. So strong compliance is not a prerequisite for regime effectiveness.

Here is a summary of the principal conclusions of the Victor *et al.* study (1998, x–xi).

- *Data availability is crucial.* Data must be full, open and accurate, verified by independent means, and constantly re-evaluated in the context of changing interpretations of the goals. We noted in the climate context that national greenhouse gas emission inventories are extremely patchy and of variable accuracy. The issue for effectiveness is whether their accuracy is improving, not whether they are in error.

- *Soft law approaches are often necessary initially.* Complex and novel agreements, and agreements that require many policies to coincide, are best begun through debate, technical aid and financial assistance, coupled with guidelines as to the best practical and voluntary compacts, rather than the rigid compliance of hard directives. The key to a successful regime is the recognition of a shared outcome, no matter how long it takes to achieve.

- *Financial transfers from rich to poor.* Normally from North to South and from West to East, these enormously assist effectiveness and verification. Making such transfers conditional on observed behaviour is increasingly used as a stick and a carrot. The major role of such transfers is capacity building, providing technical and management aid for the building up of domestic institutions that can enforce national compliance. Failure to do so makes joint implementation a potentially lost cause.

● *Public interest groups help to generate the agreement but do not monitor implementation.* NGOs are usually highly active in negotiations but find it hard to devote the resources to monitoring the outcome, except as part of wider and longer-running campaigns. This is why formal governmental structures need to be put in place, coupled with systematic community reporting.

● *Resonant support by business and by civic groups,* coupled with linked policy initiatives, also assist enormously in promoting effectiveness. This is the hearts-and-minds condition. Generating mobilisation among the organs of civil society creates nodes of self-reinforcing and self-perpetuating activity that give an agreement life in the society as a whole.

Regimes are as varied as the South African protea. What we learn from all this analysis is that regimes depend on a mutual recognition of the commons and of ethical responsibilities, and on highly effective communicators. The best regimes depend on a collective vision for a better future. That propulsion will create structures for implementation and review that should evolve as circumstances demand. Without an ethical framework, the global commons cannot be managed by cold-hearted law on its own.

Regimes also influence trading patterns, and in turn trading rules will challenge the effectiveness of regimes. This is the penultimate port of call for our commons enquiry.

Trade and the environment

The opening up of global trade under the General Agreement on Tariffs and Trades (GATT), and latterly the World Trade Organisation (WTO), has resulted in much misgiving over the rights of vulnerable peoples and economics to competition and global incursion of trade activities. The WTO principles provide for a liberalisation of trade in 'like products' without disadvantage at national borders. The only scope for national discrimination in trading regimes is based on the following:

● actions necessary to protect human, animal or plant life or health;

● actions relating to the conservation of natural resources if linked to restrictions on domestic production or consumption;

● product requirements, such as labelling and regulatory safeguards, can be deployed, but a country must show that such measures are non-discriminatory. A legal requirement of *prior informed consent* is now used by the EU to cover products prohibited in the country of export but permitted by the importer;

● process and production methods (PPMs) may form a basis of action, though the separation of process from a 'like final product' is very difficult to prove unambiguously. The USA managed to ban Mexican tuna products on the grounds that the catching techniques did not control the co-catching of dolphins, a practice banned in the USA. But energy inputs can be regarded as a process-driven approach where discrimination on the grounds of pollution control and sustainability do apply. In addition, many multilateral environmental agreements contain measures to permit trade intervention in the event of provable non-compliance;

● the imposition of economic charges on environmentally damaging products and processes is now permitted under the PPM rules, though adding or subtracting levies at the border is proving very difficult to implement without rancour;

● multilateral environmental agreements (MEAs) are increasingly interfering with trade rules. So far, the obvious ones, namely the Basle Convention on the export of hazardous substances, the Convention on the Trade of Endangered Species and the Montreal Protocol, have been allowed to determine their own trade arrangements. But the Kyoto Agreement (Protocol eventually) will cause enormous headaches for trading regimes, as the Biodiversity Convention already does.

The trade regime is handled by a WTO Committee on Trade and the Environment (CTE). This is constantly reviewing rules, regulations and practices, with a wary eye for outright discrimination under the thin guise of sustainability. But it has concern for the abuse of environment and social well-being under the equally thin veil of trade liberalisation. James Cameron (1998, 168–175) is highly critical of the CTE, arguing that its remit is too broad, its membership too narrow and its homework too sketchy. Cameron, who has closely monitored the work of the CTE for over two years, complains that the body is insufficiently interested in the purposes of the evolving environmental agenda. Furthermore, he claims, the CTE is not at all cognisant of the social and justice dimensions of sustainable development. This means it cannot address proactively the linked issues of trade as a basis for local communal uplift and trade as threat to the viability of existing community structures and environmental safeguards. Only the thoughtful intervention of a wide range of NGOs, along the lines of the stakeholders tied to the Agenda 21 process, Cameron believes, will allow for a more effective role for the CTE in trade–environment patterns. Time is running out. New and more aggressive arrangements, which are far too insensitive to the subtle aspects of community-centred sustainable development, are already being put in place.

Education for a sustainable future

What is education for sustainability? It is the creation of the sense and the practice of global citizenship for all humanity. Arguably, and hopefully, in the innermost recesses of every human conscience there is a Gaian echo. Humans are a product of life's evolution on Earth. It is logical to surmise that, deep down in an ethical well still to be tapped, we are all global citizens ready to adopt a stewardship and trustee role, as outlined in Box 1.7. It is primarily on this basis that education for sustainability has a chance to work.

Yet educational institutions are notoriously short-sighted and inflexible. Schools are driven by political norms favouring the production of good corporate citizens, adaptable to a rapidly changing technology and working experiences. Schools are expected to teach citizenship, but for the most part they do not conduct themselves as if citizenship were a way of living in the schools themselves. For instance, it would be wonderful if youngsters could carry out their own energy audit, charge an energy tax, create an energy trading regime and utilise the income for energy-saving technology. Similarly, if there were a petrol/gasoline tax in operation, parents could create a mobility charity to run a minibus on gas or solar power to take their youngsters safely and comfortably to school. An organic garden could use the school meal refuse and provide healthy food for children, who learn how to co-operate, tend and recycle, as well as what is nutritious food. Such options would be possible if curricula were even further co-ordinated, classrooms opened up into the neighbourhood and grades awarded as much on doing as thinking.

Such options are being tried out, to be fair. But the scope is limited, and too much depends on the goodwill of overworked teachers, visionary school leaders and co-operative parents. Above all, the cross-curricular co-operation is the stumbling block. Class times make it difficult to lock in companionable learning experiences. Safety regulations mean out-of-school trips are limited and expensive. The national curriculum is still obsessed with formal structures and assessments. One day, one hopes, Local Agenda 21 and the local schools will co-operate to create a common cause.

The higher education scene is hardly more imaginative. True, universities and colleges splatter 'environmental' in front of almost any science course. But this is more for the trend than the substance. The way of learning in universities remains divorced from the ethics and justice aspects of resilience, vulnerability and empowerment. Few students are genuinely encouraged to write across the disciplines, and even fewer are graded positively for acting in communities or on ethical issues. Even today, there is no policy of ethical investment for university superannuation schemes. The excuse is that pension fund managers have to go for the highest return. But there are good-stewardship funds that could be encouraged if universities and local authorities seriously addressed the issue of how to invest for sustainability (see Box 19.8).

The notion of a practical-based Masters course is beginning to evolve. In the UK, the Forum for the Future Scholarships provides 12 bright graduates with six working

Box 19.8 Higher education for sustainability

The HE21 Project has begun to develop a database for best-practice sustainability across the higher education sector in the UK. This is part of the Forum for the Future programme on best-practice databases for LA21, rural sustainability, business practice and training (www.he21.org.uk). In detail, the HE21 Project contains the following elements:

- a committed group of pilot schemes, called 'trail blazers', in universities and colleges to promote interdisciplinarity in teaching and in research, and to share experiences;

- a guide to best practice for environmental management by HE sectors, but run by students and faculty as 'hands-on' courses;

- curriculum specifications for key professions, notably business, engineering, design and teacher training at the undergraduate level;

- guidelines for community-based learning programmes;

- a manual for how to incorporate educational experience into the preparation of sustainability indicators, and *vice versa*.

Much of this effort is simply trying to get the higher education institutions to be more ecofriendly in terms of energy conservation, waste management and materials use reduction.

The present scale of pilot schemes remains modest and uninspiring. They include:

- course material in LA21 in the Open University, and in Brighton;

- courses on citizenship for sustainability in Essex, John Moores and Middlesex;

- developing sustainability leadership through a work experience Masters programme in Middlesex.

In practice, only 17 of 756 further and higher education institutions responded to the recommendations published in the Toyne Report (1993), and even in those cases, cost savings were the major incentive.

In general, higher education institutions are hopelessly unmotivated over the sustainability transition. US colleges have evolved eco-campuses where good environmental practice is part of course credit, and co-operative student ventures on recycling and leasing are awarded academic as well as financial incentives. In the UK, unless a part of higher education central budgets is held back until such schemes are in place, there is little likelihood of real action. Those who teach do not necessarily act.

Sources: Toyne (1993); Forum for the Future (1997).

experiences of six weeks each in political parties, business, local government, NGOs, the media, finance, law and regulation. Each student has to produce a practical guide as to how sustainability principles were moved on in the period of contact. Group learning is encouraged, as is distance communication through a website. Former scholars assist their counterparts of a later cohort. The spirit of co-operation is wonderful to behold. One would like to think this is a precursor of many more offshoots. There is no reason why there should not be a practical, work-experience doctorate in environmental management or sustainable development. This would be based on a 'diary' of accumulated experiences, all leading to a shift towards better practice, validated by colleagues and those assisted by the practitioner. Distance learning and selected workshops would fill in the background material and keep scholars in contact with one another. This world of working for sustainability becomes the world of learning, not just theory building and testing in the abstract.

The sustainability message throughout this book is that societal and planetary maintenance is an honourable and necessary occupation. Whether this task is done voluntarily or via some paid occupation is a matter for further debate. But it seems mad that tens of thousands of people are not encouraged to do socially supportive tasks, such as care for the uncared, restore the damaged and recreate the otherwise thrown away. This is bad enough in the developed world. It is much worse in the developing world that those who bind communities together are not so valued in aid and education. There is real scope for twinning higher degrees so that community facilitators from the groups most vulnerable to global change can work with young practitioners and benefit from the companionship of sustainability exploration. This way, two people obtain recognition and experience, and both move the world a little further towards sustainability.

Education means 'leading from within'. Here is a golden opportunity to partner the classroom with the world of resilience, vulnerability, empowerment and social learning so that these vital processes are allowed to flourish in a million ways across the globe under the cause of citizenship action.

References

Allott, P. (1990) *Eunomia: New Order for a New World.* Oxford University Press, Oxford.

Birnie, P. (1992) International environmental law: its adequacy for future needs. In A. Hurrell and B. Kingsbury (eds), *The International Politics of the Environment.* Clarendon Press, Oxford, 51–84.

Birnie, P. and Boyle, A. (1992) *International Environmental Law.* Butterworths, London.

Brown Weiss, E. (1989) *In Fairness to Future Generations: International Law, Common Patrimony and Intergenerational Equity.* Dobbs Ferry, New York.

Cameron, J. (1998) The CETE: a renewed mandate for change, or more dialogue? In D. Brack (ed.), *Trade and the Environment: Conflict or Compatibility?* Earthscan, London, 168–186.

Forum for the Future (1997) *Higher Education 21: Local Agenda 21.* Forum for the Future, London.

Fox, W. (1992) *Towards a Transpersonal Ecology: Developing New Foundations for Environmentalism.* Shambhala Press, Toronto.

Greene, O. (1996) Environmental regimes. In J. Vogler and M. Imber (eds), *The Environment and International Relations.* Routledge, London.

Haas, P. (1990) *Saving the Mediterranean.* Colombia University Press, New York.

Haas, P., Levy, M., Keohane, R.O. and Levy, M.A. (1993) *Institution for the Earth: Sources of Effective International Environmental Protection.* MIT Press, Cambridge, MA.

Harrison, W.D. and Lenton, T.M. (1998) Spora and Gaia: how microbes fly with the clouds. *Ethology, Ecology and Evolution,* **10**, 1–16.

Hobbes, T. (1651) *Leviathan.* Basil Blackwell [1989], Oxford.

Jordan, A. (1994) The politics of incremental cost financing: the evolving role of the GEF. *Environment,* **36**, 6, 12–20, 31–36.

Jordan, A (1998) The ozone endgame: the implementation of the Montreal Protocol in the United Kingdom. *Environmental Politics,* **7**(4), 23–52

Kenny, A. (1994) The earth is fine: the problem is the greens. *The Spectator,* 11 March, 3–7.

Kotov, V. and Nikitina, E. (1998) Regime and enterprise: Novilsk Nickel and transboundary air pollution. In D. Victor *et al.* (eds), *The Implementation and Effectiveness of International Environmental Agreements.* MIT Press, Cambridge, MA, 549–574.

Lenton, T. (1998) Gaia and natural selection. *Nature,* **394**, 439–447.

Levy, M. (1996) Assessing the effectiveness of international environmental institutions. *Global Environmental Change,* **6**(4), 395–397.

Lovelock, J. (1992) *Gaia: the Practical Guide to Planetary Science.* Gaia Books, London.

Naess, A. (1989) *Ecology, Community and Lifestyle: Outline of an Ecophilosophy.* Cambridge University Press, Cambridge.

O'Riordan, T. and Voisey, H. (eds) (1998) *The Transition to Sustainability: the Politics of Agenda 21 in Europe.* Earthscan, London.

Paterson, M. (1996) *Global Warming and Global Politics.* Routledge, London.

Prakesh, A. and Gupta, A. (1997) Ecologically sustainable institutions. In F. Smith (ed.), *Environmental Sustainability: Practical Global Implications.* St Lucie's Press, Boca Raton, LA.

Rittberger, T. (ed.) (1993) *Regime Theory and International Relations.* Clarendon Press, Oxford.

Rowlands, I. (1995) *The Politics of Global Atmospheric Change.* Manchester University Press, Manchester.

Sands, P. (1997) The Rio principles. In F. Dodds (ed.), *The Way Forward: Beyond Agenda 21.* Earthscan, London, 21–40.

Sbragia, A. and Damro, C. (1999) The changing international role of the European Union: institution building and the politics of climate change. *Government and Policy,* **17**(1) (in press).

Schelling, T. (1972) On the ecology of micromotives. *Public Interest,* **25**, 61–98.

Schneider, S.D.H. and Boston, P.J. (eds) (1991) *Science of Gaia.* MIT Press, Boston, MA.

Shanklin, A. (1998) *The Antarctic Zone.* British Antarctic Survey, Cambridge.

Toyne, P. (chairman) (1993) *Environmental Responsibility: an Agenda for Further and Higher Education.* Department of Education and Employment, London.

Victor, D.G., Raustiala, K. and Skolnikoff, E. (eds) (1998) *The Implementation and Effectiveness of International Environmental Commitments: Theory and Practice.* MIT Press, Cambridge, MA.

Vogler, J. (1995) *The Global Commons.* John Wiley, Chichester.

Von Miltke, K. (1996) External perspectives on climate change: a view from the United States. In T. O'Riordan and J. Jäger (eds), *Politics of Climate Change: a European Dimension.* Routledge, London, 330–346.

Williams, M. (1997) The Group of 77 and Global Environmental Politics. *Global Eenvironmental Change,* **7**(3), 295–298.

Further reading

There is now a good choice of books on international environmental politics and law as a result of the growth of interest in these topics. However, the best introductory textbooks are still:

Birnie, P. and Boyle, A. (1992) *International Environmental Law.* Oxford University Press, Oxford.

Porter, G. and Welsh Brown, J. (1996) *Global Environmental Politics* (2nd edn). Westview Press, Bolder, CO.

Otherwise, the following are recommended:

Dodds, F. (ed.) (1997) *The Way Forward: Beyond Agenda 21.* Earthscan, London.

Flanders, L. (1998) The United Nations Department on Policy Co-ordination and Sustainable Development. *Global Environmental Change*, 7(4), 391–394.

Haas, P. M., Levy, M.A. and Parson, E.A. (1992) Appraising the Earth Summit: how should we judge UNCED's success? *Environment*, **34**, (8), 6–12.

Hurrell, A. and Kingsbury, B. (1992) *The International Politics of the Environment.* Clarendon Press, Oxford.

Thomas, C. (1992) *The Environment in International Relations.* Royal Institute for International Affairs, London.

Victor, D.G., Raustiala, K. and Skolnikoff, E. (eds) (1998) *The Implementation and Effectiveness of International Environmental Commitments: Theory and Practice.* MIT Press, Cambridge, MA.

Vogler, J. (1995) *The Global Commons.* John Wiley, Chichester.

Werksman, J. (ed.) (1996) *Greening International Institutions.* Earthscan, London.

World Commission on Environment and Development (1987) *Our Common Future.* Oxford University Press, Oxford.

Index